Biohydrogen Production: Sustainability of Current Technology and Future Perspective

Anoop Singh • Dheeraj Rathore
Editors

Biohydrogen Production: Sustainability of Current Technology and Future Perspective

Springer

Editors
Anoop Singh
Department of Scientific and Industrial Research, Ministry of Science & Technology, Government of India
Technology Bhawan
New Delhi, India

Dheeraj Rathore
School of Environment and Sustainable Development
Central University of Gujarat
Gandhinagar, Gujarat, India

ISBN 978-81-322-3575-0 ISBN 978-81-322-3577-4 (eBook)
DOI 10.1007/978-81-322-3577-4

Library of Congress Control Number: 2016945986

Printed on acid-free paper

This Springer imprint is published by Springer Nature
The registered company is Springer (India) Pvt. Ltd.

Foreword

At the beginning of the twenty-first century, there are two important and interrelated global problems facing the humankind. They are the depletion of fossil fuels (coal, petroleum and natural gas) and the environmental problems caused by their utilisation. According to the best estimates, most of the fossil fuels, which can be extracted at a reasonable cost, will be depleted by the end of this century. The utilisation of fossil fuels is resulting into huge environmental problems, such as climate change, global warming, ozone layer depletion, acid rains, pollution, oxygen depletion and others. Carcinogens produced by the combustion of fossil fuels are many times more than those produced by cigarettes. Radioactivity produced by coal burning is many times greater than those produced by nuclear power plants. We are observing more frequent and stronger hurricanes, typhoons and tornadoes. We are also observing more and more droughts and floods. As global warming grows, they will also grow in size and frequency. Oceans are rising due to melting glaciers and ice caps. It is estimated that the cost of worldwide environmental and health damages caused by fossil fuels is presently 8 trillion US dollars per year and is growing.

There is an elegant solution to the above-described global problems. It is replacing the fossil fuels by hydrogen produced from clean and renewable energies. Hydrogen is the most efficient, the cleanest and the lightest fuel. It is also renewable. Once we convert to the hydrogen energy system, we shall never have to convert to another energy system, as we shall never run out of hydrogen. So long as we have the sun and renewable energy sources, we shall be able to produce clean and abundant hydrogen.

Of course, hydrogen is a synthetic fuel and it must be manufactured. There are various hydrogen manufacturing methods such as direct thermal, thermochemical, electrochemical, biological, etc. Among the hydrogen production methods, the biological method has the potential of producing the most cost-effective hydrogen. Because of this, many research groups around the world are working on biological hydrogen production. In several cases, bench-scale production systems have come up with encouraging results.

Clearly, time has arrived for a book on biohydrogen production. I congratulate the editors, Dr. Anoop Singh and Dr. Dheeraj Rathore, for seeing the need for such a book and producing it. This book entitled "Biohydrogen Production: Sustainability of Current Technology and Future Perspective" covers the biological hydrogen production authoritatively from A to Z, including raw material sources, various production technologies, sustainability problems, regional variations, economics, global trends and future perspectives.

I strongly recommend this excellent book to energy scientists, engineers and students, who are interested in hydrogen production in general and biological hydrogen production in particular, as well as to those with industrial concerns, who are looking for inexpensive hydrogen production technologies.

International Association for Hydrogen Energy
Miamim, FL, USA

T. Nejat Veziroğlu,

Preface

Presently most of the energy demand is fulfilled by the fossil fuel. Global petroleum demand has increased steadily from 57 million barrels day^{-1} in 1973 to 90 million barrels day^{-1} in 2013 and will continue to increase in line with the world's economy. The increasing energy demands will speed up the exhaustion of the finite fossil fuel. United Arab Emirates, one of the major oil export countries, would fail to meet the share in the oil and natural gas demands by 2015 and 2042, respectively. The fossil fuel resources in Egypt would be exhausted within two decades.

Using petroleum-based fuels creates atmospheric pollution during combustion. Apart from emission of the greenhouse gas (GHG) CO_2, air contaminants like NO_X, SO_X, CO, particulate matter and volatile organic compounds are also emitted which leads not only to climate change but also to deterioration of environmental and human health. Continued use of fossil fuel is now widely recognised as unsustainable. A renewable, carbon-neutral energy resource is necessary for environmental and economic sustainability. Concern for exhausting the availability of fossil fuel for fulfilling future energy demand and considering changes in global climate by conventional energy resource has diverted researchers towards exploring a way to environmentally safe and sustainable energy resources. Finding sufficient supplies of clean energy for the future is one of the most daunting challenges for humanity and is intimately linked to global stability, economic prosperity and quality of life. A rapid surge in research activities with intensive focus on alternative fuels has been seen in the past decades in order to reduce the dependency on fossil fuels, mainly by providing local energetic resources.

Biofuels are considered as the most environment friendly alternative energy source because they are renewable and also sequester carbon. Currently, biofuels are commercially produced from the food crops, developing serious ecological and socio-economical anxiety such as land use changes and food-fuel competition issue. About 1 % (14 million hectares) of the world's arable land is able to produce current biofuels, to supply 1 % of global transport fuel demand. Between 1980 and 2005, worldwide production of biofuels increased by an order of magnitude from 4.4 billion litres to 50.1 billion litres. Clearly, increasing the share, it will be impractical due to the severe impact on the world's food supply and the large areas of production land required. This is manifested by the recent increase in grain prices due to utilisation of maize at large scale as a feedstock for production of fuel ethanol in the USA. This caused riots in Mexico due to the increase in the price of tortillas, a staple food. Further, GHG saving is another constraint for developing a sustainable biofuel. The Intergovernmental Panel on Climate Change has calculated that reductions of 25–40 % of CO_2 emissions by 2020 and up to 80 % by 2050 are required to stay within temperature range, i.e. less than 2 °C, to avoid dangerous climate changes worldwide. The production of sustainable bioenergy is a challenging task in the promotion of biofuels for replacing the fossil-based fuels to get a cleaner environment and also to reduce the dependency on other countries and uncertainty of fuel price.

Among the various renewable energy sources, biohydrogen is a strong candidate for future energy source by virtue of the fact that it is renewable, does not evolve GHG and ozone layer-depleting chemicals in combustion, liberates large amount of energy per unit weight in combution and is easily converted into electricity by fuel cell. Hydrogen is also harmless to mammals and the environment. Hydrogen can be produced safely and considered as the ultimate cleanest energy carrier to be generated from renewable sources. Progress in the late 1990s contributed to a breakthrough in terms of sustainable hydrogen production. There are various technologies (direct biophotolysis, indirect biophotolysis, photo-fermentations and dark fermentation) available for the production of biohydrogen from biomass/organic wastes, and many of these technologies have some drawbacks (e.g. low yield, low production rate, etc.), which limit the practical application. Studies on the biohydrogen production have been focused on photo-decomposition of organic compounds by photosynthetic bacteria, dark fermentation from organic compounds with anaerobes and biophotolysis of water using algae and cyanobacteria. Among these technologies, metabolic engineering is presently the most promising for the production of biohydrogen as it overcomes most of the limitations in other technologies. The biohydrogen production from biomass is particularly suitable for a relatively small and decentralised system, and it can be considered as an important key for a sustainable renewable energy source.

The present book is an effort to provide an up-to-date information and knowledge on the state of the art of biohydrogen production technology by the internationally recognised experts and subject peers in different areas of biohydrogen. It is a comprehensive collection of chapters related to choices of feedstock, microbiology, biochemistry, molecular biology, enzymes and metabolic pathways involved, bioprocess engineering, waste utilisation, economics, life cycle assessment and perspectives of the biohydrogen production in different countries and regions of the world and also include scale-up and commercialisation issues. The introductory chapter (Chap. 1) gives a general background for global energy statistics, available sources for energy supply, options of renewable energy sources, benefits of adoption of biohydrogen and its sustainability and future perspectives. The following chapter (Chap. 2) reviews the potentiality of different biomass that can be utilised for biohydrogen production and also discusses various technologies for production of biohydrogen and sums up with the required further research. Chapters (3 and 4) focused on biohydrogen production from agricultural biomass and wastes to analyse their suitability for biohydrogen production and also point out the challenges for biohydrogen production from agricultural biomass and wastes.

A series of chapters (Chaps. 5, 6, 7 and 8) are concentrated on the potential of microbial biohydrogen production especially from cyanobacteria and green algae. These chapters discussed on the physiology of biohydrogen production from microbial biomass, industrial approaches for biohydrogen production by photoautotrophic microbes, characterisation and identification of algal strains, mechanism of hydrogen photoproduction by algae, design and modelling of photobioreactor for algae cultivation and biohydrogen production, algal engineering for improving photosynthetic efficiency and hydrogenase and constraints and challenges for biohydrogen production. The following chapter (Chap. 9) is an attempt to review the latest findings on hydrogenase enzyme, responsible for hydrogen production, and also enlighten the metabolic engineering to increase the enzyme production and activity. Two chapters (Chaps. 10 and 11) reviewed the present status and future perspectives of biohydrogen production in Asia and Saudi Arabia. The economics, a major limitation of biohydrogen popularity for industrial production, is also covered (Chap. 12). Life cycle assessment (LCA) techniques allow detailed analysis of material and energy fluxes on regional and global scales. LCA studies of renewable energy sources calculate the environmental impact and can relate the results against sustainability criteria. The comprehensive LCA of biohydrogen production and its comparison with other biofuels is covered in the Chap. 13 and can be a tool for sustainability assessment and policy decisions. Chapter 14 presented a global trend of biohydrogen research and its future perspectives.

This book is aimed at a wide audience, mainly researchers, energy specialists, academicians, entrepreneurs, industrialists, policymakers and others who wish to know the latest development and future perspectives of biohydrogen production, and also discusses the bottlenecks of the various processes that currently limit the scale-up and commercialisation. Each chapter begins with a fundamental explanation for general readers and ends with in-depth scientific details suitable for expert readers. The text in all the chapters is supported by numerous clear, illustrative and informative diagrams, flow charts and comprehensive tables detailing the scientific advancements, providing an opportunity to understand the process thoroughly and meticulously. Written in a lucid style, the book comprehensively covers each point to give the reader a holistic picture about biohydrogen production technology and its sustainability. The book may even be adopted as a textbook for university courses that deal with biohydrogen and renewable energy sources.

Despite the great efforts of authors and editors along with extensive checks conducted by many experts in the field of biohydrogen production, mistakes may have been made. We would appreciate if the readers could highlight mistakes and make comments or suggestions to improve and update the book contents for future editions.

New Delhi, India — Anoop Singh
Gujarat, India — Dheeraj Rathore

Acknowledgements

There are several people behind the successful completion of this book. We, the editors of this book, jointly express our gratitude towards our teacher Professor S.B. Agrawal (Department of Botany, Banaras Hindu University, Varanasi, India) for his inspiration and continuous support. We are thankful to Surabhi Shukla from Springer who first approached us with the proposal for the book and helped us crystallise our ideas on the topic. Afterwards Raman Shukla, Dr. Mamata Kapila and Hemalatha Gunasekaran took over and provided excellent support with all the administrative work. We also thank all the authors who kindly agreed to contribute the chapters and worked with us throughout the process. We are also grateful to the reviewers who took time out of their busy schedule and critically reviewed the chapters of this book and provided very valuable suggestions for their improvements. This book is a labour of love for us since we spent a lot of our weekends and free time on working on it. For this reason alone, our families deserved to be thanked for bearing with us all this while. Anoop Singh likes to thank the administration of the Department of Scientific and Industrial Research (DSIR), Ministry of Science and Technology, government of India, for their support. Dheeraj Rathore would like to thank the vice-chancellor of the Central University of Gujarat, India, and dean of the School of Environment and Sustainable Development, Central University of Gujarat, India, for their unflinching support and encouragement towards this endeavour.

Contents

List of Editors and Contributors

About the Editors

Dr. Anoop Singh has a PhD in botany (2004). Dr. Singh is a scientist at the Department of Scientific and Industrial Research (DSIR), Ministry of Science and Technology, government of India. Before joining DSIR, he worked at the Technical University of Denmark, Denmark; University College Cork, Ireland; the Energy and Resources Institute (TERI), New Delhi, India; the Indian Agricultural Research Institute (IARI), New Delhi, India; Banaras Hindu University, Varanasi, India; and VBS Purvanchal University, Jaunpur, India. He has visited several European countries and participated in several international conferences. He has published more than 50 research articles in scientific journals (h-index 22) and also published an edited book on *Life Cycle Assessment of Renewable Energy Sources* with Springer. Dr. Singh is a member of several scientific communities and serving as editorial board member for a number of journals besides being a reviewer. His research is focused on sustainable agriculture; waste management through agriculture; the utilisation of industrial, agricultural and household waste for eco-friendly energy production; and renewable energy sources and their life cycle assessment (LCA).

Dr. Dheeraj Rathore is an assistant professor in the School of Environment and Sustainable Development at the Central University of Gujarat (CUG), Gandhinagar, India. He obtained his PhD in botany from AAI-DU, Allahabad, India, and his M.Phil. in environment management from Vikram University, Ujjain, India. Before joining CUG, he worked at the University of Dodoma, Tanzania. He has published more than 20 research articles (h-index 7) and is a member of several scientific communities besides being a reviewer. His research interests are focused on stress physiology and physiology of biofuel production.

Contributors

Hidayet Argun Department of Environmental Engineering, Pamukkale University, Denizli, Turkey

Khorcheska Batyrova Département de Microbiologie, infectiologie et immunologie, Université de Montréal, Montréal, Canada

Punyasloke Bhadury Integrative Taxonomy and Microbial Ecology Research Group, Department of Biological Sciences, Indian Institute of Science Education and Research Kolkata, Nadia, West Bengal, India

Sourish Bhattacharya Process Design and Engineering Cell, Central Salt and Marine Chemicals Research Institute, Bhavnagar, Gujarat, India

S. Bhuvanesh Department of Biochemical Engineering and Biotechnology, Indian Institute of Technology Delhi, New Delhi, India

Neon J. Bordoloi Biofuel Laboratory, Department of Energy, Tezpur University, Tezpur, India

Hélène Carrère Laboratoire de Biotechnologie de l'Environnement, Narbonne, France

Lucile Chatellard Laboratoire de Biotechnologie de l'Environnement, Narbonne, France

Rahul S. Chutia Biofuel Laboratory, Department of Energy, Tezpur University, Tezpur, India

Manojit Debnath Systematics and Applied Phycology Laboratory, Postgraduate Department of Botany, Hooghly Mohsin College, Hooghly, West Bengal, India

Integrative Taxonomy and Microbial Ecology Research Group, Department of Biological Sciences, Indian Institute of Science Education and Research Kolkata, Nadia, West Bengal, India

Ramkrishna Ghosh Systematics and Applied Phycology Laboratory, Postgraduate Department of Botany, Hooghly Mohsin College, Hooghly, West Bengal, India

Lina Gogoi Biofuel Laboratory, Department of Energy, Tezpur University, Tezpur, India

Pelin Gokfiliz Institute of Applied and Natural Sciences, Department of Biotechnology, Dokuz Eylul University, Buca-İzmir, Turkey

Patrick C. Hallenbeck Département de Microbiologie, infectiologie et immunologie, Université de Montréal, Montréal, Canada

Life Sciences Research Center, Department of Biology, United States Air Force Academy, Colorado Springs, CO, USA

I.M.I. Ismail Center of Excellence in Environmental Studies (CEES), King Abdul Aziz University, Jeddah, Saudi Arabia

Martin Kaltschmitt Institute of Environmental Technology and Energy Economics, Hamburg University of Technology, Hamburg, Germany

Ilgi Karapinar Department of Environmental Engineering, Dokuz Eylul University, Buca-İzmir, Turkey

Rupam Kataki Biofuel Laboratory, Department of Energy, Tezpur University, Tezpur, India

Anubha Kaushik University School of Environmental Management, GGS Inderprastha University, New Delhi, India

M.Z. Khan Environmental Research Laboratory, Department of Chemistry, Aligarh Muslim University, Aligarh, Uttar Pradesh, India

Nicholas E. Korres Department of Crop, Soil and Environmental Sciences, University of Arkansas, Fayetteville, AR, USA

Ayse Kose Department of Bioengineering, Ege University, Izmir, Turkey

Antonella Marone Laboratoire de Biotechnologie de l'Environnement, Narbonne, France

R. Miandad Center of Excellence in Environmental Studies (CEES), King Abdul Aziz University, Jeddah, Saudi Arabia

S. Venkata Mohan Bioengineering and Environmental Sciences (BEES), CSIR-Indian Institute of Chemical Technology (CSIR-IICT), Hyderabad, India

Rumi Narzari Biofuel Laboratory, Department of Energy, Tezpur University, Tezpur, India

G.N. Nikhil Bioengineering and Environmental Sciences (BEES), CSIR-Indian Institute of Chemical Technology (CSIR-IICT), Hyderabad, India

A.S. Nizami Center of Excellence in Environmental Studies (CEES), King Abdul Aziz University, Jeddah, Saudi Arabia

Jason K. Norsworthy Department of Crop, Soil and Environmental Sciences, University of Arkansas, Fayetteville, AR, USA

Suphi S. Oncel Department of Bioengineering, Ege University, Izmir, Turkey

O.K.M. Ouda Department of Civil Engineering, Prince Mohamed Bin Fahd University, Al-Khobar, Saudi Arabia

Dheeraj Rathore School of Environment and Sustainable Development, Central University of Gujarat, Gandhinagar, Gujarat, India

Ibrahim M. Abu Reesh Department of Chemical Engineering, School of Engineering, Qatar University, Doha, Qatar

M. Rehan Center of Excellence in Environmental Studies (CEES), King Abdul Aziz University, Jeddah, Saudi Arabia

Ruprekha Saikia Biofuel Laboratory, Department of Energy, Tezpur University, Tezpur, India

Omprakash Sarkar Bioengineering and Environmental Sciences (BEES), CSIR-Indian Institute of Chemical Technology (CSIR-IICT), Hyderabad500 007, India

Surajbhan Sevda Department of Chemical Engineering, School of Engineering, Qatar University, Doha, Qatar

Department of Biochemical Engineering and Biotechnology, Indian Institute of Technology Delhi, New Delhi, India

K. Shahzad Center of Excellence in Environmental Studies (CEES), King Abdul Aziz University, Jeddah, Saudi Arabia

Mona Sharma Department of Environmental Sciences, School of Earth, Environment and Space Studies, Central University of Haryana, Mahendergarh, India

Anoop Singh Department of Scientific and Industrial Research (DSIR), Ministry of Science and Technology, Government of India, Technology Bhawan, New Delhi, India

Ratan Singh School of Environment and Sustainable Development, Central University of Gujarat, Gandhinagar, Gujarat, India

T.R. Sreekrishnan Department of Biochemical Engineering and Biotechnology, Indian Institute of Technology Delhi, New Delhi, India

Debashis Sut Biofuel Laboratory, Department of Energy, Tezpur University, Tezpur, India

Lisa Thormann Institute of Environmental Technology and Energy Economics, Hamburg University of Technology, Hamburg, Germany

Eric Trably Laboratoire de Biotechnologie de l'Environnement, Narbonne, France

Christina Wulf Institute of Environmental Technology and Energy Economics, Hamburg University of Technology, Hamburg, Germany

1 Biohydrogen: Next Generation Fuel

Anoop Singh and Dheeraj Rathore

Abstract

Concern over sustainability of fossil fuel use is raised due to depleting fuel resources and emitting greenhouse gases (GHGs) from it. Among many alternative energy sources, biofuels, natural gas, hydrogen, and synthesis gas (syngas) emerge as four strategically important sustainable energy sources. As hydrogen gas is renewable, it does not evolve GHGs, and releases large amount of energy in combustion of unit weight and hydrogen can also be easily converted into electricity by fuel cell. It could be a strong candidate for future alternate energy resource. Biological H_2 production delivers clean H_2 in sustainable manner with simple technology and more attractive potential than the current chemical production of H_2. Although present industrial hydrogen production system is based on chemical processing units, research trend on biohydrogen promises a deafening potential of industrial biohydrogen production in the near future.

1.1 Introduction

The conventional fossil-based fuels contributed major share in the global primary energy consumption, while in the present scenario, the fossil fuel use is widely considered as unsustainable fuel due to depletion of fossil resources and accelerated accumulation of greenhouse gases (GHGs) in the environment that already has exceeded the "dangerously high" threshold of 450 ppm CO_2e (Schenk et al. 2008). This contributes to different environmental challenges including global warming, climate change, biodiversity loss, receding of glaciers, sea level rise, etc.

A. Singh (✉)
Department of Scientific and Industrial Research (DSIR), Ministry of Science and Technology, Government of India, Technology Bhawan, New Delhi 110016, India
e-mail: apsinghenv@gmail.com

D. Rathore
School of Environment and Sustainable Development, Central University of Gujarat, Sector-30, Gandhinagar, Gujarat 382030, India
e-mail: dheeraj.rathore@cug.ac.in

A. Singh, D. Rathore (eds.), *Biohydrogen Production: Sustainability of Current Technology and Future Perspective*, DOI 10.1007/978-81-322-3577-4_1

(Gullison et al. 2007). The researchers had pointed out the three basic assumptions in current policy debates on climate, energy, and GHGs emissions, viz., (a) strong requirement for cleaner energy production and conservation technologies on a global scale, (b) the need for future mandates on emission reduction to be aligned with the production of clean energy and energy-conservation policies, and (c) the need to act with urgency (Subhadra and Edwards 2010).

Traditional fuels like wood, charcoal, agricultural residues, and animal wastes are major contributors to household energy supply in many of the developing countries having agrarian economies (FAO 2005; Dhanya et al. 2013). The current disposal practices for agricultural residues have caused widespread environmental concern as they represent hindrance to sustainable development in rural areas as well as to national economies (Dhanya et al. 2013; Sheehan 2009). Environmental contaminations due to faulty disposal of waste have also necessitated identification of environmentally sound and economically feasible technologies for waste management (Prasad et al. 2007a, b).

The concerns related to energy security, environmental safety, and sustainability have encouraged researchers toward alternative, renewable, sustainable, efficient, and cost-effective energy sources with lesser emissions (Singh and Olsen 2012). Renewable energy can play a decisive role at global and national levels in dealing with the concerns related to energy security, climate change, eco-friendliness, and sustainability (Singh et al. 2010a, b, 2011). Hence, renewable energy sources as an alternative to conventional fossil fuels have been depicted as the main energy supplier in the future that could increase the energy-supply security and emission reduction and render a stabilized income for farmers (Singh and Olsen 2012). The production of sustainable renewable energy is a challenging task to replace the conventional fossil fuels to get cleaner environment, to reduce the dependency on foreign countries, and to cope up with the fuel price uncertainty (Singh and Olsen 2012).

Among many alternative energy sources, biofuels, natural gas, hydrogen, and synthesis gas (syngas) emerge as four strategically important sustainable energy sources in the foreseeable future (Nigam and Singh 2011). Currently, most of the biofuel production at commercial scale are made using the food crops as raw material, developing serious ecological and socioeconomical concern, e.g., land-use changes and food vs. fuel competition (Rathore et al. 2015).Most of the issues related to energy security, production, and consumption can be solved by utilization of biohydrogen as fuel, as biohydrogen is renewable and can be utilized as fuel for electricity, heat, and transportation purposes, with some modifications to existing technologies and have potential to improve sustainability and reduce GHG emissions significantly (Rathore and Singh 2013). This chapter is an attempt to bring out the sustainability and future prospectus of utilization of biohydrogen as an energy source.

1.1.1 Global Energy Demand, Supply, and CO_2 Emission

Energy is the backbone for civilization. Development of a nation is fueled by its energy availability because one of the major inputs for economic development of every country is energy. The energy sector assumes a crucial emphasis in view of the ever-increasing energy demands necessitating big investments to meet them. According to the recently published International Energy Agency (IEA) 2015 Key World Energy Statistics (IEA 2015a), the primary energy supply of the world has grown by 122 % in 40 years, from approximately 6.10 billion tonne of oil equivalent (TOE) in 1973 to about 13.54 billion TOE in 2013. During this period, a shift of fuel source is noticed as the share of natural gas has enhanced from 16 % to 24.1 %, while coal share increased only 4.4 % and oil share is decreased by 15 %. The other/alternate sources of energy were also explored and added about 5 % additional share during this period (Fig. 1.1a). It is expected in the future that the increase in the energy demand will depend upon economic growth of emerging market countries, e.g., India, China, and the Middle East (IEA, 2007, 2008, 2015b). An estimation has

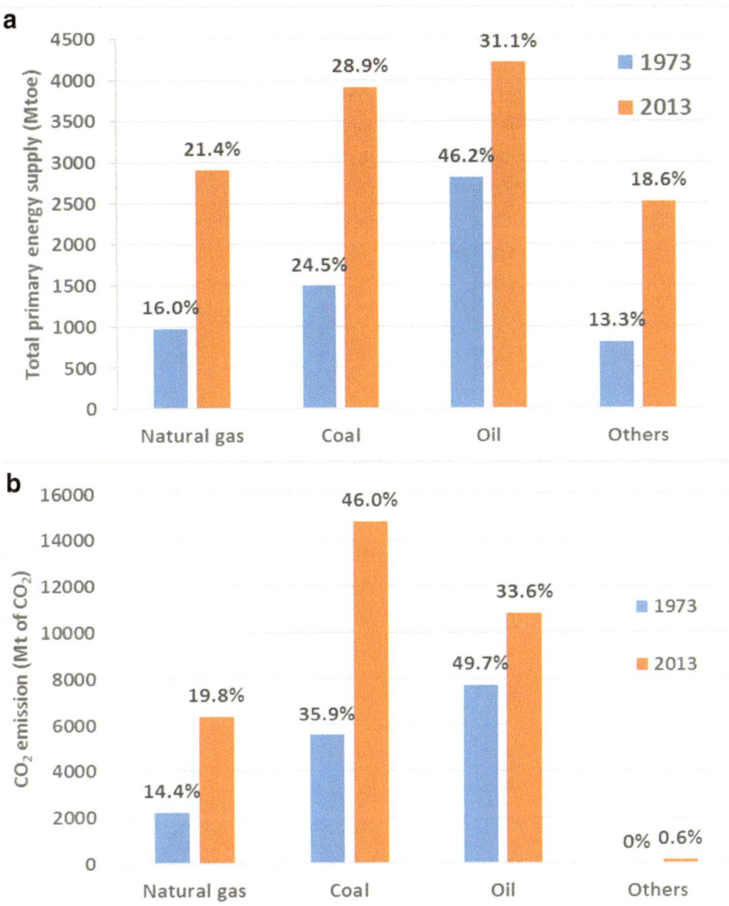

Fig. 1.1 Fuel share in total primary energy supply (**a**) and CO_2 emissions from fuel consumption (IEA 2015a)

been reported that there is an expected increase of 48 % over 25 years from about 11.43 billion TOE during 2005 to about 17 billion TOE in 2030 (Benchmarking of Biodiesel Fuel Standardization in East Asia Working Group 2010). IEA raised its forecast of global oil demand to 93.6 million barrels per day in 2015 (a gain of 1.1 million barrels a day on the year) due to increasing energy demand in India, China, and Europe and a spate of colder temperatures in the first quarter for heating purposes at homes and factories. The IEA named this hike a "notable acceleration" from 2014 growth levels of 0.7 million barrels per day (mb d^{-1}) (Gallucci 2015).

Energy production and consumption are affected by disruptions, from wars to extreme weather (BP 2015). The worldwide energy use in the central scenario of IEA is set to grow by one-third to 2040. The energy use growth is primarily driven by China, India, Africa, Southeast Asia, and the Middle East. Non-OECD (Organisation for Economic Co-operation and Development) countries account for all the increase in global energy use because demographic and structural economic trends, allied with greater efficiency, reduce collective consumption in OECD countries from the peak reached in 2007 (IEA 2015b). As projected by IEA (2007), the average annual rate of energy consumption is to grow by 3 % from 2004 to 2020 in developing countries. Energy demand in industrialized nations with mature economies and relatively low population growth is expected to be at the lower rate of 0.9 % per year, admitted from a much higher starting point. About half of the increase in global energy demand by 2030 will be for power generation and one-fifth for transport needs, mostly in the form of petroleum fuels.

IEA's Factsheet proclaimed that "rising crude oil-import needs of China and India, from the Middle East and other regions, increase their vulnerability to the implications of a possible shortfall in investment or a disruption to oil supply" (IEA 2014). Statistics of the factsheet demonstrated that natural gas share in total inter-regional fossil fuel trade rises by one quarter to more than 20 % by 2040; the increasing availability of liquefied natural gas (LNG) eased the gas security concerns. The increase in coal trade is driven by strong Asian demand and likely to grow to 40 % by 2040. The rise in world oil supply trend from 14 to 104 mb d^{-1} in 2040 hinges critically on timely investments in the Middle East. The rise in the production of natural gas at global level is in a near-linear fashion to 5,400 bcm (billion cubic meter) in 2040, with a major role for unconventional gas which increases its contribution in output from 17 % to 31 %. The coal demand at global level rises to 6,350 Mtoe in 2040 at a much lower rate (0.5 % per year) than over the last 30 years. The coal demand growth is restricted by new air pollution and climate policies in the main markets of the United States, China, and Europe. The accelerated growth in coal use continues in India. Four countries, viz., India, China, Australia, and Indonesia, alone account for more than 70 % of global coal output by 2040 that underscore Asia's importance in global coal trade and pricing.

The carbon dioxide emissions by the consumption of fuels have increased by 107 % during the period from 1973 (15,515 Mt of CO_2) to 2013 (32,190 Mt of CO_2). The emission share of natural gas increased with the similar magnitude of, while increase in the emission share of coal consumption (~10 %) more than doubled to the increase in the supply of coal (4.4 %) (Fig. 1.1b). The emission share of other fuels is very less (0.6 %) while it contributed about 18 % of global energy supply. The share of energy supply by other sources is majorly contributed by biofuels and wastes (about 10 % of global energy supply during 2013). The remaining energy supply is contributed by nuclear energy, geothermal power, hydropower, wind energy, solar energy, etc. (IEA 2015a).

Presently most of the energy supply is based on fossil fuel with a minor portion coming from renewable resources. Renewables Global Status Report published in 2014 accounted that about 19 % of global final energy consumption in 2012 rendered by renewable and continued to grow in 2013 (REN21's Renewables Global Status Report 2014). Among this 19 %, about 10 % is furnished by modern renewables and the remaining 9 % is accounted for traditional biomass. The share of heat energy from modern renewable sources in the total final energy use is about 4.2 %, while hydropower contributed about 3.8 % and about 2 % is coming from solar, wind, biomass, biofuels, and geothermal (REN21's Renewables Global Status Report 2014).

1.2 Renewable Energy Sources

Global dependence on fossil fuels has led to the release of over 1,100 $GtCO_2$ into the atmosphere since the mid-nineteenth century. Currently, energy-related GHG emissions, mainly from fossil fuel combustion for heat supply, electricity generation, and transport, account for around 70 % of total emissions including carbon dioxide, methane, and some traces of nitrous oxide (Sims et al. 2007). With the situation of increasing energy demand and energy prices and implementation of policies for global warming reduction, the sources of renewable energy have popularized. Renewable energy is not only providing the energy but also a tool to solve several other problems associated with the fossil energy, viz., improving the energy security, resolving the health and environmental anxiety, decreasing greenhouse gas emissions, and reducing poverty by increasing the employment.

1.2.1 Classification

The inexhaustible renewable energy sources include solar, wind, ocean, hydroelectric, biomass, and geothermal energy. These renewable energy sources offer many environmental benefits over to conventional energy sources. The different types of renewable energy sources have own specific advantages, which make them uniquely suited to limited applications. Almost all these renewable energy sources are not releasing gaseous or liquid pollutants during operation. In their technological development, the renewable ranges from technologies that are well established and mature to those that need further research and development (Hepbasli 2008). International Energy Agency classified renewable energy sources, viz., (i) first-generation source, the technologies have already reached up to maturity level, e.g., combustion, hydropower, and geothermal energy; (ii) second-generation source includes those technologies which are going through rapid development such as solar energy, wind power, and bioenergy; and (iii) third-generation sources, which are presently under developmental stages such as concentrating solar power, improved geothermal and ocean energy, and integrated bioenergy systems (IEA 2006).

1.2.2 Benefits of Biohydrogen Production

Biohydrogen economy has captured global consideration due to its social, economic, and environmental benefits (Sekoai and Daramola 2015). As biohydrogen can be produced by sunlight and minimal nutrients or organic waste effluents as a nutrient source, it has considerably less impact on environment and production cost. The production of biohydrogen does not have any competition with the food/fodder and it also not required fertile land, like first- and second-generation biofuels. By the virtue of the fact that hydrogen gas is renewable, does not liberate greenhouse gases, has unshackle large amount of energy per unit weight during combustion, and can easily be converted into electricity by fuel cell, it is considered as a strong participant for future energy (www.oilgee.com, 2012). Despite the existing technological constraint for industrial production of biohydrogen, its multidimensional advantages make it the most popular alternative over other renewable energy resources.

1.2.2.1 Environmental Benefit

Combustion and refining process of the finite fossil fuels cause severe environmental problems. The CO_2 generation by burning of hydrocarbon is a major cause of global warming and other greenhouse gases as well as left with the toxic compounds as in the case of coal. International Energy Agency predicted that 30 billion tons of CO_2 was emitted from hydrocarbon fuels in 2008, which is doubled since 1970 (Energy Information Administration 2011). Life Cycle Assessment studies suggested that biofuels such as biodiesel, bioethanol, and biomethane are considered as a better option for carbon saving (Rathore et al. 2013), although present biofuel feedstock develops a conflict over food and fuel. Despite the fact that meeting food demands remains the primary objective of agriculture, the promotion of energy crops for biofuel production has added an additional component to the conventional production portfolio of the agricultural sector and thus further intensifies the challenges of widespread land-use pattern and land grabbing (Venghaus and Selbmann 2014). Changing land-use pattern for production of biofuel crops resulted into distortion of ecological sustainability of the area. Hydrogen is characterized as a "clean fuel," as it produces only water vapor as the by-product after its use as an energy carrier, no emissions of toxic waste and adding no GHG to the atmosphere (Brentner et al. 2010). The hydrogen produced by physical and chemical processes does not liberate CO_2 during combustion though its production process required energy input which directly or indirectly comes from fossil fuel (Brentner et al. 2010; Lee 2014).

Biological hydrogen production is a potentially carbon neutral process that is carried out at lower temperatures and pressures and is therefore less energy intensive than thermochemical and electrochemical processes (Levin and Chahine 2010). Production of biohydrogen does not cause burden on food product as it offers potential to generate renewable H_2 from inexpensive "waste" feedstocks (Brentner et al. 2010; Ghimire et al. 2015), wastewater (Skonieczny and Yargeau 2009), sludge (Sittijunda et al. 2010), or microalgae (Rathore and Singh 2013).

1.2.2.2 Economic Benefit

Developing hydrogen economy is broadly based on the need to provide a more sustainable energy system to overcome the climate change, diminishing fossil fuel resources, dwindling supplies, and lessen reliance on foreign oil (Brentner et al. 2010). Utilization of biohydrogen will not only be encouraged by its application in transportation sector but also by the superiority of the cost and competence to other energy production technologies (Ma et al. 2013). The International Energy Agency (Maniatis 2003) stated that biohydrogen is now a weak technology but with potential to capture market. Lee and Chiu (2012) showed the effect of percentage increments in investment on the output of the biohydrogen sector in India, Japan, USA, and China and estimated the increment of US$ 2.14, 3.61, 10.42, and 12.48 billion, respectively, in the biohydrogen sector output during years 2011–2050. Baseline results of Taiwan general equilibrium model by Lee and Hung (2012) indicated that wind, biofuel, biohydrogen, and hydrogen fuel cell technologies are sensitive to external support and will perform well without external support. In case government supported to clean energy, biohydrogen and hydrogen fuel cells will lead all clean energies.

1.2.2.3 Social Benefit

Major impacts of biohydrogen on society include reduction in air pollution and global warming issues (Sørensen 2012). Due to operational costs and the high capital investment, nonthermal production of pure hydrogen, i.e., from biomass, will have significant income impacts (Claassen 2011). However, the biomass plant operation is a labor-proficient process and has limited employment opportunities. Alternatively, employment opportunities and income multipliers are sensibly high, reflecting the intensive investment in goods and services. Report of European Union on 2 MW (Megawatt) nonthermal hydrogen production from biomass plants in 2030 anticipated varied result. This report suggested more than 100,000 jobs generated over a 15-year period by construction and operation of 2,300 biohydrogen plants (http://www.hyways.de/).

1.3 Sustainability of Biohydrogen Production

Perhaps the most critical issues faced by today's society are identifying and building a sustainable energy system. Replacing our existing dependency on fossil fuel with a sustainable energy source is one of the major pieces in that system (Turner 2004). With 2.75 times greater energy yield (122 kJ/g) of hydrocarbon fuels, hydrogen is often cited as the green fuel (Das and Veziroglu 2008). Currently almost 96 % of the total production of H_2 comes from steam reforming of natural gas (48 %), partial oxidation of refinery oil (about 30 %), and coal gasification (18 %) (Holladay et al. 2009; Brentner et al. 2010; Corbo et al. 2011; Lee 2014). However, the hydrogen production process is energy intensive. It is also not environmentally friendly and unsustainable due to cost and high level of carbon emission.

Biohydrogen holds the potential for a substantial contribution to the future renewable energy demands. Biological H_2 production delivers clean H_2 in sustainable manner with simple technology and more attractive potential than the current chemical production of H_2 since it is suited for the conversion of a wide spectrum of substrate utilization such as organic wastes, industrial manufacturing process by-products, and biomass as feedstock costing almost zero (Venkata Mohan 2010; Maru 2014). However, present technology for biohydrogen production has its limitation. Development in process technology and pathways for industrial-scale biohydrogen production will make it more profitable, cost-effective sustainable energy option.

1.4 Future Perspectives

Currently hydrogen use is largely for chemical industry mainly to produce ammonia and methanol. Nonetheless, in the near future, hydrogen is expected to a fuel that will significantly improve the air quality (Kalamaras and Efstathiou 2013), provide economic stability (Lee and Hung 2012), and demonstrate social equitability (Claassen 2011; Sørensen 2012). Intensive efforts are going on throughout the globe to make hydrogen as a carbon neutral fuel by producing it via biological process (biohydrogen) and making it as a strong candidate to replace fossil fuel. Several technologies, feedstocks, and pathways have been demonstrated by researchers to produce biohydrogen, and some laboratory and pilot-scale studies for biohydrogen production by fermentations have come up with the promising results for industrial biohydrogen production (Show et al. 2011a, b).

Biohydrogen can be produced in three broad ways: by biophotolysis (using microalgae), dark fermentation, and photo fermentation (Melis and Melnicki 2006; Manish and Banerjee 2008; Sinha and Pandey 2011; Show et al. 2011a, b; Rathore and Singh 2013; Basak et al. 2014). Primarily the slow production rate and low hydrogen yield are two common challenges for the biological hydrogen-producing systems. Results from the last two decades suggested an encouraging scenario of biohydrogen production. There has been a significant improvement in the yield and volumetric production rate of hydrogen production and sanguine development in biological hydrogen production routes. However, for industrial approach that makes a sense in hydrogen economy, present production rate and hydrogen yield necessarily surpass the present achievements (Show et al. 2011a, b, 2012).

By an estimate 80 kg of hydrogen per acre per day could be produced by diverting the entire photosynthetic efficiency of the algae toward hydrogen production. In a realistic efficiency of 50 %, hydrogen production cost comes close to a \$2.80 a kilogram (Melis and Happe 2001). Though in the current scenario, below 10 % of the algae photosynthetic capacity was utilized for biohydrogen production (Show et al. 2012). Researches on biotechnological approach to improve algal photosynthetic biohydrogen production are underway and demonstrating promising result (Lay 2001; Oncel et al. 2015).

Choosing suitable process parameters such as illumination intensity, carbon to nitrogen ratio, age of inoculums, and bioreactor configuration can significantly improve the overall yield for biological hydrogen production by photo fermentation of purple non-sulfur bacteria (PNS) (Show et al. 2012; Basak et al. 2014). Several researchers favor dark fermentation or heterotrophic fermentation under anaerobic conditions since it is low cost, high rate, and high hydrogen-yielding process which can utilize various organic substrates and carbohydrate-rich wastewater (Hallenbeck and Ghosh 2009; Ghimire et al. 2015; Marone et al. 2015). In the dark fermentation, hydrogen is produced as an intermediate metabolite at the first stage and used as an electron donor at the second stage by many methanogens. It might be viable to harvest hydrogen produced in the first stage, leaving the remaining acidification products for further methanogenic process (Show et al. 2012).

Bioreactor design to improve process efficiency is another major aspect for the industrial biohydrogen production. The yield and conversion rates of biohydrogen bacteria in dark fermentation are highly dependent on the reactor type, reactor operating parameters, and media conditions. A good reactor design for biohydrogen dark fermentation should be able to operate at very low hydraulic retention time (HRT) at the same time avoiding the associated biomass washout (Arimi et al. 2015).

1.5 Conclusion

Continues increase in energy demand from the individual to the national level keeping an extra burden on exhaustible fossil fuel. The use of fossil fuel not only causes threat to the environment but also influences development of the country. The use of renewable resources could be an alternative approach to resolve the problem of energy resource. Biohydrogen could be a next generation biofuel by eliminating constraints of first- and second-generation biofuels and able to provide a sustainable option to replace current energy carrier mix. Biohydrogen is a carbon neutral process, which can be obtained from a variety of feedstocks. Nevertheless, biohydrogen is a potential candidate for future energy source, which could largely contribute to the energy security, improve air quality, and provide economic stability and social equitability.

References

Arimi MM, Knodel J, Kiprop A, Namango SS, Zhang Y, Geiben SU (2015) Strategies for improvement of biohydrogen production from organic-rich wastewater: a review. Biomass Bioenergy 75:101–118

Basak N, Jana AK, Das D, Saikia D (2014) Photofermentative molecular biohydrogen production by purple-non-sulfur (PNS) bacteria in various modes: the present progress and future perspective. Int J Hydrog Energy 39:6853–6871

Benchmarking of Biodiesel Fuel Standardization in East Asia Working Group (2010) Energy situation in the world. In: Goto S, Oguma M, Chollacoop N (eds) EAS-ERIA biodiesel fuel trade handbook: 2010. ERIA, Jakarta, pp 6–15

BP (2015) BP energy outlook 2035. British Petroleum. Available at: http://www.bp.com/content/dam/bp/pdf/energy-economics/energy-outlook-2015/bp-energy-outlook-2035-booklet.pdf. Accessed on 24 Feb 2016

Brentner LB, Peccia J, Zimmerman JB (2010) Challenges in developing biohydrogen as a sustainable energy source: implications for a research agenda. Environ Sci Technol 44(7):2243–2254

Claassen PAM (2011) Final activity report of integrated project "Non-thermal production of pure hydrogen from biomass". Available at: http://www.biohydrogen.nl/downloadattachment/22636/26107/PES_P5_Final_Report_08092011.pdf on 21 June 2015

Corbo P, Migliardini F, Veneri O (2011) Hydrogen fuel cells for road vehicles. Green energy and technology. Springer-Verlag London Limited, London, UK

Das D, Veziroglu TN (2008) Advances in biological hydrogen production processes. Int J Hydrogen Energy 33:6046–6057

Dhanya MS, Prasad S, Singh A (2013) Biogas technology for developing countries: an approach to sustainable development. In: Koress NE, O'Kiely P, Benzie JAH, West JS (eds) Bioenergy production by anaerobic digestion: using agricultural biomass and organic wastes. Routledge (Taylor and Francis Group), New York, pp 30–45

Energy Information Administration (2011) Annual energy outlook 2011 with projections to 2035. Washington, DC. Available at http://www.eia.gov/oiaf/aeo/aeoref_tab.htm

FAO (2005) Bioenergy. Sustainable Development Department, FAO, Rome, Italy. 11 Mar 2006. Available from http://www.fao.org/sd/dim_en2/en2_050402_en.htm. Accessed on 5 Jan 2012

Gallucci M (2015) Oil price outlook is still uncertain as global demand and supplies both surge, energy agency says. The International Business Times, New York. Available at: http://www.ibtimes.com/oil-price-outlook-still-uncertain-global-demand-supplies-both-surge-energy-agency-1883284 [Accessed on 30th May 2016].

Ghimire A, Frunzo L, Pirozzi F, Trably E, Escudie R, Lens PNL, Esposito G (2015) A review on dark fermentative biohydrogen production from organic biomass: process parameters and use of by-products. Appl Energy 144:73–95

Gullison RE, Frumhoff PC, Canadell JG, Field CB, Nepstad DC, Hayhoe K, Avissar R, Curran LM, Friedlingstein P, Jones CD, Nobre C (2007) Tropical forests and climate change. Science 316:985–986

Hallenbeck PC, Ghosh D (2009) Advances in fermentative biohydrogen production: the way forward? Trends Biotechnol 27:287–297

Hepbasli A (2008) A key review on exergetic analysis and assessment of renewable energy resources for a sustainable future. Renewable sustain. Energy Rev 12:593–661

Holladay JD, Hu J, King DL, Wang Y (2009) An overview of hydrogen production technologies. Catal Today 139(4):244–260

http://www.hyways.de/. Accessed on Jan 2015

IEA (2007) World energy outlook 2007 - special report - focus on energy poverty. International Energy Agency, Paris

IEA (2008) Worldwide trends in energy use and efficiency key insights from IEA indicator analysis. International Energy Agency, Paris

IEA (2014) World energy outlook 2014 factsheet. International Energy Agency, Paris

IEA (2015a) Key world energy statistics. International Energy Agency (IEA), France

IEA (2015b) World energy outlook 2015: executive summary. International Energy Agency (IEA), France

International Energy Agency (2006) Renewable energy: RD&D priorities, insights from the IEA technology programmes. IEA, Paris

Kalamaras CM, Efstathiou AM (2013) Hydrogen production technologies: current state and future developments. In: Conference papers in energy, Article ID 690627. Hindawi Publishing Corporation

Lay JJ (2001) Biohydrogen generation by mesophilic anaerobic fermentation of microcrystalline cellulose. Biotechnol Bioeng 74:280–287

Lee DH (2014) Development and environmental impact of hydrogen supply chain in Japan: assessment by the CGE-LCA method in Japan with a discussion of the importance of biohydrogen. Int J Hydrog Energy 39(33):19294–19310

Lee DH, Chiu LH (2012) Development of a biohydrogen economy in the United States, China, Japan, and India: with discussion of a chicken-and-egg debate. Int J Hydrog Energy 37:15736–15745

Lee DH, Hung CP (2012) Toward a clean energy economy: with discussion on role of hydrogen sectors. Int J Hydrog Energy 37:15753–15765

Levin DB, Chahine R (2010) Challenges for renewable hydrogen production from biomass. Int J Hydrog Energy 35:4962–4969

Ma T, Chen M, Wang C, Mao Z, Jiang M (2013) Study on the environment-resource economy comprehensive efficiency evaluation of the biohydrogen production technology. Int J Hydrog Energy 38:13062–13068

Maniatis K (2003) Pathways for the production of biohydrogen: opportunities and challenges. In: Towards hydrogen. IEA, Paris. 3 Mar 2003

Manish S, Banerjee R (2008) Comparison of biohydrogen production processes. Int J Hydrog Energy 33:279–286

Marone A, Varrone C, Fiocchetti F, Giussani B, Izzo G, Mentuccia L, Rosa S, Signorini A (2015) Optimization of substrate composition for biohydrogen production from buffalo slurry co-fermented with cheese whey and crude glycerol, using microbial mixed culture. Int J Hydrog Energy 40(1):209–218

Maru BT (2014) Sustainable production of hydrogen and chemical commodities from biodiesel waste crude glycerol and cellulose by biological and catalytic processes. PhD thesis submitted to Universitat Rovira I Virgili

Melis A, Happe T (2001) Hydrogen production. Green algae as a source of energy. Plant Physiol 127:740–748

Melis A, Melnicki MR (2006) Integrated biological hydrogen production. Int J Hydrogen Energy 31:1563–1573

Nigam PS, Singh A (2011) Production of liquid biofuels from renewable resources. Progress Energy Combust Sci 37:52–68

Oncel SS, Kose A, Faraloni C (2015) Genetic optimization of microalgae for biohydrogen production. In: Handbook of marine microalgae. Biotechnology advances. Academic, pp 383–404

Prasad S, Singh A, Joshi HC (2007a) Ethanol as an alternative fuel from agricultural, industrial and urban residues'. Resour Conserv Recycl 50:1–39

Prasad S, Singh A, Jain N, Joshi HC (2007b) Ethanol production from sweet sorghum syrup for utilization as automotive fuel in India. Energy Fuel 21(4):2415–2420

Rathore D, Singh A (2013) Biohydrogen production from microalgae. In: Gupta VK, Tuohy MG (eds) Biofuels technologies recent developments. Springer, Berlin, pp 317–333

Rathore D, Pant D, Singh A (2013) A comparison of life cycle assessment studies of different biofuels. In: Singh A, Pant D, Olsen SI (eds) Life cycle assessment of renewable energy sources. Spinger-Verlag, London, pp 269–289

Rathore D, Nigam PS, Singh A (2015) Biorefinery concept for a microalgal bioenergy production system. In: Kim SK, Lee CG (eds) Marine bioenergy: trends and developments. CRC Press, Boca Raton, pp 179–194

Schenk PM, Thomas-Hall SR, Stephens E, Marx UC, Mussgnug JH, Posten C, Kruse O, Hankamer B (2008) Second generation biofuels: high-efficiency microalgae for biodiesel production. Bioenerg Res 1:20–43

Sekoai PT, Daramola MO (2015) Biohydrogen production as a potential energy fuel in South Africa. Biofuel Res J 6:223–226

Sheehan J (2009) Biofuels and the conundrum of sustainability. Curr Opin Biotechnol 20:318–324

Show KY, Lee DJ, Jhang ZP (2011a) Production of biohydrogen: current perspectives and future prospects. In: Pandey A, Larroche C, Ricke SC, Dussap C-G, Gnansounou E (eds) Biofuels alternative feedstocks and conversion processes. Academic Press, Amsterdam, pp 467–479.

Show KY, Lee DJ, Chang JS (2011b) Bioreactor and process design for biohydrogen production. Bioresour Technol 102:8524–8533

Show KY, Lee DJ, Tay JH, Lin CY, Chang JS (2012) Biohydrogen production: current perspectives and the way forward. Int J Hydrog Energy 37:15616–15631

Sims REH, Schock RN, Adegbululgbe A, Fenhann J, Konstantinaviciute I, Moomaw W, Nimir HB, Schlamadinger B, Torres-Martínez J, Turner C, Uchiyama Y, Vuori SJV, Wamukonya N, Zhang X (2007) Energy supply. In: Metz B, Davidson OR, Bosch PR, Dave R, Meyer LA (eds) Climate change 2007: mitigation. Contribution of working group III to the fourth assessment report of the intergovernmental panel on climate change. Cambridge University Press, Cambridge, UK

Singh A, Olsen SI (2012) Key issues in life cycle assessment of biofuels. In: Gopalakrishnan K et al (eds) Sustainable bioenergy and bioproducts, green energy and technology. Springer-Verlag London Limited, London, UK, pp 213–228.

Singh A, Pant D, Korres NE, Nizami AS, Prasad S, Murphy JD (2010a) Key issues in life cycle assessment of ethanol production from lignocellulosic biomass: challenges and perspectives. Bioresour Technol 101(13):5003–5012

Singh A, Smyth BM, Murphy JD (2010b) A biofuel strategy for Ireland with an emphasis on production of biomethane and minimization of land-take. Renew Sustain Energy Rev 14(1):277–288

Singh A, Olsen SI, Nigam PS (2011) A viable technology to generate third generation biofuel. J Chem Technol Biotechnol 86(11):1349–1353

Sinha P, Pandey A (2011) An evaluation report and challenges for fermentative biohydrogen production. Int J Hydrog Energy 36:7460–7478

Sittijunda S, Reungsang A, O-thong S (2010) Biohydrogen production from dual digestion pretreatment of poultry slaughterhouse sludge by anaerobic self-fermentation. Int J Hydrog Energy 35(24):13427–13434

Skonieczny MT, Yargeau V (2009) Biohydrogen production from wastewater by Clostridium beijerinckii: effect of pH and substrate concentration. Int J Hydrog Energy 34(8):3288–3294

Sørensen B (2012) Hydrogen and fuel cells emerging technologies and applications, 2nd edn. Academic Press, Oxford

Subhadra B, Edwards M (2010) An integrated renewable energy park approach for algal biofuel production in United States. Energy Policy 38(9):4897–4902

Turner JA (2004) Sustainable hydrogen production. Science 305(5686):972–974

Venghaus S, Selbmann K (2014) Biofuel as social fuel: introducing socio environmental services as a means to reduce global inequity? Ecol Econ 97:84–92

Venkata Mohan S (2010) Waste to renewable energy: a sustainable and green approach towards production of biohydrogen by acidogenic fermentation. In: Singh OV, Harvey SP (eds) Sustainable biotechnology. Springer, Amsterdam, pp 129–164

Biohydrogen Production Potential of Different Biomass Sources

2

Hidayet Argun, Pelin Gokfiliz, and Ilgi Karapinar

Abstract

Hydrogen production from renewable energy sources has gained special attention in recent years. Especially biohydrogen production from biomass resources is accepted as an environmentally-friendly and sustainable approach. However, this process is slow due to the recalcitrant biomass structure hindering the liberation of readily fermentable sugars for fermentation. Therefore, biohydrogen production is usually integrated with a relevant biomass pretreatment process. This book chapter presents an overview of potential biomass resources, biomass pretreatment options, and fermentation processes used for biohydrogen gas production. The text focuses especially on separate and integrated dark and photofermentative hydrogen production processes by discussing principles and recent research outcomes from the literature.

2.1 Introduction

Hydrogen is a clean energy carrier with a high energy content and has a wide range of applications from transportation fuel to electricity generation. Currently, most hydrogen demand is supplied from fossil resources, such as natural gas and coal by steam reforming or gasification. However, these processes require high energy. Besides, CO_2 is produced as a by-product which is a main gas that causes greenhouse effect. In response to those problems, hydrogen production from renewable sources like biomass draws great attention.

Biomass is a general term which is called as organic material that is produced via photosynthesis by green plants including algae,

H. Argun
Department of Environmental Engineering, Pamukkale University,
Kınıklı Campus, 20070 Denizli, Turkey
e-mail: hargun@pau.edu.tr

P. Gokfiliz
Institute of Applied and Natural Sciences, Department of Biotechnology, Dokuz Eylul University,
Tinaztepe Kampus, 35160 Buca-İzmir, Turkey
e-mail: pelingokfiliz@hotmail.com

I. Karapinar (✉)
Department of Environmental Engineering, Dokuz Eylul University,
Tinaztepe Campus, 35160 Buca-İzmir, Turkey
e-mail: ilgi.karapinar@deu.edu.tr

A. Singh, D. Rathore (eds.), *Biohydrogen Production: Sustainability of Current Technology and Future Perspective*, DOI 10.1007/978-81-322-3577-4_2

trees, and crops (McKendry 2002). Utilization of biomass as feedstock for hydrogen production is not only cost-effective but also environmentally-friendly option, because the processes are carbon neutral (have net zero CO_2 emission) due to the fact that CO_2 is fixed in the atmosphere by plants during photosynthesis. Besides, agricultural crops (sugar and oilseed crops) and their waste by-products, lignocellulosic products such as wood and wood waste, aquatic plants like algae and water weeds, industrial or municipal solid wastes, and animal wastes are accepted as biomass sources (Nath and Das 2003; Caputo et al. 2005; Ni et al. 2006). Hydrogen can be produced from these biomass sources via thermochemical or biological processes. Over the last few decades, investigations on biological hydrogen production as a by-product of microorganism metabolism have accelerated for generating sustainable energy to meet increasing global energy demand (Gupta et al. 2013).

Dark fermentation and photofermentation are the major bioprocesses for hydrogen generation from carbohydrate-rich substrates. The most important criteria for raw material selection for biohydrogen production are its availability, carbohydrate content, fermentability, and cost (Kapdan and Kargi 2006). It is a fact that biomass is the most abundant biopolymer on Earth and an alternative resource to fossil fuels. Biomass sources for biohydrogen production have been categorized as first-generation biomasses (agricultural crops), second-generation biomasses (lignocellulosic wastes), and third-generation biomasses (algae). Second- and third-generation biomasses are preferred ones, since they are not food source. Recalcitrance of these biomasses is the limiting factor hindering microbial biomass degradation for hydrogen production. Therefore, fermentative hydrogen production is usually integrated with a relevant pretreatment process. Physical and chemical biomass pretreatments are mostly used processes prior to biohydrogen production. Enzymatic biomass pretreatments on the other hand are in minority due to slow conversion rates and practical difficulties. The complexity of pretreatment is directly related with the biomass content. Therefore, selection of a suitable biomass requiring less pretreatment steps is of great importance.

Production of clean energy source and utilization of biomasses make biological hydrogen production a novel and promising approach to meet the increasing energy needs as a substitute for fossil fuels. On the basis of these facts, this chapter focuses on biomass types, potential use of different biomasses as the raw material, biomass pretreatment options, biohydrogen production potentials from biomasses, bioprocessing strategies, and challenges on biohydrogen production.

2.2 Biomass Sources for Biohydrogen Production

First-generation biomasses are often edible agricultural crops which are grown for food and animal feed purposes (Sims et al. 2008; Lee and Lavoie 2013). Sugar-containing crops such as sugarcane and sugar beet; starch-containing ones such as wheat, barley, potato, and corn; and oily plants and seeds such as soybeans, sunflower, and palm are some of the examples for first-generation biomasses. A number of studies have been reported in the literature for biohydrogen production from first-generation biomasses, especially from starchy and sugar-rich biomasses due to easy fermentability attribute of these feedstocks by anaerobic organisms. Even though higher hydrogen yields are obtained from first-generation biomasses, the biggest obstacle when using these sources as feedstock is the utilization of arable land to produce energy crops instead of food production. This will lead both severe food shortages and overmuch usage of water and fertilizers (Dragone et al. 2010). For this reason, nowadays, biohydrogen production studies have been shifted from first-generation biomass to second-generation biomass.

Lignocellulosic biomasses including agricultural and forestry wastes and nonedible crop residues, as well as industrial and municipal organic wastes, wastes from food processing, and industrial effluents, constitute second-generation biomasses (Cheng et al. 2011; Singh et al. 2011). Beet molasses is a by-product of sugar industry and a common biomass that is used in fermentation processes due to its high sucrose content. Besides carrot pulp (de Vrije et al. 2010), solid organic wastes such as carbohydrate-rich wastes (apples, carrots, Jerusalem artichoke roots, maize flour, oats, potatoes, and wheat flour), protein-rich wastes (soybean milk cake, chicken meat, cow manure with straw, fish residues, and meat waste from restaurants), agro-industrial wastes (including food waste from restaurants, rapeseed oil cakes, sunflower oil cakes, grape marc, fruit peels – orange peels and banana peels – and maize cob), agricultural residues (Jerusalem artichoke leaves and stalks, giant reed stalks and leaves, maize stalks, rice straw, and sorghum stalks) (Guo et al. 2014), palm oil mill effluent (POME) (Al-Shorgani et al. 2014), distillery wastewater (Sridevi et al. 2014), and waste papers (Ntaikou et al. 2009) are second-generation biomasses that were used as substrate for biohydrogen production.

Second-generation biomasses should be considered as organic wastes of agricultural or industrial activities, in general. Therefore, these biomasses are abundant and cheap and they do not compete with food production. Researchers have been taking a great interest in biohydrogen production from second-generation biomasses especially agricultural and forestry by-products due to their low cost. Qian (2014) stated that estimated annual production of lignocellulosic biomass is over 200 billion tons on Earth. The main component of plant biomass and primary building block of plant cell walls is lignocellulose. The structure of lignocellulose is complicated because it is composed of mainly 20–45 % cellulose, 16 % and 37 % hemicellulose, and 12–26 % lignin with wide range of carbon to nitrogen ratio between C/N = 118 and C/N = 10 (Sawatdeenarunat et al. 2015). Cellulose is a linear polymer of D-glucose which is linked to each other by β-(1,4)-glycosidic bonds. Hemicellulose is a branched heteropolymer consisted of pentoses (D-xylose, L-rhamnose, and L-arabinose), hexoses (D-glucose, D-mannose, and D-galactose), and uronic acid derivatives (e.g., D-glucuronic, D-galacturonic acids). Lignin is a complex, hydrophobic polymer and is made of cross-linked phenolic monomers. These biopolymers are difficult to be fermented by anaerobic microorganisms directly for biohydrogen production. Thus, a pretreatment step is required to remove lignin and to hydrolyze complex carbohydrates into their monomers. Biohydrogen production from second-generation biomass on a large scale is a challenge due to the requirement of high pretreatment costs.

Algae are third-generation biomasses and they have been in use as feedstock for biohydrogen production due to their rich carbohydrate content. They are unicellular or multicellular organisms which can be classified as prokaryotic, like cyanobacteria (blue-green algae), or eukaryotic such as green algae, red algae, and brown algae. Generally, algae are grouped as microalgae and macroalgae, according to their morphology and size. They may grow autotrophically (use atmospheric CO_2 via photosynthesis), heterotrophically (use organic carbon), or mixotrophically (use both inorganic and organic carbon depending on the condition). Algae can store carbon in the form of starch, cellulose, and lipids. The carbohydrate source in algae is mainly starch which is deposited in the cytoplasm and cellulose in the cell wall. The carbohydrate storage type can be different according to the types of algae. For instance, cyanobacteria have glycogen, green algae and red algae have starch, and brown algae have β-glucans as a storage carbohydrate (Mollers et al. 2014). The carbohydrate content of algae can vary between 30.7 and 48.2 % (Batista et al. 2014; Yun et al. 2014; Liu and Wang 2014; Nayak et al. 2014). Algae are common biomass in bioethanol and biodiesel production. But, researchers have focused on biohydrogen production from third-generation biomass due to many advantages such as high CO_2 capture rate, ease of cultivation, rapid biomass production, and high carbohydrate content. The other advantage of using algae as substrate in biohydrogen production is the absence of lignin and hemicellulose. The pretreatment requirement for the separation of lignin and hemicellulose as applied in second-generation biomass is omitted. Thus, cost of pretreatment is reduced and formation of some toxic end products as furfurals and 5-hydroxymethylfurfural (5-HMF) during pretreatment does not occur.

2.3 Pretreatment (Hydrolysis) of Biomasses for Biohydrogen Production

Biomass is an abundant renewable resource for biological hydrogen gas production and can be found in different forms in nature as presented in Sect. 2.2. However, in most cases biomass cannot directly be used as feedstock for fermentative hydrogen production due to the presence of hardly biodegradable compounds and unsuitable nitrogen levels. Hydrogen-producing microorganisms prefer fermentable sugar monomers and low nitrogen in the substrate. Therefore, biomass is usually subjected to a convenient pretreatment in order to provide the desired microbial milieu conditions. A general schematic diagram covering various pretreatment steps for biohydrogen production is shown in Fig. 2.1.

Biomass pretreatment is related with the use of several sources like energy, water, chemicals, equipment, and by-product formations like wastes and toxic compounds as shown in Fig. 2.1. Therefore pretreatment with little resource consumption requirement and waste production under simple operation conditions with high product formation yields should be preferred. An ideal pretreatment process is identified as a process that produces a disrupted, hydrated substrate that is easily hydrolyzed with no toxic by-product formation (Agbor et al. 2011).

Feedstock preparation from biomass involves a number of steps prior to fermentation as shown in Fig. 2.1. Therefore, the selection of a proper biomass is of great importance. Feedstock preparation requirements for hydrogen production from green or brown colored biomass, for example, are different. Green biomasses such as grass leaves or cabbage residues are rich in hemicellulose, nitrogen, and lignin (Ruggeri and Tommasi 2012). Therefore green biomass requires more pretreatment steps when compared to brown biomass (corn, wheat straw, corn stover, etc.) that is generally rich in cellulose but poor in nitrogen (CalRecycle 2015). Thus the decision on starting with a convenient biomass will significantly

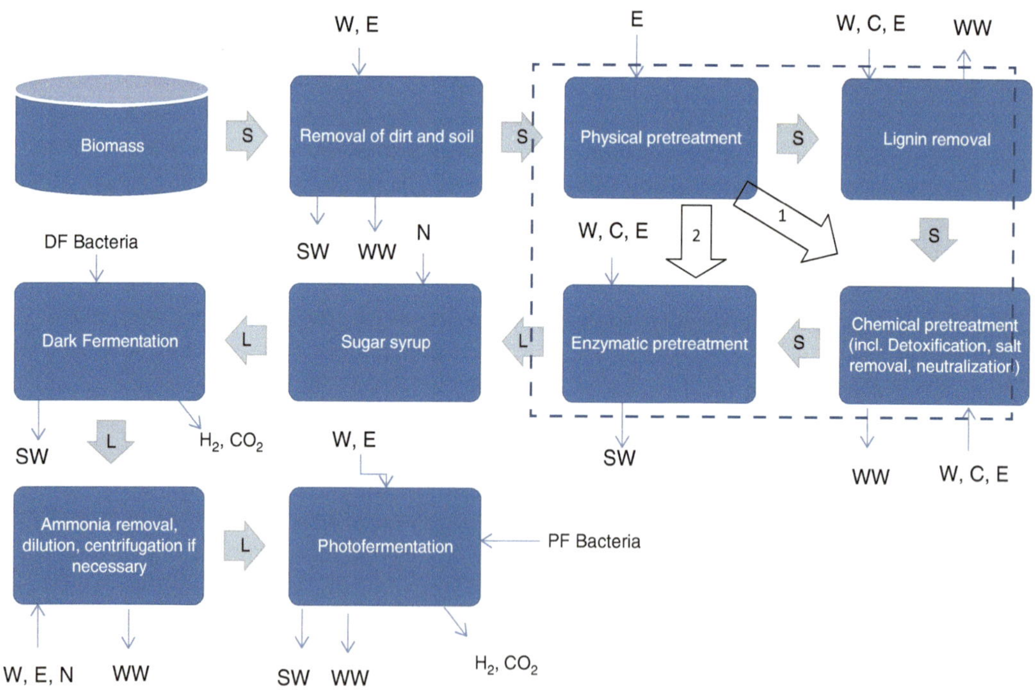

Fig. 2.1 Fermentative hydrogen gas production from biomass (*E* energy, *S* solid, *L* liquid, *C* chemicals, *W* water, *SW* solid waste, *WW* wastewater, *1* 1st alternative route, *2* 2nd alternative route, *area within dashed line* biomass pretreatment processes, *DF* dark fermentation, *PF* photofermentation, *N* nutrient)

affect required pretreatment stages and operation costs in hydrogen production.

After starting with a proper biomass, the next step is the removal of dirt and soil. Soil may contain hydrogen-consuming methanogens that may negatively affect hydrogen production performance. On the other hand, any coarse material within the biomass might cause damage during further operation phases. Dirt and soil removal can be accomplished by washing or by hand sorting. Usually washing is applied if there is dirt or soil on biomass. Wet biomass cannot be stored due to microbial attack and degradation of fungi. Therefore, it has to be dried by solar or oven heating if storage is considered (Velázquez-Martí et al. 2011). Drying at high temperature could cause collapse of cellulose pores which is an irreversible process limiting microbial degradation (Dowe and Nrel 2001).

If biomass is considered to be used as feedstock in fermentation, the lignins and hemicelluloses that hinder or block enzymatic attack and the crystallinity of cellulose have to be reduced by suitable pretreatment processes. Depending on the biomass structure, pretreatment can be applied in single or multisteps (Agbor et al. 2011; Alvira et al. 2010; Mosier et al. 2005). Biomass pretreatment processes are classified in general as physical, chemical, and biological processes (FitzPatrick et al. 2010). Physical pretreatment is related with size reduction or the contribution of a physical force to decompose the biomass structure to some extent. Chemical pretreatment is usually applied at high temperatures under severe acidic or alkaline conditions. Biological pretreatments, on the other hand, can be accomplished at ambient operation conditions with lower conversion rates and yield of complex carbohydrates to their monomers. The objectives in all pretreatment steps are to enable microbial access to fermentable sugars within the biomass. Factors limiting this access are the cellulose crystallinity, degree of polymerization, lignin and hemicellulose structure, available surface area for enzymatic attack, particle size, and porosity of biomass (Alvira et al. 2010). The effects of different pretreatment processes on biomass structure are summarized in Table 2.1, and the principles are briefly explained in the following sections.

2.3.1 Physical Pretreatment

Physical pretreatment involves the treatment of biomass by physical forces without using any chemicals or microorganisms (Zheng et al. 2014). Comminution, steam explosion, liquid hot water, extrusion, and irradiation are the most used physical pretreatment processes (Zheng et al. 2014). The main goal in physical pretreatment is to enhance the accessible biomass surface area by decreasing the particle size, cellulose crystallinity, and degree of polymerization (Alvira et al. 2010). Achieving these goals enables a more efficient chemical or microbial hydrolysis of the biomass matrix and decreases hydrolytic enzyme limitations (Chandra et al. 2007; Mansfield et al. 1999). Most of the studies in the literature relating to biohydrogen production from biomass usually report the application of physical treatment prior to fermentation, and mechanical size reduction is the most popular method applied.

Size reduction by mechanical comminution is an energy-intensive process and can be accomplished by different equipment such as ball or hammer millers, shredders, and grinders (Alvira et al. 2010). The energy consumption of this process increases depending on the biomass structure, the moisture, and the particle size that is to be obtained. For example, a particle size less than 2 mm is desired prior to delignification of different types of biomass by the alkaline peroxide pretreatment process (Sun et al. 2000; Gould 1985; Banerjee et al. 2011). Size reduction after chemical pretreatment was suggested as an option to decrease energy consumption and process costs. However, this is not applicable for all kind of biomass types (Zhu et al. 2010).

Physical pretreatment of biomass can also be accomplished without using mechanical size reduction. Steam explosion, for example, enables a direct pretreatment of biomass. This technique was reported to be one of the most cost-effective

Table 2.1 Effects of different pretreatment methods on biomass structure

	Pretreatment	Increase of accessible surface area	Decrystallization of cellulose	Solubilization of hemicellulose	Solubilization of lignin	Alteration of lignin structure	Formation of toxic products like furfural/hydroxymethylfurfural (HMF)
Physical	Mechanical	●	●				
	Irradiation	●	○	○			○
	Steam explosion	●		●	○	●	●
	Liquid hot water	●	ND	●	○	○	○
Chemical	Catalyzed steam explosion	●		●	●/○	●/○	●
	Acid	●		●	○	●	●
	Alkaline	●		○	●/○	●	○
	Oxidative	●	ND		●/○	●	○
	Ionic liquids	●	●	○			
	Thermal acid	●	ND	●			●
	Thermal alkaline	●	ND	○	●/○	●	○
	Thermal oxidative	●	ND	○	●/○	●	○
	Ammonia fiber explosion	●	●	○	●	●	○
Biological	Enzymatic	●	ND	●	●	●	

Adapted from Mosier et al. (2005) and Zheng et al. (2014)

options for agricultural biomass pretreatment (Prasad et al. 2007). During steam explosion, biomass is subjected to a quick pressure release under high temperature (Han et al. 2010), which causes cell wall disruption due to heating and shearing forces and the hydrolysis of glycosidic bonds by formed organic acids (Jacquet et al. 2011). However this process still suffers from problems such as toxic and inhibitory by-product (e.g., 5-hydroxymethylfurfural (5-HMF) and furfurals) formation (Cantarella et al. 2004) and incomplete removal of lignin (Shevchenko et al. 1999; Martín-Sampedro et al. 2012). Factors affecting the efficiency in steam explosion are particle size, biomass humidity, temperature, pressure, and residence time (Jeoh 1998). Liquid hot water pretreatment is a similar process with steam explosion, however, with higher retention times and no sudden pressure release. During liquid hot water treatment, biomass fractions are exposed to hot water resulting in cellulose hydrolysis, hemicellulose solubilization, and delignification to some extent due to penetration of hot water into the cell structure (Zheng et al. 2014). Different from steam explosion, liquid hot water treatment retention times vary from minutes up to hours (Zheng et al. 2014). Steam explosion and liquid hot water treatments can be preceded in temperature- and pressure-resistant reactors.

Another way of physical treatment is extrusion where the biomass is exposed to heating, mixing, and shearing while passing through an extruder (Agbor et al. 2011). In this process the biomass is forced to move through a barrel with the torque of single or twin screws (Zheng and Rehmann 2014). This movement produces compression and expansion within the zones of the barrel, disrupting the biomass structure into shorter fibers with broken-down cell walls (Karunanithy et al. 2012). Temperature, screw speed, retention time, and biomass moisture are critical parameters affecting the extrusion process (Karunanithy and Muthukumarappan 2012).

Biomass disruption by physical treatment could also be accomplished by irradiation methods like microwave, gamma ray, ultrasound, or electron beam applications (Zheng et al. 2014. When biomass is irradiated with microwaves, macromolecules and water within the cells absorb the wave energy resulting in enormous heat generation and consequently cell wall degradation in a short period of time (Chaturvedi and Verma 2013). Biomass irradiation by gamma rays is another option that causes glycosidic bond breakage and cell degradation (Orozco et al. 2012). When biomass is subjected to ultrasonication, a phenomenon called bubble collapse occurs by cavitation, resulting in high pressure and heat formation which disrupts the cell wall (Luo et al. 2014). Electron beam irradiation is a pretreatment process that induces a cleavage and depolymerization mechanism within the lignocellulosic biomass structure (Bak 2014). This process can proceed under normal pressure and temperature which is accepted as an advantage (Mehnert 1995).

2.3.2 Chemical Pretreatment

Chemical pretreatment (CPT) is the process where chemicals are used to depolymerize the biomass (Harmsen et al. 2010). The objective of CPT is to enable enzymatic access to fermentable sugars by breaking down the lignin and hemicellulose structures or the related chemical bonds (Badiei et al. 2014). Acid, alkaline, organosolv, ammonia fiber explosion (AFEX), ionic liquids (IL), oxidation, or any combination of those processes is the most applied CPT methods (Harmsen et al. 2010). There exist many studies in the literature regarding CPT of biomass for various biofuel productions and the mechanisms are explained as well (Harmsen et al. 2010; Ulhoff HET 2012). Therefore, this section will focus on some intensively used CPT technologies that were applied for biohydrogen production after a short description of each process (see Table 2.1).

Most of the studies on chemical biomass pretreatment in the literature report dilute acid hydrolysis (AH). Among several acids like HCl, HNO_3, H_3PO_4, and some volatile fatty acids, H_2SO_4 is preferred since it is less corrosive and relatively cheaper (Jeihanipour et al. 2011). The acid used in hydrolysis can be in liquid or solid form (Guo et al. 2012). Dilute AH can be accomplished at 100–250 °C, in 0.5–30 min with 0.5–3 % acid concentrations. Concentrated AH usually takes place at room temperature with acid concentrations above 30 % (Taherzadeh and Karimi 2007a). The main disadvantages in AH are toxic by-product (Harmsen et al. 2010) and excess salt formations (Liu et al. 2013). Also corrosion, temperature, and pressure-resistant reactors required are other factors to be considered for process simplicity (Taherzadeh and Karimi 2007a). Toxic compounds like 5-HMF and furfural are produced more during dilute AH due to dehydration of hexose (C_6) and pentose (C_5) sugars at high reaction temperatures (Mosier et al. 2005). Biomass dissolution can proceed faster in concentrated AH (Jung et al. 2013); however, the need of high alkaline dosages for neutralization results in more salt formation than that in dilute AH. During AH, biomass depolymerization occurs due to hemicellulose dissolution and breakage of glycosidic bonds (Taherzadeh and Karimi 2007a). Hydrogen can be produced from the obtained C_5 and C_6 sugar monomers after AH, though C_6 sugars like glucose are preferred by the majority of microorganisms (Lai et al. 2014). The efficiency of AH can be enhanced by optimizing pretreatment operation condition parameters like temperature, pressure, retention time, solid/liquid ratio, and acid dosage (Panagiotopoulos et al. 2011).

Biomass depolymerization can also be done by alkaline pretreatment where NaOH, KOH, and $Ca(OH)_2$ (lime) are used as chemicals (Chaturvedi and Verma 2013). High-temperature alkaline pretreatment, which is also known as the Kraft pulping process, has been in use for years in the pulp and paper industry (Perlack et al. 2005). Lignin and hemicellulose can effectively be removed at hot and alkaline conditions resulting in a recovery of the cellulose for paper production (Sixta and Rutkowska 2006). The recovered cellulose in this way has also been used for biofuel production (Jeihanipour and Taherzadeh 2009). The alkaline treatment is usually applied at high temperature where cellulose cleavage occurs due to scission and cleavage reactions (Knill and Kennedy 2002). This, however, may result in unfavorable toxic 5-HMF and furfural productions due to dehydration of C_5 and C_6 sugars at elevated temperatures (Yin et al. 2011) Those unfavorable conditions could be prevented by integrating the usage of hydrogen peroxide which is known as the alkaline peroxide (ALPER) process that can be operated at room temperature and does not result in toxic by-product formation (Karagöz et al. 2012; Gould and Freer 1984).

2.3.3 Enzymatic Pretreatment

It has been clearly stated that biomass can be converted into fermentable feedstock for biohydrogen production after a suitable physical or chemical pretreatment process. However, it should be realized that physical or chemical biomass pretreatments may require intensive energy, chemicals, and sometimes severe operation conditions resulting in wastewater and toxic by-product formations.

Therefore, selection of an environmentally-friendly and sustainable process is of great importance. In this context, enzymatic pretreatment could be considered as an alternative option to physical and chemical biomass pretreatments.

Microorganisms such as fungi (*Trichoderma reesei*, *Phanerochaete chrysosporium*) and some bacteria (*Clostridium thermocellum*, *Ruminococcus albus*) can produce special proteins that can degrade biomass to liberate fermentable sugars which is known as enzymatic biomass degradation (Rodrigo de Souza 2013). Enzymatic reactions can proceed under mild operation conditions without toxic by-product formation (Verardi et al. 2012). However, this process is slower and more expensive compared with physical and chemical pretreatment processes (Taherzadeh and Karimi 2007b). Enzymes that take part in biomass degradation are diverse. In general cellulose-, hemicellulose-, and lignin-degrading enzymes are cellulases, xylanases, and lignin peroxidases and laccases, respectively (Perez et al. 2002). Cellulases can degrade β-(1,4)-glycosidic bonds of cellulose resulting in glucose and cellobiose formation which can further be used as feedstock for biohydrogen gas production (Carere et al. 2008). Xylanases, on the other hand, can degrade hemicelluloses into C_5 sugars like xylose, arabinose, or mannose (De Menezes et al. 2010). Mostly, enzymatic hydrolysis is used for C_6 production since the majority of microorganisms can metabolize C_6 sugars rather than C_5 sugars for hydrogen production (Chen et al. 2013). Enzymes can be used in pure form or could be generated by microorganisms spontaneously during the fermentation process. The latter is generally preferred in hydrogen production due to economic and practical considerations.

Table 2.2 summarizes different pretreatments applied to first-, second-, and third-generation biomass resources prior to fermentative hydrogen production. The biomass and hydrolysate content along with experimental pretreatment conditions are presented as well. As can be seen from Table 2.2, there is no single way to treat biomass for hydrogen production. Usually pretreatment starts with physical size reduction followed with diverse combinations of chemical and enzymatic treatments. Also detoxification is usually applied after chemical pretreatments in order to remove toxic by-products like 5-HMF, furfural, salts, and volatile fatty acids from the pretreatment effluents. Detoxification stage is not required in enzymatic pretreatment. Activated carbon adsorption, ion-exchange resin, and lime treatment are the most intensively used detoxification processes among several detoxification methods.

2.4 Biohydrogen Production from Biomass

Biological processes for hydrogen production from carbon-containing materials are achieved mainly by dark fermentation and photofermentation (light fermentation). The major substrates required in the dark fermentation are simple sugars like glucose and sucrose. The substrate for photofermentation is organic acids such as acetic, butyric, and lactic acids. Sequential dark fermentation and photofermentation or combined dark fermentation and photofermentation processes are two-stage fermentation approaches used in order to increase the yield of production. The simple sugars are converted to organic acids in dark fermentation, and then these products are used as substrate in photofermentation.

The main problem in biohydrogen production is the cost of the substrate when pure simple sugars are used for this purpose. In order to solve this problem, the recent trend is the utilization of carbon-rich waste materials or biomasses. Organic wastes must be somehow converted into environmentally acceptable form. The conventional approach to achieve this goal is applying one of the waste treatment or disposal technology in order to convert carbon content of these wastes into CO_2. However, these methods are cost intensive and they are not successful all the time for the complete conversion of their carbon content all the way to CO_2. Using biomass for energy production like hydrogen helps to overcome both substrate requirement of biohydrogen and intensive cost requirement of waste disposal.

Two main stages in biohydrogen production from biomass are hydrolysis and fermentation. The approaches for the realization of these stages are (i) separate hydrolysis and fermentation (SHF), (ii) simultaneous saccharification and fermentation (SSF), and (iii) direct fermentation (DF) or also known as consolidated bioprocessing (CBP).

Separate hydrolysis and fermentation is the process in which biomass is first hydrolyzed by pretreatment methods as mentioned in Sect. 2.3 to obtain fermentable sugars and to remove nonfermentable lignin in a separate unit. Then, the sugar solution is subjected to dark fermentation or photofermentation in a bioreactor for hydrogen production. SHF provides certain advantages. Hydrolysis of complex substrates is the rate-limiting step in most of the bioprocesses, and conducting hydrolysis in a separate process overcomes this problem. The operating conditions in pretreatment can be optimized to reach maximum biomass to sugar solution conversion yield. The desired hydrolysis products as lignin and toxic substance-free sugar solution can be obtained by using appropriate pretreatment methods. Finally, required or optimized sugar concentration with addition of other nutrients, which may be required

Table 2.2 Biomass pretreatment processes for fermentative hydrogen production

Biomass and its composition	Physical pretreatment	Chemical pretreatment	Enzymatic pretreatment	Detoxification	Pretreated product composition	Reference
Rice straw	Shredding and crushing	Concentrated acid hydrolysis	–	Sulfate removal and neutralization by $Ca(OH)_2$ treatment for 20 h	51.73 g total sugar/L	Liu et al. (2013)
	200 μm	(55 % H_2SO_4, 40 °C, 2 h)				
Cornstalk	Milling	–	1st step:	–	6 g/L glucose	Zhao et al. (2014)
	<40 mesh size	–	Lignin and holocellulose removal by *Phanerochaete chrysosporium* for 21 days	–	3.5 g/L xylose	
					1.1 g/L arabinose	
			2nd step:	–	54 mg/L acetic acid	
			Enzymatic saccharification by enzymes of *Trichoderma viride* for 4 days		No toxic product formation	
Water hyacinth	Pulverization	Acid pretreatment by microwave heating (1 % H_2SO_4, 140 °C for 15 min)	Cellulase treatment for 120 h (0.15 g, P10 U/mg)	Detoxification by 1.67 g/L activated carbon (37 °C, 180 rpm, 20 min)	17.24 g/L glucose	Cheng et al. (2015)
(24.9 % cellulose, 23.2 % hemicellulose,					1.56 g/L acetic acid	
10.1 % lignin					1.21 g/L furfural	
					0.51 g/L 5-HMF	
20.4 % protein					0.076 g/L vanillin	
Mixture of cassava starch and *Chlorella pyrenoidosa*	–	Steam heating with dilute acid (1 % H_2SO_4, 135 °C for 15 min)	–	–	13.7 g/L reducing sugar (glucose equivalent)	Xia et al. (2014)

Wheat straw	Hammer milling <0.5 mm	Alkaline pretreatment (7.4 % $Ca(OH)_2$, 20 °C, 48 h)	–	–	0.685 g total carbohydrates/g vs untreated straw	Reilly et al. (2014)
Corn stover		1st step: acid pretreatment	–	Detoxification by 100 g/L activated charcoal treatment (30 °C, 200 rpm, 60 min)	13.14 g/L glucose	Datar et al. (2007)
42.2 % glucan		1.2 % H_2SO_4, 2 h	–	Detoxification by 100 g/L activated charcoal treatment (30 °C, 200 rpm, 60 min)	32.32 g/L xylose	
25.7 % xylan 15.6 % lignin 11.0 % extractives, 1.9 % protein, and 3.6 % ash		2nd step: steam explosion (200 °C, 1 min)	–		2.85 g/L-galactose 4.63 g/L-arabinose 1.97 g/L-mannose	
Wetland biomass mixture of water arum (*Phragmites* (sp.)) and green reed (*Thalia dealbata*) (37–43 % cellulose, 23–33 % hemicelluloses, and 13–14 % lignin)	Cutting and milling 60 mesh	1st step: steam explosion 2nd step: ionic liquid pretreatment with 1-butyl-3-methyl-imidazolium chloride as ionic liquid ([Bmim]Cl)	Cellulase treatment for 24 h (20 FPU/g cellulase/substrate)	–	72.11–88.8 % glucose (%w/w biomass) 0.62–1.27 % xylose (%w/w biomass)	Peng et al (2014)
Red algae	Blender milling	Dilute acid pretreatment	–	Detoxification by 100 g/L activated carbon (35 °C, 150 rpm, 6 h)	15 g/L total carbohydrate by	Park et al. (2013)

(continued)

Table 2.2 (continued)

Biomass and its composition	Physical pretreatment	Chemical pretreatment	Enzymatic pretreatment	Detoxification	Pretreated product composition	Reference
(*Gelidium amansii*) (2.4 % agar, 14.9 % cellulose, 15.6 % protein, and 5.7 % ash)	<0.3 mm	1.30 % H_2SO_4 159 °C, 15 min			69.7 % hydrolysis efficiency	
Miscanthus 38.2 % cellulose 24.2 % hemicellulose 25 % lignin 1.3 % protein 4.2 % solvent extract 1.4 % hot water extract 2 % ash	1st step: chopping to 0.5–5 cm 2nd step: extrusion by twin extruder	Alkaline treatment During extrusion (12 % NaOH, 70 °C, 4 h)	Cellulase treatment for 72 h (10 FPU/g glucan)	–	10 g glucose 3.7 g xylose+ Arabinose	de Vrije et al. (2002)
Soybean straw 39.93 % cellulose 14.45 % hemicellulose 23.31 % lignin	Comminution to get 20 mesh	Alkaline peroxide treatment 16 % H_2O_2 and 0.5 % NaOH, 30 min	–	–	18 % total reducing sugar 33 % cellulose 5.5 % hemicellulose 8 % lignin	Han et al. (2012)

Corncob 32 % cellulose 41.8 % hemicellulose 8.6 % lignin and 0.9 % ash	Knife milling 2 mm	Acid steam explosion 0.5 % H_2SO_4, 121 °C, 20 min	Cellulase treatment for 72 h	–	5.1 g/L reducing sugar	Tang et al. (2013)
Wheat straw 35.1 % cellulose 24.8 % hemicellulose and 20.4 % lignin	Milling <2 mm	Ozonation	–	–	564 mg sugar/g dry straw	Wu et al. (2013)
Arthrospira platensis (*Spirulina*) (CYA-1)	–	Ultrasonication (200 W ultrasound, 15 min)	1st step: α-amylase (activity ≥ 50 U/mg) treatment for 2 h (60 °C, 120 rpm) 2nd step: glucoamylase (activity 70 U/mg) treatment for 12 h (60 °C, 120 rpm)	–	407 mg sugar/g dry weight	Cheng et al. (2012)
Cornstalk (37.1 % cellulose, 21.5 % hemicelluloses, and 18.5 % lignin)	Milling 0.5–2 mm	Lime treatment 0.1 g/g biomass, 23 ± 2 °C, 96 h	–	–	40.1 % cellulose 18.5 % hemicellulose 16.7 % lignin	Cao et al. (2012)
Cotton (85 % cellulose, 0 % lignin)	–	Concentrated acid hydrolysis (55 % H_2SO_4, 40 °C, 2 h)	–	Sulfate removal by ion-exchange resin	51.73 g reducing sugar/L	Chu et al. (2011)

for fermentation, can be controlled. The disadvantage of SHF is the cost of pretreatment used for hydrolysis. Selection of a pretreatment method with high yield of sugar formation and free of toxic substances, hydrogen production potentials depending on pretreatment methods, and biomass are the major interests in SHF

In simultaneous saccharification and fermentation, hydrolysis of biomass to sugar solution by corresponding enzymes (see Sect. 2.3.3) and fermentation of that sugar by hydrogen-producing organisms are realized in a single unit. The problem in this approach is the difference in the rate of hydrolysis and fermentation. Low rate of hydrolysis will cause low concentration of sugar formation which could then reduce the rate of fermentation and thus hydrogen formation. Another problem is adjusting the necessary conditions for maximum activity of enzyme in hydrolysis and growth or activity of microorganisms for biohydrogen production. In case raw biomass is used, there is a possibility of the occurrence or existence of some inhibitory substance that could lower the activity of enzyme. A physical or chemical pretreatment stage may be required prior to SSF to make the biomass suitable for hydrolysis reaction. The selection of enzyme with high activity and pretreatment methods are the major concerns in SSF studies.

Direct microbial conversion is achieved in a single unit by using single microorganism culture or consortium of microorganisms which are capable of hydrolyzing and then fermenting the biomass to hydrogen gas. The process is more cost-effective with respect to SSF and SHF since no chemical consumption and external enzyme additions are required for pretreatment and hydrolysis of biomass, respectively. The process would be ideal if only hydrolysis and fermentation are achieved by single organisms without a rate-limiting step. Nevertheless, the problem about the difference in the rate of hydrolysis and fermentation still exists in this approach. The optimal growth conditions could be different when the consortium of organisms as hydrolyzing and fermenting is used. Then, control of operating condition with respect to organism type could be a nuisance which may affect both hydrolysis and fermentation rates. Moreover, sugars obtained from microbial hydrolysis can be consumed by hydrolyzing organisms for growth instead of hydrogen-producing organisms with the result of decreasing hydrogen production potential. One of the strategies developed in order to divert sugar consumption to hydrogen production rather than growth of hydrolyzing culture was "temperature shift" (Lo et al. 2011). Hydrolysis was conducted at 35 °C to obtain sufficient amount of cellulase/xylanase production by *Cellulomonas uda* E3-01, and then temperature was raised to 45 °C to increase the rate of hydrolysis and to prevent the sugar consumption by the hydrolyzing organisms. But, results indicated that sugar was used for growth by hydrolyzing organism even though an effective temperature shift was achieved (Nasirian et al. 2011). Direct microbial conversion studies concentrate on isolation and selection of microbial cultures capable of producing high concentration of hydrolytic enzyme and hydrogen. Cocultures or consortium of either mesophilic or thermophilic organism has been identified and used for direct conversion of lignocellulosic biomass to biohydrogen. It has been stated that production by cocultures is higher than that of monocultures (Ho et al. 2012; Li and Liu 2012; Ren et al. 2008; Liu et al. 2008; Geng et al. 2010).

2.4.1 Dark Fermentative Hydrogen Production from Biomass

Hydrogen production by dark fermentation is realized under the acidogenic phase of the anaerobic process. Methanogenic phase is inhibited to prevent the consumption of hydrogen. In other words, substrate is converted to organic acids preferably acetic and butyric acid, and then no further conversion of these organic acids to methane or CO_2 is desired. Hydrogen production potential of dark fermentation process is evaluated through the yield of formation as mol H_2/mol glucose. Theoretically conversion of 1 mol of glucose yields 12 mol of hydrogen gas (H_2). According to the reaction stoichiometry, bioconversion of 1 mol of glucose into acetate yields 4 mol H_2/mol glucose, but only 2 mol H_2/mol glucose is formed when butyrate is the end product.

Ethanol and lactate are non-hydrogen-forming and propionic acid is a hydrogen-consuming end product of fermentation. The observed yields from glucose are in general around 2.0–2.5 mol H_2/mol glucose (Kapdan and Kargi 2006) which is lower than theoretical estimations. The reasons of low yields are the formation of butyric and acetic acid mixture which closes the yield around 2.5 mol H_2/mol glucose, utilization of substrate as an energy source for bacterial growth rather than desired organic acid generation and hence hydrogen formation, or generation of non-hydrogen-forming or hydrogen-consuming end products. The effects of environmental conditions, media compositions, and microbial culture type to reach the maximum theoretical yield of hydrogen formation from simple sugars have been evaluated in detail (Kapdan and Kargi 2006). The recent challenge in dark fermentative hydrogen production is to develop process technologies for biomass.

Hydrogen production can occur under mesophilic, thermophilic, and hyperthermophilic conditions. The well-known mesophilic hydrogen-producing microorganism species is *Clostridium* sp. which is a Gram-negative, spore-forming bacterium (Chong et al. 2009; Kapdan and Kargi 2006). Pure *Clostridium* sp. can be used in the production. However, anaerobic sludge is pretreated by acid, heat, or chemicals for the formation of *Clostridium* spore and elimination of methanogens. The most common method is the heat treatment in which anaerobic sludge is exposed to high temperature (90–100 °C) for a certain period of time (Ho et al. 2012; Panagiotopoulos et al. 2010; Saraphirom and Reungsang 2010; Fang et al. 2006; Argun et al. 2008a). Then, *Clostridium* spores are activated by adjusting the media composition and environmental conditions for hydrogen production. The recent trend is the utilization of thermophilic and hyperthermophilic cultures for hydrogen production. Some of those used for this purpose are *Caldicellulosiruptor* sp. (Pawar et al. 2013; Panagiotopoulos et al. 2010), *Thermoanaerobacter* sp. (Brynjarsdottir et al. 2013; Cao et al. 2009; Hniman et al. 2011; Phummala et al. 2014), and *Thermotoga* sp. (Nguyen et al. 2008; de Vrije et al. 2002).

2.4.1.1 Biohydrogen Production from First-Generation Biomass

First-generation biomasses are mainly starch and sugar-rich agricultural products such as sweet sorghum, sugar beet, potato, wheat, pumpkin, etc., and their residues after being processed (Ghimire et al. 2015; Argun et al. 2008a; Oztekin et al. 2008; Panagiotopoulos et al. 2010; Saraphirom and Reungsang 2010). Lignin and hemicellulose contents of those biomasses are lower compared with those of second-generation biomasses. Therefore, pretreatment of these biomasses is more effective in obtaining high concentration of fermentable sugar solution. Moreover, formation of fermentation inhibitory substances such as furfural and 5-HMF is not generally encountered.

Hydrogen production potentials of first-generation biomass have been widely studied. Sugar beet is supposed to be a promising energy crop because of its high sucrose and water content. Biohydrogen production by batch dark fermentation of sugar beet juice by *Caldicellulo-siruptor saccharolyticus* under thermophilic condition that resulted in 3 mol H_2/mol hexose was obtained which is comparable with theoretical yield of 4 mol H_2/mol glucose. It was mentioned that sugar beet juice contains some required nutrients such as amino acids (e.g., glutamine, glutamic acid, asparaginic acid, leucine, isoleucine, alanine), organic acids (mainly lactic acid), and inorganic acids (mainly phosphoric acid) for cell growth and hydrogen production (Panagiotopoulos et al. 2010). Another sugar-rich biomass that attracted attention for biohydrogen production is sweet sorghum. Batch dark fermentation of sweet sorghum syrup by anaerobic mixed cultures resulted in maximum hydrogen production potential of 6864 mL H_2/L medium and hydrogen yield of 2.22 mol H_2/mol hexose (Saraphirom and Reungsang 2010). Potato is a starch-rich first-generation biomass. Xie et al. (2008) used potato in two-phase anaerobic fermentation in which hydrogen was produced in the first stage, and methane was produced in the second stage. The effect of enzyme hydrolysis was investigated in this study. The maximum hydrogen yield of 271.2 mL/g TVS was achieved with α-amylase and glucoamylase hydrolyzed potato feedstock. On the other hand, Mars et al. (2010) studied hydrogen production by the extreme thermophiles *Caldicellulosiruptor saccharolyticus* and *Thermotoga neapolitana* from potato steam peel (PSP) which is a coproduct of the potato processing industry. Untreated PSP and enzyme hydrolyzed PSP were used as substrate for batch fermentation. Amyloglucosidase-hydrolyzed PSP resulted in higher sugar concentration compared to that of α-amylase. The yields from amyloglucosidase-hydrolyzed PSP were 3.4 mol H_2/mol glucose by *C. saccharolyticus* and 3.3 mol H_2/mol glucose by *T. neapolitana*. However, they were 3.5 mol H_2/mol glucose by *C. saccharolyticus* and 3.8 mol H_2/mol glucose by *T. neapolitana* for untreated PSP. Similarly, hydrogen production yield from batch dark fermentation of untreated sweet potato was obtained as 1.24 mol H_2/mol hexose by mix cultures of *Ruminococcus schinkii* and *Lactovum miscens* (Lay et al. 2012). These results indicated that pre-hydrolysis stage is not essential for potato and its peels when hydrolyzing microbial cultures are available during fermentation.

Kargi and his research group investigated hydrogen production from acid-hydrolyzed and partially hydrolyzed wheat powder by boiling. The effects of factors such as substrate concentration (Argun et al. 2008a), media composition (Oztekin et al. 2008; Argun et al. 2008b), fermentation temperature (Cakır et al. 2010), inoculum type (Argun et al. 2009c), bioreactor operation mode in continuous processes (Kargi and Pamukoglu 2009; Sagnak et al. 2010) on biohydrogen yield, and production rate were evaluated. The cumulative biohydrogen formation, hydrogen yield, and formation rate were at maximum at the heat-treated WP concentration of 20 g/L. Higher WP concentrations resulted in lower fermentation performance probably due to substrate and product (VFA) inhibition (Argun et al. 2008a). Wheat powder was nitrogen and phosphorus limited for dark fermentative hydrogen production and required external nutrient addition. The optimal nutrient ratios for boiled wheat powder were determined as C/N = 200 and C/P = 1000 for the maximum yield of formation of 281 mL H_2/g starch (Argun et al. 2008b). When acid-hydrolyzed wheat powder solution was used, the optimum nutrient ratios reaching the maximum hydrogen yield (Y=2.84 mol H_2 mol/glucose) were C/N = 0.02, P/C = 0.008, and Fe(II)/C = 0.015 (Oztekin et al. 2008). Dark fermentation of acid-hydrolyzed ground wheat under thermophilic conditions (55 °C) was proven to be more beneficial as compared to mesophilic or thermophilic fermentation of boiled wheat powder (Cakır et al. 2010). Another approach used for hydrogen production from wheat was to apply enzyme for hydrolysis (Han et al. 2015a, b). The hydrolysate was fermented at 37 °C by *Biohydrogenbacterium* R3 and anaerobic sludge. The yields of hydrogen formation were 2.34 mol H_2/mol glucose and 1.9 mol H_2/mol glucose, respectively.

2.4.1.2 Biohydrogen Production from Second-Generation Biomass

The hydrogen production potential from lignocellulosic materials depends on applied pretreatment method, microbial culture, and biomass type (Nissila et al. 2014). Pretreatment of lignocellulosic materials increases the porosity, removes lignin, and reduces the cellulose crystalline structure. As a result, cellulose becomes more accessible to be converted into simple sugars through enzymatic hydrolysis during fermentation (Nasirian et al. 2011). It is evident that applying pretreatment or hydrolysis stage enhances hydrogen production (Cui et al. 2009, 2010; Brynjarsdottir et al. 2013; Han et al. 2012; Cao et al. 2012; Sekoai et al. 2013). The yield of production with pretreatment application prior to fermentation could reach up to 3 mol/mol which is 75 % of the theoretical yield (Nissila et al. 2014).

Some of the lignocellulosic materials used for hydrogen production are bagasse (Lo et al. 2011), cornstalk (Cao et al. 2012; Sawatdeenarunat et al. 2015), rice straw (Liu et al. 2013, 2014a, b; He et al. 2014), wheat straw (Nasirian et al. 2011), hemp (*Cannabis sativa*), barley straw, Timothy grass (*P. pratense*) (Brynjarsdottir et al. 2013), soybean straw (Han et al. 2012), oil palm trunk (Hniman et al. 2011), and poplar leaves (Cui et al. 2010) and even mixture of agricultural waste and organic fraction of municipal wastes (Sekoai et al. 2013). High cellulose and hemicellulose contents rather than lignin are required. Those carbohydrates are easily converted to fermentable sugars as glucose and xylose through hydrolysis. Since lignin is the non-fermentable part of the lignocellulosic materials, hydrogen is not generated from lignin (Ho et al. 2012; Klimiuk et al. 2010; Thomsen et al. 2008; Cheng et al. 2010). Hydrolysis of lignocellulosic materials will result in mixture of glucose and xylose after lignin separation. Although glucose is known as the most preferred carbon source by hydrogen-producing organisms, it was reported that xylose is the preferred sugar over glucose due to higher attainable hydrogen yields per substrate (Prakasham et al. 2009, 2010). The optimal ratio for maximum hydrogen production was suggested as glucose/xylose: two-third (Prakasham et al. 2009, 2010). High xylose utilization for hydrogen production was attributed to metabolic and biochemical nature of inoculum (buffalo dung) which may contain microbial consortia that are well adapted to digest lignocellulosic materials as food. Xylose utilization was improved by using cocultures of *C. beijerinckii* and *Geobacter metallireducens*. The hydrolysates with the lowest glucose/xylose ratio of 1:10 resulted in the highest xylose utilization, hydrogen production, and no accumulation of acetate which could be inhibitory to hydrogen production through metabolic shift from hydrogen generation to solvent generation as ethanol (Zhang et al. 2013). Patel et al. (2015) reported that xylose resulted in the highest hydrogen production and organic acid generation among other carbon sources as cellulose, cellobiose, starch, and sucrose. On the other hand, no significant difference was observed in hydrogen production when xylose, glucose, and xylose–glucose mixture was used as substrate under thermophilic fermentation condition (Hniman et al. 2011). The theoretical yield of hydrogen formation from xylose is 3.33 mol H_2/mol xylose. The observed yield of formations varies depending on the bacterial cultures used. Some of the observed yields from xylose were 0.45 mol H_2/mol xylose with *Bacillus firmus* (Sinha and Pandey 2014) and 0.56 mol/mol by the mix culture of *Caldoanaerobacter subterraneus* and *Caloramator fervidus* (Yokoyama et al. 2007). Better yields from xylose were obtained under thermophilic fermentation conditions such as 1.62 mol/mol by extreme thermophilic mixed culture (Kongjan et al. 2009), 2.09 mol/mol by *Thermoanaerobacterium thermosaccharolyticum* (Khamtib and Reungsang 2012), and 2.24 mol/mol by *Caldicellulosiruptor saccharolyticus* (Kádár et al. 2004).

Acid and alkaline pretreatments are the most common hydrolysis methods applied to second-generation biomass. The hydrolysate content and hydrogen production potential vary depending on acid or alkali pretreatment. Therefore, it is not possible to suggest one of these pretreatment methods superior to the other. The use of cornstalk as substrate for H_2 production has gained special attention due to its abundance. Cellulose, hemicellulose, and lignin content of fresh cornstalk are around 37 %, 17 %, and 20 %, respectively. Alkaline pretreatment with NaOH provided increase in cellulose content to 40 % and hemicellulose to 27 % with the decrease in lignin content to 10 % and the yield of hydrogen was 82.5 mL/g substrate (Amutha and Murugesan 2013). Lime pretreatment of cornstalk provided more hemicellulose and lignin removal compared with cellulose, and maximum yield of hydrogen formation was achieved as 170 mL/g TVS at 0.10 g lime/g raw biomass (Cao et al. 2012). Acid treatment of soybean straw was more effective in obtaining high sugar content with decreasing hemicellulose and lignin than alkaline treatment, and 11-fold increase in hydrogen production was achieved with acid-treated soybean straw compared to that of raw one (Han et al. 2012). Hydrogen production from acid hydrolysis of bagasse mainly by H_3PO_4 was higher than alkaline treatment (Lo et al. 2011). Acid treatment of beer lees was more effective on hydrolysis of hemicellulose than cellulose or lignin, and cumulative hydrogen yield was 17-fold more compared to that of raw beer lees (Cui et al. 2009). Similarly, the increase in the cumulative hydrogen yield of formation from acid-hydrolyzed wheat straw was 136-fold higher than that of raw wheat straw (Fan et al. 2006). Higher hydrogen formation yields after acid or alkaline pretreatment is due to substantial decrease in mainly lignin content and increase in accessibility of organisms to cellulose for hydrogen production.

Enzymatic hydrolysis of chemically treated lignocellulosic biomass enhances hydrogen production. Nasirian et al. (2011) compared the hydrogen production from fresh wheat straw, with dilute acid pretreatment, by simultaneous saccharification and fermentation (SSF) and by separate hydrolysis and fermentation (SHF). No hydrogen was produced from fermentation of fresh wheat straw. The highest yields and volumes of production were obtained from enzymatic SSF and from fermentation of acid-hydrolyzed wheat straw. Similar results were observed when lime-treated wheat straw was subjected to enzymatic hydrolysis with celluloses, hemicelluloses, and beta-glucosidases in SSF (Reilly et al. 2014). Enzymatic SSF resulted in higher yield of formation (58.78 ± 4.02 mL H_2/g TVS) compared to single-stage alkaline treatment (43.28 ± 3.77 mL H_2/g TVS). The findings from hydrogen production from poplar leaves support these results. Cumulative hydrogen yield from enzymatic hydrolysis was threefold higher than that of raw poplar leaves and 1.34-fold higher than that of substrate pretreated with HCl (Cui et al. 2010).

Utilization of thermophilic organisms for hydrogen production from lignocellulosic materials has recently received more attention. It has been stated that the enzymatic activity is slow at mesophilic fermentation condition which causes slow rate of biomass hydrolysis and low yield of sugar formation. Thermophilic organisms have thermostable cellulose and xylanase enzymes which exert high activity and rate of hydrolysis at fermentation temperatures above 50 °C. On the other hand, low amount of enzyme production is the limitation in thermophilic fermentation (Bhalla et al. 2013). Therefore, pretreatment could be required before fermentation to provide at least partial hydrolysis of biomass. Dark fermentation of acid-pretreated corn stover hydrolysate by thermophilic organism, *T. thermosaccharolyticum*, resulted in 2.24 mol H_2/mol sugar hydrogen formation yield (Cao et al. 2009). Sequential alkali and enzymatic treatment of wooden chopsticks resulted in 195 mL H_2/g total sugar under extreme thermophilic conditions with *Thermoanaerobacterium* sp. (Phummala et al. 2014). Hydrogen production studies from hemp (*Cannabis sativa*), barley straw, and Timothy grass (*P. Pratense*) by *Thermoanaerobacter* sp. showed lower yields on biomass without chemical treatment as compared with acid or alkali-pretreated substrates. The yield obtained from barley straw was the highest and almost ten times more yield was achieved by alkali-treated straw (Brynjarsdottir et al. 2013). Pretreatment stage can be eliminated by using cocultures of thermophilic cellulolytic and hydrogen-producing organisms. These cultures form a cellulose–enzyme–microbe complex (Lu et al. 2006). Cellulolytic bacteria hydrolyze the cellulose to sugars for hydrogen-producing bacteria in the coculture. Some of the studies conducted on coculture hydrogen production revealed that hydrogen production was significantly improved. In most cases, pretreatment of biomass is not required. Direct fermentation of cornstalk powder without any pretreatment was achieved by using thermophilic coculture of *Clostridium thermocellum* and *Clostridium thermosaccharolyticum*. The yield of hydrogen formation with coculture was 94 % more than that of monoculture (Li and Liu 2012). Similarly, a twofold increase in the yield (1.8 mol H_2/mol glucose) was obtained when coculture of *Clostridium thermocellum* and *Thermoanaerobacterium thermosaccharolyticum* (Liu et al. 2008) was used. Cellobiose and glucose accumulation was observed during fermentation with monoculture of *C. thermocellum* indicating that culture cannot completely utilize cellobiose and glucose produced by the cellulose degradation. Utilization of coculture increased cellobiose and glucose consumption, cell division, and organic acid production. This result was supported by coculture of *Clostridium thermocellum* and *Clostridium thermopalmarium* which depend on cellulose and soluble sugar for growth, respectively. The soluble sugars produced by *C. thermocellum* from hydrolysis of cellulose were consumed by

C. thermopalmarium and twofold increase was obtained in hydrogen production compared to monoculture fermentation. However alkali treatment was required at high substrate concentrations (Geng et al. 2010). Ren et al. (2008) identified mesophilic cocultures of *Clostridium acetobutylicum* and *Ethanoigenens harbinense* in hydrogen production. Coculture provided 20 % more cellulose hydrolysis and 2.7 more hydrogen production than that of monoculture of *C. acetobutylicum*.

Particle size of the biomass could be a factor in hydrogen generation by dark fermentation. Hydrogen and organic acid concentrations increased with decreasing particle size of raw wheat straw from 10 to 1 mm (Yuan et al. 2011). On the other hand, Song et al. (2014) used cornstalk as biomass without pretreatment and investigated the effect of particle size on biohydrogen yield by using *Clostridium* sp. FS3 which was isolated from anaerobic acclimated sludge. It was concluded that particle size didn't substantially affect hydrogen yield. The maximum yield was 92.9 mL H_2/g cornstalk under optimized media composition.

Hydrolysis of lignocellulosic materials generates some by-products such as furfural, 5-HMF, phenols, formate, acetate, and other unknown ones (Monlau et al. 2013, 2015). Furfural, 5-HMF, and phenolic compounds are known as inhibitory on hydrogen-producing organisms. Severe alkaline and acid treatments could increase the amount of inhibitory phenolic compounds in the hydrolysate and thereby adversely affect the hydrogen production potential (Cui et al. 2009; Phummala et al. 2014). Detoxification of hydrolysate through resin adsorption could substantially improve the hydrogen production (Rai et al. 2014). Monlau et al. (2015) used with and without alkaline-pretreated sunflower stalks as feedstock in hydrogen production. The yield of formation from fresh stalk (7.1 mL H_2/g TVS) was higher than that of pretreated one (6 ml H_2/g TVS). Lower yield formation from alkaline-pretreated stalks was attributed to inhibition effect of phenolic compounds released from lignin degradation. Similar result was observed when dilute acid hydrolysis was applied to sunflower stalk (Monlau et al. 2013). Hydrogen production yield substantially decreased when hydrolysate volume in the media was 7.5 % at which by-product concentrations in the fermentation media reached to 61 mg/l acetate, 45 mg/L formate, 86.2 mg/L furfural, 9.5 mg/L 5-HMF, and 1.5 mg/L phenolic compounds. A shift of dominant microbial culture from *Clostridium* sp. to *Sporolactobacillus* sp. was observed which indicates specific inhibition of biohydrogen-producing bacteria by hydrolysate. Depending on this population shift, a metabolic shift from acetate/butyrate to non-hydrogen-generating ethanol/lactate was also observed. On the other hand, Liu et al. (2015) reported that although furfural inhibits hydrogen production, 5-HMF enhances. Forty percent higher hydrogen production was obtained from 100 mg/L HMF containing steam-exploded cornstalk hydrolysate in comparison to control experiment without addition of HMF. It was also observed that volatile fatty acid production increased, and 90 % HMF was degraded up to 1000 mg/L HMF concentration during hydrogen formation. Microbial community analysis indicated that HMF stimulated higher proportion of hydrogen-producing *Clostridium* and cellulose-hydrolyzing rumen bacteria *Ruminococcaceae*. It was stated that furfural inhibits hydrogen generation, VFA concentration is not affected by furfural concentration, and it is degraded by anaerobic culture. Similarly, no inhibition effects on hydrogen production up to 0.14 g/L HMF and 1.09 g/L furfural concentrations were observed in pretreated corncob (Nasr et al. 2014).

Most of the hydrogen production studies from lignocellulosic biomass were conducted in small-scale batch fermentation. A summary of hydrogen yields and rates obtained from batch fermentation of different second-generation biomasses with applied pretreatment method is given in Table 2.3. Recently, attentions were diverted to continuously operated bioprocesses to determine optimal operating conditions such as initial substrate concentration, hydraulic retention time (HRT), and organic loading rates for the used second-generation biomass and bioreactor. Hydrogen production potentials of continuously

Table 2.3 Hydrogen production yields and rates obtained from batch dark fermentation of different second-generation biomasses with applied pretreatment method

Biomass	Pretreatment	Culture	T^b C	H_2 yield	H_2 rate	Reference
Corn stover	Acid	Heat-treated river sediment	37	4.17 mmol/g sugar	0.78 mmol/L/h	Zhang et al. (2014a, b)
Corn stover	Acid	*Thermoanaerobacterium thermosaccharolyticum*	60	2.24 mol/mol sugar		Cao et al. (2009)
Cornstalk	Alkaline	*Bacillus licheniformis* MSU AGM 2	37	80 ml/g substrate	0.14 ml/h	Amutha and Murugesan (2013)
Cornstalk	Wet stream explosion	Anaerobic digested sludge		12 ml/g TVS	–	Liu et al. (2014a, b)
Corncob	Acid	Anaerobic digested sludge	35	265 ml/g COD	–	Nasr et al. (2014)
Corncob	High-pressure autohydrolysis	Anaerobic digested sludge	35	31 ml/g COD	–	Nasr et al. (2014
Wheat straw	Acid and enzyme	Mixed culture	35	1 mol/mol glucose	–	Nasirian et al. (2011)
Wheat straw	Acid and stream explosion	*Clostridium* sp. IODB03	37	2.52 mol/mol sugar	153 mol/L/h	Patel et al. (2015)
Wheat straw	Acid, stream explosion, and enzyme	*Clostridium* sp. IODB03	37	2.62 mol/mol sugar	136 mol/L/h	Patel et al. (2015)
Wheat straw	Acid and microwave heating	Cow dung	36	64 ml/g TVS	–	Fan et al. (2006)
Wheat straw	Acid	Mixed culture	35	1.19 mol/mol glucose	–	Nasirian et al. (2011)
Wood chopsticks	Alkaline and enzyme	*Thermoanaerobacterium*	50	195 ml/g TVS	116 ml/d	Phummala et al. (2014)
Rice straw	Acid	Heat-treated sludge	37	0.44 mol/mol sugar	6.43 ml/h	Liu et al. (2013)

Soybean straw	Acid	*Clostridium buytricum*	35	47.65 ml/g substrate	2.26 ml/h	Han et al. (2012)
Potato and pumpkin	–	BESA-treated anaerobic sludge	35	171 ml/g TVS	–	Ghimire et al. (2015)
Oil palm trunk	Acid and alkaline	*Thermoanaerobacterium*, *Thermoanaerobacter*, *Caloramator*	60	301 ml/g sugar	–	Hniman et al. (2011)
Poplar leaves	Enzyme	*Clostridium pasteurianum*	35	44.9 ml/g poplar leaves	2.17 ml/h	Cui et al. (2010)
Sugarcane leaves	Acid	Anaerobic sludge	37	248.05 ml/g	–	Moodley and Kana (2015)
Waste paper	–	*Ruminococcus albus*	37	282 ml/g paper		Ntaikou et al. (2009)
Miscanthus	Hydrothermal	*Clostridium beijerinckii*, *Geobacter metallireducens*	30		2.43 mmol/L/h	Zhang X et al. (2013)
Beer lees	Acid	*Clostridium pasteurianum*	35	53 ml/g dry beer lees	6.68 ml/h	Cui et al (2009)
Raw cornstalk	–	*C. sartagoforme FZ11*	35	87.2 mL/g cornstalk	6.2 mL/g/h	Zhang et al. (2015)
Sugarcane bagasse	Acid	*Enterobacter aerogenes MTCC 2822*	30		8.3 mL/h	Rai et al. (2014)
Coffee mucilage	–	Swine manure		77.6 mL/g COD	7.6 Nl/L/d	Hernández et al. (2014)
Water hyacinth	Acid	Anaerobic digester sludge	37	134.9 mL/g TVS	90 mL/L/h	Cheng et al. (2015)

stirred tank reactor (CSTR) (Zhao et al. 2013; Liu et al. 2013; Pawar et al. 2013), anaerobic contact filter (Vijayaraghavan et al. 2006), continuously external circulating bioreactor (Liu et al. 2014a, b), upflow anaerobic sludge blanket (UASB) reactor, anaerobic filter (Kongjan and Angelidaki 2010), and anaerobic biotrickling filter (Arriaga et al. 2011; Vargas et al. 2014) from different second-generation biomasses have been investigated. Hydrogen production rates and yields at optimal operating conditions of these continuously operated bioreactors, used substrate, and applied pretreatment method were given in Table 2.4. HRT is one of the most significant operating parameters that should be optimized in a bioprocess to obtain the highest product yield. The optimal HRT of a bioprocess is the factor of substrate type, substrate concentration, environmental conditions, microorganism type, and certainly bioreactor type as immobilized and suspended growth systems. The advantage of CSTR is the presence of homogenous conditions which eliminate substrate limitation and provide easy control of environmental conditions within the reactor. On the other hand, HRTs for CSTR are generally longer than that of immobilized systems due to difficulty in holding slow-growing organism in bioreactor at short retention times and consequently washing out of organism. Accumulation of VFAs, decrease in pH, lower glucose and xylose utilization, more lactic acid production, and washout of organisms are the problems encountered in hydrogen production in CSTR at short HRTs (Zhao et al. 2013; Liu et al. 2014a, b; Kongjan and Angelidaki 2010). Liu et al. (2014a, b) developed continuously external circulating bioreactor (CECBR) with volumetric circulation rate of 9.6 L/min, and a substantial decrease in optimal HRT to 4 h was provided with external recirculation of effluent. Immobilized systems are more advantageous in terms of high biomass holding capacity and ability to operate at lower HRTs without washing out of organisms. There are limited studies on hydrogen production in immobilized systems. Kongjan and Angelidaki (2010) studied UASB and anaerobic filter (AF) system with hydrothermal-treated wheat straw. It was observed that hydrogen production in immobilized bioprocesses can easily be recovered when system was operated in optimal conditions after inappropriate operation period. Efficient toxic substance removal was obtained, a stable long-term acetic acid production was achieved under optimal operation conditions, but lactic acid accumulation occurred when the systems were operated at shorter HRT (Kongjan and Angelidaki 2010). Vargas et al. (2014) examined hydrogen production from acid and enzymatic hydrolysate of oat straw in trickling biofilter. Although there is efficient hydrogen production from enzymatic hydrolysate, production is totally suppressed when acid hydrolysate was used. The reason for the suppression was explained as sugar gap occurred between enzymatic and acid hydrolyses rather than the presence of 5-HMF or furfural in the hydrolysate. On the other hand, no hydrogen production suppression was observed for acid hydrolysate of oat straw in trickling biofilter (Arriaga et al. 2011). The difference in these two studies could be acid hydrolysis condition which results in different sugar contents. By considering the advantages of immobilized systems over CSTR, more studies are needed to evaluate the performance of immobilized systems in hydrogen production from lignocellulosic biomasses.

2.4.1.3 Biohydrogen Production from Third-Generation Biomass

Algae are the main source of third-generation biomass. Low-cost cultivation of algae is possible with no requirement for expensive substrate and energy input (Park et al. 2011). Utilization of algae as substrate for hydrogen production provides CO_2 reduction besides clean and sustainable fuel production. Algae are a CO_2-fixing organism and hydrogen production will require massive cultivation of this microorganism. Therefore, large-scale production of algae offers massive CO_2 reduction from the atmosphere.

Algae can fix CO_2 in the form of carbohydrate, such as glycogen, starch, or other cellular storage materials in cytoplasm and in the form of lipids in cell membrane (Nayak et al. 2014). Direct fermentation of algae biomass is possible (Shi et al. 2011). But, it was stated that biodegradation

Table 2.4 Hydrogen production yields and rates obtained from batch dark fermentation of third-generation biomasses

Biomass	Pretreatment	Culture	T[a] °C	H_2 yield	H_2 rate	References
Chlorella vulgaris	Acid	*Clostridium buytricum* CGS5	37	1.15 mol/mol reducing sugar	246 ml/L/h	Liu et al. (2012)
Chlorella vulgaris	–	Heat-treated anaerobic sludge	60	19 ± 2.94 ml/g TVS	0.123 ml/h	Wieczorek et al. (2014)
Chlorella vulgaris	Enzyme	Heat-treated anaerobic sludge	60	135 ± 3.11 ml/g TVS	3.14 ml/h	Wieczorek et al. (2014)
Chlorella vulgaris	–	TC60 from compost pile	60	4.2 mmol/g TVS	–	Carver et al. (2011)
Chlorella vulgaris FSP-E	Acid	*Clostridium butyricum* CGS5	35	1.42 mol/mol reducing sugar	176.9 mL/L/h	Chen et al. (2015)
Chlorella sp.	–	Heat-treated anaerobic digested sludge	35	7.13 ml/g TVS	0.417 ml/h	Sun et al. (2011)
Chlorella sorokinicna	Acid	Thermophilic mixed culture (*Thermoanaerobacterium thermosaccharolyticum*)	60	2.68 mol/mol hexose	330 ml/L/h	Roy et al. (2014)
Chlorella sorokinicna	Acid	*Enterobacter cloacae* IIT-BT 08	37	9 ± 2 mol/kg COD	148 ml/L/h	Kumar et al. (2013)
Gelidium amansii	Acid	Anaerobic digested sludge	35	1.07 mol/mol sugar added	420 ml/h	Park et al. (2011)
Scenedesmus obliquus	Grinding	Coculture of anaerobic sludge and *Clostridium butyricum*	37	2.74 mol/mol sugar added	154 ml/h	Ortigueira et al. (2015)
Scenedesmus obliquus	Acid	*C. butyricum* DSM 10702	37	2.9 mol/mol sugars	–	Ferreira et al. (2013)
Laminaria japonica	–	*Clostridium* sp.	35	0.92 mol/mol hexose (120 ml/g TVS)	–	Yan et al. (2011)
Anabaena PCC 7120	Enzyme	Thermophilic mixed culture	60	2.68 mol/mol hexose	–	Nayak et al. (2014)
Dunaliella tertiolecta	–	TC 60 from compost pile	60	2.1 mmol/g TVS	–	Carver et al. (2011)
Taihu blue algae	Alkaline	Heat-treated anaerobic granular sludge	35	105 mL/g TVS	–	Yan et al. (2010)

[a]Fermentation temperature

of intact macroalgae cell membranes and walls could be slow and prolong fermentative hydrogen production. For an efficient fermentation, the embedded carbohydrate within the cell has to be liberated properly (Liu and Wang 2014). Therefore, pretreatment of algae or third-generation biomass is the first step in hydrogen production as it was the case for hydrogen production from the first- and second-generation biomasses. Enzymatic, heat, sonication, alkaline, and acid–heat treatments are some of the pretreatment methods that have been applied to algae biomass for hydrolysis purpose (Nayak et al. 2014; Yun et al. 2014; Batista et al. 2014; Liu and Wang 2014; Park et al. 2011).

Hydrogen production from different algae biomasses has been presented in the literature. Nayak et al. (2014) studied thermophilic dark fermentative hydrogen production using pretreated *Anabaena* PCC 7120 as substrate by mixed microflora. *Anabaena* PCC 7120 is a cyanobacterium and the carbon content of it was found to be 48.2 % on dry-weight basis.

It was mentioned that the carbohydrate content of *Anabaena* sp. PCC 7120 was mostly composed of glycogen stored in cytoplasm and several polysaccharides associated with the cell envelope. No direct fermentation of this type of biomass has been accomplished yet. Enzymatic (α-amylase), autoclaving, sonication, and acid–heat pretreatments were applied to algae biomass, and the maximum yield was obtained as 2.68 mol H_2/mol of hexose (6.42 mmol H_2/g COD reduced), by α-amylase pretreatment. Another algae biomass used for hydrogen production is *Chlorella vulgaris*. It is a well-known microalgae and it was supposed to be a potential feedstock for hydrogen production due to its starch content in the cell wall which should be hydrolyzed to make fermentation feasible. A crude hydrolytic extracellular enzyme solution extracted from the H_2-fermented effluent of food waste was used instead of a commercial enzyme for a cost-effective hydrolysis. Carbohydrate content of microalgal biomass was 38.8 g/100 g of cell, and the reducing sugar concentration was ca.6 g/L after the hydrolysis of biomass with enzyme solution. The highest H_2 yield and production rate were 43.1 mL H_2/g dry cell weight and 21.8 mL H_2/L/h, respectively (Yun et al. 2014). Liu et al. (2012) reported 100 % recovery of glucose and xylose by acid hydrolysis of *C. vulgaris* with the maximum hydrogen yield of 1.15 mol H_2/mol sugar. But, alkaline treatment provided only breakdown of the cell wall, and enzymes were required for further hydrolysis of the components to fermentable sugars.

Laminaria japonica, which is a brown macroalgae, was used as substrate for dark fermentative hydrogen production by Liu and Wang (2014). The carbohydrate content of *L. japonica* was 47.6 % and the main carbohydrate types were laminarin (β-glucan) and alginate. Heat, acid, alkaline, and ultrasonication pretreatment methods were applied to *L. japonica*, and the highest cumulative hydrogen volume of 66.7 ml H_2/g was obtained by heat pretreatment, which was approximately sixfold greater than that of raw *L. japonica* (10.0 mL/g). *Gelidium amansii* is a red algae which mainly contains glucose and galactose as the main sugar monomers. Acid-hydrolyzed red algae yielded maximum hydrogen production of 2.8 mol H_2/mol galactose and 1.5 mol H_2/mol glucose (Park et al. 2011). Combined pretreatment of *Chlorella pyrenoidosa* by steam heating with diluted acid and by microwave heating with diluted acid enhanced the dark fermentative hydrogen yield about tenfold compared to single pretreatments like steam heating, microwave heating, and ultrasonication (Xia et al. 2013). *C. pyrenoidosa* was reported to be composed of 26.2 % carbohydrates, 35.2 % proteins, 16.3 % lipids, 9.2 % moisture, and 13.1 % ash, and its low C/N ratio was considered as a limiting factor in dark fermentative hydrogen production. Steam heating with dilute acid-pretreated *Chlorella pyrenoidosa* (CP) was mixed with cassava starch (CS) to increase the C/N ratio. The maximum dark fermentative hydrogen yield of 276.2 mL/g total volatile solids (TVS) from the mixed biomass at C/N molar ratio of 25.3 showed 3.7-fold and 1.8-fold increases, respectively, compared with those from only CP and only CS (Xia et al. 2014). Batista et al. (2014) compared batch fermentative hydrogen production by *Enterobacter aerogenes* ATCC 13048 and

Clostridium butyricum DSM 10702 from dried (5 % moisture) and wet (69 % moisture) *Scenedesmus obliquus*. The main carbohydrate storage of *S. obliquus* was determined as starch and its sugar content was around 30.7 % on dry-weight basis. H_2 yields obtained from batch fermentation by *C. butyricum* were considerably higher than those obtained by *E. aerogenes*, and the highest H_2 yield was 113.1 ml H_2/g TVS algae without any pretreatment of algae. Formation of high hydrogen yields from raw algae was attributed to the fact that *C. butyricum* produces amylases able to hydrolyze starch, which is not possible by *E. aerogenes*. No significant difference in hydrogen production potentials from wet and dried algae biomass was observed for both cultures. It was stated that algal hydrogen production could be cost-effective when the high-energy-demanding drying process could be eliminated. Table 2.4 summarizes applied pretreatment, hydrogen production yields, and rates obtained from batch fermentation of third-generation biomasses.

2.4.2 Hydrogen Production from Biomass by Photofermentation

Hydrogen-producing photofermentative microorganisms are purple nonsulfur bacteria (PNS). *Rhodobacter sphaeroides*, *Rhodobacter capsulatus*, *Rhodospirillum rubrum*, and *Rhodopseudomonas palustris* are some of those bacteria used for hydrogen production (Kars and Ceylan 2013; Kapdan et al. 2009; Argun et al. 2008c). The main substrates for hydrogen production in photofermentation are organic acids such as acetic, butyric, and lactic acids (Kars and Ceylan 2013; Keskin et al. 2011). Photofermentation requires strict environmental conditions and media compositions. Light intensity, light source type, and lighting regime are the major environmental conditions needed to be optimized for maximum hydrogen production by photofermentation (Argun and Kargi 2010a). Fe and Mo are essential elements for hydrogen production and they should be externally added into the fermentation media (Kars and Ceylan 2013). Although theoretically high yield of hydrogen can be obtained from conversion of organic acids to hydrogen, low light conversion efficiency and production volumes are reported obstacles in photofermentative hydrogen production (Azwar et al. 2014).

Photofermentation has a complementary role of dark fermentative hydrogen production despite its strict process requirements and low conversion hydrogen rates. As it was mentioned in Sect. 2.4.1.1, the theoretical yield for glucose conversion to hydrogen is 12 mol H_2/mol glucose. Dark fermentation stage can achieve only, e.g., 4 mol H_2/mol glucose which is one-third of theoretical yield if acetate is the end product. The theoretical yield of hydrogen formation from acetate is 4 mol H_2/mol acetate. Two moles of acetate obtained from dark fermentation of glucose are further converted to H_2 and CO_2 by photofermentation with a total yield of 8 mol H_2/mol. Therefore, photofermentative hydrogen production is considered to be the next stage of dark fermentation in order to enhance the yield for hydrogen formation. The two main bioprocess approaches for this purpose are sequential dark fermentation and photofermentation and combined dark fermentation and photofermentation. On the other hand, there are reports about photofermentation of simple sugars and organic acids obtained from hydrolysis of biomass for hydrogen production. In this case, single-stage photofermentation of biomass hydrolysate is another option in hydrogen production.

2.4.2.1 Single-Stage Photofermentation

Photofermentative organisms cannot use polysaccharides directly as substrate. Therefore, utilization of biomass for photofermentative hydrogen production requires a pretreatment step for the hydrolysis of polysaccharides into monosaccharides. Lignocellulosic biomass hydrolysate may contain organic acids such as acetic, lactic, formic, propionic, and citric acid as well as sugars (Kars and Ceylan 2013; Patel et al. 2015; Pattanamanee et al. 2012; Pattra et al. 2008).

Hydrogen production potential of single-stage photofermentation of biomass hydrolysate has not been studied in detail, yet. There are limited numbers of studies about direct photofermentation of biomass hydrolysate to hydrogen production in literature. In one of those, acid-hydrolyzed waste ground wheat was used as substrate for direct photofermentation by pure culture of *Rhodobacter sphaeroides* cultures. The maximum hydrogen yield of 1.23 mol H_2/mol glucose was obtained with *Rhodobacter sphaeroides* RV at a sugar concentration of 5 g/L (Kapdan et al. 2009). In another study waste barley was acid hydrolyzed and the hydrolysate was subjected to photofermentation directly by *Rhodobacter sphaeroides* O.U.001 without a dark fermentation stage.

The starch content of waste barley in that study was 37 % and other components were 2 % fat and 11 % protein. Acid hydrolysate of waste barley contained mainly acetic, lactic, formic, propionic, and citric acid besides sugar. The maximum volumetric hydrogen production was 0.4 L H_2/L culture. The ammonia concentration in the hydrolysate was lower than inhibitory concentration, but the addition of Fe and Mo was required for an effective fermentation (Kars and Ceylan 2013). Zhang et al. (2014a, b) reported hydrogen production from different types of pretreated milled agricultural residues like corn stover, corncobs, sorghum stover, soybean stalks, cotton stalks, and rice straw by photofermentation. After the enzymatic hydrolysis of feedstock with Solarbio cellulase, agricultural residue hydrolysates were used as substrate in photofermentation by a mix inoculum of *Rhodospirillum rubrum*, *Rhodobacter capsulatus*, and *Rhodopseudomonas palustris*. Maximum cumulative hydrogen yield of 229 mmol H_2/L culture and hydrogen production rate of 5.97 mmol H_2/L/h were obtained with corncob. Hydrolysis of oil palm with acid treatment resulted in a hydrolysate containing mainly glucose, xylose, and acetic acid. Hydrogen production from hydrolysate by direct photofermentation with *R. sphaeroides* S10 was highly affected by media composition such as Fe, Mo, and yeast extract. The maximum production rate obtained at optimized media composition was 22.4 ml H_2/L/h (Pattanamanee et al. 2012).

2.4.2.2 Sequential Dark and Photofermentation

Organic acids required for photofermentative hydrogen production can be generated by dark fermentation of mono- and polysaccharide containing biomass. The process may require a pretreatment like acid, alkaline, and enzymatic hydrolysis for polysaccharides to monosaccharides and lignin separation from biomass as the first step. The second step is dark fermentation of sugars for hydrogen organic acid production. This step is followed with the utilization of dark fermentation effluent as substrate for photofermentative hydrogen production.

The optimal growth and hydrogen production conditions for dark and photofermentative organisms are different. Running the process in two different reactors provides better control of corresponding optimal conditions. For example, optimal hydrogen production in photofermentation occurs at 30 °C. However, dark fermentation requires higher temperatures like 37 °C or 50 °C. A cooling stage of dark fermentation effluent is required especially in thermophilic dark fermentation. Initial organic acid requirement of photofermentative organisms is generally lower than the amount of organic acids produced in dark fermentation. High organic acid concentrations generated during dark fermentation could be inhibitory to photofermentative organisms (Oh et al. 2004; Rai et al. 2014; Ozkan et al. 2012). The effluent of dark fermentation can be diluted to the required level of organic acids for photofermentative organism to eliminate organic acid inhibition. Similarly, high ammonia concentration could inhibit photofermentative hydrogen production. Argun et al. (2008c) reported the optimum TVFA and NH_4–N concentrations in dark fermentation effluent of ground wheat as 2350 and 47 mg/L. Özgür and Peksel (2013) suggested the maximum acetic and ammonia concentrations as 30–40 mM and 2 mM, respectively. Ammonia concentration can be decreased to certain levels by applying ammonia separation techniques or by diluting the effluent before photofermentation.

Sequential dark fermentation and photofermentation of first- and second-generation biomasses have been widely studied. Pretreated barley straw was first subjected to dark fermentation with hyperthermophilic bacteria *Caldicellulosiruptor saccharolyticus* and then photofermentation with *Rhodobacter capsulatus*. Dark fermentation effluent was composed of mainly acetate with relatively low concentrations of lactate and formate. It was reported that external additions of Fe and Mo are crucial in photofermentative hydrogen production (Özgür and Peksel 2013). Hydrogen production from beet molasses by sequential dark fermentation and photofermentation resulted in 13.7 mol of H_2/mol of sucrose production yield which corresponds to 57 % of the theoretical yield of 24 mol of H_2/mol of sucrose. The effluent of dark fermentation effluent was ammonia, Fe, and Mo deficient and external addition of those elements were required for photofermentation (Özgür et al. 2010). Laurinavichene et al. (2010) reported total yield of 5.6 mol H_2/mol glucose from potato homogenate without pretreatment where the contribution of dark fermentation and photofermentation was 0.7 mol H_2/mol glucose and 4.9 mol H_2/mol glucose, respectively. The effluent of dark fermentation was supplemented with Fe/Mg/phosphate nutrients and diluted (≤10 % in water) to decrease organic acid and nitrogen to non-inhibitory level for photofermentative organisms. Sugar beet was first subjected to thermophilic dark fermentation and then to photofermentation in an outdoor process. Dark fermentation effluent was cooled to 30 °C, diluted three times to adjust the acetate to non-inhibitory concentration of 40 mM, and supplemented with potassium phosphate, Fe, and Mo. The total yield obtained in the two-stage process was 77 % of the theoretical yield (Ozkan et al. 2012). Rai et al. (2014) studied biohydrogen production from sugarcane bagasse (SCB) by integrating dark fermentation and photofermentation. Acid-hydrolyzed SCB contained glucose, xylose, arabinose, and inhibitors such as furfural, 5-HMF, and ferulic compounds. Total volatile fatty acid-rich media obtained from dark fermentation of acid hydrolysate was used as substrate for photofermentation by *Rhodopseudomonas* BHU 01. The total hydrogen production potential was 1753 mL H_2/L.

2.4.2.3 Combined Dark Fermentation and Photofermentation

Hydrogen production by combined dark fermentation and photofermentation from biomass is realized in a single reactor which contains both dark fermentative and photofermentative organisms. The theoretical background of this approach can be explained as that organic acids produced in dark fermentation are simultaneously converted to hydrogen gas by photofermentative organisms. The process offers distinct advantages over single-stage dark fermentation or photofermentation due to the prevention of organic acid accumulation by photofermentative bacteria. However, the process has to be operated at convenient fermentation conditions both satisfying dark fermentative and photofermentative microorganisms. Some important parameters affecting hydrogen production performance in combined dark fermentation and photofermentation are the proper selection of microbial culture, dark fermentative/photofermentative biomass ratio, initial biomass and substrate concentration, light source, lighting regime, and light intensity. The effects of those parameters on hydrogen gas production from waste wheat powder, for example, have been reported in the literature. Among various combinations of dark fermentative and photofermentative microorganism, compositions using heat-treated anaerobic sludge with *Rhodobacter sphaeroides* NRLL-1727 reported the most convenient culture combination for hydrogen production from waste wheat hydrolysate by combined fermentation (Ozmihci and Kargi 2010). Argun et al. used a combination of heat-treated anaerobic sludge with four types of photofermentative bacteria *Rhodobacter sphaeroides* (NRLL-1727), *Rhodobacter sphaeroides* (DSMZ-158), *Rhodopseudomonas palustris* (DSMZ-127), and *Rhodobacter sphaeroides* (RV) for combined fermentation of wheat powder solution and reported that the optimum initial dark fermentative/photofermentative biomass ratio as 1:7 (Argun et al. 2009a). In another study, the same group investigated the effects of initial biomass and wheat powder concentrations on hydrogen formation yield and rate by using the same culture composition. Optimum initial biomass and wheat powder concentrations resulting most convenient hydrogen yield and rate were determined as 5 g wheat powder/L and 1.1 g total biomass/L, respectively (Argun et al. 2009b). Same authors reported that using halogen lamp as light source with 670 w/m2 continuous light illumination was the most effective option compared with other light sources such as tungsten, incandescent, fluorescent, and infrared lamps (Argun and Kargi 2010a). Hydrogen gas production from waste wheat powder solution by combined fermentation was also produced in a continuously operated annular hybrid reactor where *Clostridium beijerinckii* DSM-791 and *Rhodobacter sphaeroides* RV were used as inoculum (Argun and Kargi 2010b). The reactor was continuously illuminated with halogen and fluorescent lamps at light intensity of 10 klux resulting in 85 mLH_2/g starch (Argun and Kargi 2010b). The highest hydrogen formation yields and rates from waste wheat powder by the above-mentioned studies were 1.45 mol H_2/mol glucose and 50.26 ml H_2/g cells/h, respectively. However combined fermentation studies with yields up to 7.1 mol H_2/mol glucose have been reported in the literature (Asada et al. 2006).

2.4.3 Challenges in Biohydrogen Production from Biomass

Hydrogen is one of the best choices in terms of renewable energy production. The method used for hydrogen production should provide high productivity and high purity at an affordable cost. Physicochemical methods are highly efficient in productivity and in purity of hydrogen but not cost-effective due to high energy requirement in the production line. Biological methods for hydrogen production have received considerable attention in the last decades since bioprocesses operate under mild conditions (25–35 °C, 1 atm) which make the process cost-effective. The other factor that attracts attentions in biohydrogen production is the utilization of organic residues which need high technologies to handle safely or to convert into environmentally acceptable form. The limitations in this approach are somewhat lower rate and yields of production as compared to the other methods. Therefore, the challenges in biohydrogen production are to develop strategies in order to increase the yield and rate. The main obstacles to achieve high rates and yields have been summarized as partial pressure of hydrogen gas in the produced gas mixture, competing reactions, research needs to develop bioprocess technology, insufficient active hydrogenase enzyme, and efficient hydrogen-producing culture (Hallenbeck 2009). *Clostridium* sp. produces two moles of more hydrogen during glycolysis of glucose by reoxidizing NADH. Redox potential difference between hydrogen and NADH/NAD couple makes this reaction unfavorable. Therefore, one of the solutions is to reduce partial pressure of hydrogen appreciably to derive electrons form NADH (Hallenbeck 2009). The simplest method to decrease partial pressure is to sparge inert gas such as argon and nitrogen. However, more cost-effective or technological methods should be developed to overcome this problem.

The reactions that reduce hydrogen production are metabolic shift from acetic acid generation to solvent or hydrogen-consuming organic acid generation and consumption of hydrogen by uptake hydrogenase and homoacetogens (Hallenbeck 2009; Mathews and Wang 2009). Elimination of uptake hydrogenase has minor effect on hydrogen production. Therefore, studies should be devoted to the development of methods to eliminate homoacetogens and enhance acetic acid production by controlling process conditions or through metabolic engineering.

Commercial production of hydrogen from lignocellulosic biomasses can be achieved by the development of bioreactors. It is evident that immobilized systems are superior over suspended growth systems in hydrogen production (Azbar and Kapdan 2012). The studies on hydrogen production in immobilized system should be diverted to the selection of microbial support particle type or immobilization methods. Support particle characteristics should include high surface area, high porosity, high biomass holding capacity, no toxic component, and low cost. The main immobilization techniques are entrapment in a porous material, attachment on a surface, containment, and self-aggregation. Both support particle type and immobilization technique should be suitable for effective substrate diffusion from liquid phase into biofilm and hydrogen separation from liquid phase to gas phase. Optimization of operating parameters specific to immobilization method or support particle and substrate type is essential. HRT, organic loading rate, and substrate concentration are the major operating parameters. Bioreactors should provide high yield and rate for hydrogen production with efficient substrate removal at short HRTs, high substrate concentrations, or organic loadings for the sake of process economy. Characteristic of lignocellulosic hydrolysate after applied pretreatment is another factor that affects hydrogen production potential of a bioreactor. The effect of glucose, xylose, arabinose, or toxic substance concentrations (5-HMF, furfural, or phenols) and glucose to xylose ratio in the hydrolysate on hydrogen production in studied bioreactor should be considered. Another factor is bioprocess development to convert substrate into hydrogen for target hydrogen yield of 12 mol/mol glucose. Two-stage biohydrogen production as dark fermentation and photofermentation is the first approach to achieve this yield. However, photofermentation needs extensive studies to overcome the problems that limit the expected performance of the process to reach its theoretical potential. Therefore, biohydrogen process needs development of an efficient and low-cost second-stage fermentation process to valorize dark fermentation effluent for hydrogen production and to obtain processed water which does not require further treatment before discharging to receiving media. Although it is desired to have a biological process as second stage, alternatively, methods such as electrohydrolysis can be applied to dark fermentation effluent to obtain high-purity and high-volume hydrogen (Tuna et al. 2009).

Isolation of hydrogen-producing microorganisms with high hydrogen production ability is the major challenge. Mesophilic cultures might be preferred due to lower heating requirement and hence consumption of less energy for the process. However hydrogen production ability of thermophilic organisms is the utilization of thermophilic hydrogen-producing cultures. Thermophilic organisms have thermostable cellulose and xylanase enzymes which exert high activity and rate of hydrolysis at fermentation temperatures above 50 °C. On the other hand, although the yield of thermophilic hydrogen production is higher than mesophilic ones, volumetric productivities are comparable. Therefore, more researches on thermophilic hydrogen production are needed to evaluate yields and rates if it is worth to spend energy for heating to obtain relatively higher yields by thermophilic fermentation.

Another problem in biohydrogen production is the purity of hydrogen in gas phase. Hydrogen purity varies between 30 % and 60 % according to available literature (see Table 2.5). Sparging the head space with an inert gas to reduce hydrogen partial pressure causes further dilution in hydrogen percentage. Simultaneous production and separation of hydrogen by using selective membranes could help both reducing hydrogen partial pressure and increasing purity of hydrogen (Neves et al. 2009).

Pretreatment of lignocellulosic biomasses is the major cost-intensive stage of biohydrogen production. Although there are different methods, each has its own merit and limits. Acid, enzyme, and hydrothermal pretreatments are still the most common methods used for lignocellulosic biomasses. Recent trend is the utilization of ionic liquids for pretreatment. Effective separation of lignin, obtaining pure cellulose and hemicellulose, no toxic substance generation, recovery, and repeated use of liquids without substantial decrease in the separation of cellulose and hemicellulose are the major advantages of ionic liquids. One of the disadvantages of ionic liquids is their cost due to small amount of production. Large-scale utilization of these chemicals, for example, for pretreatment of lignocellulosic biomass for hydrogen production will help to commercialize them and reduce their cost. Another approach is to eliminate pretreatment stage completely. Direct fermentation of lignocellulosic material by coculture and solid

Table 2.5 Hydrogen production yields and rates obtained from dark fermentation of different second-generation biomasses by continuously operated bioreactors

Biomass	Pretreatment	Bioprocess	Culture	T[b] C	Substrate conc.	HRT h	% H_2	H_2 rate	H_2 yield	Reference
Rice straw	Acid treatment	CECBR	Heat-treated anaerobic sludge	40	20 g TS/L	8		5.52 L/L/d	0.72 mol/mol hexose	Liu et al. (2014a, b)
Rice straw	Acid treatment	CECBR	Heat-treated anaerobic sludge	40	20 g TS/L	4		16.32 L/L/d	1.02 mol/mol hexose	Liu et al. (2014a, b)
Cornstalk	Fungal and enzymatic	CSTR	*T. thermosaccharolyticum* W16	60	6.9 g glucose/L 3 g xylose/L	50	60.6	8.4 mmol/L/h	1.9 mol/mol sugar	Zhao et al. (2013)
Wheat straw	Steam–acid pretreatment and enzymatic hydrolysis	CSTR	*C. saccharolyticus* DSM 8903	37	6.7 g glucose/L 3.7 g xylose/L 0.4 g arabinose/L	20		5.09 L/L/d	3.43 mol/mol sugar	Pawar et al. (2013)
Wheat straw[a]	Acid treatment	CSTR	Heat-treated anaerobic sludge	37	20 g TS/L	4	30–40	10 L/L/d	0.69 mol/mol hexose	
Wheat straw	Hydrothermal	CSTR	Mixed extreme thermophilic culture	70	3.9 g glucose/L	72	41	241 ml/L/d	188 ml/g sugar	
Wheat straw	Hydrothermal	UASB	Sterilized methanogenic granules	70	3.9 g glucose/L	24	43	821 ml/L/d	212 ml/g sugar	Kongjan and Angelidaki (2010)
Wheat straw	Hydrothermal	AF	Mixed extreme thermophilic culture	70	3.9 g/glucose/L	24		488 ml/L/d	126 ml/g sugar	
Oat straw	Acid	TBF	Heat-treated anaerobic granular sludge	28	1.2 g COD/L	24	45	81.4 mL/L/h	2.9 mol/mol hexose	Arriaga et al. (2011)
Oat straw	Enzyme	TBF	Heat-treated triticale silage	35	5 g COD/L	12		26 ml/L/h	2.3 mol/mol sugar	Vargas et al. (2014)
Oat straw	Acid						66	71 ml/L/h	0.59 mol/mol sugar	Vargas et al. (2013)
Oat straw	Enzymatic	ASBR	Heat-treated anaerobic granular sludge	35	4.5 g COD/L	8	50	29.6 ml/L/h	0.81 mol/mol sugar	
Oat straw	Acid and enzyme						58	27 ml/L/h	0.38 mol/mol sugar	

TS total sugar, *CECBR* continuously external circulating bioreactor, *CSTR* continuously stirred tank reactor, *UASB* upflow anaerobic sludge blanket reactor, *AF* anaerobic filter, *TBF* trickling biofilter, *ASBR* anaerobic sequencing batch reactor

[a]Wheat straw hydrolysate was mixed with food industry wastewater

[b]Fermentation temperature

state fermentation are two strategies for this purpose to reduce the cost of production. These approaches need detailed investigations to determine process control parameters and to select microorganism types.

2.5 Conclusions

Hydrogen is certainly accepted as an energy carrier and alternative to fossil fuels. The development of economical large-scale hydrogen production is required to replace hydrogen with fossil fuels as one of the main energy sources. It is evident that cost-effective hydrogen production cannot be achieved by energy-intensive processes and by using fossil fuels as raw material which are already limited and highly competitive for the generation of other energy sources. Sustainable, renewable, and green energy demands address biohydrogen production from biomasses. Generation of biohydrogen from, especially, second- and third-generation biomasses combines these three concepts along with providing economical process. Therefore, hydrogen production from biomass by fermentative processes is a viable alternative to existing physical and chemical methods. Process limitations or obstacles are well known, and intensive studies are conducted by researchers to overcome all these obstacles and to make biohydrogen competitive with the conventionally produced ones. Other areas that need extensive effort are commercialization and adaptation of hydrogen sector to biohydrogen. It is necessary to transfer research outcomes into practice. Interdisciplinary studies with participants from sectors could accelerate application of biohydrogen technology in large scale.

References

Agbor VB, Cicek N, Sparling R et al (2011) Biomass pretreatment: fundamentals toward application. Biotechnol Adv 29:675–685

Al-Shorgani NKN, Tibin EM, Ali E et al (2014) Biohydrogen production from agroindustrial wastes via *Clostridium saccharoperbutylacetonicum* N1-4 (ATCC 13564). Clean Technol Environ 16:11–21

Alvira P, Ballesteros M, Negro MJ (2010) Pretreatment technologies for an efficient bioethanol production process based on enzymatic hydrolysis: a review. Bioresour Technol 101:4851–4861

Amutha KB, Murugesan AG (2013) Biohydrogen production using corn stalk employing *Bacillus licheniformis* MSU AGM 2 strain. Renew Energy 50:621–627

Argun H, Kargi F (2010a) Effects of light source, intensity and lighting regime on bio-hydrogen production from ground wheat starch by combined dark and photo-fermentations. Int J Hydrog Energy 35:1604–1612

Argun H, Kargi F (2010b) Bio-hydrogen production from ground wheat starch by continuous combined fermentation using annular-hybrid bioreactor. Renew Energy 35:6170–6178

Argun H, Kargi F, Kapdan IK (2008a) Light fermentation of dark fermentation effluent for biohydrogen production by different *Rhodobacter* species at different initial volatile fatty acid (VFA) concentrations. Int J Hydrog Energy 33:7405–7412

Argun H, Kargi F, Kapdan IK et al (2008b) Batch dark fermentation of powdered wheat starch to hydrogen gas: effects of the initial substrate and biomass concentrations. Int J Hydrog Energy 33:6109–6115

Argun H, Kargi F, Kapdan IK et al (2008c) Biohydrogen production by dark fermentation of wheat powder solution: effects of C/N and C/P ratio on hydrogen yield and formation rate. Int J Hydrog Energy 33:1813–1819

Argun H, Kargi F, Kapdan IK (2009a) Hydrogen production by combined dark and light fermentation of ground wheat solution. Int J Hydrog Energy 34:4305–4311

Argun H, Kargi F, Kapdan IK (2009b) Effects of the substrate and cell concentration on bio-hydrogen production from ground wheat by combined dark and photo-fermentation. Int J Hydrog Energy 34:6181–6188

Argun H, Kargi F, Kapdan IK (2009c) Microbial culture selection for bio-hydrogen production from waste ground wheat by dark fermentation. Int J Hydrog Energy 34:2195–2200

Arriaga S, Rosas I, Mondragón FA et al (2011) Continuous production of hydrogen from oat straw hydrolysate in a biotrickling filter. Int J Hydrog Energy 36:3442–3449

Asada Y, Tokumoto M, Aihara Y et al (2006) Hydrogen production by co-cultures of *Lactobacillus* and a photosynthetic bacterium, *Rhodobacter sphaeroides* RV. Int J Hydrog Energy 31:1509–1513

Azbar N, Kapdan IK (2012) Use of immobilized cell systems in biohydrogen production. In: Azbar N, Levin D (eds) State of the art and progress in production of biohydrogen. Bentham Science Publishers. ISBN 978-1-60805-411-4

Azwar MY, Hussain MA, Abdul-Wahab AK (2014) Development of biohydrogen production by photobiological, fermentation and electrochemical processes: a review. Renew Sust Energ Rev 31:58–173

Badiei M, Asim N, Jahim JM et al (2014) Comparison of chemical pretreatment methods for cellulosic biomass. APCBEE Procedia 9:170–174

Bak JS (2014) Electron beam irradiation enhances the digestibility and fermentation yield of water-soaked lignocellulosic biomass. Biotechnol Rep 4:30–33

Banerjee G, Car S, Scott-craig JS et al (2011) Alkaline peroxide pretreatment of corn stover : effects of biomass, peroxide, and enzyme loading and composition on yields of glucose and xylose. Biotechnol Biofuels 4:16. http://www.biotechnologyforbiofuels.com/content/4/1/16

Batista AP, Moura P, Marques PASS et al (2014) *Scenedesmus obliquus* as feedstock for biohydrogen production by *Enterobacter aerogenes* and *Clostridium butyricum*. Fuel 117:537–543

Bhalla A, Bansal N, Kumar S et al (2013) Improved lignocellulose conversion to biofuels with thermophilic bacteria and thermostable enzymes. Bioresour Technol 128(2013):751–759

Brynjarsdottir H, Scully SM, Orlygsson J (2013) Production of biohydrogen from sugars and lignocellulosic biomass using *Thermoanaerobacter* GHL15. Int J Hydrog Energy 38:14467–14475

Cakır A, Ozmihci S, Kargi F (2010) Comparison of biohydrogen production from hydrolyzed wheat starch by mesophilic and thermophilic dark fermentation. Int J Hydrog Energy 35:13214–13218

CalRecycle. http://www.calrecycle.ca.gov/organics/Homecompost/. California's Dep Resour Recycl Recover. Accessed 2 Sept 2015

Cantarella M, Cantarella L, Alberto Gallifuoco AS et al (2004) Effect of inhibitors released during steam-explosion treatment of poplar wood on subsequent enzymatic hydrolysis and SSF. Biotechnol Prog 20:200–206

Cao G, Ren N, Wang A, Lee DJ et al (2009) Acid hydrolysis of corn stover for biohydrogen production using *Thermoanaerobacterium thermosaccharolyticum* W16. Int J Hydrog Energy 34:7182–7188

Cao G-L, Guo W-Q, Wang A-J et al (2012) Enhanced cellulosic hydrogen production from lime-treated cornstalk wastes using thermophilic anaerobic microflora. Int J Hydrog Energy 37:13161–13166

Caputo AC, Palumbo M, Pelagagge PM et al (2005) Economics of biomass energy utilization in combustion and gasification plants: effects of logistic variables. Biomass Bioenerg 28:35–51

Carere CR, Sparling R, Cicek N et al (2008) Third generation biofuels via direct cellulose fermentation. Int J Mol Sci 9:1342–1360

Carver SM, Hulatt CJ, Thomas DN et al (2011) Thermophilic, anaerobic co-digestion of microalgal biomass and cellulose for H_2 production. Biodegradation 22:805–814

Chandra RP, Bura R, Mabee WE et al (2007) Substrate pretreatment: the key to effective enzymatic hydrolysis of lignocellulosics. Adv Biochem Eng Biotechnol 108:67–93

Chaturvedi V, Verma P (2013) An overview of key pretreatment processes employed for bioconversion of lignocellulosic biomass into biofuels and value added products. Biotech 3:415–431

Chen R, Wang YZ, Liao Q et al (2013) Hydrolysates of lignocellulosic materials for biohydrogen production. BMB Rep 46:244–521

Chen CY, Chang HY, Chang JS (2015) Producing carbohydrate-rich microalgal biomass grown under mixotrophic conditions as feedstock for biohydrogen production. Int J Hydrog Energy. doi:10.1016/j.ijhydene.2015.05.163

Cheng J, Xie B, Zhou J et al (2010) Cogeneration of H_2 and CH_4 from water hyacinth by two-step anaerobic fermentation. Int J Hydrog Energy 35:3029–3035

Cheng CL, Lo YC, Lee KS et al (2011) Biohydrogen production from lignocellulosic feedstock. Bioresour Technol 102:8514–8523

Cheng J, Xia A, Song W et al (2012) Comparison between heterofermentation and autofermentation in hydrogen production from *Arthrospira* (*Spirulina*) *platensis* wet biomass. Int J Hydrog Energy 37:6536–6544

Cheng J, Lin R, Song W et al (2015) Enhancement of fermentative hydrogen production from hydrolyzed water hyacinth with activated carbon detoxification and bacteria domestication. Int J Hydrog Energy 40:2545–2551

Chong ML, Sabaratnam V, Shirai Y et al (2009) Biohydrogen production from biomass and industrial wastes by dark fermentation. Int J Hydrog Energy 34:3277–3287

Chu C-Y, Wu S-Y, Tsai C-Y et al (2011) Kinetics of cotton cellulose hydrolysis using concentrated acid and fermentative hydrogen production from hydrolysate. Int J Hydrog Energy 36:8743–8750

Cui M, Yuan Z, Zhi X et al (2009) Optimization of biohydrogen production from beer lees using anaerobic mixed bacteria. Int J Hydrog Energy 34:7971–7978

Cui M, Yuan Z, Zhi X et al (2010) Biohydrogen production from poplar leaves pretreated by different methods using anaerobic mixed bacteria. Int J Hydrog Energy 35:4041–4047

Datar R, Huang J, Maness PC et al (2007) Hydrogen production from the fermentation of corn stover biomass pretreated with a steam-explosion process. Int J Hydrog Energy 32:932–939

De Menezes CR, Silva ÍS, Pavarina ÉC et al (2010) Production of xylooligosaccharides from enzymatic hydrolysis of xylan by white-rot fungi *Pleurotus*. Acta Sci Technol 32:37–42

de Vrije T, de Haas GG, Tan GB et al (2002) Pretreatment of miscanthus for hydrogen production by *Thermotoga* elfii. Int J Hydrog Energy 27:1381–1390

de Vrije T, Budde MAW, Lips SJ et al (2010) Hydrogen production from carrot pulp by the extreme thermophiles *Caldicellulosiruptor saccharolyticus* and *Thermotoga neapolitana*. Int J Hydrog Energy 35:13206–13213

Dowe N, Nrel JMSSF (2001) Experimental protocols — lignocellulosic biomass hydrolysis and fermentation laboratory analytical procedure (LAP)

Dragone G, Fernandes B, Vicente AA et al (2010) Third generation biofuels from microalgae. In: Mendez-Vilas A (ed) Current research, technology and education topics in applied microbiology and microbial biotechnology. Formatex Research Center, Spain, pp 1355–1366

Fan YT, Zhang YH, Zhang SF et al (2006) Efficient conversion of wheat straw wastes into biohydrogen gas by cow dung compost. Bioresour Technol 97:500–505

Fang HHP, Li C, Zhang T (2006) Acidophilic biohydrogen production from rice slurry. Int J Hydrog Energy 31:683–692

Ferreira AF, Ortigueira J, Alves L et al (2013) Energy requirement and CO_2 emissions of bioH_2 production from microalgal biomass. Biomass Bioenerg 49:249–259

FitzPatrick M, Champagne P, Cunningham MF et al (2010) A biorefinery processing perspective: treatment of lignocellulosic materials for the production of value-added products. Bioresour Technol 101:8915–8922

Geng A, He Y, Qian C et al (2010) Effect of key factors on hydrogen production from cellulose in a co-culture of *Clostridium thermocellum* and *Clostridium thermopalmarium*. Bioresour Technol 101:4029–4033

Ghimire A, Frunzo L, Pontoni L et al (2015) Dark fermentation of complex waste biomass for biohydrogen production by pretreated thermophilic anaerobic digestate. J Environ Manag 152:43–48

Gould JM (1985) Enhanced polysaccharide recovery from agricultural residues and perennial grasses treated with alkaline hydrogen peroxide. Biotechnol Bioeng 27:893–896

Gould JM, Freer SN (1984) High-efficiency ethanol production from lignocellulosic residues pretreated with alkaline H_2O_2. Biotechnol Bioeng 26:628–631

Guo F, Fang Z, Xu CC et al (2012) Solid acid mediated hydrolysis of biomass for producing biofuels. Prog Energy Combust Sci 38:672–690

Guo XM, Trably E, Latrille E et al (2014) Predictive and explicative models of fermentative hydrogen production from solid organic waste: role of butyrate and lactate pathways. Int J Hydrog Energy 39:7476–7485

Gupta SK, Kumari S, Reddy K et al (2013) Trends in biohydrogen production: major challenges and state-of-the-art developments. Environ Technol 34:1653–1670

Hallenbeck PC (2009) Fermentative hydrogen production: principles, progress, and prognosis. Int J Hydrog Energy 34:7379–7389

Han G, Deng J, Zhang S et al (2010) Effect of steam explosion treatment on characteristics of wheat straw. Ind Crop Prod 31:28–33

Han H, Wei L, Liu B et al (2012) Optimization of biohydrogen production from soybean straw using anaerobic mixed bacteria. Int J Hydrog Energy 37:13200–13208

Han W, Liu DN, Li YF et al (2015a) Utilization of wheat for biohydrogen production by a combination of solid-state fermentation and batch fermentation. Int J Hydrog Energy 40:5849–5855. doi:10.1016/j.ijhydene.2015.03.036

Han W, Wang X, Ye L et al (2015b) Fermentative hydrogen production using wheat flour hydrolysate by mixed culture. Int J Hydrog Energy 40:4474–4480. doi:10.1016/j.ijhydene.2015.02.016

Harmsen P, Huijgen W, Bermudez L, Bakker RRC (2010) Literature review of physical and chemical pretreatment processes for lignocellulosic biomass. BioSynergy Project,Wageningen University & Research centre – Food & Biobased Research (WUR-FBR, NL), Energy Research Centre of the Netherlands (ECN, NL), Abengoa Bioenergía Nuevas Tecnologías (ABNT, ES)

He L, Huang H, Lei Z (2014) Enhanced hydrogen production from anaerobic fermentation of rice straw pretreated by hydrothermal technology. Bioresour Technol 171:145–151

Hernández MA, Rodríguez SM, Andres Y (2014) Use of coffee mucilage as a new substrate for hydrogen production in anaerobic co-digestion with swine manure. Bioresour Technol 168:112–118. doi:10.1016/j.biortech.2014.02.101

Hniman A, O-Thong S, Prasertsan P (2011) Developing a thermophilic hydrogen-producing microbial consortia from geothermal spring for efficient utilization of xylose and glucose mixed substrates and oil palm trunk hydrolysate. Int J Hydrog Energy 36:8785–8793

Ho KL, Lee DJ, Suc A et al (2012) Biohydrogen from lignocellulosic feedstock via one-step process. Int J Hydrog Energy 37:15569–15574

Jacquet N, Quiévy N, Vanderghem C et al (2011) Influence of steam explosion on the thermal stability of cellulose fibres. Polym Degrad Stab 96:1582–1588

Jeihanipour A, Taherzadeh MJ (2009) Ethanol production from cotton-based waste textiles. Bioresour Technol 100:1007–1010

Jeihanipour A, Karimi K, Taherzadeh MJ (2011) Acid hydrolysis of cellulose-based waste textiles. In: 7th International Chemical Engineering Congress and Exhibition. IChEC, Kish, Iran. pp 21–24

Jeoh T (1998) Steam explosion pretreatment of cotton gin waste for fuel ethanol production. Msc Thesis Virginia Polytechnic Institute and State University

Jung JY, Choi MS, Yang JK (2013) Optimization of concentrated acid hydrolysis of waste paper using response surface methodology. J Kor Wood Sci Technol 41:87–99

Kádár Z, De Vrije T, Noorden GEV et al (2004) Yields from glucose, xylose, and paper sludge hydrolysate during hydrogen production by the extreme thermophile *Caldicellulosiruptor saccharolyticus*. Appl Biochem Biotechnol 113–116:497–508

Kapdan IK, Kargi F (2006) Bio-hydrogen production from waste materials. Enzym Microbiol Tech 38:569–582

Kapdan IK, Kargi F, Oztekin R et al (2009) Biohydrogen production from hydrolyzed wheat starch by photofermentation. Int J Hydrog Energy 34:2201–2207

Karagöz P, Rocha IV, Özkan M et al (2012) Alkaline peroxide pretreatment of rapeseed straw for enhancing bioethanol production by same vessel saccharification and co-fermentation. Bioresour Technol 104:349–357

Kargi F, Pamukoglu MY (2009) Dark fermentation of ground wheat starch for bio-hydrogen production by fed-batch operation. Int J Hydrog Energy 34:2940–2946

Kars G, Ceylan A (2013) Biohydrogen and 5-aminolevulinic acid production from waste barley by *Rhodobacter sphaeroides* O.U.001 in a biorefinery concept. Int J Hydrog Energy 38:5573–5579

Karunanithy C, Muthukumarappan KA (2012) Comparative study on torque requirement during extrusion pretreatment of different feedstocks. Bioenergy Res 5:263–276

Karunanithy C, Muthukumarappan K, Gibbons WR (2012) Extrusion pretreatment of pine wood chips. Appl Biochem Biotechnol 167:81–99

Keskin T, Abo-Hashesh M, Hallenbeck PC (2011) Photofermentative hydrogen production from wastes. Bioresour Technol 102:8557–8568

Khamtib S, Reungsang A (2012) Biohydrogen production from xylose by *Thermoanaerobacterium thermosaccharolyticum* KKU19 isolated from hot spring sediment. Int J Hydrog Energy 37:12219–12228

Klimiuk E, Pokoj T, Budzyniski W et al (2010) Theoretical and observed biogas production from plant biomass of different fibre contents. Bioresour Technol 101:9527–9535

Knill CJ, Kennedy JF (2002) Degradation of cellulose under alkaline conditions. Carbohydr Polym 51:281–300

Kongjan P, Angelidaki I (2010) Extreme thermophilic biohydrogen production from wheat straw hydrolysate using mixed culture fermentation: effect of reactor configuration. Bioresour Technol 101:7789–7796

Kongjan P, Min B, Angelidaki I (2009) Biohydrogen production from xylose at extreme thermophilic temperatures (70 °C) by mixed culture fermentation. Water Res 43:1414–1424

Kumar K, Roy S, Das D (2013) Continuous mode of carbon dioxide sequestration by *C. sorokiniana* and subsequent use of its biomass for hydrogen production by *E. cloacae* IIT-BT 08. Bioresour Technol 145:116–122

Lai Z, Zhu M, Yang X et al (2014) Optimization of key factors affecting hydrogen production from sugarcane bagasse by a thermophilic anaerobic pure culture. Biotechnol Biofuels:1–11. http://www.biotechnologyforbiofuels.com/content/7/1/119

Laurinavichene TV, Belokopytov BF, Laurinavichius KS et al (2010) Towards the integration of dark- and photo-fermentative waste treatment. Potato as substrate for sequential dark fermentation and light-driven H_2 production. Int J Hydrog Energy 35:8536–8543

Lay HC, Lin HC, Sen B et al (2012) Simultaneous hydrogen and ethanol production from sweet potato via dark fermentation. J Clean Prod 27:155–164

Lee RA, Lavoie JM (2013) From first- to third-generation biofuels: challenges of producing a commodity from a biomass of increasing complexity. Anim Front 3:6–11

Li Q, Liu CZ (2012) Co-culture of *Clostridium thermocellum* and *Clostridium thermosaccharolyticum* for enhancing hydrogen production via thermophilic fermentation of cornstalk waste. Int J Hydrog Energy 37:10648–10654

Liu H, Wang G (2014) Fermentative hydrogen production from macroalgae Laminaria japonica using anaerobic mixed bacteria. Int J Hydrog Energy 39:9012–9017

Liu Y, Yu P, Song X, Qu Y (2008) Hydrogen production from cellulose by co-culture of *Clostridium thermocellum* JN4 and *Thermoanaerobacterium thermosaccharolyticum* GD17. Int J Hydrog Energy 33:2927–2933

Liu CH, Chang CY, Cheng CL et al (2012) Fermentative hydrogen production by *Clostridium butyricum* CGS5 using carbohydrate-rich microalgal biomass as feedstock. Int J Hydrog Energy 37:15458–15464

Liu CM, Chu CY, Lee WY et al (2013) Biohydrogen production evaluation from rice straw hydrolysate by concentrated acid pre-treatment in both batch and continuous systems. Int J Hydrog Energy 38:15823–15829

Liu CM, Wu SY, Chu CY et al (2014a) Biohydrogen production from rice straw hydrolyzate in a continuously external circulating Bioreactor. Int J Hydrog Energy 39:19317–19322

Liu Z, Li Q, Zhang C et al (2014b) Effects of operating parameters on hydrogen production from raw wet steam-exploded cornstalk and two-stage fermentation potential for biohythane production. Biochem Eng J 90:234–238

Liu Z, Zhang C, Wang L et al (2015) Effects of furan derivatives on biohydrogen fermentation from wet steam-exploded cornstalk and its microbial community. Bioresour Technol 175:152–159

Lo YC, Su YC, Cheng CL et al (2011) Biohydrogen production from pure and natural lignocellulosic feedstock with chemical pretreatment and bacterial hydrolysis. Int J Hydrog Energy 36:3955–3963

Lu YP, Zhang YHP, Lynd LR (2006) Enzyme-microbe synergy during cellulose hydrolysis by Clostridium thermocellum. Proc Natl Acad Sci U S A 103: 16165–16169

Luo J, Fang Z, Smith RL (2014) Ultrasound-enhanced conversion of biomass to biofuels. Prog Energy Combust Sci 41:56–93

Mansfield SD, Mooney C, Saddler JN (1999) Substrate and enzyme characteristics that limit cellulose hydrolysis. Biotechnol Prog 15:804–816

Mars AE, Veuskens T, Budde MAW et al (2010) Biohydrogen production from untreated and hydrolyzed potato steam peels by the extreme thermophiles *Caldicellulosiruptor saccharolyticus* and *Thermotoga neapolitana*. Int J Hydrog Energy 35:7730–7737

Martín-Sampedro R, Eugenio ME, García JC et al (2012) Steam explosion and enzymatic pre-treatments as an approach to improve the enzymatic hydrolysis of *Eucalyptus globulus*. Biomass Bioenerg 42:97–106

Mathews J, Wang G (2009) Metabolic pathway engineering for enhanced biohydrogen production. Int J Hydrogen Energy 34:7404–7416

McKendry P (2002) Energy production from biomass (part 1): overview of biomass. Bioresour Technol 83:37–46

Mehnert R (1995) Electron beams in research and technology. Nucl Instrum Methods Phys Res B 105:348–358

Mollers KB, Cannella D, Jorgensen H et al (2014) Cyanobacterial biomass as carbohydrate and nutrient feedstock for bioethanol production by yeast fermentation. Biotechnol Biofuels 64:1–11

Monlau F, Aemig Q, Trably E et al (2013) Specific inhibition of biohydrogen-producing Clostridium sp. after dilute-acid pretreatment of sunflower stalks. Int J Hydrog Energy 38:12273–12282

Monlau F, Kaparaju P, Trably E et al (2015) Alkaline pretreatment to enhance one-stage CH_4 and two-stage H2/CH4 production from sunflower stalks: mass, energy and economical balances. Chem Eng J 260:377–385

Moodley P, Kana EBG (2015) Optimization of xylose and glucose production from sugarcane leaves (Saccharum officinarum) using hybrid pretreatment techniques and assessment for hydrogen generation at semi-pilot scale. Int J Hydrog Energy 40:3859–3867. doi:10.1016/j.ijhydene.2015.01.087

Mosier N, Wyman C, Dale B et al (2005) Features of promising technologies for pretreatment of lignocellulosic biomass. Bioresour Technol 96:673–686

Nasirian N, Almassi M, Minaei S et al (2011) Development of a method for biohydrogen production from wheat straw by dark fermentation. Int J Hydrog Energy 36:411–420

Nasr N, Gupta M, Elbeshbishy E et al (2014) M. Biohydrogen production from pretreated corn cobs. Int J Hydrog Energy 39:19921–19927

Nath K, Das D (2003) Hydrogen from biomass. Curr Sci 85:265–271

Nayak BK, Roy S, Das D (2014) Biohydrogen production from algal biomass (*Anabaena sp.* PCC 7120) cultivated in airlift photobioreactor. Int J Hydrog Energy 39:7553–7560

Neves LA, Nemestóthy N, Alves VD et al (2009) Separation of biohydrogen by supported ionic liquid membranes. Desalination 240:1–3, 311–315

Nguyen TAD, Kim YP, Kim MS et al (2008) Optimization of hydrogen production by hyperthermophilic eubacteria, *Thermotoga maritima* and *Thermotoga neapolitana* in batch fermentation. Int J Hydrog Energy 33:1483–1488

Ni M, Leung DYC, Leung MKH et al (2006) An overview of hydrogen production from biomass. Fuel Process Technol 87:461–472

Nissila ME, Lay CH, Puhakka JA (2014) Dark fermentative hydrogen production from lignocellulosic hydrolyzates: a review. Biomass Bioenergy 67:145–159

Ntaikou I, Koutros E, Kornaros M (2009) Valorisation of wastepaper using the fibrolytic/hydrogen producing bacterium *Ruminococcus albus*. Bioresour Technol 100:5928–5933

Oh YK, Seol EH, Kim MS et al (2004) Photo-production of hydrogen from acetate by a chemoheterotrophic bacterium *Rhodopseudomonas palustris* P4. Int J Hydrog Energy 29:1115–1121

Orozco RS, Hernández PB, Ramírez NF et al (2012) Gamma irradiation induced degradation of orange peels. Energies 5:3051–3063

Ortigueira J, Alves L, Gouveia L, Moura P (2015) Third generation biohydrogen production by *Clostridium butyricum* and adapted mixed cultures from *Scenedesmus obliquus* microalga biomass. Fuel 153:128–134. doi:10.1016/j.fuel.2015.02.093

Özgür E, Peksel B (2013) Biohydrogen production from barley straw hydrolysate through sequential dark and photofermentation. J Clean Prod 52:14–20

Özgür E, Mars AE, Peksel B et al (2010) Biohydrogen production from beet molasses by sequential dark and photofermentation. Int J Hydrog Energy 35:511–517

Ozkan E, Uyar B, Ozgur E et al (2012) Photofermentative hydrogen production using dark fermentation effluent of sugar beet thick juice in outdoor conditions. Int J Hydrog Energy 37:2044–2049

Ozmihci S, Kargi F (2010) Comparison of different mixed cultures for bio-hydrogen production from ground wheat starch by combined dark and light fermentation. J Ind Microbiol Biotechnol 37:341–347

Oztekin R, Kapdan IK, Kargi F et al (2008) Optimization of media composition for hydrogen gas production from hydrolyzed wheat starch by dark fermentation. Int J Hydrog Energy 33:4083–4090

Panagiotopoulos IA, Bakker RR, de Vrije T et al (2010) Prospects of utilization of sugar beet carbohydrates for biological hydrogen production in the EU. J Clean Prod 18:9–14

Panagiotopoulos I, Bakker RR, de Vrije T et al (2011) Effect of pretreatment severity on the conversion of barley straw to fermentable substrates and the release of inhibitory compounds. Bioresour Technol 102:11204–11211

Park JH, Yoon JJ, Park HD et al (2011) Feasibility of biohydrogen production from *Gelidium amansii*. Int J Hydrog Energy 36:13997–14003

Park JH, Cheon HC, Yoon JJ et al (2013) Optimization of batch dilute-acid hydrolysis for biohydrogen production from red algal biomass. Int J Hydrog Energy 38:6130–6136

Patel AK, Debroy A, Sharma S et al (2015) Biohydrogen production from a novel alkalophilic isolate *Clostridium sp.* IODB-O3. Bioresour Technol 175:291–297

Pattanamanee W, Choorit W, Deesan C et al (2012) Photofermentive production of biohydrogen from oil palm waste hydrolysate. Int J Hydrog Energy 37:4077–4087

Pattra S, Sangyoka S, Boonmee M et al (2008) Biohydrogen production from the fermentation of sugarcane bagasse hydrolysate by *Clostridium butyricum*. Int J Hydrog Energy 33:5256–5265

Pawar SS, Nkemka N, Zeidan AA et al (2013) Biohydrogen production from wheat straw hydrolysate using *Caldicellulosiruptor saccharolyticus* followed by biogas production in a two-step uncoupled process. Int J Hydrog Energy 38:9121–9130

Peng H, Gao L, Li M, Shen Y et al (2014) Steam explosion-ionic liquid pretreatments on wetland lignocellulosic biomasses of *Phragmites* (sp.) and *Thalia dealbata* for $BioH_2$ conversion. RSC Adv 4:36603

Pérez J, Muñoz-Dorado J, De La Rubia T et al (2002) Biodegradation and biological treatments of cellulose, hemicellulose and lignin: an overview. Int Microbiol 5:53–63

Perlack RD, Wright LL, Turhollow AF et al (2005) Biomass as feedstock for a bioenergy and bioproducts ındustry : the technical feasibility of a billion-ton annual supply Agriculture DOE/GO-102. p 78

Phummala K, Imai T, Reungsang A et al (2014) Delignification of disposable wooden chopsticks waste for fermentative hydrogen production by an enriched culture from a hot spring. J Environ Sci 26(2014):1361–1368

Prakasham RS, Brahmaiah P, Sathish T et al (2009) Fermentative biohydrogen production by mixed anaerobic consortia: impact of glucose to xylose ratio. Int J Hydrog Energy 34:9354–9361

Prakasham RS, Sathish T, Brahmaiah P (2010) Biohydrogen production process optimization using anaerobic mixed consortia: a prelude study for use of agro-industrial material hydrolysate as substrate. Bioresour Technol 101:5708–5711

Prasad S, Singh A, Joshi HC (2007) Ethanol production from sweet sorghum syrup for utilization as automotive fuel in India. Energy Fuel 21:2415–2420

Qian EW (2014) Pretreatment and saccharification of lignocellulosic biomass. In: Tojo S (ed) Research approaches to sustainable biomass systems. Elsevier, Oxford, pp 181–204

Rai PK, Singh SP, Asthana RK et al (2014) Biohydrogen production from sugarcane bagasse by integrating dark- and photo-fermentation. Bioresour Technol 152:140–146

Reilly M, Dinsdale R, Guwy A (2014) Mesophilic biohydrogen production from calcium hydroxide treated wheat straw. Int J Hydrog Energy 39:16891–16901

Ren N, Wang A, Gao L et al (2008) Bioaugmented hydrogen production from carboxymethyl cellulose and partially delignified corn stalks using isolated cultures. Int J Hydrog Energy 33:5250–5255

Rodrigo DSW (2013) Microbial degradation of lignocellulosic biomass, sustainable degradation of lignocellulosic biomass – techniques, applications and commercialization, Dr. Anuj Chandel (Ed.). ISBN: 978-953-51-1119-1, InTech. doi:10.5772/54325. http://www.intechopen.com/books/sustainable-degradation-of-lignocellulosic-biomass-techniques-applications-and-commercialization/microbial-degradation-of-lignocellulosic-biomass

Roy S, Kumar K, Ghosh S et al (2014) Thermophilic biohydrogen production using pre-treated algal biomass as substrate. Biomass Bioenerg 61:157–166

Ruggeri B, Tommasi T (2012) Efficiency and efficacy of pre-treatment and bioreaction for bio-H2 energy production from organic waste. Int J Hydrog Energy 37:6491–6502

Sagnak R, Kapdan IK, Kargi F (2010) Dark fermentation of acid hydrolyzed ground wheat starch for biohydrogen production by periodic feeding and effluent removal. Int J Hydrog Energy 35:9630–9636

Saraphirom P, Reungsang A (2010) Optimization of biohydrogen production from sweet sorghum syrup using statistical methods. Int J Hydrog Energy 35:13435–13444

Sawatdeenarunat C, Surendra KC, Takara D et al (2015) Anaerobic digestion of lignocellulosic biomass: challenges and opportunities. Bioresour Technol 178:178–186

Sekoai PT, Gueguim EB, Kana A (2013) Two-stage modelling and optimization of biohydrogen production from a mixture of agro-municipal waste. Int J Hydrog Energy 38:8657–8663

Shevchenko SM, Beatson RP, Saddler JN (1999) The nature of lignin from steam explosion/enzymatic hydrolysis of softwood. Appl Biochem Biotechnol 77–79:867–876

Shi X, Jung KW, Kim DH et al (2011) Direct fermentation of Laminaria japonica for biohydrogen production by anaerobic mixed cultures. Int J Hydrog Energy 36:5857–5864

Sims R, Taylor M, Saddler J, Mabee W (2008) From first to second generation biofuel technologies. An overview of current industry and RD&D activities. International Energy Agency, France. https://www.iea.org/publications/freepublications/publication/2nd_Biofuel_Gen.pdf

Singh A, Olsena SI, Nigamb PS (2011) A viable technology to generate third-generation biofuel. J Chem Technol Biotechnol 86:1349–1353

Sinha P, Pandey A (2014) Biohydrogen production from various feedstocks by *Bacillus firmus* NMBL-03. Int J Hydrog Energy 39:7518–7525

Sixta H, Rutkowska EW (2006) Comprehensive kinetic study of delignification, carbohydrate degradation, cellulose chain scissions, and hexenuronic acid reactions during kraft pulping of eucalyptus globulus. Lenzinger Ber 86:32–45

Song ZX, Li XH, Li WW et al (2014) Direct bioconversion of raw corn stalk to hydrogen by a new strain Clostridium sp. FS3. Bioresour Technol 157:91–97

Sridevi K, Sivaraman E, Mullai P (2014) Back propagation neural network modelling of biodegradation and fermentative biohydrogen production using distillery wastewater in a hybrid upflow anaerobic sludge blanket reactor. Bioresour Technol 165:233–240

Sun R, Tomkinson J, Wang S et al (2000) Characterization of lignins from wheat straw by alkaline peroxide treatment. Polym Degrad Stab 67:101–109

Sun J, Yuan X, Shi X et al (2011) Fermentation of *Chlorella* sp. for anaerobic bio-hydrogen production: influences of inoculum–substrate ratio, volatile fatty acids and NADH. Bioresour Technol 102:10480–10485

Taherzadeh K, Karimi M (2007a) Acid-based hydrolysis process for ethanol from lignocellulosic materials: a review. Bioresources 2:472–499

Taherzadeh MJ, Karimi K (2007b) Enzyme-based hydrolysis processes for ethanol from lignocellulosic materials: a review. Bioresources 2:707–738

Tang X, Ren N, Xu J (2013) Evaluation of hydrogen production from corn cob with the mesophilic bacterium *Clostridium hydrogeniproducens* HR-1. Int J Hydrog Energy 38:9104–9110

Thomsen MH, Thygesen A, Thomsen AB (2008) Hydrothermal treatment of wheat straw at pilot plant scale using a three-step reactor system aiming at high hemicellulose recovery, high cellulose digestibility and low lignin hydrolysis. Bioresour Technol 99:4221–4228

Tuna E, Kargi F, Argun H (2009) Hydrogen gas production by electrohydrolysis of volatile fatty acid (VFA) containing dark fermentation effluent. Int J Hydrog Energy 34:262–269

Ulhoff HET, Berlin HCS (2012) Carbides. Ullmann's Encycl Ind Chem 6:565–582

Vargas AJ, Celis LB, Buitron G et al (2013) Hydrogen production from acid and enzymatic oat straw hydrolysate in an anaerobic sequencing batch reactor: performance and microbial population analysis. Int J Hydrog Energy 38:13884–13894

Vargas AJ, Mondragon FA, Celis LB et al (2014) Continuous hydrogen production in trickling bed reactor by using triticale silage as inoculum: effect of simple complex substrates. J Chem Technol Biotechnol 90:1062–1069

Velázquez-Martí B, Fernández-González E, López-Cortés I et al (2011) Quantification of the residual biomass obtained from pruning of vineyards in Mediterranean area. Biomass Bioenerg 35:3453–3464

Verardi A, Bari I De, Ricca E et al (2012) Hydrolysis of lignocellulosic biomass: current status of processes and technologies and future perspectives, bioethanol. In: Lima MAP (Ed.). ISBN: 978-953-51-0008-9, InTech. doi:10.5772/23987. Available from: http://www.intechopen.com/books/bioethanol/hydrolysis-of-lignocellulosic-biomass-current-status-of-processes-and-technologies-and-future-perspective

Vijayaraghavan K, Ahmad D, Ibrahim KMB (2006) Biohydrogen generation from jackfruit peel using anaerobic contact filter. Int J Hydrog Energy 31:569–579

Wieczorek N, Kucuker MA, Kuchta K (2014) Fermentative hydrogen and methane production from microalgal biomass (*Chlorella vulgaris*) in a two-stage combined process. Appl Energy 132:108–117

Wu J, Upreti S, Ein-Mozaffari F (2013) Ozone pretreatment of wheat straw for enhanced biohydrogen production. Int J Hydrog Energy 38:10270–10276

Xia A, Cheng J, Ding L et al (2013) Improvement of the energy conversion efficiency of *Chlorella pyrenoidosa* biomass by a three-stage process comprising dark fermentation, photofermentation, and methanogenesis. Bioresour Technol 146:436–443

Xia A, Cheng J, Ding L et al (2014) Enhancement of energy production efficiency from mixed biomass of *Chlorella pyrenoidosa* and cassava starch through combined hydrogen fermentation and methanogenesis. Appl Energy 20:23–30

Xie B, Cheng J, Zhou J et al (2008) Production of hydrogen and methane from potatoes by two-phase anaerobic fermentation. Bioresour Technol 99:5942–5946

Yan Q, Zhao M, Miao H et al (2010) Coupling of the hydrogen and polyhydroxyalkanoates (PHA) production through anaerobic digestion from Taihu blue algae. Bioresour Technol 101:4508–4512

Yan Q, Wang A, Yu C et al (2011) Enzymatic characterization of acid tolerance response (ATR) during the enhanced biohydrogen production process from Taihu cyanobacteria via anaerobic digestion. Int J Hydrog Energy 36:405–410

Yin S, Mehrotra AK, Tan Z (2011) Alkaline hydrothermal conversion of cellulose to bio-oil: influence of alkalinity on reaction pathway change. Bioresour Technol 102:6605–6610

Yokoyama H, Moriya N, Ohmori H et al (2007) Community analysis of hydrogen-producing extreme thermophilic anaerobic microflora enriched from cow manure with five substrates. Appl Microbiol Biotechnol 77:213–222

Yuan X, Shi X, Zhang P et al (2011) Anaerobic biohydrogen production from wheat stalk by mixed microflora: kinetic model and particle size influence. Bioresour Technol 102:9007–9012

Yun YM, Kim DH, Oh YK et al (2014) Application of a novel enzymatic pretreatment using crude hydrolytic extracellular enzyme solution to microalgal biomass for dark fermentative hydrogen production. Bioresour Technol 159:365–372

Zhang X, Ye X, Guoa B et al (2013) Lignocellulosic hydrolysates and extracellular electron shuttles for H2 production using co-culture fermentation with *Clostridium beijerinckii* and *Geobacter metallireducens*. Bioresour Technol 147:89–95

Zhang K, Ren NQ, Wang AJ (2014a) Enhanced biohydrogen production from corn stover hydrolyzate by pretreatment of two typical seed sludges. Int J Hydrogen Energy 39:14653–14662

Zhang Z, Yue J, Zhou X et al (2014b) Photo-fermentative bio-hydrogen production from agricultural residue enzymatic hydrolyzate and the enzyme reuse. Bioresources 9:2299–2310

Zhang JN, Li YH, Zheng HQ et al (2015) Direct degradation of cellulosic biomass to bio-hydrogen from a newly isolated strain *Clostridium sartagoforme* FZ11. Bioresour Technol 192:60–67

Zhao L, Cao GL, Wang AJ et al (2013) Evaluation of continuous biohydrogen production from enzymatically treated cornstalk hydrolysate. Int J Hydrog Energy 38:15100–15104

Zhao L, Cao GL, Wang AJ et al (2014) An anaerobic sequential batch reactor for enhanced continuous hydrogen production from fungal pretreated cornstalk hydrolysate. Int J Hydrog Energy 39:19311–19316

Zheng J, Rehmann L (2014) Extrusion pretreatment of lignocellulosic biomass: a review. Int J Mol Sci 15:18967–18984

Zheng Y, Zhao J, Xu F et al (2014) Pretreatment of lignocellulosic biomass for enhanced biogas production. Prog Energy Combust Sci 42:35–53

Zhu JY, Pan X, Zalesny RS (2010) Pretreatment of woody biomass for biofuel production: energy efficiency, technologies, and recalcitrance. Appl Microbiol Biotechnol 87:847–857

Biohydrogen Production from Agricultural Biomass and Organic Wastes

3

Nicholas E. Korres and Jason K. Norsworthy

Abstract

With increasing demand for energy, depleting primary energy sources (i.e., coal and oil), and deteriorating environment, the efficient use and conservation of existing resources along with the production of energy from alternative nonconventional sources such as biomass have become essential. Hydrogen is a clean energy form with no emissions that can be used for generation of electricity or as fuel for transportation purposes. The production of hydrogen from lignocellulosic biomass or biohydrogen is a very promising energy form, and its role as an energy carrier in the future energy market is considered of major importance. Biohydrogen production from organic sources such as energy crops, agricultural residues, or organic waste seems very promising as it fulfills the basic sustainability criteria compared to conventional energy type. The technology for biomass conversion to biohydrogen has advanced in recent years and ensures relatively high volumes of production. Nevertheless, the production of biohydrogen shows a wide variation which can be largely explained by differences in substrate characteristics and operational conditions. Hence, attention needs to focus on sustainability issues concerning the production of the biomass, the standardization of the operational parameters, and the logistics of biohydrogen cycle production.

3.1 Introduction

With an increasing demand for energy in combination with depleting primary energy sources (i.e., coal and oil) and deteriorating environment, the efficient use and conservation of existing resources along with the production of energy from alternative nonconventional sources such as biomass have become essential. Hydrogen is a

N.E. Korres (✉) • J.K. Norsworthy
Department of Crop, Soil and Environmental Sciences, University of Arkansas,
1366 W. Altheimer Drive, Fayetteville, AR 72704, USA
e-mail: korres@uark.edu; nkorres@yahoo.co.uk

A. Singh, D. Rathore (eds.), *Biohydrogen Production: Sustainability of Current Technology and Future Perspective*, DOI 10.1007/978-81-322-3577-4_3

clean fuel with zero carbon dioxide (CO_2) emissions and its use is easily suited in fuel cells for electricity production. When burned or oxidized, it generates small amounts of pollutants such as water and nitrous oxides which can be easily controlled to minimum levels as the only emissions (Abedi et al. 2001). Its high energy yield of 122 kJ g^{-1}, which is 2.75 times greater than known hydrocarbon fuels, allows its use as a fuel for transportation, an option, which is highly considered due to many technical advantages and policy measures (Cherry 2004; Gemici et al. 2009). In addition, it can be stored either chemically or physiochemically in various solid and liquid composites, e.g., metal hydrides, carbon nanostructures, alanates, borohydrides, methane, methanol, or light hydrocarbons (Shakya et al. 2005). As stated by Winter (2005), 50 million tons of hydrogen are traded each year worldwide with an annual growth rate of approximately 10 %. Based on the National Hydrogen Program of the United States, the contribution of hydrogen to total energy market will reach 8–10 % by 2025 (Armor 1999). Additionally, as reported by the US Department of Energy, hydrogen power for use in transportation will be available in all regions of the United States by the year 2040 (US-DOE 2004). Hence, the development of efficient and cost-effective hydrogen production technologies, caused by the increasing need for hydrogen energy, has gained considerable attention.

It has been projected that a significant part of the European renewable energy will originate from agriculture and forestry (Korres et al. 2013). Lignocellulosic biomass, as a source of renewable forms of energy, is the most investigated type of feedstock, as it is one of the most abundant resources with wide availability in most countries (Korres et al. 2013). Hydrogen production from lignocellulosic biomass or biohydrogen is one of the most promising energy forms (Balat and Kirtay 2010) since its role as an energy carrier is of high importance (Balat 2008a, b; Veziroglou and Sahin 2008).

The aim of this chapter is to (1) highlight and discuss the potential of the most important lignocellulosic biomass sources that can be used as primary feedstock for biohydrogen production and (2) to discuss the most appropriate techniques and processes to convert these sources into biohydrogen.

3.2 Types of Feedstock for Biohydrogen Production

Hydrogen can be produced from a wide-ranging variety of primary energy sources and different production technologies (Balat and Kirtay 2010) although currently most of it is produced from nonrenewable feedstock, e.g., oil, natural gas, and coal (Balat and Kirtay 2010). More particularly, half of the hydrogen is produced from thermo-catalytic and gasification processes using natural gas, heavy oils, naphtha (Abedi et al. 2001), and coal (Mohan et al. 2007) as starting materials. The amount of fossil-derived CO_2 that is released into the atmosphere by these technologies contributes greatly to air pollution (Abedi et al. 2001) through greenhouse gas (GHG) emissions and consequently reduces the sustainability of the hydrogen-based energy production. Alternatively, hydrogen energy can be produced from renewable sources such as solar-hydrogen electrolysis process or from natural gas in a stream-methane reformation process (Kazim and Veziroglou 2001). The use of biomass is an attractive alternative to conventional feedstocks mainly because of the zero net CO_2 effect (Das et al. 2008; Anonymous 1995).

3.2.1 Biomass for Biohydrogen Production

Biomass is considered as any organic material that is renewable, including agricultural crops and trees, wood and timber residues, plants (either of terrestrial or aquatic origin), grasses, animal originating wastes (e.g., manure or slurry), or municipal wastes (Perlack et al. 2005; Jae-Hwa et al. 2010). The use of biomass for the production of hydrogen is of major importance (Chong et al. 2009). This is because the amount of CO_2 production when biomass is transformed into hydrogen energy counteracts the amount of CO_2 consumed during the growing cycle while biomass is produced. Hence the net CO_2 contribution from biomass-derived fuels is considerably less than that from fossil-derived fuels, a characteristic known as biomass carbon neutrality. In addition, producing biomass on a sustainable basis by growing energy crops, for example, will support the agricultural sector (Anonymous 1995) and increase agricultural diversification (Korres et al. 2010). Nevertheless, the ultimate impact of the biomass-to-hydrogen process depends on the overall economics since the major problem in utilization of biohydrogen as energy source is its unavailability in nature and the need for low-cost production methods (Kapdan and Kargi 2006). Moreover, feedstock issues, such as supply, cost, and logistics, are major destabilizing factors of the biohydrogen production process (Abedi et al. 2001).

Two types of biomass are suitable for biohydrogen production, namely, (a) bioenergy crops and (b) agricultural/wood-processing wastes (NRC 2004). These, based on their origin, can be classified as plant- or animal-oriented biomass (Table 3.1).

Theoretically any organic feedstock rich in carbohydrates, fats, or proteins can be considered as a possible source for biohydrogen production via biochemical transformation processes (Ntaikou

Table 3.1 Various biomass types suitable for biohydrogen energy production classified based on their origin and generation source

	Biomass origin		Main conversion processes
Generation	Plant	Animal	
Energy crops	Agriculture (maize, sweet sorghum, sugarcane, grass, etc.)		Microbiological fermentation[a]
	Herbaceous and woody (eucalyptus, miscanthus, etc.)		Microbiological fermentation, pyrolysis[b]
	Marine (algae, water hyacinth, etc.)		
Agricultural residues	Agricultural origin (rice husk, straw, stalk, nut shell, olive husk, grain residue, etc.)		Pyrolysis, steam gasification, supercritical fluid extraction[c]
	Forestry (sawdust, woodchip, bark, etc.)		Gasification (methanation included)[d]
	Fishery		
Wastes	Municipal waste (food, food processing waste, pulp and paper, etc.)		Microbiological fermentation, supercritical water extraction
	Agricultural waste (straw, maize stalk, fruit peel wastes, olive pulp, rice waste, etc.)		Pyrolysis
		Animal excrements (cattle, pig, chicken, etc.)	Microbiological fermentation

[a]Subjected to feedstock, e.g., olive and crop straw are converted into biohydrogen mainly by pyrolysis, nut shell by steam gasification, grain residue by supercritical fluid extraction (Balat 2008a, b; Balat and Kırtay 2010), energy crops by microbiological fermentation (Ntaikou et al. 2010)
[b]Subjected to feedstock, e.g., olive and crop straw are converted into biohydrogen mainly by pyrolysis, nut shell by steam gasification, grain residue by supercritical fluid extraction (Balat 2008a, b; Balat and Kırtay 2010), energy crops by microbiological fermentation (Ntaikou et al. 2010)
[c]Subjected to feedstock, e.g., olive and crop straw are converted into biohydrogen mainly by pyrolysis, nut shell by steam gasification, grain residue by supercritical fluid extraction (Balat 2008a, b; Balat and Kırtay 2010), energy crops by microbiological fermentation (Ntaikou et al. 2010)
[d]Subjected to feedstock, e.g., olive and crop straw are converted into biohydrogen mainly by pyrolysis, nut shell by steam gasification, grain residue by supercritical fluid extraction (Balat 2008a, b; Balat and Kırtay 2010), energy crops by microbiological fermentation (Ntaikou et al. 2010)

et al. 2010). However, among macromolecular organic compounds, carbohydrates are the major source of biochemical hydrogen production; thus biomass and wastes with high content of soluble carbohydrates are considered most suitable feedstocks (Kapdan and Kargi 2006; Lay et al. 2003).

3.2.1.1 Energy Crops

The use of energy crops along with forest and/or orchard residues and municipal solid wastes (Powlson et al. 2005; Demirbas 2008) using an appropriate technology (Petersen 2008) can greatly decrease GHG emissions while decoupling food and biofuel production (Kaparaju et al. 2009). Energy crops refer to certain plants that are cultivated solely for the exploitation of their biomass as feedstock for energy production. These can be directly utilized for their energy content via combustion or biotransformation to biofuels (Ntaikou et al. 2010). Crops such as maize or sweet sorghum and herbaceous and woody crops such as short rotation coppice, miscanthus (*Miscanthus × giganteus*), switchgrass (*Panicum virgatum*), kenaf (*Hibiscus cannabinus*), reed canarygrass (*Phalaris arundinacea*) or giant reed (*Arundo donax*), and eucalyptus (*Eucalyptus globulus*) are considered suitable crops for the production of renewable energy (Sims et al. 2006; Monti et al. 2009) such as biohydrogen (Antonopoulou et al. 2008; de Vrije et al. 2009). Nevertheless, it has been argued that crops such as maize or sugar are not economical for hydrogen production compared to by-products from agricultural crops or wastes from industrial processes (Benemann 1996). Hydrogen yield, from energy crops, exhibits wide variation (Table 3.2), which is mostly explained by differences in operational conditions (pH, inoculum source, temperature, loading rate, etc.) and the characteristics of the substrate (Pakarinen 2011).

Some of the research on biohydrogen production from energy crops has been focused on the

Table 3.2 Various energy crops and their potential biohydrogen yield

Energy crop	H_2 production rate	Units	H_2 yield[a]	Units	References
Miscanthus	–	–	1.1	mol H_2/mol sugars	de Vrije et al. (2002)
Miscanthus	13.1	mmol H_2 L^{-1} h^{-1}	3.2	mol H_2 mol^{-1} sugars	de Vrije et al. (2009)
Miscanthus	12.6	mmol H_2 L^{-1} h^{-1}	3.4	mol H_2 mol^{-1} sugars	de Vrije et al. (2009)
Water hyacinth	–	–	52	ml H_2 g^{-1}TVS	Cheng et al (2010)
Ryegrass	–	–	75	ml H_2 g^{-1} TS added[b]	Kyazze et al. (2008)
Ryegrass			22	ml H_2 g^{-1} TS added	Kyazze et al. (2008)
Sugar beet juice	2.2	ml m^{-1} L^{-1}	1.9	mol H_2 mol^{-1} sugars	Hussy et al. (2005)
Sweet sorghum extract			10.4	ml H_2 g^{-1} sweet sorghum or 0.86 mol H_2 mol^{-1} glucose consumed	Antonopoulou et al. (2009)
Sweet sorghum extract	8.52	L H_2 L^{-1} day^{-1}	0.86	mol H_2 mol^{-1} hexose converted	Antonopoulou et al. (2008)
Sugarcane			170 ml or 78.5 mmol g^{-1} VS		Hafner (2007)

TVS total volatile solids, *TS* total solids, *VS* volatile solids
[a]Feedstock has undergone various pretreatments, before it is exposed to a wide range of microorganism cultures during various operational modes and operational parameters (e.g., temperature, pH, loading rate, etc.)

pretreatment of the raw material such as the sugar-rich extract, which was extracted after water extraction (Antonopoulou et al. 2011) or after steam explosion (Datar et al. 2007; Kongjan et al. 2010). In addition, mixed or pure microbial cultures were used for bioconversion of energy crops to hydrogen production, whereas batch operational modes used widely in earlier studies have been recently substituted by continuous hydrogen production experiments (Pakarinen 2011).

3.2.1.2 Wastes

Waste materials can be subdivided as agricultural waste, industrial waste, municipal waste, and other hazardous wastes. Based on their origin, these can be further classified as organic waste materials such as food processing waste, organic wastes generated by industries, crop residues, or animal manures generated from agricultural practices or as organic fraction of municipal wastes (Korres et al. 2013). Biomass from organic wastes is an inexpensive and environmentally friendly way for biohydrogen production (Ni et al. 2006).

Agricultural Wastes

Different types of lignocellulosic residues originating either from the agricultural production per se (e.g., wheat straw, maize cob-stover, etc.) or as by-products from the processing of the agricultural commodity (e.g., sugarcane and sweet sorghum bagasse, barley hull, wheat bran, rice husks, rice washing drainage, etc.) (Singh et al. 2010) or forestry residues such as wood trimmings are potential sources for biohydrogen production (Ntaikou et al. 2010). Despite their abundance and their low cost, agricultural and forestry wastes contain complex carbohydrate polymers such as cellulose and hemicellulose, which are tightly joined with lignin and resistant to soluble carbohydrates released during fermentation. Consequently, their biotransformation to hydrogen, in most cases, is a difficult task despite the use of cellulolytic microorganisms during the fermentative hydrogen production (Ntaikou et al. 2010). Therefore, a pretreatment is required either using mechanical, chemical, or biological means (Korres and Nizami 2013). This will facilitate their delignification and subsequent loosening of the cellulose and hemicellulose structure followed by the release of soluble carbohydrates and their process either to hydrogen gas or organic acid production (Panagiotopoulos et al. 2009; Kapdan and Kargi 2006). The most important agricultural wastes along with their potential hydrogen yield are presented in Table 3.3.

Additionally, the highly degradable and widely available fruits and vegetables wastes have shown to produce high biohydrogen yields (Tenca et al. 2011). High biohydrogen yields were obtained by the fermentation of fruit waste such as jackfruit peel and sweet lime peeling extracts (Vijayaraghavan et al. 2006; Venkata Mohan et al. 2009) that equal to 76 $mLH_2\ g^{-1}$ COD and 198 $mLH_2\ g^{-1}$ VS, respectively. Vegetable pulp from cabbage and carrots also resulted in relatively high yields of biohydrogen equal to 62 and 71 $mLH_2\ g^{-1}$ VS, respectively (Okamoto et al. 2000). Additionally, wastes from the process of lettuce and potato yielded 50 and 106 mL $H_2\ g^{-1}$ VS, respectively (Dong et al. 2009), while a mixture of various vegetables resulted in 89 mL $H_2\ gCOD^{-1}$ (Venkata Mohan et al. 2009).

The quantity of forestry-based feedstocks and their by-products which are used for energy production is estimated by analyzing the actual firewood production. About 1.4 billion m^3 of firewood is used each year in the tropics resulting in approximately 40 million metric tons of charcoal production. Based on a wood-to-charcoal conversion rate (i.e., 8 and 17), the global charcoal supply in this region is between a quarter and a half of the firewood supply (May-Tobin 2011). Nevertheless, land use change along with efficient operational procedures needs to be taken under consideration for sustainable hydrogen production based on forestry products and wastes.

Table 3.3 Potential biohydrogen yield of various agriculture-based wastes

Feedstock	H_2 yield[a]	Units	References
Apple processing	0.9	L H_2 L^{-1} medium (0.1 L H_2 g^{-1} COD	Van Ginkel et al. (2005)
Maize stalk	20–176	ml H_2 g^{-1} VS	Fan et al. (2008)
Maize stalk	3–150	ml H_2 g^{-1}TVS	Zhang et al. (2007)
Fodder maize	62	ml H_2 g^{-1} TS added	Kyazze et al. (2008)
Fruit peel waste	459	Ml H_2 g^{-1}VS destroyed	Vijayaraghavan et al. (2007, 2009)
Olive pulp	0.19	Mmol H_2 g^{-1} TS	Koutrouli et al. (2009)
Olive pulp	0.321–1.6	mmol H_2 g^{-1} TS added	Gavala et al. (2005)
Pineapple waste	5.92	mmol H_2 g^{-1} COD	Wang et al. (2006)
Poplar leaves	15–44.9	ml H_2 g^{-1} TS	Cui et al. (2010)
Potato processing	2.1	L H_2 L^{-1} medium or 0.1 L H_2 g^{-1}COD	Van Ginkel et al. (2005)
Potato waste	30	ml H_2 g^{-1} TS	Zhu et al. (2008)
Wheat straw	0.5–68.1	ml H_2 g^{-1} TS	Fan et al. (2006)
Sugar wastewater	2.6	mol H_2 mol^{-1} hexose	Ueno et al. (1996)
Rice winery	2.14	mol H_2 mol^{-1} hexose	Yu et al. (2002)
Rice slurry	346	ml H_2 g^{-1} carbohydrate	Fang et al. (2006)
Starch wastewater	92	ml H_2 g^{-1} starch	Zhang et al. (2003)
POME (palm oil mill effluent)	2.3–6.33	L H_2 L^{-1} raw POME	Atif et al. (2005) and Thong et al. (2007)
POME	0.42	L H_2 g^{-1} COD reduced	Vijayaraghavan and Ahmad (2006)

[a]Feedstock has undergone various pretreatments, before it is exposed to a wide range of microorganism cultures during various operational modes and operational parameters (e.g., temperature, pH, supplements, etc.)

3.2.2 Animal-Generated Wastes for Biohydrogen Production

3.2.2.1 Livestock Manure

Livestock raised in large restrained animal feeding operations produce excessive concentrations of manure (and slurry), a suitable source for bioenergy production, but also a frequent water polluter. Fortunately, on the smaller end of the livestock production scale, farmers can use fermentation techniques to convert manure and slurry into biohydrogen and securing economic and environmental benefits (UCS 2012a, b). An analysis by the Union of Concerned Scientists (UCS) reported that the United States can tap almost 60 million tons of manure to produce bioenergy by 2030 (UCS 2012a, b). Nevertheless, the use of livestock residues as a feedstock for biohydrogen production via fermentation is not a suitable solution due to their chemical characteristics (e.g., alkaline pH, very low carbohydrate content). Swine manure, for example, as sole feedstock for biohydrogen production yields low biohydrogen volumes by fermentation at both mesophilic (between 30 and 40 °C) (Wagner et al. 2009) and hyperthermophilic (above 50–60 °C) temperatures with production lower than 4 LH_2/kg VS (Kotsopoulos et al. 2009). When hexose was added in the manure, the hydrogen yield raised up to 200 LH_2/kg hexose (Wu et al. 2009; Zhu et al. 2009) indicating the suitability of livestock residues as co-substrates to a carbohydrate-rich feedstock mixture for fermentative hydrogen production. According to Holm-Nielsen and Al Seadi (2004), this approach seems particularly advantageous and practical as it has been demonstrated in European conventional anaerobic digestion process plants. In these establishments agricultural crops and other organic wastes were co-digested with manure because of its positive influence, as a source of essential trace elements and beneficial buffering substances, to anaerobic digestion process (Tenca et al. 2011). This is particularly important regarding fruit and vegetable wastes, which are widely available and represent a form of highly degradable feedstock for use in biohydrogen production industry.

3.2.2.2 Livestock Waste

Wastewater from cheese and dairy industry, which is rich in lactose, can be used for the production of biohydrogen (Ferchichi et al. 2005a, b; Feng et al. 2005; Yang et al. 2005; Nath et al. 2008; Hasson 2009). Several studies have revealed that pH and operational mode or organic loading rates can affect the production of hydrogen from crude cheese (Collet et al. 2004) or dairy wastewater (Venkata Mohan et al. 2007).

3.2.3 Municipal and Industrial Wastes for Biohydrogen Production

Wastes originating from industrial and municipal sources should be favored for bioenergy production as this will relieve the pressure on natural habitats and the conflicts about food availability and conservation of biodiversity.

3.2.3.1 Organic Fraction of Municipal Solid Wastes

Municipal solid wastes (MSW) refer to the stream of garbage collected through community sanitation services (Anonymous 2003). MSW consists of items such as product packaging, grass clippings, furniture, clothing, bottles, household wastes, food scraps, newspapers, appliances, and batteries (Anonymous 2003). The combustion of MSW for the production of electricity and heat is an effective use of this type of feedstock as it eases the problem of their disposal (Tenca 2011). Nevertheless, combustion of MSW contributes to increases of GHG emissions (Tenca 2011). Alternatively, the 70 % of the organic fraction of municipal solid waste (OFMSW) that includes paper, garden residues, food wastes, and wood, and represents about 70 % of the total MSW produced, can be bioprocessed for hydrogen production. The feasibility of the process has been investigated by many workers. Lay et al. (1999) used a wide range of anaerobic microbes at mesophilic conditions for the digestion of OFMSW with varying "food-to-microorganisms" ratio. High hydrogenic activity was recorded at 0.4 g OFMSW g^{-1} bacterial biomass with rate of biohydrogen production rising at 43 LH_2 kg^{-1}VSS h^{-1} and production potential equal to 125 LH_2 kg^{-1}VS h^{-1}. Similarly, Valdez-Vazquez et al. (2005) recorded a yield of 95 LH_2 kg^{-1}VS added to the treating municipal organic waste. In addition, Okamoto et al. (2000) reported a hydrogen production of 19.3–96.0 LH_2 kg^{-1} VS added by mesophilic batch fermentation of OFMSW which contained rice and carrots. There is a great variation of various agricultural, food, or industrial waste feedstocks regarding their yield and production rate of biohydrogen (Table 3.4).

Table 3.4 Potential biohydrogen production of various food, animal, and industrial wastes

Feedstock	H_2 yield[a]	Units	References
OFMSW	0.15	L H_2 L g^{-1} OFMSW	Lay et al. (1999)
Food waste and sewage sludge	122.9	ml H_2 g^{-1} COD carbohydrate	Kim et al. (2004)
Waste paper	2.29	mol H_2 mol^{-1} hexose	Ntaikou et al. (2009)
Cheese whey	7.89	mmol H_2 g^{-1} lactose	Ferchichi et al. (2005a)
Cheese-processing wastewater	2.4	mmol H_2 g^{-1} COD	Yang et al. (2007)
Rice slurry	346	ml H_2 g^{-1} carbohydrate	Fang et al. (2006)
Dairy wastewater	1.59	mmol H_2 L^{-1} day^{-1}	Venkata Mohan et al. (2007)
Starch water	92	ml H_2 g^{-1} starch	Zhang et al. (2003)

[a]Feedstock has undergone various pretreatments, before it is exposed to a wide range of microorganism cultures during various operational modes and operational parameters (e.g., temperature, pH, supplements, etc.)

3.2.3.2 Sewage and Paper Sludge

Hason (2009) reported that a catalytic gasification of biomass in supercritical water (water at high temperature and pressure) has been developed for the bioconversion of sewage and paper sludge into biohydrogen. This process produced a gas at high pressure (>22 MPa) that was unusually rich in hydrogen (Hawaii Renewable Resources Program 2000).

3.2.3.3 Glycerol Waste

Glycerol is a by-product of biodiesel production (Chong et al. 2009). Four and half liters of crude glycerol are produced by the production of 45 l of biodiesel through the transesterification of vegetable oils or animal fats (Yazdami and Gonzalez 2007). Glycerol is used widely in cosmetics industry, but a further increase in the production of biodiesel fuels would most probably cause difficulties on the efficient treatment of glycerol-containing wastes. Nevertheless, the production of 1,3-propanediol using glycerol from biodiesel industry attracts the attention of stakeholders since it can be used as a basic ingredient for the production of polyesters (Petitdemange et al. 1995). A glycerol treatment under anaerobic conditions and thermophilic or mesophilic temperatures using an immobilized bioreactor packed with polyurethane was successfully developed by Yang et al. (2008). The syntrophic coexistence of hydrogenotrophic and *Methanobacterium* spp. bacteria was also recorded (Yang et al. 2008). Therefore, the production of biohydrogen from glycerol wastes with indigenous hydrogenotrophic bacteria is possible. According to Ito et al. (2005) the addition of yeast extract and tryptone could accelerate the biohydrogen production from biodiesel manufacturing process wastes by using *Enterobacter aerogenes* HU-101. Yeast extract, NH_4Cl, KCl, and $CaCl_2$ have been reported as the most important substances that influence hydrogen production from glycerol (Liu and Fang 2007). Sakai and Yagishita (2007) reported the biohydrogen production in bioelectrochemical cells using thionine as an extrinsic electron transfer mediator.

3.3 Processes for Biohydrogen Production

Hydrogen is considered as an important alternative energy vector and a conduit for future sustainable energy (Balat and Kirtay 2010). Hydrogen is not a primary energy source existing freely in nature but a secondary form of energy that has to be produced (Herzog and Tatsutani 2005). A wide variety of primary energy sources and different production technologies are available for hydrogen production. As mentioned earlier half of all the hydrogen is produced from thermo-catalytic and gasification processes using natural gas, heavy oils, naphtha, and coal (Abedi et al. 2001; Herzog and Tatsutani 2005; Mohan et al. 2007). Biomass as an alternative option for biohydrogen has a large potential to provide sustainable energy quantities in the future (Perlack et al. 2005). The two major pathways for the biomass conversion to energy are the biochemical (i.e., anaerobic digestion or fermentation) (Korres et al. 2013) and thermochemical (i.e., combustion and gasification) conversion (Chen et al. 2014).

Anaerobic digestion uses bacteria in an oxygen-free environment to break down a wide range of feedstocks (e.g., agricultural crops, animal manure, organic wastes, etc.) to produce methane-rich biogas which can be used to generate heat and electricity (Al Seadi et al. 2008) or can be upgraded to biomethane and used as transport fuels (Korres et al. 2010).

3.3.1 Biological Hydrogen Production

Biological hydrogen production involves the enzymatic breakdown of macromolecular substances by microorganisms, e.g., yeasts, bacteria, or enzymes, at low pressure under a wide range of temperature.

More particularly, three microbial groups with the potential for hydrogen production in a biochemical reaction have been identified. The

Box 3.1

Direct photolysis (a) and direct photolysis with respiratory O_2 uptake (b) for biohydrogen production by cyanobacteria and green algae (PSII=photosystem II; PS I=photosystem I; H_2ase=hydrogenase; hν=light energy in terms of electromagnetic wave frequency; (ν) and Planck's constant (=6.55×10^{-34} J·s))

(a)

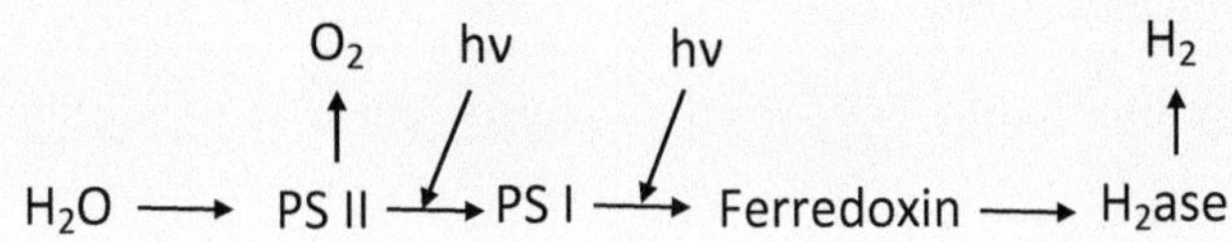

(b)

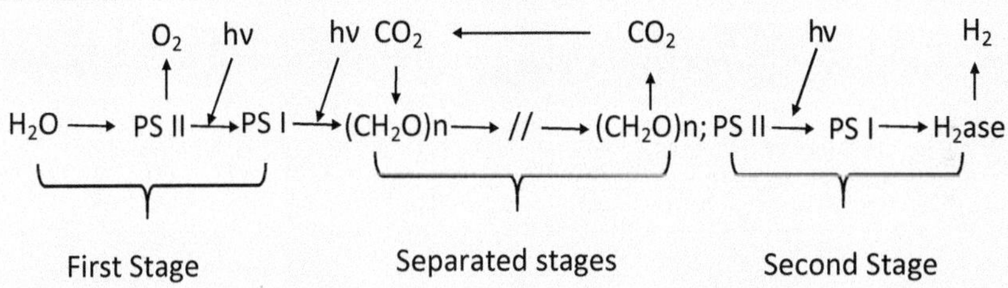

first group includes the autotrophic photosynthetic green algae and cyanobacteria that, in the presence of light, split water to molecular hydrogen and oxygen directly (Hallenbeck and Benemann 2002). This biological reaction is classified into two categories, namely, direct photolysis by cyanobacteria or green algae (Box 3.1) and indirect photolysis by cyanobacteria or microalgae (Box 3.2) (Levin et al. 2004).

Hydrogenase, the enzyme responsible for this production process (Box 3.1), catalyzes the reaction $2H^+ + 2X_{reduced} \rightarrow 6\ H_2 + 2X_{oxidized}$ where water acts as an electron donor and ferredoxin, X, as an electron carrier. Due to ferredoxin reduction, by this photochemical reaction, green algae are theoretically considered as water-splitting microorganisms (Gaffron and Rubin 1942). Nevertheless, hydrogenase is too oxygen labile for sustainable hydrogen production; light-dependent hydrogen production ceases within a few to several tens of minutes since photosynthetically produced oxygen inhibits or inactivates hydrogenase (Miyamoto 1997). More recent studies have determined that the reducing power of hydrogenase does not originate exclusively from water but sometimes is due to organic compounds and depends on algal species and feedstock conditions (Miyamoto 1997).

Nitrogenase is responsible for nitrogen fixation (Miyake et al. 1989), and it is found mainly within prokaryotes, including cyanobacteria, but not in eukaryotes, such as microalgae. Catalytic hydrogen production by nitrogenase (Box 3.2) occurs as a side reaction at a rate of one-third to one-fourth than that of nitrogen fixation (Miyamoto 1997). Cyanobacteria have developed mechanisms to protect nitrogenase from oxygen gas, since it is a substance extremely oxygen labile (Miyamoto 1997). The most successful mechanism is the localization of nitrogenase in the heterocysts of filamentous cyanobacteria in which vegetative cells perform oxygenic photosynthesis. Organic compounds produced by CO_2 reduction are transferred into heterocysts and are decomposed to provide nitrogenase with reducing power. ATP can be pro-

Box 3.2

Indirect single-stage photolysis by cyanobacteria (a) and indirect two-stage photolysis by microalgae (b) for biohydrogen production (based on Hallenbeck and Benemann, undated) (PSII=photosystem II; PS I=photosystem I; Na_2ase=nitrogenase; H_2ase=hydrogenase; hν=light energy in terms of electromagnetic wave frequency; (ν) and Planck's constant ($=6.55\times10^{-34}$ J·s))

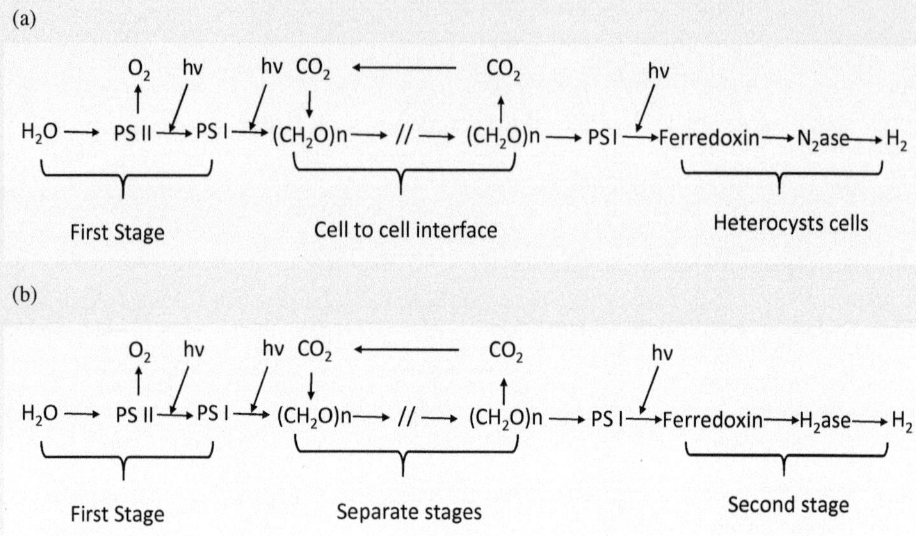

vided by PSI and anoxygenic photosynthesis within heterocysts (Miyamoto 1997).

Although these reactions are attractive from environmental perspective, the limitations for the photolysis impose difficulties for its application. As stated previously, the inhibition of biohydrogen production by the presence of oxygen during direct photolysis is the main limitation found by using green algae (Nath and Das 2004). In addition, most of the cyanobacteria investigated exhibited lower photochemical efficiency, due to the complicated reaction systems needed to overcome the large Gibbs free energy (+237 kJ/mol hydrogen) requirements (Miyake 1998).

3.3.2 Fermentation

The fermentative hydrogen production by anaerobic bacteria is recognized as one of the most promising biological routes due to the ability of these organisms to convert biomass and organic wastes into valuable biohydrogen energy (Chuang et al. undated). There are various types of fermentation such as dark, combined sequential dark and photo-, and photofermentation. A plethora of microorganisms have been reported for the biotransformation of biomass to hydrogen for each of dark, combined dark and photo-, and photofermentation technologies. This approach seems to endorse great potential for commercial use.

3.3.2.1 Biohydrogen Production by Dark Fermentation

Hydrogen can be produced by anaerobic bacteria through dark fermentation (Chong et al. 2009). The gram-positive bacteria of the genus *Clostridium* are promising because they exhibit high hydrogen production rate. They are fast growing and capable of forming endospores, which make them easy to handle for industrial applications (Ferchichi et al. 2005a, b; Krupp and Widmann 2009).

Anaerobic bacteria capable of hydrogen production also include species of *Enterobacter* (Nath et al. 2006), *Bacillus* (Kotay and Das 2007), and *Thermotoga* (Schroder et al. 1994). Dark fermentation can use renewable biomass including agriculture waste (Hussy et al. 2005; Logan et al. 2002), municipal waste (Wang et al. 2003), and food processing waste (Van Ginkel et al. 2005). The biohydrogen production cycle by dark fermentation is shown in Fig. 3.1.

Saccharolytic *Clostridium* species metabolize simple sugars using the Embden-Meyerhof-Parnas glycolytic pathway, by which glucose is converted to pyruvate (Schroder et al. 1994). This is oxidized by the enzyme pyruvate-ferredoxin oxidoreductase to yield acetyl CoA, CO_2, and reduced ferredoxin (Hallenbeck and Benemann 2002). The reoxidation of the reduced ferredoxin is catalyzed by the enzyme hydrogenase through which hydrogen gas is produced.

The amount of hydrogen production from glucose by bacteria culture is affected by metabolic pathway and end products. Hydrogen-producing microbes have been found in environments with a wide range of temperature and include mesophiles (25–40 °C) (Kotay and Das 2007; Shin et al. 2007), thermophiles (40–65 °C), extreme thermophiles (65–80 °C), or hyperthermophiles (>80 °C) (Jannasch et al. 1988).

3.3.2.2 Biohydrogen Production by Photofermentation

Photoheterotrophic bacteria, under anaerobic conditions, are capable of converting organic acids (e.g., acetic, lactic, and butyric) to hydrogen and CO_2 in the presence of light (Fig. 3.2). Thus, the organic acids that are produced during the acidogenic phase of anaerobic digestion of organic feedstock can be converted to H_2 and CO_2 by these photoheterotrophic anaerobic microorganisms (Kapdan and Kargi 2006).

Cyanobacteria are capable of hydrogen production by oxygenic photosynthesis (Lee et al.

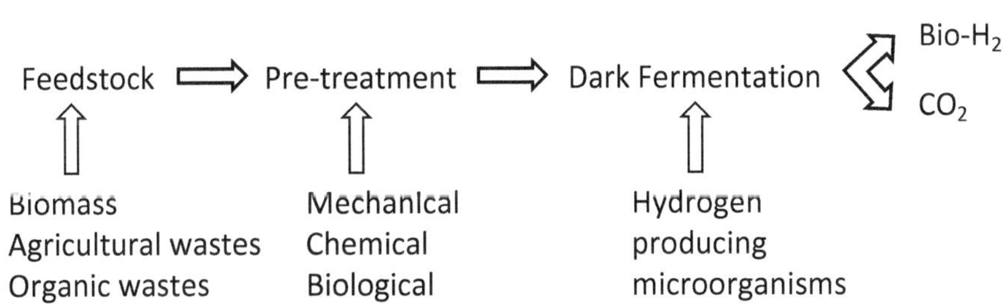

Fig. 3.1 Biohydrogen production cycle by dark fermentation

Fig. 3.2 Biohydrogen production by photofermentation. *fd* ferredoxin, *Na$_2$ase* nitrogenase

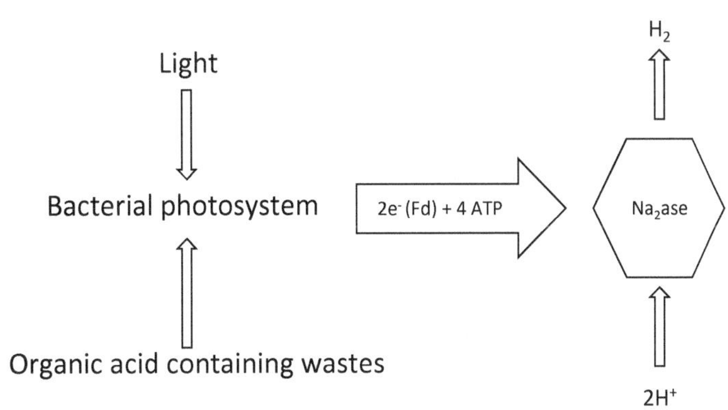

2010) although the purple non-sulfur bacteria (e.g., genus *Rhodobacter*) seem more promising for hydrogen production by anoxygenic photosynthesis and photofermentation (Rai Pankaj et al. 2012; Yongzhen et al. 2007; Koku et al. 2003; Kondo et al. 2002; Kim et al. 2006; Zhu et al. 2002).

Several studies have reported the capability of *Rhodobacter sphaeroides* for hydrogen production in the presence of organic acids (Rai Pankaj et al. 2012; Yongzhen et al. 2007).

Some purple photosynthetic bacteria capable for biohydrogen production include the species of *Rhodobacter sphaeroides*, *Rhodobacter capsulatus* (Zhu et al. 1999; He et al. 2005; Fang et al. 2005; Shi and Yu 2005), *Rhodovulum sulfidophilum* (Matsunaga et al. 2000; Maeda et al. 2003), and *Rhodopseudomonas palustris* (Barbosa et al. 2001).

The biohydrogen production occurs under anaerobic conditions with light illumination and optimum growth temperature between 30 and 35 °C and pH 7.0 (Zhu et al. 2002; He et al. 2005; Kim et al. 2006). The microbes prefer organic acids as carbon source such as acetic (Barbosa et al. 2001; Najafpour et al. 2004; Oh et al. 2004; Fang et al. 2005), butyric (He et al. 2005), propionic (Shi and Yu 2004), lactic (Miyake et al. 1999; Kondo et al. 2002; He et al. 2005;) and malic (Eroglu et al. 1999). Nevertheless, many workers have reported that other carbohydrates than those mentioned above and industrial effluents can also be used for hydrogen gas production by photosynthetic bacteria (Ike et al. 1999; Yetis et al. 2000; Zhu et al. 2002; Maeda et al. 2003). Hydrogen production rates vary depending on the light intensity, carbon source, and the type of microbial culture. Nitrogenase is the key enzyme that catalyzes hydrogen gas production by photosynthetic bacteria (Fig. 3.2). Factors that can affect enzymatic activity include oxygen, ammonia, or high N/C ratios (Koku et al. 2003). Light intensity is another factor that affects the performance of photofermentation. Increasing light intensity enhances the biohydrogen yield and production rate (Barbosa et al. 2001; Shi and Yu 2005). Light intensity affects the consumption rates of organic acids as they exhibit varied requirements, e.g., efficient butyrate consumption involves higher light intensities in comparison with acetate and propionate. Usually, hydrogen yield under dark conditions is lower than that of the illuminated conditions (Oh et al. 2004) although alternating 14 /10 light/dark hours of light/dark cycles yielded slightly higher hydrogen production rates in comparison to continuous illumination (Koku et al. 2003).

There is limited experience with photofermentation at industrial scale. The distribution of light within the photofermenter needs to be designed to minimize shading. Therefore any externally illuminated photo-bioreactor must have a high ratio of surface area to volume resulting in constructions which are material intensive and costly.

3.3.2.3 Biohydrogen Production by Sequential Dark and Photofermentation

Sequential dark and photofermentation is a relatively new approach in biological hydrogen gas production (Kapdan and Kargi 2006). The biohydrogen yield is usually higher when dark fermentation and photofermentation are combined since the effluent from dark fermentation is sufficient in providing organic acids for photofermentation (Yokoi et al. 2002; Kapdan and Kargi 2006. However, the operational conditions of the system should be controlled since ammonia concentration and C/N ratio in the effluent from dark fermentation should not be at the inhibitory levels for the photosynthetic bacteria (Lee et al. 2002). Microbe coculture of *C. butryricum* and *Rhodobacter* spp. yielded higher biohydrogen volumes compared to single-phase dark fermentation (Yokoi et al. 1998). Comparisons of the various biological processes for hydrogen production are listed in the following table (Table 3.5).

3.3.3 Thermochemical Biohydrogen Production

Gasification (i.e., the conversion of feedstock into fuel gases under high temperature and pres-

Table 3.5 Advantages and disadvantages of various biological processes for hydrogen production

Process	Microorganism	Advantages	Disadvantages
Direct biophotolysis	Green algae	(a) Hydrogen is produced directly from water and sunlight	(a) High light intensity is required
		(b) Solar conversion energy increased by tenfold in comparison to crops and trees	(b) Oxygen can be dangerous for the system
Indirect biophotolysis	Cyanobacteria	(a) Produce hydrogen from water	(a) Removal of hydrogenase enzymes to avoid degradation of hydrogen
		(b) Ability to fix nitrogen from atmosphere	(b) Lower photochemical efficiency
			(c) Oxygen presence at 30 %
			(d) Inhibitory effect of oxygen on nitrogenase
Dark fermentation	Fermentative bacteria	(a) Wide variety of feedstock	(a) Gas mixture needs cleaning and updated from the presence of CO2
		(b) Hydrogen production under dark	(b) Relatively low biohydrogen yields
		(c) Metabolites produce added value products	
Photofermentation	Photosynthetic bacteria	(a) Use of wide-spectrum light energy	(a) Low light conversion efficiency
		(b) Wide variety of feedstock	(b) Presence of oxygen inhibits hydrogenase

sure in a low-oxygen environment) and pyrolysis (i.e., a similar process to gasification that is conducted under anaerobic conditions and produces liquids as the primary product) are the main thermochemical conversion processes. These processes are characterized by faster conversion rates and best suited for lower-moisture feedstock at high temperatures. Thermochemical routes can convert the entire organic (carbon) portion of the feedstock under treatment into energy. On the contrary, the inorganic fraction (ash) is fouling the high-temperature equipment, increases nutrient loading in wastewater treatment and disposal facilities, and adds disposal cost. Inorganic constituents may also accelerate some of the conversion reactions (Barrows 2001).

3.3.3.1 Gasification for Biohydrogen Production

Biomass gasification has been under intensive development over the last two decades Zuideveld (2001) although the difficulties regarding its employment were mainly financial rather than technical due to high capital and operating costs. Also, environmental concern is another obstacle that needs to be solved if biomass gasification is to become sustainable (Herzog and Tatsutani 2005). Nevertheless, an increasing demand for biohydrogen energy will most probably reduce the cost of biomass gasification in the near future significantly. Another major cost-effective method of hydrogen production is through thermochemical gasification of biomass from agricultural residues, urban waste, woodchips from biomass plantation, and animal residues. The gasification output consists mainly of H_2, CO, and CH_4 which is also converted into H_2 and CO through steam reforming process (Zuideveld 2001). The hydrogen content in the hydrogen-enriched product gas contains around 15–20 % and it can be increased to 30–40 % if part of the air needed for gasification is replaced by steam. A high overall electrical efficiency can be attained in the range between 25 % and 50 %, if hydrogen is used in fuel cells for power generation. Hence, it is extremely important to have high hydrogen content in the produced gas from the thermochemical process (EIA 2001).

3.3.3.2 Pyrolysis

Pyrolysis is a thermophilic conversion of biomass into energy at elevated temperatures (~600 °C) in oxygen-free environment. The process recovers approximately 80 % of the stored energy in biomass (Hogg 2007). Different types of reactors have been used. In a typical two-stage reactor, the first chamber operates at low temperature, while the second chamber runs at high temperatures, in which the combustion of the feedstock is completed (Chen et al. 2014).

The pyrolysis capacity has grown from 45,000 MWth in 2004 to 84,500 MWth in 2013 with an increase of about 88 % worldwide. The expected growth of this capacity by 2021 has been estimated at 140,157 MWth (Childress 2007; Higman and Tam 2014).

Pyrolysis is considered to be economically profitable in large scales, which minimizes environmental concerns such as carbon sequestration, soil amendment, energy/heat supply, and value-added chemicals (Higman and Tam 2014). It should be mentioned that increased operating and maintenance costs can be a restrictive factor for the application of this technology.

3.3.3.3 Integrated Process for Biohydrogen Production

Biomass can be converted to hydrogen by the integration of two distinct thermochemical techniques, i.e., (a) gasification followed by shift conversion and (b) pyrolysis of biomass to form a bio-oil that can be subsequently converted to hydrogen via catalytic steam reforming and shift conversion. The latter approach is potentially cost competitive compared to current commercial processes for hydrogen production (Mann 1995) as it was demonstrated in bench-scale applications using model compounds and the carbohydrate-derived fraction of bio-oil (Wang et al. 1998). The advantages of this approach compared to traditional gasification technology are numerous as the bio-oil is easily transportable; hence the second phase (i.e., steam reforming) can be carried out at a different location that facilitates the logistics of biohydrogen production (Abedi et al. 2001). Additionally, the production and recovery of added value coproducts from bio-oil can significantly impact the economics of the entire process.

3.4 Conclusions

Various types of biomass can be used for sustainable production of biohydrogen. Technological advantages in bioconversion of biomass and various wastes with high soluble carbohydrate content have been developed to ensure high biohydrogen yields. Biohydrogen production using algae and cyanobacteria is a promising approach that merits further consideration. Additionally, integrated thermochemical or biological processes appear very promising for higher biohydrogen yields. However, attention needs to focus on sustainability issues concerning the production and suitability of the biomass for hydrogen production, the standardization of the operational parameters, and the logistics of biohydrogen cycle production. Based on the evidence presented in this chapter, ryegrass and extracts from sweet sorghum along with maize stalks, fodder maize, fruit peel wastes, rice slurry, food wastes, and sewage slurry are the most suitable feedstock concerning biohydrogen production.

References

Abedi J, Yeboah YD, Realff M, McGee D, Howard J, Bota KB (2001) An integrated approach to hydrogen production from agricultural residues for use in urban transportation. In: Proceedings of the 2001 DOE Hydrogen Program Review, NREL/CP-570-30535, Baltimore, MD, April 17–19, 2001, pp 160–186

Al Seadi T, Rutz D, Prassl H, Kottner M, Finsterwalder T, Volk S, Janssen R (2008) The biogas handbook. University of Southern Denmark, Esbjerg

Anonymous (1995) The green hydrogen report: the 1995 Progress Report of the Secretary of Energy's Hydrogen Technical Advisory Panel, May 1995, DOE/GO-10095-179 DE95009213

Anonymous (2003) Municipal Solid Waste in the United States: 2001 facts and figures. Office of Solid Waste and Emergency Response (5305W) EPA530-R-03-011. www.epa.gov, October 2003. Accessed Jan 2015

Antonopoulou G, Gavala HN, Skiadas IV, Angelopoulos K, Lyberatos G (2008) Biofuels generation from sweet sorghum: fermentative hydrogen production and anaerobic digestion of the remaining biomass. Bioresour Technol 99:110–119

Antonopoulou G, Gavala HN, Skiadas IV, Lyberatos G (2011) Effect of substrate concentration on fermentative hydrogen production from sweet sorghum extract. Int J Hydrog Energy 36:4843–4851

Armor JN (1999) The multiple roles for catalysis in the production of H_2. Appl Catal A Gen 176:159–176

Atif AAY, Fakhrul-Razi A, Ngan MA, Morimoto M, Iyuke SE, Veziroglu NT (2005) Fed batch production of hydrogen from palm oil mill effluent using anaerobic microflora. Int J Hydrog Energy 30:1393–1397

Balat M (2008a) Hydrogen-rich gas production from biomass via pyrolysis and gasification processes and effects of catalyst on hydrogen yield. Energy Sources Part A 30:552–564

Balat M (2008b) Potential importance of hydrogen as a future solution to environmental and transportation problems. Int J Hydr Energy 33:4013–4029

Balat H, Kırtay E (2010) Hydrogen from biomass. Present scenario and future prospects. Int J Hydrog Energy 35:7416–7426

Barbosa MJ, Rocha JMS, Tramper J, Wijffels RH (2001) Acetate as a carbon source for hydrogen production by photosynthetic bacteria. J Biotechnol 85:25–33

Barrows B (2001) Briefing paper: what are "Conversion Technologies"? State of Oregon, Dept. of Environmental Quality, November 2011

Benemann J (1996) Hydrogen biotechnology: progress and prospects. Nat Biotechnol 14:1101–1103

Chen D, Yin L, Wang H, He P (2014) Pyrolysis technologies for municipal solid waste: a review. Waste Manag 34:2466–2486

Cheng J, Xie B, Zhou J, Song W, Cen K (2010) Cogeneration of H_2 and CH_4 from water hyacinth by two-step anaerobic fermentation. Int J Hydrog Energy 29:173–185

Cherry RS (2004) A hydrogen utopia. Int J Hydrog Energy 29:125–129

Childress J (2007) The Gasification industry: status and forecast, COAL-GEN Mega Session August 3, 2007 Milwaukee, WI

Chong ML, Sabaratnamb V, Shiraic Y, Hassan MA (2009) Biohydrogen production from biomass and industrial wastes by dark fermentation. Int J Hydrog Energy 34:3277–3287

Chuang YS, Liang YC, Huang CY, Chang FY, Lay CH, Lin CY (undated) Biohydrogen production from agricultural wastes using anaerobic mixed microflora

Collet C, Adler N, Schwitzguebel JP, Peringer P (2004) Hydrogen production by Clostridium thermolacticum during continuous fermentation of lactose. Int J Hydrog Energy 29:1479–1485

Cui M, Yuan Z, Zhi X, Wei L, Shen J (2010) Biohydrogen production from poplar leaves pretreated by different methods using anaerobic mixed bacteria. Int J Hydrog Energy 35:4041–4047

Das D, Khanna N, Veziroglu TN (2008) Recent developments in biological hydrogen production processes. Chem Ind Chem Eng Q 14:57–67

Datar R, Huang J, Maness PC, Mohagheghi A, Czernik S, Chornet E (2007) Hydrogen production from the fermentation of corn stover biomass pretreated with a steam-explosion process. Int J Hydrog Energy 32:932–939

de Vrije T, de Haas GG, Tan GB, Keijsers ERP, Claassen PAM (2002) Pretreatment of Miscanthus for hydrogen production by *Thermotoga elfii*. Int J Hydrog Energy 27:1381–1390

de Vrije T, Bakker RR, Budde MAW, Lai MH, Mars AE, Claassen PAM (2009) Efficient hydrogen production from the lignocellulosic energy crop *Miscanthus* by the extreme thermophilic bacteria *Caldicellulosiruptor saccharolyticus* and *Thermotoga neapolitana*. Biotechnol Biofuels 2:12. doi:10.1186/1754-6834-2-12

Demirbas A (2008) Products from lignocellulosic materials via degradation processes. Energy Sources A 30:27–37

DOE (2004) United States Department of Energy. An integrated research, development and demonstration plan. Hydrogen Pasteur Plan 2004

Dong L, Zhenhong Y, Yongming S, Xiaoying K, Yu Z (2009) Hydrogen production characteristics of the organic fraction of municipal solid wastes by anaerobic mixed culture fermentation. Int J Hydrog Energy 34:812–820

EIA-Energy Information Administration (2001) International energy outlook 2001. DOE/EIA, Washington, DC

Hogg R (2007) Energy from waste by pyrolysis and gasification the experience and performance of an operational plant. In: Proceedings of the international conference on sustainable solid waste management. Chennai, India, 5–7 September, 385–392

Eroglu I, Aslan K, Gunduz U, Yucel M, Turker L (1999) Substrate consumption rate for hydrogen production by *Rhodobacter sphaeroides* in a column photobioreactor. J Biotechnol 70:103–113

Fan YT, Zhang YH, Zhang SF, Hou HW, Ren BZ (2006) Efficient conversion of wheat straw wastes into biohydrogen gas by cow dung compost. Bioresour Technol 97:500–505

Fan YT, Xing Y, Ma HC, Pan CM, Hou HW (2008) Enhanced cellulose hydrogen production from corn stalk by lesser panda manure. Int J Hydrog Energy 33:6058–6065

Fang HHP, Liu H, Zhang T (2005) Phototrophic hydrogen production from acetate and butyrate in wastewater. Int J Hydrog Energy 30:785–793

Fang HHP, Li CL, Zhang T (2006) Acidophilic biohydrogen production from rice slurry. Int J Hydrog Energy 31:683–692

Feng GL, Letey J, Chang AC, Campbell M (2005) Simulating dairy liquid waste management options as

a nitrogen source for crops. Agri Ecosys Environ 110:219–229
Ferchichi M, Crabbe E, Gil GH, Hintz W, Almadidy A (2005a) Influence of initial pH on hydrogen production from cheese whey. J Biotechnol 120:402–409
Ferchichi M, Crabbe E, Hintz W, Gil GH, Almadidy A (2005b) Influence of culture parameters on biological hydrogen production by *Clostridium saccharoperbutylacetonicum* ATCC 27021. World J Microbiol Biotechnol 21:855–862
Gaffron H, Rubin JJ (1942) Fermentative and photochemical production of hydrogen in algae. Gen Physiol 26:219–240
Gavala HN, Skiadas LV, Ahring BK, Lyberatos G (2005) Potential for biohydrogen and methane production from olive pulp. Water Sci Technol 52(1–2):209–215
Gemici Z, Ulusarslan T, Teke I (2009) Currency of district cooling systems and alternative energy sources. Energy Educ Sci Technol A 23:31–53
Hafner SD (2007) Biological hydrogen production from nitrogen-deficient substrates. Biotechnol Bioeng 97:435–437
Hallenbeck PC, Benemann JR (2002) Biological hydrogen production; fundamentals and limiting processes. Int J Hydrog Energy 27:1185–1193
Hasson A (2009) Biohydrogen production from industrial organic wastes. J Appl Sci Environ Sanitation 4(1):7–10
Hawaii Renewable Resources Program (2000) University of Hawaii's Department of Chemistry
He D, Bultel Y, Magnin JP, Roux C, Willison JC (2005) Hydrogen photosynthesis by *Rhodobacter capsulatus* and its coupling to PEM fuel cell. J Power Sources 141:19–23
Herzog A, Tatsutani M (2005) A hydrogen future? An economic and environmental assessment of hydrogen production pathways. Natural Resources Defense Council, Issue Paper, November 2005
Higman CAA, Tam S (2014) Advances in coal gasification, hydrogenation and gas treating for the production of chemicals and fuels. Chem Rev 114:1673–1708
Holm-Nielsen JB, Al Seadi T (2004) Manure based biogas systems. In: Lens P, Hamelers B, Hoitink H, Bidlingmaier W (eds) Resource recovery and reuse in organic solid waste management. IWA Publishing, London, pp 377–394
Hussy I, Hawkes FR, Dinsdale R, Hawkes DL (2005) Continuous fermentative hydrogen production from sucrose and sugarbeet. Int J Hydrog Energy 30:471–483
Ike A, Murakawa T, Kawaguchi H, Hirata K, Miyamoto K (1999) Photoproduction of hydrogen from raw starch using a halophilic bacterial community. J Biosci Bioeng 88:72–77
Ito T, Nakashimada Y, Senba K, Matsui T, Nishio N (2005) Hydrogen and ethanol production from glycerol containing wastes discharged after biodiesel manufacturing process. J Biosci Bioeng 100:260–265
Jae-Hwa L, Dong-Geun L, Jae-Il P, Ji-Youn K (2010) Biohydrogen production from a marine brown algae and its bacterial diversity. Korean J Chem Eng 27:187–192
Jannasch HW, Huber R, Belkin S, Stetter KO (1988) *Thermotoga neapolitana* sp. nov. of the extremely thermophilic, eubacterial genus *Thermotoga*. Arch Microbiol 103–104
Kaparaju P, Serrano M, Thomsen AB, Kongjan P, Angelidaki I (2009) Bioethanol, biohydrogen and biogas production from wheat straw in a biorefinery concept. Bioresour Technol 100:2562–2568
Kapdan IK, Kargi F (2006) Bio-hydrogen production from waste materials. Enzym Microb Technol 38:569–582
Kazim A, Veziroglou TN (2001) Utilization of solar-hydrogen energy in the UAE to maintain its shape in the world energy market for the 21st century. Renew Energy 24:259–274
Kim SH, Han SK, Shin HS (2004) Feasibility of biohydrogen production by anaerobic co-digestion of food waste and sewage sludge. Int J Hydrog Energy 29:1607–1616
Kim MS, Baek JS, Lee JK (2006) Comparison of H2 accumulation by *Rhodobacter sphaeroides* KD131 and its uptake hydrogenase and PHB synthase deficient mutant. Int J Hydrog Energy 31:121–127
Koku H, Eroglu I, Gunduz U, Yucel M, Turker L (2003) Kinetics of biohydrogen production by the photosynthetic bacterium *Rhodobacter spheroids* O.U. 001. Int J Hydrog Energy 28:381–388
Kondo T, Arakawa M, Hiral T, Wakayama T, Hara M, Miyake J (2002) Enhancement of hydrogen production by a photosynthetic bacterium mutant with reduced pigment. J Biosci Bioeng 93:145–150
Kongjan P, Thong S, Kotay M, Min B, Angelidaki I (2010) Biohydrogen production from wheat straw hydrolysate by dark fermentation using extreme thermophilic mixed culture. Biotechnol Bioeng 105(5):899–908
Korres NE, Nizami AS (2013) Variation in anaerobic digestion. Need for process monitoring. In: Korres NE, O'Kiely P, Benzie JAH, West JS (2013) Bioenergy production by anaerobic digestion. Using agricultural biomass and organic waste. Pub Earthscan from Routledge/Taylor and Francis Pub. Group, pp 194–230
Korres NE, Singh A, Nizami AS, Murphy JD (2010) Is grass biomethane a sustainable transport biofuel? Biofuels Bioprod Biorefin 4:310–325
Korres NE, O'Kiely P, Benzie JAH, West JS (2013) Bioenergy production by anaerobic digestion. Using agricultural biomass and organic waste. Pub Earthscan from Routledge/Taylor and Francis Pub. Group
Kotay SM, Das D (2007) Microbial hydrogen production with Bacillus coagulans IIT-BT S1 isolated from anaerobic sewage sludge. Bioresource Technol 98:1183–1190
Kotsopoulos TA, Fotidis IA, Tsolakis N, Martzopoulos GG (2009) Biohydrogen production from pig slurry in a CSTR reactor system with mixed cultures under hyperthermophilic temperature. Biomass Bioenergy 33:1168–1174

Koutrouli EC, Kalfas H, Gavala HN, Skiadas IV, Stamatelatou K, Lyberatos G (2009) Hydrogen and methane production through two-stage mesophilic anaerobic digestion of olive pulp. Bioresour Technol 100:3718–3723

Krupp M, Widmann R (2009) Biohydrogen production by dark fermentation: experiences of continuous operation in large lab scale. Int J Hydrog Energy 34:4509–4516

Kyazze G, Dinsdale R, Hawkes FR, Guwy AJ, Premier GC, Donnison IS (2008) Direct fermentation of fodder maize, chicory fructans and perennial ryegrass to hydrogen using mixed microflora. Bioresour Technol 99:8833–8839

Lay JJ, Lee YJ, Noike T (1999) Feasibility of biological hydrogen production from organic fraction of municipal solid waste. Water Res 33:2579–2586

Lay JJ, Fan KS, Chang IJ, Ku CH (2003) Influence of chemical nature of organic wastes on their conversion to hydrogen by heat-shock digested sludge. Int J Hydrog Energy 28:1361–1367

Lee CM, Chen PC, Wang CC, Tung YC (2002) Photohydrogen production using purple nonsulfur bacteria with hydrogen fermentation reactor effluent. Int J Hydrog Energy 27:1309–1313

Lee H, Vermaas WFJ, Rittmann BE (2010) Biological hydrogen production: prospects and challenges. Trends Biotechnol 28:262–271

Levin DB, Pitt L, Love M (2004) Biohydrogen production: prospects and limitations to practical application. Int J Hydrog Energy 29:173–185

Liu F, Fang B (2007) Optimization of biohydrogen production from biodiesel wastes by *Klebsiella pneumoniae*. Biotech J 2:374–380

Logan BE, Oh SE, Kim IS, Van Ginkel S (2002) Biological hydrogen production measured in batch anaerobic respirometers. Environ Sci Technol 36:2530–2535

Maeda I, Miyasaka H, Umeda F, Kawase M, Yagi K (2003) Maximization of hydrogen production ability in high-density suspension of *Rhodovulum sulfidophilum* cells using intracellular poly(3 hydroxybutyrate) as sole substrate. Biotechnol Bioeng 81:474–481

Mann MK (1995) Technical and economic analyses of hydrogen production via indirectly heated gasification and pyrolysis. In: Proceedings of the 1995 Hydrogen Program Review, Vol. 1, NREL/CP-430-20036-Vol. 1, pp 205–236

Matsunaga T, Hatano T, Yamada A, Matsumato M (2000) Microaerobic hydrogen production by photosynthetic bacteria in a double phase photobioreactor. Biotechnol Bioeng 68:647–651

May-Tobin C (2011) Wood for fuel. In: Boucher D, Elias P, Lininger K, May-Tobin C, Roquemore S, Saxon E (eds) (2011). The root of the problem. What's driving tropical deforestation today? Union of Concerned Scientists, pp 79–86

Miyake J (1998) The science of biohydrogen. An energetic view. In: Zaborsky OR (ed) Biohydrogen. Plenum Press, New York, pp 7–18

Miyake J, Asada Y, Kawamura S (1989) Nitrogenase. In: Hall CW, Kitani O (eds) Biomass Handbook Gorton and Breach. Science Publishers, New York, pp 362–370

Miyake J, Wakayama T, Schnackenberg JS, Arai T, Asada Y (1999) Simulation of daily sunlight illumination pattern for bacterial photo-hydrogen production. J Biosci Bioeng 88:659–663

Miyamoto K (1997) Renewable biological systems for alternative sustainable energy production. FAO Agricultural Services Bulletin – 128. Food and Agriculture Organization of the United Nations, Rome

Mohan SV, Bhaskar YV, Sarma PN (2007) Biohydrogen production from chemical wastewater treatment in biofilm configured reactor operated in periodic discontinuous batch mode by selectively enriched anaerobic mixed consortia. Water Res 41:2652–2664

Monti A, Fazio S, Venturi G (2009) Cradle-to-farm gate life cycle assessment in perennial energy crops. Eur J Agron 31:77–84

Najafpour G, Younesi H, Mohammed AR (2004) Effect of organic substrate on hydrogen production from synthesis gas using *Rhodospirillum rubrum* in batch culture. Biochem Eng J 21:123–130

Nath K, Das D (2004) Improvement of fermentative hydrogen production: various approaches. Appl Microbiol Biotechnol 65:520–529

Nath K, Kumar A, Das D (2006) Effect of some environmental parameters on fermentative hydrogen production by Enterobacter cloacae DM11. Can J Microbiol 52:525–532

Nath A, Dixit M, Bandiya A, Chavda S, Desai AJ (2008) Enhanced PHB production and scale up studies using cheese whey in fed batch culture of Methylobacterium sp. ZP24. Biores Technol 99:5749–5755

Ni M, Leung MKH, Sumathy K, Leung DYC (2006) Potential of renewable hydrogen production for energy supply in Hong Kong. Int J Hydrog Energy 31:1401–1412

NRC-National Research Council (U.S) (2004) The hydrogen economy: opportunities, costs, barriers, and R&D needs. National Academies Press, Washington, DC

Ntaikou I, Koutros E, Kornaros M (2009) Valorization of wastepaper using the fibrolytic/hydrogen bacterium Ruminococcus albus. Bioresour Technnol 100(15):5928–5933

Ntaikou I, Antonopoulou G, Lyberatos G (2010) Biohydrogen production from biomass and wastes via dark fermentation: a review. Waste Biomass Valor 1:21–39

Oh YK, Scol EH, Kim MS, Park S (2004) Photoproduction of hydrogen from acetate by a chemoheterotrophic bacterium *Rhodopseudomonas palustris* P4. Int J Hydrog Energy 29:1115–1121

Okamoto M, Miyahara T, Mizuno O, Noike T (2000) Biological hydrogen potential of materials characteristic of the organic fraction of municipal solid wastes. Water Sci Technol 41(3):25–32

Pakarinen O (2011) Methane and hydrogen production from crop biomass through anaerobic digestion. Jyvaskyla studies in biological and environmental science. University of Jyvaskyla, Jyvaskyla

Panagiotopoulos IA, Bakker RR, Budde MAW, de Vrije T, Claassen PAM, Koukios EG (2009) Fermentative hydrogen production from pretreated biomass: a comparative study. Bioresour Technol 100:6331–6338

Perlack RD, Wright LL, Turhollow AF, Graham RL, Stokes BJ, Erbach DC (2005) Biomass as feedstock for a bioenergy and Bioproducts industry: the technical feasibility of a billion-ton annual supply. Joint Study Sponsored by U.S. Department of Energy and U.S. Department of Agriculture, May 2005. DOE/GO-102005-2135 ORNL/TM-2005/66

Petersen JE (2008) Energy production with agricultural biomass: environmental implications and analytical challenges. Eur Rev Agric Econ 35(3):385–408

Petitdemange G, Durr C, Abbad Andaloussi S, Raval G (1995) Fermentation of raw glycerol to 1,3-propanediol by new strains of Clostridium butyricum. J Ind Microbiol 15:498–502

Powlson DS, Riche AB, Shield I (2005) Biofuels and other approaches for decreasing fossil fuel emissions from agriculture. Anal Appl Biol 146:193–201

Rai Pankaj K, Singh SP, Asthana RK (2012) Biohydrogen production from cheese whey wastewater in a two-step anaerobic process. Appl Biochem Biotechnol 167:1540–1549

Sakai S, Yagishita T (2007) Microbial production of hydrogen and ethanol from glycerol-containing wastes discharged from a biodiesel fuel production plant in a bioelectrochemical reactor with thionine. Biotech Bioeng 98:340–348

Schroder C, Selig M, Schonheit P (1994) Glucose fermentation to acetate, Co_2 and H_2 in the anaerobic hyperthermophilic eubacterium Thermotoga maritima – involvement of the embden-meyerhof pathway. Arch Microbiol 161:460–470

Shakya BD, Aye L, Musgrave P (2005) Technical feasibility and financial analysis of hybrid wind-photovoltaic system with hydrogen storage for cooma. Int J Hydrog Energy 30:9–20

Shi XY, Yu HQ (2004) Hydrogen production from propionate by *Rhodopseudomonas capsulate*. Appl Biochem Biotechnol 117:143–154

Shi XY, Yu HQ (2005) Response surface analysis on the effect of cell concentration and light intensity on hydrogen production by *Rhodopseudomonas capsulate*. Process Biochem 40:2475–2481

Shin JH, Yoon JH, Ahn EK, Kim MS, Sim SJ, Park TH (2007) Fermentative hydrogen production by the newly isolated *Enterobacter asburiae* SNU-1. Int J Hydrog Energy 32:192–199

Sims REH, Hastings A, Schlamadinger B, Taylor G, Smith P (2006) Energy crops: current status and future prospects. Glob Chang Biol 12:20542076

Singh A, Pant D, Korres NE, Nizami ASA, Prasad S, Murphy J (2010) Key issues in life cycle assessment of ethanol production from lignocellulosic biomass. Challenges and perspectives. Bioresour Technol 101:5003–5012

Tenca QA (2011) Biohydrogen production from agricultural and livestock residues within an integrated bioenergy concept. Ph.D. Thesis Università degli Studi di Milano, Facoltà di Agraria, Department of Agricultural Engineering. Via Celoria 2, 20133, Milano (MI), Italy

Tenca A, Schievano A, Perazzolo F, Adani F, Oberti R (2011) Biohydrogen from thermophilic co-fermentation of swine manure with fruit and vegetable waste: maximizing stable production without pH control. Bioresour Technol 102:8582–8588

Thong S, Prasertsan P, Intrasungkha N, Dhamwichukorn S, Birkeland NK (2007) Improvement of biohydrogen production and treatment efficiency on palm oil mill effluent with nutrient supplementation at thermophilic condition using an anaerobic sequencing batch reactor. Enzy Microb Technol 41:583–590

UCS (Union of Concerned Scientists) (2012a) Turning agricultural residues and manure into bioenergy. Fact sheets. Union of concerned Scientists. https://s3.amazonaws.com/ucs-documents/clean-vehicles/Agricultural-Residue-Ranking.pdf. Accessed Jan 2015

UCS (Union of Concerned Scientists) (2012b) The promise of biomass: clean power and fuel—if handled right. (Union of Concerned Scientists). Cambridge, MA. http://www.ucsusa.org/assets/documents/clean_vehicles/BiomassResource-Assessment.pdf. Accessed Jan 2015

Ueno Y, Otsuka S, Morimoto M (1996) Hydrogen production from industrial wastewater by anaerobic microflora in chemostat culture. J Ferment Bioeng 82:194–197

Valdez-Vazquez I, Rios-Leal E, Esparza-Garcia F, Cecchi F, Poggi-Varaldo HM (2005) Semi-continuous solid substrate anaerobic reactors for H2 production from organic waste: mesophilic versus thermophilic regime. Int J Hydrog Energy 30:1383–1391

Van Ginkel SW, Oh SE, Logan BE (2005) Biohydrogen gas production from food processing and domestic wastewaters. Int J Hydrog Energy 30:1535–1542

Venkata Mohan S, Lalit Babu V, Sarma PN (2007) Anaerobic biohydrogen production from dairy wastewater treatment in sequencing batch reactor (AnSBR): effect of organic loading rate. Enzym Microb Technol 41:506–515

Venkata Mohan S, Babu LM, Reddy V, Mohanakrishna G, Sarma PN (2009a) Harnessing of biohydrogen by acidogenic fermentation of Citrus limetta peelings: effect of extraction procedure and pretreatment of biocatalyst. Int J Hydrog Energy 34:6149–6156

Venkata Mohan S, Mohanakrishna G, Kannaiah Goud R, Sarma PN (2009b) Acidogenic fermentation of vegetable based market waste to harness biohydrogen with simultaneous stabilization. Bioresour Technol 100:3061–3068

Veziroglu TN, Sahin S (2008) 21st century's energy: hydrogen energy system. Energy Convers Manag 49:1820–1831

Vijayaraghavan K, Ahmad D (2006) Biohydrogen generation from palm oil mill effluent using anaerobic contact filter. Int J Hydrog Energy 31:1284–1291

Vijayaraghavan K, Ahmad D, Bin Ibrahim MK (2006) Biohydrogen generation from jackfruit peel using anaerobic contact filter. Int J Hydrog Energy 31:569–579

Vijayaraghavan K, Ahmad D, Soning C (2007) Biohydrogen generation from mixed fruit peel waste using anaerobic contact filter. Int J Hydrog Energy 32:4754–4760

Wagner RC, Regan JM, Oh SE, Zuo Y, Logan BE (2009) Hydrogen and methane production from swine wastewater using microbial electrolysis cells. Water Res 43:1480–1488

Wang D, Czernik S, Chornet E (1998) Production of hydrogen from biomass by Catalytic Steam Reforming of Fast Pyrolysis Oil. Energy Fuel 12:19–24

Wang CC, Chang CW, Chu CP, Lee DJ, Chang BV, Liao CS, Tay JH (2003) Using filtrate of waste biosolids to effectively produce biohydrogen by anaerobic fermentation. Water Res 37:2789–2793

Wang CH, Lin PJ, Chang JS (2006) Fermentative conversion of sucrose and pineapple waste into hydrogen gas in phosphate-buffered culture seeded with municipal sewage sludge. Process Biochem 41:1353–1358

Winter CJ (2005) Into the hydrogen energy economy-milestones. Int J Hydrog Energy 30:681–685

Wu X, Zhu J, Dong C, Miller C, Li Y, Wang L, Yao W (2009) Continuous biohydrogen production from liquid swine manure supplemented with glucose using an anaerobic sequencing batch reactor. Int J Hydrog Energy 34:6636–6645

Yang P, Ruihong Z, Mcgarvey JA, Benemann J (2005) Hydrogen production from waste by anaerobic fermentation. American Society of Agricultural Engineers, July 17–20, 2005, Tampa, Fl. Paper No. 056019 Continuous hydrogen production from organic waste

Yang PL, Zhang RH, McGarvey JA, Benemann JR (2007) Biohydrogen production from cheese processing wastewater by anaerobic fermentation using mixed microbial communities. Int J Hydrog Energy 32:4761–4771

Yang Y, Tsukahara K, Sawayama S (2008) Biodegradation and methane production from glycerol-containing synthetic wastes with fixed-bed bioreactor under mesophilic and thermophilic anaerobic conditions. Proc Biochem 43:362–367

Yazdani SS, Gonzalez R (2007) Anaerobic fermentation of glycerol: a path to economic viability from the biofuels industry. Curr Opin Biotech 18:213–219

Yetis M, Gunduz U, Eroglu I, Yucel M, Turker L (2000) Photoproduction of hydrogen from sugar refinery wastewater by *Rhodobacter sphaeroides* O.U.001. Int J Hydrog Energy 25:1035–1041

Yokoi H, Mori S, Hirose J, Hayashi S, Takasaki Y (1998) H_2 production from starch by mixed culture of *Clostridium butyricum* and *Rhodobacter* sp M-19. Biotechnol Lett 20:895–899

Yokoi H, Maki R, Hirose J, Hayashi S (2002) Microbial production of hydrogen from starch manufacturing wastes. Biomass Bioenergy 22:89–95

Yongzhen T, Yang C, Yongqiang W, Yanling H, Zhihua Z (2007) High hydrogen yield from a two-step process of dark- and photo-fermentation of sucrose. Int J Hydrog Energy 32:200–206

Yu H, Zhu Z, Hu W, Zhang H (2002) Hydrogen production from rice winery wastewater in an upflow anaerobic reactor by using mixed anaerobic cultures. Int J Hydrog Energy 27(11–12):1359–1365

Zhang T, Liu H, Fang HHP (2003) Biohydrogen production from starch in wastewater under thermophilic condition. J Environ Mang 69:149–156

Zhang ML, Fan YT, Xing Y, Pan CM, Zhang GS, Lay JJ (2007) Enhanced biohydrogen production from cornstalk wastes with acidification pretreatment by mixed anaerobic cultures. Biomass Bioenergy 31:250–254

Zhu H, Suzuki T, Tsygankov AA, Asada Y, Miyake J (1999) Hydrogen production from tofu wastewater by *Rhodobacter sphaeroides* immobilized agar gels. Int J Hydrog Energy 24:305–310

Zhu H, Ueda S, Asada Y, Miyake J (2002) Hydrogen production as a novel process of wastewater treatment-studies on tofu wastewater with entrapped *R. sphaeroides* and mutagenesis. Int J Hydrog Energy 27:1349–1357

Zhu H, Stadnyk A, Béland M, Seto P (2008) Co-production of hydrogen and methane from potato waste using a two-stage anaerobic digestion process. Bioresour Technol 99:5078–5084

Zhu J, Li Y, Wu X, Miller C, Chen P, Ruan R (2009) Swine manure fermentation for hydrogen production. Bioresour Technol 100:5472–5477

Zuideveld PL (2001) Overview of shell gasification projects. EPRI gasification Technology. Gasification Technology Conference, San Francisco, October 7–10 2001

Trends and Challenges in Biohydrogen Production from Agricultural Waste

4

Lucile Chatellard, Antonella Marone, Hélène Carrère, and Eric Trably

Abstract

Over the past decade, increasing interest has been given to anaerobic fermentative processes for hydrogen production and other high-value by-products. The development of technologies dedicated to energy production from biomass has recently emerged. Indeed, agricultural residues, such as agricultural waste or energy crops, have become economically and technologically attractive for their low-cost and carbohydrate-rich substrates. Moreover, dark fermentation methods present an ingenious solution to process them. However, low hydrogen production yields are often reported because of their rather low biodegradability due to the presence of complex polymers recalcitrant to biodegradation, such as lignocellulose. Hydrogen potentials range between less than 1 ml $H_2.g^{-1}$ of dry matter for complex lignocellulosic residues and 240 ml $H_2.g^{-1}$ of dry matter for purified polymers such as starch. Many solutions for increasing hydrogen potential have been proposed such as microbial consortium selection, substrate pretreatment and process parameter optimisation. Consequently, higher hydrogen yields have recently been obtained, reaching 150 ml $H_2.g_{TVS}^{-1}$ for pretreated rice straw. Nevertheless, the only manner to reach viable industrialisation of dark fermentation processes would be to combine this process with other biological energy production techniques such as photofermentation, bioelectrochemically assisted hydrogen production and anaerobic digestion, in a so-called environmental biorefinery concept.

4.1 Introduction

Green economy, defined as an ecologically based approach of the economic market (Costanza et al. 1997), implies that research efforts need to focus on new sources and modes of production of

L. Chatellard • A. Marone • H. Carrère
E. Trably (✉)
LBE, INRA, 102 Avenue des Etangs, 11100 Narbonne, France
e-mail: eric.trably@supagro.inra.fr

A. Singh, D. Rathore (eds.), *Biohydrogen Production: Sustainability of Current Technology and Future Perspective*, DOI 10.1007/978-81-322-3577-4_4

renewable energy. In this view, biomass potentially represents an abundant resource for producing a renewable and clean energy. Its biological conversion into hydrogen is an attractive concept, as this gas is characterised by a clean combustion process and is an efficient energy carrier with a considerable calorific value (122 MJ/kg). Among the technologies producing energy from biomass, anaerobic dark fermentative processes have gained a widespread increasing interest as they produce hydrogen efficiently, in addition to high-value by-products (Chong et al. 2009; Das 2009; Das and Veziroä 2001; Hallenbeck 2009; Hawkes et al. 2007).

Currently, biomass residues and organic effluents released from human activities, e.g. forestry, agriculture and livestock management, are not well managed, causing significant environmental problems (Levin et al. 2007). Considering that agricultural waste is made up of complex substrates and can be degraded biologically by complex microbial ecosystems, dark fermentation represents a key technology for the treatment and, by extension, the production of hydrogen from these residues. In particular, biohydrogen production from agricultural waste is very advantageous since these biomasses are abundant, cheap, renewable and biodegradable (Guo et al. 2010).

Dark fermentative hydrogen production is a ubiquitous phenomenon that occurs in natural environments under anoxic or anaerobic conditions and is carried out by a wide variety of bacteria. Usually, biological processes that convert biomass into more versatile energy carriers are dedicated to the production of methane. Indeed methane is the only energy carrier produced from the conversion of organic matter in nature. As part of the anaerobic digestion process, hydrogen corresponds to a by-product of the first acidogenic conversion of organic matter into organic acids. In nature, the gas is not released because numerous other bacteria consume the hydrogen as a source of reducing power (Marone 2012). Technologically, the main advantage in investigating the production of hydrogen instead of methane is the higher economic value of hydrogen, owing to its wider range of applications in the chemical industry (Li and Fang 2007). When processes are developed aiming at hydrogen production, a physical separation is required within the anaerobic digestion processing chain, exploiting a specific environment in which hydrogen-producing microorganisms flourish and other H_2 consumers perish (Marone 2012). Hence, successful biological hydrogen production requires the inhibition of numerous hydrogen-using microorganisms, such as homoacetogens and methanogens. Inhibition is commonly accomplished by heat treatment of the inoculum to kill all microorganisms except spore-forming, H_2-producing fermenting bacteria. Other inhibition methods include a refined control of reactor operation parameters with high dilution rates, low pH and maintenance of low hydrogen partial pressure through nitrogen gas sparging (De Vrije and Claassen 2003).

In dark fermentation, organic biomass is used as a substrate and produces not only hydrogen in the gas phase but also organic acids in the liquid phase. Effluents generated after dark fermentation processes therefore contain high-energy elements that can be extracted as valuable chemicals or used in other bioprocesses when integrated in a cascade process in a context of environmental biorefinery. This makes dark fermentation processes particularly interesting for applications in decentralised areas with high production of organic waste, such as in areas with intense agriculture and animal production (Bruni 2010). As support of the acidogenic degradation mechanism, the fermentative degrading microorganisms can be issued either from the substrate itself in a so-called self-fermentation process or from external sources after inoculation, mainly with leachate, manure, rumen or sewage sludge. Although microorganisms are important actors, the substrate composition is also a crucial element to consider since hydrogen is mainly produced from carbohydrates, proteins and lipids being degraded to other by-products such as ethanol (Guo et al. 2014; Monlau et al. 2012). This type of carbohydrate-rich substrate is present in wastewater effluents from food manufacturing (cassava, brewery, tofu, etc.), agri-industrial waste (food waste from restaurant, fruit and vegetables wastes, etc.) and agricultural end products (leaves, stalks, straw, etc.).

The use of low-cost feedstock is essential to obtain cost-effective biotechnologies for biohydrogen and biogas production (Ni and Sun 2009; Rabelo et al. 2009). Unfortunately, low-cost material often results from low biodegradability. Similar to other bioprocesses, dark fermentation does not efficiently degrade the lignocellulose from agricultural residues. Such low digestibility causes a loss of hydrogen production, consequently limiting the overall energetic efficiency of the process (Bruni 2010). Indeed, lignocellulosic material is a polymer comprising a complex matrix formed by three main fractions: cellulose, hemicellulose and lignin. These fractions are linked together to form a rigid barrier for plants against environmental threats (Menon and Rao 2012). Due to their carbohydrate content, cellulose and hemicelluloses are the most degradable fractions of the lignocellulosic matrix that can be used by anaerobic flora to produce biohydrogen. As they are part of a complex matrix, their degradation is limited by their availability and nowadays still represents a challenge. Monlau et al. (2012) showed that only the soluble sugars in lignocellulosic biomass lead to significant hydrogen production. They also suggested that the hydrolysis step is the main limiting factor when using lignocellulosic biomass as a substrate. Hence, hydrolysis pretreatment has been investigated in order to increase the performances of dark fermentation and, thus, the hydrogen yields.

Overall, research and development in dark fermentation have been focusing on three main fields: bioreactor engineering, operational optimisation and metabolic engineering with aims to enhance chemical, biomolecular and physiological aspects of hydrogen metabolism; optimise pretreatment and bioreactor conditions to improve substrate degradability; favour hydrogen-producing consortia and maintain a constant H_2 production; maximise microbial cell concentration to improve substrate utilisation and increase H_2 yield; reduce competition for electrons with competitive bacteria present in the reactor; identify available and cheap gas separation technologies for hydrogen recovery, like membrane systems; and improve the quality of the substrate for biogas production (Massi 2012).

Considering that agricultural wastes are complex substrates and can be degraded biologically by complex microbial ecosystems, the present chapter focuses on dark fermentation as a key technology for producing hydrogen from agricultural residues, energy crops and livestock. Based on recent findings, this chapter deals with the question about how to optimise the dark fermentation process to get as much energy as possible out of the agricultural waste.

In the following chapter, the currently adopted key operational parameters, the potentially obtainable hydrogen production and the possible strategies to enhance the biohydrogen production are discussed so as to improve the understanding of recent findings and the development of further research in this field. The integration of dark fermentation within a biorefinery concept is also reviewed, since dark fermentation does not entirely convert organic matter and cannot be dissociated into other processes for treating the remaining by-products.

4.2 Agricultural Residues

Agriculture is one of the most common economic activities in the world. Each year it generates a significant amount of waste, evaluated at more than 220 billion tons per year and mainly composed of carbohydrates in simple and polymeric forms (Ren et al. 2009). With the constantly increasing quantity of crop residues issued from agricultural activities arises an opportunity for developing new valuable processes such as the production of biohydrogen by dark fermentation. Agricultural residues do not compete with food crops for energy generation; however, their fermentative degradation, due to their large content in lignocellulose, remains a technological challenge.

4.2.1 Hydrogen Production Potential from Agricultural Residues

According to a Food and Agriculture Organization of the United Nations (FAO) analysis, about 20 % of 1640 × 10^6 t of fruits and vegetables are lost or wasted during harvest (FAO 2013; FAO 2011). Three hundred and twenty-eight million tons of unmarketable vegetables could therefore be potentially used as biomass for dark fermentation. As they are mostly composed of simple sugars, they are ideal candidates for fermentative hydrogen production.

Furthermore, a yield of 17–18.6 ml $H_2.g^{-1}$ of total solid (TS) has been obtained from the self-fermentation of leaf-shaped vegetable refuses, thus demonstrating that vegetable wastes can be used as efficient substrate as well as source of microorganisms for hydrogen production (Marone et al. 2014). By adding other sources of inoculum to the substrate, such as anaerobic digested sludge, a higher hydrogen production was obtained with 48, 62, 71 and 102 ml $H_2.g^{-1}$ of TS from lettuce, cabbage, carrot and potatoes, respectively (Dong et al. 2009; Okamoto et al. 2000). These results support that the hydrogen yield is closely related to component degradability, since potatoes are mainly composed of starch, while lettuce, closer to leaves, mostly contains lignocellulose with a low lignin content. Venkata Mohan et al. (2009) made similar observations, by producing more hydrogen with a juice of mixed vegetables rather than with the pulp (poorest in free sugars), the hydrogen yield reaching 124 versus 105 ml $H_2.g^{-1}$ of equivalent sugars, respectively. Hydrogen yields from beet pulp have also been reported to reach highest values with about 90.1 ml $H_2.g^{-1}$ of chemical oxygen demand (COD) without any pretreatment (Ozkan et al. 2011).

The low biodegradability of lignocellulosic waste presents a higher impact especially when nonfood crops residues are investigated. As an illustration, low biodegradable straws contain about 80 % of lignocellulose (Guo et al. 2010). Using grinding as a sole pretreatment, Han et al. (2012) showed a production of only 5.46 ml $H_2.g^{-1}$ of dry soybean straw in mesophilic conditions, while Monlau et al. (2013b) obtained 2.3 ml $H_2.g^{-1}$ of volatile solids (VS) with sunflower stalks. Hydrogen production from wheat straws was not greater, with yields ranging from 5 to 37 ml $H_2.g^{-1}$ of VS (Nasirian et al. 2011; Quéméneur et al. 2012; Li and Chen 2007; Chu et al. 2011). In these latter cases, hydrolysis was the limiting step and needed to be improved in order to increase the hydrogen yields of these substrates (Monlau et al. 2013c). Table 4.1 presents the maximal hydrogen yields obtained from different crop wastes and according to the inoculum and process parameters.

4.2.2 Processes for Enhancing the Biohydrogen Production from Agricultural Residues

Although crop residues are good candidates as substrate for producing biohydrogen by dark fermentation, processes need to be optimised in order to convert a greater amount of compounds constituting the biomass, so as to consequently reach higher hydrogen yields. For this, the microbial inoculum must be carefully selected so as to be perfectly adapted to the substrate. Optimal process parameters are strongly dependent on substrate composition as well as on the type of inoculum.

The origin of the inoculum, as well as pretreatments applied on the biomass, directly impacts the microbial community and, subsequently, the hydrogen potentials. Pan et al. (2008) observed that hydrogen yields from wheat bran fermentation can double between two digested sludges originating from two different paper mills and used as microbial inoculum. By comparing different sources of inoculum, these authors found that compost inoculum led to lower hydrogen yields than activated sludge with 17.7 ml $H_2.g^{-1}$ of total volatile solids (TVS) versus 50.6 ml $H_2.g^{-1}$ of TVS, respectively. In contrast, Chu et al. (2011) found higher hydrogen production by using digested dairy manure as inoculum with a production of 37.0 ml $H_2.g^{-1}$ of VS from wheat stalks against 23.0 ml $H_2.g^{-1}$ of VS with anaerobic digested activated sludge.

Table 4.1 Biohydrogen potential from different agricultural residues

Substrate	Type of inoculum/ pretreatment	Process conditions ($g.l^{-1}$/°C/pH)	Type of process	H_2 production (ml $H_2.g^{-1}$)	References
Beet pulp	Seed sludge	20/35/6.0	Batch	90.1[a]	Ozkan et al. (2011)
Leaf shape of mixed vegetable and potatoes	Indigenous microflora	–/37/6.7	Batch	19[b]	Marone et al. (2014)
Lettuce	Heat-treated anaerobic sludge	20/37/5.5	Batch	48[b]	Dong et al. (2009)
Potatoes	Heat-treated anaerobic sludge	20/37/5.5	Batch	102[b]	Dong et al. (2009)
Rice straw	Anaerobic sludge	90/55/6.5	Batch	24.8[b]	Chen et al. (2012)
Soybean straw	Cracked cereal acclimated in continuous stirred tank reactor (CSTR)	–/35/7.0	Batch	5.46[c]	Han et al. (2012)
Sunflower stalks	Anaerobic digested sludge	5/35/5.5	Batch	2.3[c]	Monlau et al. (2013c)
Wheat bran	Activated sludge paper mill	100/36/7.0	Batch	50.6[d]	Pan et al. (2008)
Wheat bran	Digested sludge paper	100/36/7.0	Batch	28.6[d]	Pan et al. (2008)
Wheat bran	Mill cornstalk compost	100/36/7.0	Batch	22.6[d]	Pan et al. (2008)
Wheat bran	Wheat straw compost	100/36/7.0	Batch	17.7[d]	Pan et al. (2008)
Wheat stalks	Anaerobic digested activated sludge	60/35/6.5	Batch	23[c]	Chu et al. (2011)
Wheat straw	Seed sludge from H2-producing CSTR	6/35/–	Batch	5.69[c]	Nasirian et al. (2011)
Wheat straw	Mesophilic anaerobically digested sludge	4/37/5.5	Batch	10.52[c]	Quéméneur et al. (2012)
Wheat straw	*Clostridium butyricum*	40/35/7.2	Batch	9[e]	Li and Chen (2007)
Wheat stalks	Anaerobic digested dairy manure	60/35/6.5	Batch	37[c]	Chu et al. (2011)
Wheat straw	Cow dung compost	25/36/7.0	Batch	0.5[d]	Fan et al. (2006)

[a]ml $H_2.g_{COD}^{-1}$
[b]ml $H_2.g_{TS}^{-1}$
[c]ml $H_2.g_{VS}^{-1}$
[d]ml $H_2.g_{TVS}^{-1}$
[e]ml $H_2.g^{-1}_{substrate}$

The low hydrogen yields from lignocellulosic residues are not only mostly due to a low accessibility of carbohydrate compounds but also to biomass recalcitrance to biodegradation. Hence, to improve biodegradability and optimise hydrogen yields, several types of pretreatment can be applied to the original biomass. The main types of pretreatments used are physical, physico-chemical and biological and have been widely reviewed (Monlau et al. 2013b; Mosier et al. 2005; Nissilä et al. 2014).

Physical pretreatment techniques aim at decomposing organic material in order to decrease particle size. This leads to an increase in surface area and to a reduction in the degree of crystallinity (Monlau et al. 2013b). Microorganisms can thus reach initially non-accessible areas and subsequently increase the chances of degrading the substrate. When submitted to chemical treatment, such as acid or alkaline attacks, or to biological treatment (enzyme or microorganisms), part of the organic matter is solubilised and further degraded by hydrolysis. After pretreatment, the medium, called hydrolysate, can be used directly in dark fermentation, alone or added to the remaining solid.

With this approach, a 136-fold production of 68 ml $H_2.g^{-1}$ of TVS from wheat straw pretreated with acid has been observed, instead of the 0.5 ml $H_2.g^{-1}$ of TVS from straw that was submitted to fermentation with a mesophilic cow dung compost (36 °C) (Fan et al. 2006). This kind of treatment increases the biodegradability in high proportions as confirmed with soybean straws for which other authors reached 98.87 % of degradation efficiency after fermentation by microorganisms issued from cracked cereals (Han et al. 2012). To solubilise more biomass, acid can also be used in combination with other types of pretreatments. Wheat straw grinding at 1 mm followed by 4 % sulphuric acid pretreatment during 90 min at 120 °C led to a production of 37.11 ml $H_2.g^{-1}$ of VS (Nasirian et al. 2011). Combined with steam explosion, acid pretreatment of corn stover led to an interesting hydrogen yield of 3 mol $H_2.mol^{-1}$ of sugars, with anaerobic digested sludge as inoculum (Datar et al. 2007).

As lignocellulosic materials comprise cleaved ester bonds of phenolic-carbohydrate linkage, alkaline pretreatment has also proved to be very useful for improving overall biodegradability. Hence, efficient degradability of 92.14 % was observed after alkaline treatment with 4 % NaOH (Han et al. 2012). The association between microwaves and alkali can severely improve hydrogen yields as has been observed on rice straw hydrolysate which reached a production of 155 ml $H_2.g^{-1}$ of TVS (Cheng et al. 2011a). Ozone gas has also been suggested as an efficient pretreatment technique, as it also acts on delignification and therefore improves hydrogen production from 107 to 166 %. This range depends on the ozonation period (15–45 min) and on the type of substrate, i.e. wheat or barley straws (Wu et al. 2013a, b).

As for biological pretreatments, hydrolytic enzymes can specifically increase hydrogen yields by improving the hydrolytic fermentation step. In Quéméneur et al. (2012), an enzymatic cocktail secreted by an engineered *Trichoderma* strain was directly added to the fermentative process, leading to a twofold increase in the hydrogen yield from wheat straw. However, results did warn that precautions need to be taken when enzymatic pretreatment is separately carried out prior to fermentation. In this case the substrate should be sterilised to avoid direct sugar consumption by indigenous microflora (Quéméneur et al. 2012).

Similarly, by combining this pretreatment step with other techniques, higher hydrogen yields can yet be obtained. As an illustration, the addition of a concentrated solution of cellulase issued from *Penicillium decumbens* after steam explosion increased the hydrogen yield from corn straw sixfold, by reaching 63 ml $H_2.g^{-1}$ of substrate with a pure culture of fermenting *Clostridium butyricum* (Li and Chen 2007). Combining acid treatment with enzyme hydrolysis, Nasirian et al. (2011) observed a yield of 125.11 ml $H_2.g^{-1}$of VS, corresponding to a greater than 3.5-fold increase. On barley straw, the combination of enzymatic hydrolysis with a mixture of cellulase complex, endoxylanase and beta-glucosidase with 45-min ozonation led to a production of 93.40 ml $H_2.g^{-1}$ of TVS against 35.05 ml $H_2.g^{-1}$ of TVS without ozone (Wu et al. 2013a). Using an alkaline attack with other pretreatments can lead to even higher yields. Thermo-alkaline pretreatment followed by enzymatic hydrolysis of sunflower stalks has allowed the hydrogen yields to increase by a factor of 25 with 59.5 ml H_2/g of initial VS against 2.3 ml $H_2.g^{-1}$ of initial VS with raw substrate (Monlau et al. 2013c).

As stated in the introduction, the release of soluble sugars during the pretreatment should be efficiently achieved with enzyme hydrolysis of lignocellulosic biomass. Enzyme hydrolysis can be significantly enhanced thanks to thermochemical pretreatments such as alkali which increase enzyme accessibility to holocellulose by partial delignification (Monlau et al. 2013b). Sugars can be solubilised by thermo-acid pretreatment of lignocellulosic biomass (Han et al. 2012) or to a smaller extent by a thermo-alkali attack. However, such pretreatments lead to the release of various by-products such as furanic compounds originating from sugar degradation (mainly during acid pretreatment at high temperature) and phenolic compounds originating from lignin degradation (favoured by alkali pretreatments). Depending on their concentration, these by-products are capable of inhibiting hydrogen production as reviewed by Monlau et al. (2014). For example, the hydrolysate from acid pretreatment of sunflower stalks with 4 g HCl.100 g^{-1} of TS for 1 h at 170 °C totally inhibited glucose fermentation when added at volumes higher than 15 % (Monlau et al. 2013a). In this case, a detoxification step has been proposed (Mussatto and Roberto 2004).

After inoculum selection and substrate pretreatment optimisation, process parameters can also be improved, such as pH, temperature or hydraulic retention time (HRT). Working in thermophilic conditions, i.e. 55 °C, substantially increased the hydrogen production with a selection of specific hydrolytic microorganisms such as *C. pasteurianum*, *C. stercorarium* or *Thermoanaerobacterium saccharolyticum* (Chen et al. 2012). Using statistical analysis, Shanmugam et al. (2014) found that the optimal process parameters for continuous fermentation of steam-exploded cornstalk hydrolysate in an Anaerobic Sequencing Batch Reactor (ASBR) were 53 °C, pH 4.5 and 9.5 h of HRT, with a production of 98 ml H_2.g^{-1} of TVS against an average of 10 ml H_2 from raw stalks (inoculum from mixed culture from a brewery wastewater treatment facility).

Table 4.2 presents the optimal operating conditions determined for various lignocellulosic biomasses after pretreatment.

4.3 Energy Crops

Energy crops comprise any kind of plant that is specifically cultivated and harvested for fuel production, especially when characterised by easy cultivation and high carbohydrate contents. As carbohydrates are readily biodegradable, first-generation crops, such as cassava, sugar cane or wheat grains, present the highest hydrogen conversion potentials. However the use of this type of biomass is a controversial issue since it competes with food production for agricultural land. Further expectations now rest on catch crops, i.e. nonfood plants cultivated between two growing periods of food plants, which are generally used to enrich soils (short crop sorghum or grass). While energy and catch crops are mostly used for bioethanol production, few studies have considered their degradation to hydrogen by dark fermentation.

4.3.1 Characteristics and Hydrogen Production Potential of Energy Crops

Energy crops are lignocellulosic biomass with different physico-chemical properties depending on the species and part of plant studied (Menon and Rao 2012). Hence, their biodegradability and consequently the hydrogen yield strongly depend on the degree of entrapment of carbohydrates within the lignocellulosic matrix.

Starch which mainly composes roots, tubercles and grains is a glucose polymer with β-(1-4) linkage and can be considered as one of the most biodegradable substrates. Hydrogen production from starch is one of the most effective with a hydrogen yield reaching 199 240 ml H_2.g^{-1} of starch in batch reactor. In continuous experiments, yields of about 2.8 mol H_2.mol^{-1} of glucose have been observed (Zong et al. 2009; Su et al. 2009; Oztekin et al. 2008).

In comparison, more complex substrates, such as grass, present lower potentials. Hydrogen yields from grass fermentation ranged from 0.2 to 11.5 ml H_2.g^{-1} of VS (Cui and Shen 2012; Lakaniemi et al. 2011; Pakarinen 2008; Pakarinen et al. 2009). Another energy crop widely used for biofuel production, sorghum, can lead to higher H_2 potentials. Indeed, the hydrogen yield from stalks by dark fermentation reached 52 ml H_2.g^{-1} of TVS using mixed cultures in mesophilic conditions (Shi et al. 2010).

Similar to highly biodegradable and very reactive substrates, the storage duration between harvesting and dark fermentation directly impacts

Table 4.2 Enhancement of biohydrogen production from different agricultural residues and operative conditions

Substrate/*substrate pretreatment*	Type of inoculum/pretreatment	Process conditions ($g.l^{-1}$/°C/pH)	Type of process	H_2 production (ml $H_2.g^{-1}$)	References
Barley straw/*45-min ozonation + enzyme hydrolysis*	Cow manure and pond sediment	50/35/6.0	Batch	93.4[b]	Wu et al. (2013a)
Barley straw/*enzyme hydrolysis*	Cow manure and pond sediment	50/35/6.0	Batch	35.05[b]	Wu et al. (2013a)
Beet pulp/*2M NaOH, pH = 12 30 min*	Seed sludge	20/35/6.0	Batch	108.2[c]	Ozkan et al. (2011)
Beet pulp/*2M NaOH, pH = 12 30 min + microwave 700 W 170 °C 30 min*	Seed sludge	20/35/6.0	Batch	115.6[c]	Ozkan et al. (2011)
Beet pulp/*2M NaOH, pH = 12 30 min + 121 °C 30 min*	Seed sludge	20/35/6.0	Batch	66.7[c]	Ozkan et al. (2011)
Cornstalk/*steam explosion*	Brewery waste water treatment	–/53/4.5	ASBR (HRT 9.5 h)	98[d]	Shanmugam et al. (2014)
Corn stover/*1.2 %H_2SO_4 + steam explosion*	Anaerobic digester sludge	27.5 mM/35/5.5	Batch	3.21 mol $H_2.mol^{-1}$ gluc	Datar et al. (2007)
Rice straw/*microwave heating 15 min at 140 °C in 0.5 % NaOH*	Anaerobic activated sludge	60/35/6.5	Batch	150[d]	Cheng et al. (2011a)
Soybean straw/*4 %HCl 30 min*	Cracked cereal acclimated in CSTR	–/35/7.0	Batch	20.71[b]	Han et al. (2012)
Soybean straw/*0.5 % NaOH 30 min*	Cracked cereal acclimated in CSTR	–/35/7.0	Batch	10.47[b]	Han et al. (2012)
Soybean straw/16 %H_2O_2 30 min	Cracked cereal acclimated in CSTR	–/35/7.0	Batch	23.00[b]	Han et al. (2012)
Sunflower stalks/	Anaerobic digested sludge	5/35/5.5	Batch	4.4[e]	Monlau et al. (2013c)
−55 °C 24 h 4 % NaOH				59.9[e]	
+enzymes[a]				20.6[e]	
−170 °C 1 h 4 % NaOH				80.9[e]	
+enzymes[a]					

Wheat straw/*2 %H_2SO_4 120 °C 90 min*	Seed sludge from H2-producing CSTR	6/35/–	Batch	37.11[b]	Nasirian et al. (2011)
Wheat straw/*2 %H_2SO_4 120 °C 90 min + cellulases, xylanases, and β-glucanases* ***before*** *fermentation*	Seed sludge from H2-producing CSTR	6/35/–	Batch	47.89[b]	Nasirian et al. (2011)
Wheat straw/*2 %H_2SO_4 120 °C 90 min + cellulases, xylanases, and β-glucanases* ***during*** *fermentation*	Seed sludge from H2-producing CSTR	6/35/–	Batch	125.11[b]	Nasirian et al. (2011)
Wheat straw/*enzymatic cocktail (Trichoderma strain)* ***during*** *fermentation*	Mesophilic anaerobically digested sludge	4/37/5.5	Batch	21.61[b]	Quéméneur et al. (2012)
Wheat straw/*sterilisation + enzymatic cocktail (Trichoderma strain)* ***before*** *fermentation*	Mesophilic anaerobically digested sludge	4/37/5.5	Batch	18.13[b]	Quéméneur et al. (2012)
Wheat straw/*steam explosion + enzymatic hydrolysis 25 IU/g SECS*	*Clostridium butyricum*	40/35/7.2	Batch	68[e]	Li and Chen (2007)
Wheat straw/*2 %HCl + 8-min microwaves*	Cow dung compost	25/36/7.0	Batch	68.1[d]	Fan et al. (2006)
Wheat straw/*45-min ozonation + enzyme hydrolysis*	Cow manure and pond sediment	50/35/6.0	Batch	87.35[b]	Wu et al. (2013b)

[a]Enzymes applied after alkali treatment: cellulase, β-glucosidase and xylanase: 50, 25 and 50 units.g^{-1} of TS

[b]ml $H_2.g_{VS}^{-1}$

[c]ml $H_2.g_{COD}^{-1}$

[d]ml $H_2.g_{TVS}^{-1}$

[e]ml $H_2.g^{-1}_{substrate}$

the hydrogen potential as has been reported by Kyazze et al. (2008). Indigenous microorganisms start their hydrolytic activity during the storage period, releasing high quantities of soluble sugars. For example, a hydrogen yield of only 21.8 ml $H_2.g^{-1}$ dry weight from freshly cut biomass compared to 75.8 ml $H_2.g^{-1}$ of dry weight from wilted ryegrass illustrates the likely higher availability of soluble sugars in the latter case (Kyazze et al. 2008).

Table 4.3 presents hydrogen yields obtained from different energy crops, according to the inoculum and process parameters.

4.3.2 Enhancement in Biohydrogen Production from Energy Crops

Although energy crops are efficient sources of hydrogen, challenges now lie ahead in determining how to maximise the biomass biodegradability.

As for crop residues, microbial inoculum selection is very important since it provides fermentative microorganisms with different metabolic capabilities. Depending on their origins, microbial consortia can lead to a doubling of the hydrogen yields. Akutsu et al. (2008) tested six kinds of inoculum, i.e. night soil, organic fraction of municipal solid waste, thermophilically digested activated sludge, digested cattle manure, acidified potatoes and digested organic fraction of municipal solid waste. The authors observed the lowest hydrogen yield with the organic fraction of municipal solid waste, 1.38 mol $H_2.mol^{-1}$ of hexose, and the highest with activated sludge, 2.32 mol $H_2.mol^{-1}$ of hexose. Using pure strains of the hyperthermophilic bacteria *Caldicellulosiruptor saccharolyticus*, Ivanova et al. (2009) and Claassen et al. (2004) reached 20–30 ml $H_2.g^{-1}$ of VS from untreated sweet sorghum stalks. In the latter case, a process temperature of 70 °C was adapted to the microorganisms,

Table 4.3 Biohydrogen production from different energy crops and operative conditions

Substrate	Type of inoculum	Process conditions ($g.l^{-1}$/°C/pH)	Type of process	H_2 production (ml $H_2.g^{-1}$)	References
Grass silage	Mesophilic farm biogas digestate	–/**35**/6.2	Batch	11.5[a]	Pakarinen (2008)
		–/**70**/6.0		16.0[a]	
Grass	Cracked cereal	5/35/7.0	Batch	4.39[b]	Cui and Shen (2012)
Perennial ryegrass (fresh)	Anaerobically digested sludge inoculum	20/35/5.2	Batch	21.8[c]	Kyazze et al. (2008)
Perennial ryegrass (wilted)	Anaerobically digested sludge inoculum	20/35/5.2	Batch	75.6[c]	Kyazze et al. (2008)
Reed canary grass	Sludge	–/35/–	Batch	1.3[d]	Lakaniemi et al. (2011)
Starch (cassava)	Cattle dung compost	18/37/6.8	Batch	199[b]	Zong et al. (2009)
Starch (cassava)	Anaerobic activated sludge	10/35/7.0	Batch	240[b]	Su et al. (2009)
Starch	Digested and acidogenic sludges	20/55/–	CSTR	2.82 mol $H_2.mol^{-1}$ gluc	Akutsu et al. (2009)
Sweet sorghum stalks	Anaerobic sludge	–/36/–	Batch	52.1[e]	Shi et al. (2010)

[a]ml $H_2.g_{VS}^{-1}$
[b]ml $H_2.g^{-1}_{substrate}$
[c]ml $H_2.g_{TS}^{-1}$
[d]ml $H_2.g_{COD}^{-1}$
[e]ml $H_2.g_{TVS}^{-1}$

thus avoiding contamination by indigenous microorganisms.

One of the most studied parameters for biological conversion of energy crops is the type of substrate pretreatment. This procedure aims to release sugars and to pre-hydrolyse the organic material prior to fermentation. In the case of materials rich in soluble sugars such as sweet sorghum, a simple extraction is sufficient. For example, sugar extraction from sorghum was carried out with a hot water treatment on a previously crushed fraction (Antonopoulou et al. 2008; Saraphirom and Reungsang 2011). The final "juice" was rich in simple sugars, whereas the solid residue contained the remaining lignocellulosic constituents. However, yields were still less than 0.8 mol $H_2.mol^{-1}$ of sugars, equivalent to 10 ml $H_2.g^{-1}$ of sugars.

In the case of sugar polymers, the hydrolytic pretreatment should be more invasive involving the use of enzymes and chemicals. For this purpose, enzymatic hydrolysis greatly improved starch degradation. Using alpha-amylase and glucoamylase on cassava starch, hydrogen yield was two and threefold higher than for raw substrate (Su et al. 2009). Phowan and Danvirutai (2014) used sulphuric acid to hydrolyse cassava pulp and release degradable sugars. High H_2 yields were reported with about 250 ml $H_2.g^{-1}$ of initial COD from the hydrolysate.

Finally, when lignocellulosic biomass is concerned, the same pretreatment as for agricultural residues can be applied. These are mainly thermochemical and eventually combined with enzymatic hydrolysis. Thereby, alkaline treatment on sorghum stalks leads to a production of 127.26 ml H_2/g of TVS after 24 h of 0.4 % NaOH pretreatment at room temperature (Shi et al. 2010). Other authors also reported a fivefold increase in the H_2 yield with grass silage treated with 4 % NaOH for 24 h (Pakarinen et al. 2009). Comparing alkaline and acidic pretreatment, Cui and Shen (2012) showed that treatment with 4 % HCl was more efficient than 0.5 % NaOH, producing 72.21 and 19.25 ml $H_2.g^{-1}$ of substrate, respectively. Acid pretreatment with 3 % HCl, 121 °C, and 90 min of reed canary grass led to a higher hydrogen yield of 39 ml $H_2.g^{-1}$ of initial COD (Lakaniemi et al. 2011).

Because energy crops are specifically harvested for energy production and not for food needs, the focus has been put on genetic mutation aiming at providing more biodegradable plants. Prakasham et al. (2012) studied the fermentation of various types of sorghum with different lignin contents and succeeded in producing 72 ml $H_2.g^{-1}$ of TVS, i.e. sevenfold higher than previously reported values.

In Table 4.4 increases in hydrogen yields subsequent to pretreatment on energy crops are presented.

4.4 Livestock

Over the past years, the rapid growth of animal husbandry worldwide has increased the amount of manure issued from animal production activities. Nowadays, manure can be considered to be the largest waste stream from the livestock industry. More than 1500 million tons of animal manure are produced yearly, including 128 million tons of cattle manure and 295 million tons of pig manure across the 27 member states of the European Union. Three main types of animal manure can be distinguished: urinary waste, i.e. slurry or liquid manure from livestock or poultry, solid manure or farmyard manure and wastewaters which are a collection of processed water from farms, feedlot run-off, silage juices, beddings, disinfectants and liquid manure. In particular, animal effluents represent one of the most polluting agro-industrial wastewaters. The issue of achieving adequate effluent management has appeared only recently with the development of intensive farming and the concentration of livestock on limited operating space which produce large amounts of manure. Due to its high content in organic matter, nitrogen and phosphorous, manure can contribute substantially to negative environmental and human health impacts when improperly stored. Therefore, at present and for the future, the primary issue with dairy manure is the development of management systems that use

Table 4.4 Enhancement of biohydrogen production from different energy crops and operation conditions

Substrate/*pretreatment*	Type of inoculum	Process conditions ($g.l^{-1}$/°C/pH)	Type of process	H_2 production (ml $H_2.g^{-1}$)	References
Cassava starch/*α-amylase + glucoamylase – heat gelatinising*	Anaerobic activated sludge	10/35/7.0	Batch	–258.5[a] –276.1[a]	Su et al. (2009)
Grass/*comminuted 20 mesh +3 h 105 °C + 0.5 % NaOH*	Cracked cereal	–/35/7.0	Batch	19.25[a]	Cui and Shen (2012)
Grass/*comminuted 20 mesh +3 h 105 °C + 4 % HCl*	Cracked cereal	–/35/7.0	Batch	72.21[a]	Cui and Shen (2012)
Grass sill; age/*4 % NaOH 24 h 20 °C*	Heat-treated mixed microflora	–/55/6.0	Batch	6.46[b]	Pakarinen et al. (2009)
Reed canary grass/*3 % of HCl, 121 °C, 90 min*	Untreated sludge	–/35/–	Batch	39[c]	Lakaniemi et al. (2011)
Sweet sorghum plant/*air-dried + sterilised*	*Caldicellulosiruptor saccharolyticus*	–/70 °C/7.2	Batch	30.17[a]	Ivanova et al. (2009)
Sweet sorghum plant/*air-dried + sterilised*	*Caldicellulosiruptor saccharolyticus*	–/70/7.2	Batch	30.17[a]	Ivanova et al. (2009)
Sweet sorghum stalks/*24 h 0.4 % NaOH*	Anaerobic sludge	–/36/–	Batch	127.3[d]	Shi et al. (2010)
Cassava starch/*121 °C 0.5 % H_2SO_4*	Upflow anaerobic sludge blanket (UASB) granule	25/35/5.5	Batch	250[c]	Phowan and Danvirutai (2014)
Sweet sorghum extract/*crushing + hot water solubilisation*	Indigenous flora	–/35/4.7–5.5	CSTR (12 h HRT)	0.86 mol $H_2.mol^{-1}$ gluc	Antonopoulou et al. (2008)
Sweet sorghum extract/*crushing + hot water solubilisation + concentration*	Anaerobic seed sludge	25/30/5.0	ASBR (HRT 24 h)	0.68 mol $H_2.mol^{-1}$ gluc	Saraphirom and Reungsang (2011)
Sorghum/*mutant with low lignin content (DRT07K6 bmr3 derivative)*	Buffalo dung compost	20/37/–	Batch	72[d]	Prakasham et al. (2012)
Wheat starch/*heat and acid hydrolysis*	Acidogenic anaerobic sludge	15/37/7.0	Batch	2.84 mol $H_2.mol^{-1}$ gluc	Oztekin et al. (2008)

[a]ml $H_2.g^{-1}{}_{substrate}$
[b]ml $H_2.g_{VS}{}^{-1}$
[c]ml $H_2.g_{COD}{}^{-1}$
[d]ml $H_2.g_{TVS}{}^{-1}$

the resource without associated adverse environmental impacts (Hubbard and Lowrance 1998).

4.4.1 Characteristics and Hydrogen Production Potential from Animal Waste

To date, studies dealing with the use of manure as feedstock for biohydrogen production are scarce, as the main suitable substrates for dark fermentation are typically rich in carbohydrates.

Generally, very little biohydrogen is recovered from manure fermentation, with production yields ranging between 0.7 ml $H_2.g^{-1}$ of VS (Yokoyama et al. 2007) and 37.7 ± 1.81 ml $H_2.g^{-1}$ of VS (Concetti et al. 2006) whatever the operating parameters, the fermentation temperature or the type of pretreatment applied, as presented in Table 4.5.

Table 4.5 Biohydrogen production from different manures and operative conditions

Substrate/*substrate pretreatment*	Type of inoculum	Process conditions ($g.l^{-1}$/°C/pH)	Type of process	H_2 production (ml $H_2.g_{VS}^{-1}$)	References
Cow slurry/*No*	Selected from slurry	13.4/37/8.2	Batch	0.7	Yokoyama et al. (2007)
Buffalo manure/*No*	Selected from lagoon sediments	8/37/6.7	Batch	10,4 ± 1	Concetti et al. (2006)
Buffalo slurry/*sterilisation*	Selected from lagoon sediments	2.06 %VS/37/6.5	Batch	6.57 ± 0.01	Marone et al. (2015
Dairy manure/*acid pretreatment*	Selected from manure	50.0/36/7.0	Batch	18.1 ± 0.6	Xing et al. (2010)
Buffalo manure/*sterilisation*	Selected from lagoon sediments	8/37/6.7	Batch	37,7 ± 1,81	Concetti et al. (2006)
Cow manure slurry/*minced to 1-mm mesh size and passed through a sieve*	Cow manure slurry	6 %VS/60/5.2	Semi-CSTR	10.25 ± 4.96	Wang et al. (2013)
Swine manure/*shredding and filtration 1-mm mesh size*	Sludge	3.32 %VS/70/6.7	Batch	3.65	Kotsopoulos et al. (2009)

In mesophilic conditions (37 °C), Yokoyama et al. (2007) observed a maximum of 0.7 ml $H_2.g^{-1}$ of VS from cow slurry in batch reactors, by using the indigenous microflora of the slurry and without applying any pretreatment. By increasing the operating temperature to 60 °C, the hydrogen yield increased up to 29 ml $H_2.g^{-1}$ of VS although the authors also observed a simultaneous increase in hydrogen consumption.

From the same substrate and inoculum, at a thermophilic temperature (60 °C), Wang et al. (2013) reported H_2 yields up to 10.25 ± 1.96 ml $H_2.g^{-1}$ of VS in a semi-continuous stirred tank reactor (semi-CSTR). They worked with operational parameters previously optimised during a batch experiment, by integrating the Taguchi method and the response surface methodology (RSM) to predict and optimise fermentative hydrogen production (pH 5, 6 and 7 and temperatures of 50, 60 and 70 °C, respectively). As predicted, a further increase of the fermentation temperature did not improve the hydrogen production yield from manure. By operating the fermentation in such hyperthermophilic conditions (70 °C), Yokoyama et al. (2007) observed a yield of 18 ml $H_2.g^{-1}$ of VS, which is lower compared to the 60 °C conditions. At this temperature (70 °C), the maximum hydrogen yield obtained from swine manure, inoculated with sludge obtained from a mesophilic methanogenic reactor, was only 3.65 ml $H_2.g^{-1}$ of VS (Kotsopoulos et al. 2009). As hydrogen yields obtained from these substrates are low, even at high temperature, optimisation procedures still need to be developed, through pretreatment, for instance.

Indeed, in the case of livestock waste fermentation, pretreatment can help in improving the hydrogen yield. They can be applied to microorganisms present in the feedstock (e.g. sterilisation) or help release sugars from manure (e.g. acidic pretreatment). Concetti et al. (2006) observed an increase in the hydrogen yield from 10 ± 1 to 38 ± 2 ml $H_2.g^{-1}$ of VS in batch fermentation at 37 °C, by heat treating a buffalo manure and using, as inoculum, a mixed culture produced from lagoon sediments. After testing different dairy manure pretreatments (acidic pretreatment, NaOH pretreatment and infrared radiation), Xing et al. (2010) found a maximal yield of 18.1 ± 0.6 ml $H_2.g^{-1}$ of VS after acidic pretreatment, in batch reactors at 37 °C, with an inoculum selected from manure.

4.4.2 Enhancement in Biohydrogen Production from Animal Waste (Improving Routes: Co-fermentation)

Although, at present, animal manure cannot yet be considered as the ideal substrate for dark fermentation, it can be used as a co-substrate with carbohydrate-rich and promptly degradable materials in order to improve the hydrogen yield (Marone et al. 2015). Indeed, supplementation of an adequate amount of pH buffer and minerals is often required to obtain effective hydrogen production. From a techno-economic point of view, the cost of such synthetic growth media in dark fermentation can readily increase the cost of hydrogen production. Therefore, less expensive materials should be evaluated as possible supplementary nutrients for bioconversion processes (Sarma et al. 2013).

A possible solution could be the co-fermentation of several wastes sharing complementary characteristics. Co-digestion is a widely applied procedure for increasing biogas production, since it provides balanced amounts of nutrients and a required buffering capacity, thereby reducing the cost for pH control or nutritional supplements (Esposito et al. 2012). Moreover, animal waste is equally a source of digestive bacteria that are capable of hydrolysing organic matter (Zhu et al. 2009).

Considering the possibility of co-digestion, the use of livestock waste for fermentative hydrogen production should be reconsidered. It has been recently demonstrated that cattle manure addition to carbohydrate-rich substrates increases hydrogen production (Lateef et al. 2012; Perera and Nirmalakhandan 2010). Addition of manure provides macro and micronutrients (NH_3, P, K, metals) required for bacterial growth, a minimum buffering capacity necessary to maintain an optimal pH of 4.5–6 for dark fermentation, and also serves as co-substrate, due to its organic content (Tenca et al. 2011).

In addition, Marone et al. (2015) observed that the use of buffalo slurry in co-digestion with other substrates entailed several beneficial effects on the fermentative process, such as a reduction in the lag phase in batch fermentation and an increase in the degradation efficiency of other substrates.

Several authors also reported that manure is a source of efficient hydrogen-producing bacteria. Perera and Nirmalakhandan (2010) pointed to the feasibility of using swine manure as source of hydrogen-producing microorganisms, while Prakasham et al. (2009) evidenced the efficient fermentation of untreated mixed renewable agricultural waste using heat-treated buffalo dung compost as inoculum.

In this perspective, cattle manure can supply a renewable source of biomass for efficient hydrogen fermentation (Marone et al. 2015). Wu et al. (2010) observed a high H_2 yield reaching 1.5 mol $H_2.mol^{-1}$ glucose at mesophilic temperature (37 °C). This corresponds to a maximum hydrogen production rate of 2.25 L $H_2.L^{-1}.d^{-1}$ obtained from the fermentation of swine manure with glucose in an ASBR reactor, and using heat-treated anaerobic sludge pre-incubated with glucose as microbial inoculum (Wu et al. 2010). Perera and Nirmalakhandan (2010) also showed that cattle manure supplementation in sucrose fermentation increased the hydrogen yield by 10 %. They observed H_2 yields in the range of 3.8–4.7 mol $H_2.mol^{-1}$ of sucrose added in batch reactors at 25 °C only. Later, the same authors found hydrogen yields at ambient temperature reaching 5.3 mol $H_2.mol^{-1}$ of sucrose. Based on analysis of the net energy gain, they concluded in positive net energy gains when nonthermal pretreatment is applied (Perera and Nirmalakhandan 2011).

Nevertheless, in our opinion, it is more interesting to examine reports on co-digestion of animal manure and other types of waste/feedstock, avoiding the use of expensive materials such as pure sugars, even though animal manure has often been considered as a supplement enhancing hydrogen production, rather than as a substrate.

Dareioti et al. (2014) found many advantages in the addition of a minimal amount of liquid cow manure, corresponding to 5 % of substrate volume, for the co-digestion of cheese whey (44 % v/v) together with olive mill wastewater (55 % v/v). This resulted in a higher stability of the process, improvement of the yield and a higher feasibility of the process due to a minimisation of nutrient addition. Furthermore, additional pretreatment would no longer be required. At 37 °C, they observed production yields of 0.64 mol H_2.mol^{-1} of glucose equiv., corresponding to 24 ml.g^{-1} of VS in batch process (Dareioti et al. 2014). They also reported a remarkable yield of 0.54 mol H_2.mol^{-1} of glucose equiv., in continuous process (Dareioti and Kornaros 2014).

Hydrogen yields obtained from co-fermentation of manure with carbohydrate-rich substrates typically range from 24 to 126 ml H_2.g^{-1} of VS, depending on the complexity of the substrate, the consequent biomass biodegradability and the bioreactor operation mode. Several types of substrate have already been used in co-fermentation together with animal waste for producing H_2 such as cheese whey, waste milk, molasses, fruit and vegetable waste or coffee mucilage. The choice of a proper substrate depends on the seasonality of the waste production or on the availability of the "waste" at regional/local scales. Co-digestion thus offers a significant advantage for the creation of biorefineries, where multifunctional bioprocess combinations can be made in order to generate decentralised power from different types of waste. Here, the use of locally available biomass resources is expected to ensure additional benefits by supporting rural development.

Whatever the available substrate, the principal question to be dealt with is the way the different sources of waste are to be mixed in order to ensure the efficiency of the fermentative process. Marone et al. (2015) proposed a mixture design method to find the optimal assemblage of three substrate components (in %), including buffalo slurry, cheese whey and crude glycerol. They investigated the effect of the mixing ratio on biohydrogen yields. These authors found an optimal mixing ratio (VS%) of substrate at 66 % of cheese whey and 33 % buffalo slurry. More accurately, the model estimated the maximum hydrogen yield around 117 ml H_2.g^{-1} of VS_{added}, while the maximum experimental yield was 112 ± 22 ml H_2.g^{-1} of VS_{added}, in batch reactor operated at 37 °C (Marone et al. 2015). To date, the highest hydrogen production yield observed at mesophilic temperature (37 °C) is 170 ml H_2.g^{-1} of VS from the co-fermentation of buffalo manure (20%VS) with low protein cheese whey (70% VS) and crude glycerol (10% VS), after sterilisation of the substrates and inoculation with a mixed culture enriched from sediment from a coastal lake (Concetti et al. 2006). Meanwhile, the highest H_2 productivity in a semi-continuous reactor reached 3.3 L H_2.L^{-1}.d^{-1} (equivalent to 126 ml H_2.g^{-1} of VS added) from a substrate composed of fruit and vegetable waste (35% w/w) mixed with swine manure (65% w/w) at thermophilic temperature (55 °C) with a seed issued from a 10 L laboratory-scale hydrogen-producing reactor (Tenca et al. 2011).

Undoubtedly, the comparison of results generated from different studies is not always an easy task, mainly because of the numerous differences occurring between one study and the others, such as substrate concentration, its types and composition, reactor type and size, inoculum, use of additional nutrients, pretreatments, etc. It is therefore always crucial to consider the different operational conditions (Varrone et al. 2012).

Table 4.6 summarises some of the main parameters and best achieved results for hydrogen production from recent studies dealing with co-fermentation of different manure types with other waste substrates.

4.5 Future Prospects: Integration of Dark Fermentation Process into Environmental Biorefineries

Increasing hydrogen yields from dark fermentation processes remains one of the challenges for the development of large-scale hydrogen production. To date, no full-scale biohydrogen plant yet exists due to a low economic profitability and to the generally poor performances of the process. In particular, fermentative hydrogen-producing processes operated with mixed cultures present a hydrogen conversion primarily into methane through interspecies transfer which still needs to be prevented. Parameters such as short hydraulic

Table 4.6 Comparison of optimised conditions and results of recent studies dealing with biohydrogen production from co-digestion of manure with other substrates

Substrate/(composition %) *Substrate pretreatment*	Type of inoculum	Process conditions (g.l^{-1}/°C/pH)	Type of process	H_2 production (ml H_2/g_{VS}^{-1})	References
Buffalo manure/LPCW*/crude glycerol/(20/70/10) *Sterilisation*	Selected from lagoon sediments	8/37/6.7	Batch	170±3	Concetti et al. (2006)
Buffalo slurry/cheese whey/(33VS/67VS) *Sterilisation*	Selected from lagoon sediments	2.06%VS/37/6.5	Batch	117±22	Marone et al. (2015)
Liquid swine manure/beet molasses/(0.75/10 g/L sugar) *Sieved and boiled for 30 min*	Liquid swine manure	12/37/5.4	Sequencing batch	1.57 mol H_2.mol^{-1} sugar	Wu et al. (2013c)
Liquid cow manure/cheese whey/olive mill wastewater/(5/40/55) *No*	Anaerobic sludge from acclimatised substrates	63.52/37/6	Batch	23.8 (0.64 mol H_2.mol^{-1} gluc)	Dareioti et al. (2014)
Liquid cow manure/cheese whey/olive mill wastewater/(5/40/55) *No*	Anaerobic sludge from acclimatised substrates	84.69/37/6	Continuous	0.54 mol H_2.mol^{-1} gluc	Dareioti and Kornaros (2014)
Cattle manure/slaughterhouse risk material/(90 (wt dry matter)/10 (wt dry matter)) *Heated at 90 °C for 3 h*	Cattle manure	40/55/7.1	Batch	33	Gilroyed et al. (2010)
Cow manure/milk waste/(30/70) *Sieved and heat-treated/–*	Cow manure	40/55/6.5	Batch	59.5	Lateef et al. (2012)
Swine manure/fruit-vegetable waste/(35(w/w)/65(w/w)) *Sieved/shredded in a blender*	Seed from a 10 L H_2-producing reactor	20/55/5.45	Semi-continuous	126	Tenca et al. (2011)

*low protein cheese whey

retention times (HRT of few hours), low pH (5.5–6) and high organic loading rates are typically optimised to prevent methanogenesis and even more so when considering industrial-scale development. However, a short HRT cannot be applied when using complex organic materials such as agricultural waste and manure, as the hydrolysis step for complex organic materials is generally poor and relatively slow (Liu et al. 2008). As reviewed by Hallenbeck and Ghosh (2009), various techniques have been investigated such as the alteration of reactor configurations, the use of mixed cultures, metabolic engineering of existing pathways, modelling and optimisation in order to improve both the technical efficiency (based on hydrogen yield) and economic efficiency (based on hydrogen production rate) of the biological process.

The major limitation of hydrogen production by fermentative bacteria is the maximum molar yield of biomass conversion. Owing to biochemical and thermodynamic limitations, the maximum theoretical number of moles of hydrogen that can be generated per mole of glucose is four. This can be obtained by glucose oxidation to 2 mol of acetate. Although complete oxidation to carbon dioxide and hydrogen should theoretically produce 12 mol of hydrogen per mole of hexose, such a yield has never been reached in in vivo biological systems due to thermodynamic and physiological constraints. Present yields are even lower than the maximum of 4 mol of hydrogen and usually range from 1 to 2.5 mol hydrogen per mole hexose. On the whole, hydrogen fermentative production only results in a partial oxidation of the organic substrate. Indeed, the main carbon and hydrogen contents (theoretically two-third) of the substrate are converted to microbial metabolic unused by-products, which represent both a disposal burden and a waste of energy. Environmental sustainability and economic feasibility can only be reached if H_2 yields on dissolved organic material can approach 60–80 %. Therefore, dark biohydrogen production is only likely to be industrially viable if fermentation bioprocesses could be integrated into a combination of processes that are capable of utilising metabolic end products (Hawkes et al. 2007).

This involves the development of a new approach where diversified biomass is considered as part of a concept of environmental "biorefinery": such a system combines traditional elements from environmental biotechnology in terms of waste stream treatment with industrial biotechnology purposes in terms of product maximisation.

Studies dealing with the integration of fermentative hydrogen-producing bioprocesses into environmental biorefinery usually investigate a two-step process. The fermentation end products such as volatile fatty acids are either converted to methane by anaerobic digestion or hydrogen and carbon dioxide in a phototrophic process. Extra amounts of H_2 are also generated in bioelectrochemical systems (BES), in which hydrogen is produced at the cathode after application of a low voltage to the circuit, i.e. 0.2 V instead of 2 V (vs. standard hydrogen electrode, SHE) for classical electrolysis. Alternatively products from the fermentation process can be used to produce electricity in microbial fuel cells (MFC).

Therefore, biological H_2 production processes of dark fermentation can be considered as a primary step in a larger bioenergy or biochemical production concept, as proposed in Fig. 4.1. Although all bioprocesses could be coupled, any partial process design solutions are possible with a minimum of two-stage processes including one step of H_2 production. From agricultural waste, the two-stage process H_2+CH_4 is the most likely process to be rapidly exploited since present technologies are almost ready to be upscaled to the industrial scale, while several two-step anaerobic processes have already been applied worldwide.

4.5.1 Coupling Dark Fermentation and Bioelectrochemically Assisted Hydrogen Production

Among other biotechnologies that could utilise metabolic by-products generated by dark fermentation (DF) processes, microbial electrolysis cell (MEC) is a very promising technology based on the coupling between H_2 production and organic material degradation (Rozendal et al. 2006). MEC has recently emerged as a technology where bacteria referred to as exoelectrogens oxidise a substrate and release electrons to the anode (Logan and Regan 2006). A low voltage (0.2–0.6 V vs. SHE) applied to the circuit allows H_2 production at the cathode through the reduction of protons. The energy required for this operation is theoretically ten times lower than the 1.8–2 V (vs. SHE) used for H_2 production via traditional water electrolysis.

However, lignocellulosic biomass cannot be directly utilised by microorganisms in MECs for hydrogen generation. Biomass first has to be converted into monosaccharides or other low-molecular-weight compounds (Kadier et al. 2014). Since the effluents of DF are suitable to be utilised as influents in MEC, combining DF with MEC in a cascade two-step process can result in an optimal exploitation of agricultural waste, which can concurrently maximise the energy recovery and effluent depollution.

Up to now, only a few studies have dealt with H_2 production in DF processes coupled to MEC. Mostly, single-carbon sources have been used as model substrates (Hawkes et al. 2007; Lu et al. 2009). Although relatively high H_2 yields

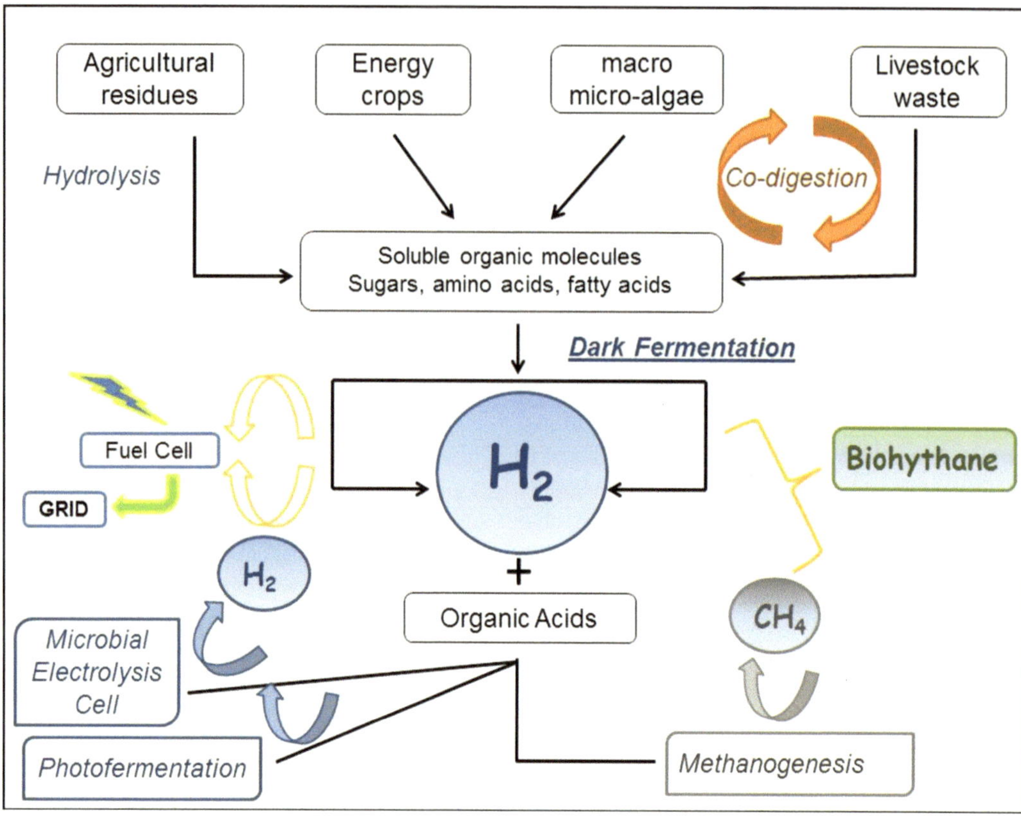

Fig. 4.1 Possible bioprocess associations in biohydrogen-based refineries

have been reported, they do not reflect a genuine coupling of bioprocesses, neither the productivity that can be expected from organic waste. Moreover, the contamination of the processes by methanogenic bacteria has been rather poorly documented for long-term running of the coupling. Indeed although the production of methane may occur spontaneously at the potential applied for hydrogen production, methane can be also produced through hydrogen uptake by hydrogenotrophic archaea. Nonetheless, more complex (fermentable) substrates, such as cellulose (Ullery and Logan 2015; Wang et al. 2011), have also been investigated. So far, only Lalaurette et al. (2009) and, more recently, Li et al. (2014) investigated a two-stage dark fermentation and electrohydrogenesis (MEC) system from lignocellulosic waste, i.e. acid-pretreated corn stover and stalks, respectively. Compared to single-stage dark fermentation, higher H_2 yields were obtained for the two-stage system (Lalaurette et al. 2009). Interestingly, the first fermentation produced 1.67 mol $H_2.mol^{-1}$ of glucose at a rate of 0.25 L $H_2.L^{-1}.d^{-1}$, using a pure culture of *Clostridium thermocellum*. The fermentation effluent primarily consisted of ethanol, acetic, lactic, succinic and formic acids. In the second MEC stage, the rate of hydrogen production was 1.00 ± 0.19 L $H_2.L^{-1}.d^{-1}$, with an energy efficiency based on electricity needed for the MEC of about $230 \pm 50\,\%$ (on the basis of the electrical energy brought to the process, therefore its values can be higher than 100 %). However, the authors also reported a high CH_4 contamination. Apparently, the duration of the batch cycle (3–4 days) was the main factor affecting cell performances. These performances were reduced compared to the ones obtained for a single synthetic substrate which required batch cycles of only 1–2 days.

By combining dark fermentation and MECs, Li et al. (2014) observed an overall H_2 yield of

387 ml $H_2.g^{-1}$ of cornstalks which was nearly three times higher than for dark fermentation alone. In the dark fermentation stage, a H_2 yield of 130 ml $H_2.g^{-1}$ of cornstalk and an average H_2 production rate of 1.73 m^3 $H_2.m^{-3}.d^{-1}$ were obtained with 20 $g.L^{-1}$ of cornstalk. Meanwhile, a H_2 yield of 257.3 ml $H_2.g^{-1}$ of cornstalk and a hydrogen production rate of 3.43±0.12 m^3 $H_2/m^{-3}.d^{-1}$ were reached in the MEC stage, with an effluent COD concentration of about 4 $g.L^{-1}$ and under an applied voltage of 0.8 V (vs. normal hydrogen electrode, NHE).

A comprehensive route for biorefinery was recently proposed by inserting a hydrogen-producing process fed with crude glycerol as by-product of the biodiesel industry, coupled to electricity production using a microbial fuel cell (MFC) (Chookaew et al. 2014). In this study, a maximum H_2 rate of 6.9 ml $H_2.L^{-1}.h^{-1}$ and a yield of 0.55 mol $H_2.mol^{-1}$ of glycerol were achieved. About 20 % of organic matter was removed in the first stage and the overall yield reached 106 ml $H_2.g^{-1}$ of COD with a coulombic efficiency (CE) of 24 % from the diluted influent in the two-chamber MEC.

Although these preliminary studies seem very promising, bench-scale reactors can still be upgraded to develop economically feasible applications. For this, many hurdles that limit MEC performances still exist and need to be overcome. These include the cost of reactor material (membrane, electrodes, etc.), the reactor design or the production of methane which occurs particularly when single-chamber systems are used. Although great efforts have been made to develop this technology and successful and promising results have been obtained, laboratory-scale equipment coupling both technologies is still in its infancy, with reported experiences limited to a sub-litre scale (Clauwaert et al. 2008).

4.5.2 Integrated Fermentative and Photofermentative Hydrogen Production

As an alternative, purple non-sulphur bacteria can also utilise the residual volatile fatty acids (VFAs) from the dark fermentation outlet effluent as electron donors. This process, called photofermentation, is catalysed by nitrogenases when bacteria are grown under nitrogen starvation (Das and Veziroä 2001). Purple non-sulphur bacteria exhibit a number of traits that make them attractive for biologically producing hydrogen from biomass, including high hydrogen yields, no oxygen production and the ability to utilise a wide range of the light spectrum.

All these features have led to an obvious combination between the dark fermentation process and a bacterial photosynthetic hydrogen production process converting VFA to H_2 and CO_2. The theoretical yield of the photofermentation process is about 8 mol H_2 per mol of acetate. Thus, theoretically, a combination between dark and light fermentation should be capable of producing the maximum possible yield of 12 mol of hydrogen per mole of hexose. Numerous experiments have been carried out on photofermentative H_2 production either with different types of waste as a direct substrate or in series following a dark fermentation reactor (Claassen et al. 2005; Lee 2002). However, coupling dark and photofermentation systems only seem possible for substrates that are characterised by a high C/N ratio (Gómez et al. 2011). Furthermore, phototrophic H_2 yields obtained with dark fermentation effluents could be adversely affected by high butyrate concentrations, greater than 2500 mg $COD_{equiv}.L^{-1}$ (Chen et al. 2008).

Although certain constraints still need to be overcome, a sequential process with dark fermentation and photofermentation stages does allow a significant increase in H_2 yields simultaneously with the processing of metabolic end products for the production of a high-quality outlet stream (Claassen et al. 2009).

To increase the potential biomass conversion, Zhu et al. (2010) investigated the fermentation of an enzymatically pretreated corn stover and then used the effluent for photofermentation by *Rhodobacter sphaeroides* ZX-5. They obtained a hydrogen production more than twofold greater than that from a synthetic medium when operated with a similar range of organic acid concentrations. These authors observed a synergy among organic acids facilitating the cell growth and hydrogen production in the photofermentative stage, although some organic acids did not directly support the hydrogen production. Similar results were obtained by Yang et al. (2010) using acidic-pretreated corncobs. The maximum bio-hydrogen yield and rate from corncobs in the first dark fermentative step were 120±5 ml $H_2.g^{-1}$ of corncob and 150 ml $H_2.L^{-1}.h^{-1}$, respectively. In the second step, a hydrogen yield of 714±44 ml $H_2.g^{-1}$ of COD was obtained from the effluent of dark fermentation by photosynthetic bacteria. In total, a 90 % COD removal from the two-stage process was observed with a consumption of 98.6±0.1 % of acetate, 99.3±0.1 % of butyrate and 100 % of propionate (Yang et al. 2010).

Nonetheless, the cost of the input light energy for these two-stage processes must be considered, especially when calculating the overall energy conversion efficiency. Su et al. (2009) studied the combination of dark and photofermentation from cassava starch by testing different raw cassava starch concentrations (10–25 $g.L^{-1}$). The maximum hydrogen yield dramatically increased from 240.4 ml $H_2.g^{-1}$ of starch with the sole dark fermentation to 402.3 ml $H_2.g^{-1}$ of starch in the combined dark and photofermentation. The associated energy conversion efficiency increased from 17.5–18.6 % to 26.4–27.1 % when considering the heat value of cassava starch only as input energy. Taking into account the additional input of light energy for photofermentation, the whole energy conversion efficiency only reached 4.46–6.04 % (Su et al. 2009).

Urbaniec et al. (2014) designed the concept of an autonomous hydrogen factory applying fermentative and photofermentative conversion of potato residues with an energy supply from combined heat and power generation in a plant based on solid oxide fuel cells (SOFC). The production plant comprised a pretreatment unit for starchy raw material, a bioreactor for dark fermentation, a photobioreactor for photofermentation and several gas upgrading and compression units. The plant was supplied with the necessary heat and power generated from the power plant itself. The demand for process heat was covered by steam generation in the boiler and oxidiser, and the power demand was covered by electricity generation in the fuel cells. Assuming that the factory produced low-purity hydrogen gas, the idea of covering the entire energy demand by a part of the energy from self-produced hydrogen does not seem practical especially when considering that the long-term operation of a fuel cell-based CHP plant supplied with a mixture of H_2 and CO_2 can become problematic. Indeed, the ratio of energy demand to the energy value of the stream of produced hydrogen only ranged from 44 to 50 % depending of the mixture of hydrogen and carbon dioxide obtained by fermentation. Nevertheless, the authors proposed a different scenario in which H_2 would be produced by an add-on factory located in an industrial site where waste heat from another factory would be available for use in the fermentation and gas upgrading stages of the production process. By supplying waste heat at a temperature level of about 80 °C, the total energy demand for hydrogen production could be covered by only 10 % of the energy value of produced hydrogen.

The main disadvantage of light-dependent processes is that it requires a more complex design of reactors, allowing for an appropriate proportioning between reactor surface area and volume to be maintained when scaling-up (Gómez et al. 2011). In addition, when experiments are to be carried out under natural sunlight, the processes are obviously subjected to light and dark (diel) cycles. This aspect might cause a switch towards another type of fermentative metabolism, leading to the consumption of substrate with no hydrogen production, either during dark periods or in the dark zones of the reactor (Koku et al. 2002). Photofermentative hydrogen production using current state of the art organisms and technology has, so far, been considered economically unrealistic by many authors (Levin 2004).

4.5.3 Integrated Fermentative Hydrogen Production and Methane Production

It is clear that among the two-stage process technologies, the combination of hydrogen and methane production in a two-stage anaerobic digestion process seems to be the most probable and feasible possibility in the near future, since two-stage processes have already been largely used for methane production at industrial scales (Vollmer and Scholz 1985). Moreover, the mixture of hydrogen and methane, so-called hythane or biohythane if biologically produced, has already proved to act as a more efficient fuel for natural gas vehicles than methane when using an internal combustion engine. Moreover, the use of hythane as fuel is more environmentally friendly than methane since it reduces the emission of hydrocarbon pollutants, carbon monoxide (CO) and nitrogen oxides (NOx) (Alavandi and Agrawal 2008; Gattrell et al. 2007).

When considering the treatment of different kinds of organic waste, the first acidogenic stage of a two-stage anaerobic digestion process has been usually employed to enhance the production of organic acids for further methane production in a second methanogenic stage (Demirel and Yenigün 2002). This kind of two-stage process has been reported to achieve enhanced stability, robustness and higher loading capacities for the methanogenesis process when compared to a traditional one-stage process (Ke et al. 2005). In a literature review on two-stage hydrogen-methane processes, Hawkes et al. (2007) observed that most publications reported a higher total efficiency on waste treatment and energy recovery than the traditional one-stage process. Additionally, this type of process has a further advantage compared to other two-stage processes, as the minerals and nutrients discharged from the anaerobic digestion of organic wastes can be reused as fertilisers in agricultural production (Angelidaki and Ellegaard 2003).

Nevertheless, only few studies have been carried out on agricultural waste for continuous production of both biohydrogen and biomethane fermentation, mainly because of the technical difficulties in operating lab-scale reactors with the whole fraction of cellulosic biomass (Liu et al. 2013). To overcome these obstacles, many studies are now focused on the production of hydrogen and methane from hydrolysate or the liquid fraction of cellulosic biomass only since these substrates can allow for continuous mode and lab-scale processes.

Antonopoulou et al. (2008) showed a two-stage hydrogen and methane system functioning continuously at a mesophilic temperature on sweet sorghum liquid extract obtained by hydrothermal hydrolysis. They observed H_2 yields of 10.4 L H_2.kg^{-1} of VS and 29 L CH_4.kg^{-1} of VS. In addition, the methane fermentation of the remaining sorghum solid extract led to an extra methane yield of 78 L.kg^{-1} of VS of sweet sorghum in batch reactor. Hemicellulose-rich liquid from wheat straw after ethanol fermentation (Kaparaju et al. 2009) and hydrolysate of hydrothermal pretreated wheat straw (Kongjan et al. 2011) were also used for efficient continuous production of hydrogen and methane under thermophilic conditions.

A biorefinery framework was proposed by Kaparaju et al. (2009), combining the production of bioethanol, biohydrogen and biogas from wheat straw, pretreated hydrothermally and hydrolysised with enzymes. The yields of hydrogen and methane from the hemicellulose-rich liquid in a CSTR reactor were 178 L H_2.kg^{-1} of sugars and 324 L CH_4.kg^{-1} of VS, respectively (Kaparaju et al. 2009). The gas yields from the hydrolysate of wheat straw in an upflow anaerobic sludge blanket (UASB) reactor were 89 L H_2.kg^{-1} of VS and 307 L CH_4.kg^{-1} of VS, respectively (Kongjan et al. 2011), which corresponded to a total TS reduction of 81 % and a total energy recovery of 87.5 %. This was 80 % higher than for the single hydrogen process.

To overcome the limits of processing solids in two-stage lab-scale systems, many studies have been carried out in batch mode. Monlau et al. (2015) proposed the combination of batch dark fermentation and continuous anaerobic digestion of sunflower stalks. Substrate pretreatment (4 g NaOH for 100 g of TS, 55 °C, 24 h) was optimised for enhancing methane production (195 L CH_4.kg^{-1} of VS with pretreatment versus 151 L CH_4.kg^{-1} of VS without pretreatment), but hydrogen yields were low (7 L H_2.kg^{-1} of VS, with or without alkali pretreatment). Lu et al. (2009) investigated the effects of the design of a microbial consortium and the operating temperature on the performances of batch hydrogen and methane production from cornstalk. *Enterobacter aerogenes* and *Clostridium paraputrificum* were introduced into the hydrogen stage to enhance the natural sludge community, and *C. paraputrificum* exhibited a very efficient capability of H_2 production. These authors operated a three-stage process consisting of the improved hydrolysis and hydrogen and methane fermentation. They found that the introduction of cellulose-degrading *Clostridium thermocellum* under thermophilic (55 °C) and acidic (pH 5.0) conditions enhanced the yield of soluble sugar release (0.482 $kg.kg^{-1}$ of cornstalk). Highest yields of hydrogen and methane were obtained with 64 and 115 $L.kg^{-1}$ of TS, respectively, which corresponded to a total energy recovery of 54 % from cornstalk.

Interestingly, alkali-pretreated water hyacinth (leaves, stems, roots) was also used as substrate for mesophilic hydrogen and methane production in batch experiments by Cheng et al. (2010). Highest yields were observed from leaves which had the highest carbohydrate and protein contents with 51.7 and 143.4 $L.kg^{-1}$ of VS for hydrogen and methane, respectively. The same group (Cheng et al. 2011b) achieved higher yields of hydrogen and methane by fermenting *Arthrospira maxima* biomass. To improve the fermentation performance, they performed microbial domestication and enzymatic hydrolysis with glucoamylase.

Lakaniemi et al. (2011) tested reed canary grass, a substrate with a high content of lignocellulosic material, for hydrogen and methane production at 35 °C. They found that acid hydrolysis was necessary for hydrogen, but not for methane production.

Wieczorek et al. (2014) investigated hydrogen and methane production in a two-stage process from microalgae (*Chlorella vulgaris*). Hydrogen production from *C. vulgaris* ranged from 1.75 ± 1.5 to 19 ± 3 ml $H_2.g^{-1}$ of VS at different concentrations with no enzymatic pretreatment. In this case, enzymatic pretreatment of the microalgae favoured the process with a sevenfold increase in H_2 production yields (19 ± 3–135 ± 3 ml $H_2.g^{-1}$ of VS) and the methane yields increased from 245 ± 2 to 414 ± 2 ml $CH_4.g^{-1}$ of VS.

Other authors (Yang et al. 2011) obtained 46 L $H_2.kg^{-1}$ of VS and 394 L $CH_4.kg^{-1}$ of VS from lipid-extracted microalgal biomass residues, which corresponded to a total energy recovery of 65 %, i.e. 14 % higher than the single-stage methane process.

Certain projects are developed worldwide in order to upscale two-step anaerobic processes. The University of Glamorgan South Wales, United Kingdom, associates a 1.25 m^3 fermenter to a 10 m^3 anaerobic digester to produce hydrogen and methane through the degradation of wheat feed pellets. Hydrogen was produced with a productivity of 0,024 L $H_2.L^{-1}.h^{-1}$ and a yield of 10 ml $H_2.g^{-1}$ of VS, representing 15–20 % of the total biogas (Massanet-Nicolau et al. 2013). With a similar objective, the University of Verona, Italy, succeeded in producing biohythane at 13 % H_2 by digesting food waste (Cavinato et al. 2012). They demonstrated that the process was stable over more than 90 days with a biohythane flow of 1.6 m^3 (H_2/CH_4).j^{-1} (excluding CO_2).

Coupling biological processes therefore appears to favour the increase of overall yields of biomass conversion. This involves the operation of two-phase systems that physically separate the acidogenic and methanogenic steps into two different reactors in series (Azbar and Speece 2001). This configuration not only enhances the overall process performances, but two-phase processes are also less sensitive to product inhibition as the acidogenic step is more robust towards substrate variability in organic load and composition. Nevertheless, process adjustment of the two-phase systems still needs to be particularly optimised. In particular the effluent issued from the first phase may require pH adjustments prior to the second phase.

4.6 Conclusion

Hydrogen production from agricultural waste by dark fermentation has a promising future since the hydrogen potential and the amounts of feedstock are substantial. Improvement of H_2 yields concerns biomass pretreatment technologies. These should increase substrate biodegradability and, as a consequence, biomass conversion and hydrogen production. The selection of pretreatment method operating conditions used for dark fermentation mainly depends on substrate composition. The simultaneous use of different pretreatment methods on biomass has led to large improvements in biodegradability along with higher hydrogen production.

Substrates with lower carbohydrate contents can also be revalued in dark fermentation as nutrient supplements or in co-fermentation with other biomasses.

In order to reach full exploitation of the residual biomass, dark fermentation processes should be associated to other biological energy production processes such as microbial electrolysis cell, biophotolysis, photofermentation and anaerobic digestion. Biorefinery is a concept which could lead to the industrial development of biological hydrogen production. At present it represents one of the cleanest solutions which associate waste treatment and non polluting energy production. The challenge now lies in the development of this clean technology at an industrial scale.

Acknowledgements A. Marone's postdoctoral program was funded by the Marie Curie Intra European Fellowship WASTE2BIOHY (FP7-MC- IEF-326974) under the 7th Framework Programme of the European Community.

References

Akutsu Y, Li Y, Tandukar M, Kubota K, Harada H (2008) Effects of seed sludge on fermentative characteristics and microbial community structures in thermophilic hydrogen fermentation of starch. Int J Hydrog Energy 33:6541–6548. doi:10.1016/j.ijhydene.2008.08.038

Akutsu Y, Li Y-Y, Harada H, Yu H-Q (2009) Effects of temperature and substrate concentration on biological hydrogen production from starch. Int J Hydrog Energy 34:2558–2566. doi:10.1016/j.ijhydene.2009.01.048

Alavandi SK, Agrawal AK (2008) Experimental study of combustion of hydrogen-syngas/methane fuel mixtures in a porous burner. Int J Hydrog Energy 33:1407–1415. doi:10.1016/j.ijhydene.2007.12.005

Angelidaki I, Ellegaard L (2003) Codigestion of manure and organic wastes in centralized biogas plants: status and future trends. Appl Biochem Biotechnol 109:95–105. doi:10.1385/ABAB:109:1–3:95

Antonopoulou G, Gavala HN, Skiadas IV, Angelopoulos K, Lyberatos G (2008) Biofuels generation from sweet sorghum: fermentative hydrogen production and anaerobic digestion of the remaining biomass. Bioresour Technol 99:110–119. doi:10.1016/j.biortech.2006.11.048

Azbar N, Speece RE (2001) Two-phase, two-stage, and single-stage anaerobic process comparison. J Environ Eng 127:240–248. doi:10.1061/(ASCE)0733-9372(2001)127:3(240)

Bruni E (2010) Improved anaerobic digestion of energy crops and agricultural residues. Ph.D. thesis, Technical University of Denmark. http://orbit.dtu.dk/fedora/objects/orbit:82849/datastreams/file_5116515/content

Cavinato C, Giuliano A, Bolzonella D, Pavan P, Cecchi F (2012) Bio-hythane production from food waste by dark fermentation coupled with anaerobic digestion process: a long-term pilot scale experience. Int J Hydrog Energy 37:11549–11555, doi: http://dx.doi.org/10.1016/j.ijhydene.2012.03.065

Chen C-Y, Lu W-B, Liu C-H, Chang J-S (2008) Improved phototrophic H2 production with Rhodopseudomonas palustris WP3-5 using acetate and butyrate as dual carbon substrates. Bioresour Technol 99:3609–3616. doi:10.1016/j.biortech.2007.07.037

Chen C-C, Chuang Y-S, Lin C-Y, Lay C-H, Sen B (2012) Thermophilic dark fermentation of untreated rice straw using mixed cultures for hydrogen production. Int J Hydrog Energy 37:15540–15546. doi:10.1016/j.ijhydene.2012.01.036

Cheng J, Xie B, Zhou J, Song W, Cen K (2010) Cogeneration of H2 and CH4 from water hyacinth by two-step anaerobic fermentation. Int J Hydrog Energy 35:3029–3035. doi:10.1016/j.ijhydene.2009.07.012

Cheng J, Su H, Zhou J, Song W, Cen K (2011a) Microwave-assisted alkali pretreatment of rice straw to promote enzymatic hydrolysis and hydrogen production in dark- and photo-fermentation. Int J Hydrog Energy 36:2093–2101. doi:10.1016/j.ijhydene.2010.11.021

Cheng J, Zhang M, Song W, Xia A, Zhou J, Cen K (2011b) Cogeneration of hydrogen and methane from Arthrospira maxima biomass with bacteria domestication and enzymatic hydrolysis. Int J Hydrog Energy 36:1474–1481. doi:10.1016/j.ijhydene.2010.11.009

Chong M-L, Sabaratnam V, Shirai Y, Hassan MA (2009) Biohydrogen production from biomass and industrial wastes by dark fermentation. Int J Hydrog Energy 34:3277–3287. doi:10.1016/j.ijhydene.2009.02.010

Chookaew T, Prasertsan P, Ren ZJ (2014) Two-stage conversion of crude glycerol to energy using dark fermentation linked with microbial fuel cell or microbial electrolysis cell. N Biotechnol 31:179–184. doi:10.1016/j.nbt.2013.12.004

Chu Y, Wei Y, Yuan X, Shi X (2011) Bioconversion of wheat stalk to hydrogen by dark fermentation: effect of different mixed microflora on hydrogen yield and cellulose solubilisation. Bioresour Technol 102:3805–3809. doi:10.1016/j.biortech.2010.11.092

Claassen PAM, Vrije T De, Budde MAW (2004) Biological hydrogen production from sweet sorghum by thermophilic bacteria. 2nd world conference on biomass for energy, Industry and Climate Protection, pp 1522–1525

Claassen AM, Budde MAW, Van Niel EWJ, De Vrije T (2005) Utilization of biomass for hydrogen fermentation. In: Lens P, Westermann P, Haberbauer M, Moreno A (eds) Biofuels for fuel cells: renewable energy from biomass fermentation. IWA Publisihing, London, pp 221–230

Claassen PAM, De Vrije T, Urbaniec K (2009) Non-thermal production of pure hydrogen from biomass: HYVOLUTION. Chem Eng Trans 18:333–338. doi:10.3303/CET0918053

Clauwaert P, Aelterman P, Pham TH, De Schamphelaire L, Carballa M, Rabaey K, Verstraete W (2008) Minimizing losses in bio-electrochemical systems: the road to applications. Appl Microbiol Biotechnol 79:901–913. doi:10.1007/s00253-008-1522-2

Concetti S, Chiariotti A, Patriarca C, Marone A, Contò G, Calì M, Signorini A (2006) Biohydrogen production from buffalo manure codigested with agroindustrial by-products in an anaerobic reactor.

Costanza R, D'Arge R, De Groot R, Farber S, Grasso M, Hannon B, Limburg K, Naeem S, O'Neill RV, Paruelo J, Raskin RG, Sutton P, Van Den Belt M (1997) The value of the world's ecosystem services and natural capital. Nature 387:253–260. doi:10.1038/387253a0

Cui M, Shen J (2012) Effects of acid and alkaline pretreatments on the biohydrogen production from grass by anaerobic dark fermentation. Int J Hydrog Energy 37:1120–1124. doi:10.1016/j.ijhydene.2011.02.078

Dareioti MA, Kornaros M (2014) Effect of hydraulic retention time (HRT) on the anaerobic co-digestion of agro-industrial wastes in a two-stage CSTR system. Bioresour Technol 167:407–415. doi:10.1016/j.biortech.2014.06.045

Dareioti MA, Vavouraki AI, Kornaros M (2014) Effect of pH on the anaerobic acidogenesis of agroindustrial wastewaters for maximization of bio-hydrogen production: a lab-scale evaluation using batch tests. Bioresour Technol 162:218–227. doi:10.1016/j.biortech.2014.03.149

Das D (2009) Advances in biohydrogen production processes: an approach towards commercialization. Int J Hydrog Energy 34:7349–7357. doi:10.1016/j.ijhydene.2008.12.013

Das D, Veziroä TN (2001) Hydrogen production by biological processes : a survey of literature. Int J Hydrog Energy 26:13–28

Datar R, Huang J, Maness P, Mohagheghi A, Czernik S, Chornet E (2007) Hydrogen production from the fermentation of corn stover biomass pretreated with a steam-explosion process. Int J Hydrog Energy 32:932–939. doi:10.1016/j.ijhydene.2006.09.027

De Vrije T, Claassen PAM (2003) Dark hydrogen fermentations. In: Reith JH, Wijffels RH, Barten H (eds) Bio-methane & bio-hydrogen, Status and perspectives of biological methane and hydrogen production. Dutch Biological Hydrogen Fundation, The Hague, pp 103–123

Demirel B, Yenigün O (2002) Two-phase anaerobic digestion processes: a review. J Chem Technol Biotechnol 77:743–755. doi:10.1002/jctb.630

Dong L, Zhenhong Y, Yongming S, Xiaoying K, Yu Z (2009) Hydrogen production characteristics of the organic fraction of municipal solid wastes by anaerobic mixed culture fermentation. Int J Hydrog Energy 34:812–820. doi:10.1016/j.ijhydene.2008.11.031

Erica M (2012) Anaerobic digestion. Fuel cells waste-to-energy chain. Springer, London, pp 47–63

Esposito G, Frunzo L, Giordano A, Liotta F, Panico A, Pirozzi F (2012) Anaerobic co-digestion of organic wastes. Rev Environ Sci Biol Technol. doi:10.1007/s11157-012-9277-8

Fan Y-T, Zhang Y-H, Zhang S-F, Hou H-W, Ren B-Z (2006) Efficient conversion of wheat straw wastes into biohydrogen gas by cow dung compost. Bioresour Technol 97:500–505. doi:10.1016/j.biortech.2005.02.049

FAO (2011) Global food losses and food waste – extent, causes and prevention. FAO, Rome

FAO (2013) FAO statistical yearbook 2013 : world food and agriculture. FAO, Rome

Gattrell M, Gupta N, Co A (2007) Electrochemical reduction of CO2 to hydrocarbons to store renewable electrical energy and upgrade biogas. Energy Convers Manag 48:1255–1265. doi:10.1016/j.enconman.2006.09.019

Gilroyed BH, Li C, Hao X, Chu A, McAllister TA (2010) Biohydrogen production from specified risk materials co-digested with cattle manure. Int J Hydrog Energy 35:1099–1105. doi:10.1016/j.ijhydene.2009.11.072

Gómez X, Fernández C, Fierro J, Sánchez ME, Escapa A, Morán A (2011) Hydrogen production: two stage processes for waste degradation. Bioresour Technol 102:8621–8627. doi:10.1016/j.biortech.2011.03.055

Guo XM, Trably E, Latrille E, Carrère H, Steyer J-P (2010) Hydrogen production from agricultural waste by dark fermentation: a review. Int J Hydrog Energy 35:10660–10673. doi:10.1016/j.ijhydene.2010.03.008

Guo XM, Trably E, Latrille E, Carrere H, Steyer J-P (2014) Predictive and explicative models of fermentative hydrogen production from solid organic waste: role of butyrate and lactate pathways. Int J Hydrog Energy 39:7476–7485. doi:10.1016/j.ijhydene.2013.08.079

Hallenbeck PC (2009) Fermentative hydrogen production: principles, progress, and prognosis. Int J Hydrog Energy 34:7379–7389. doi:10.1016/j.ijhydene.2008.12.080

Hallenbeck PC, Ghosh D (2009) Advances in fermentative biohydrogen production: the way forward? Trends Biotechnol 27:287–297. doi:10.1016/j.tibtech.2009.02.004

Han H, Wei L, Liu B, Yang H, Shen J (2012) Optimization of biohydrogen production from soybean straw using anaerobic mixed bacteria. Int J Hydrog Energy 37:13200–13208. doi:10.1016/j.ijhydene.2012.03.073

Hawkes F, Hussy I, Kyazze G, Dinsdale R, Hawkes D (2007) Continuous dark fermentative hydrogen production by mesophilic microflora: principles and progress. Int J Hydrog Energy 32:172–184. doi:10.1016/j.ijhydene.2006.08.014

Hubbard RK, Lowrance RR (1998) Management of dairy cattle manure. Agric. Uses Munic. Anim. Ind. Byprod. pp 91–102

Ivanova G, Rákhely G, Kovács KL (2009) Thermophilic biohydrogen production from energy plants by

Caldicellulosiruptor saccharolyticus and comparison with related studies. Int J Hydrog Energy 34:3659–3670. doi:10.1016/j.ijhydene.2009.02.082

Kadier A, Simayi Y, Sahaid M, Abdeshahian P, Abdul A (2014) A review of the substrates used in microbial electrolysis cells (MECs) for producing sustainable and clean hydrogen gas. Renew Energy 71:466–472. doi:10.1016/j.renene.2014.05.052

Kaparaju P, Serrano M, Thomsen AB, Kongjan P, Angelidaki I (2009) Bioethanol, biohydrogen and biogas production from wheat straw in a biorefinery concept. Bioresour Technol 100:2562–2568. doi:10.1016/j.biortech.2008.11.011

Ke S, Shi Z, Fang HHP (2005) Applications of two-phase anaerobic degradation in industrial wastewater treatment Shuizhou Ke * and Zhou Shi. Int J Environ Pollut 23:65–80

Koku H, Eroglu I, Gündpuz U, Yücel M, Türker L (2002) Aspects of the metabolism of hydrogen production by Rhodobacter sphaeroides. Int J Hydrog Energy 27:1315–1329. doi:10.1016/S0360-3199(02)00127-1

Kongjan P, O-Thong S, Angelidaki I (2011) Performance and microbial community analysis of two-stage process with extreme thermophilic hydrogen and thermophilic methane production from hydrolysate in UASB reactors. Bioresour Technol 102:4028–4035. doi:10.1016/j.biortech.2010.12.009

Kotsopoulos TA, Fotidis IA, Tsolakis N, Martzopoulos GG (2009) Biohydrogen production from pig slurry in a CSTR reactor system with mixed cultures under hyper-thermophilic temperature (70 °C). Biomass Bioenergy 33:1168–1174. doi:10.1016/j.biombioe.2009.05.001

Kyazze G, Dinsdale R, Hawkes FR, Guwy AJ, Premier GC, Donnison IS (2008) Direct fermentation of fodder maize, chicory fructans and perennial ryegrass to hydrogen using mixed microflora. Bioresour Technol 99:8833–8839. doi:10.1016/j.biortech.2008.04.047

Lakaniemi A-M, Koskinen PEP, Nevatalo LM, Kaksonen AH, Puhakka JA (2011) Biogenic hydrogen and methane production from reed canary grass. Biomass Bioenergy 35:773–780. doi:10.1016/j.biombioe.2010.10.032

Lalaurette E, Thammannagowda S, Mohagheghi A, Maness P-C, Logan BE (2009) Hydrogen production from cellulose in a two-stage process combining fermentation and electrohydrogenesis. Int J Hydrog Energy 34:6201–6210. doi:10.1016/j.ijhydene.2009.05.112

Lateef SA, Beneragama N, Yamashiro T, Iwasaki M, Ying C, Umetsu K (2012) Biohydrogen production from co-digestion of cow manure and waste milk under thermophilic temperature. Bioresour Technol 110:251–257. doi:10.1016/j.biortech.2012.01.102

Lee C (2002) Photohydrogen production using purple nonsulfur bacteria with hydrogen fermentation reactor effluent. Int J Hydrog Energy 27:1309–1313. doi:10.1016/S0360-3199(02)00102-7

Levin D (2004) Biohydrogen production: prospects and limitations to practical application. Int J Hydrog Energy 29:173–185. doi:10.1016/S0360-3199(03)00094-6

Levin DB, Zhu H, Beland M, Cicek N, Holbein BE (2007) Potential for hydrogen and methane production from biomass residues in Canada. Bioresour Technol 98:654–660. doi:10.1016/j.biortech.2006.02.027

Li D, Chen H (2007) Biological hydrogen production from steam-exploded straw by simultaneous saccharification and fermentation. Int J Hydrog Energy 32:1742–1748. doi:10.1016/j.ijhydene.2006.12.011

Li C, Fang HHP (2007) Fermentative hydrogen production from wastewater and solid wastes by mixed cultures. Crit Rev Environ Sci Technol 37:1–39. doi:10.1080/10643380600729071

Li X-H, Liang D-W, Bai Y-X, Fan Y-T, Hou H-W (2014) Enhanced H2 production from corn stalk by integrating dark fermentation and single chamber microbial electrolysis cells with double anode arrangement. Int J Hydrog Energy 39:8977–8982. doi:10.1016/j.ijhydene.2014.03.065

Liu D, Angelidaki I, Zeng R, Min B (2008) Bio-hydrogen production by dark fermentation from organic wastes and residues

Liu Z, Zhang C, Lu Y, Wu X, Wang L, Wang L, Han B, Xing X-H (2013) States and challenges for high-value biohythane production from waste biomass by dark fermentation technology. Bioresour Technol 135:292–303. doi:10.1016/j.biortech.2012.10.027

Logan BE, Regan JM (2006) Electricity-producing bacterial communities in microbial fuel cells. Trends Microbiol 14:512–518. doi:10.1016/j.tim.2006.10.003

Lu L, Ren N, Xing D, Logan BE (2009) Hydrogen production with effluent from an ethanol-H2-coproducing fermentation reactor using a single-chamber microbial electrolysis cell. Biosens Bioelectron 24:3055–3060. doi:10.1016/j.bios.2009.03.024

Marone A (2012) Bio-hydrogen production by self fermentation of vegetable waste: from screening of microbial diversity to bioaugmentation of indigenous fermentative communities. Ph.D. thesis, Università degli Studi della Tuscia, Archivio delle tesi di dottorato di ricerca. http://hdl.handle.net/2067/2529

Marone A, Izzo G, Mentuccia L, Massini G, Paganin P, Rosa S, Varrone C, Signorini A (2014) Vegetable waste as substrate and source of suitable microflora for bio-hydrogen production. Renew Energy 68:6–13. doi:10.1016/j.renene.2014.01.013

Marone A, Varrone C, Fiocchetti F, Giussani B, Izzo G, Mentuccia L, Rosa S, Signorini A (2015) Optimization of substrate composition for biohydrogen production from buffalo slurry co-fermented with cheese whey and crude glycerol, using microbial mixed culture. Int J Hydrog Energy 40(1):209–218

Massanet-Nicolau J, Dinsdale R, Guwy A, Shipley G (2013) Use of real time gas production data for more accurate comparison of continuous single-stage and two-stage fermentation. Bioresour Technol 129:561–567. doi:10.1016/j.biortech.2012.11.102

Menon V, Rao M (2012) Trends in bioconversion of lignocellulose: biofuels, platform chemicals & biorefinery concept. Prog Energy Combust Sci 38:522–550. doi:10.1016/j.pecs.2012.02.002

Monlau F, Sambusiti C, Barakat A, Guo XM, Latrille E, Trably E, Steyer J-P, Carrere H (2012) Predictive models of biohydrogen and biomethane production based on the compositional and structural features of lignocellulosic materials. Environ Sci Technol 46:12217–12225. doi:10.1021/es303132t

Monlau F, Aemig Q, Trably E (2013a) Specific inhibition of biohydrogen-producing Clostridium sp. after dilute-acid pretreatment of sunflower stalks. Int J Hydrog Energy 38:12273–12282

Monlau F, Barakat A, Trably E, Dumas C, Steyer J-P, Carrère H (2013b) Lignocellulosic materials into biohydrogen and biomethane: impact of structural features and pretreatment. Crit Rev Environ Sci Technol 43:260–322. doi:10.1080/10643389.2011.604258

Monlau F, Trably E, Barakat A, Hamelin J, Steyer J-P, Carrere H (2013c) Two-stage alkaline-enzymatic pretreatments to enhance biohydrogen production from sunflower stalks. Environ Sci Technol. doi:10.1021/es402863v

Monlau F, Sambusiti C, Barakat A, Quéméneur M, Trably E, Steyer J-P, Carrère H (2014) Do furanic and phenolic compounds of lignocellulosic and algae biomass hydrolyzate inhibit anaerobic mixed cultures? A comprehensive review. Biotechnol Adv 32:934–951. doi:10.1016/j.biotechadv.2014.04.007

Monlau F, Kaparaju P, Trably E, Steyer JP, Carrere H (2015) Alkaline pretreatment to enhance one-stage CH4 and two-stage H2/CH4 production from sunflower stalks: mass, energy and economical balances. Chem Eng J 260:377–385. doi:10.1016/j.cej.2014.08.108

Mosier N, Wyman C, Dale B, Elander R, Lee YY, Holtzapple M, Ladisch M (2005) Features of promising technologies for pretreatment of lignocellulosic biomass. Bioresour Technol 96:673–686. doi:10.1016/j.biortech.2004.06.025

Mussatto SI, Roberto IC (2004) Alternatives for detoxification of diluted-acid lignocellulosic hydrolyzates for use in fermentative processes: a review. Bioresour Technol 93:1–10. doi:10.1016/j.biortech.2003.10.005

Nasirian N, Almassi M, Minaei S, Widmann R (2011) Development of a method for biohydrogen production from wheat straw by dark fermentation. Int J Hydrog Energy 36:411–420. doi:10.1016/j.ijhydene.2010.09.073

Ni Y, Sun Z (2009) Recent progress on industrial fermentative production of acetone-butanol-ethanol by Clostridium acetobutylicum in China. Appl Microbiol Biotechnol 83:415–423. doi:10.1007/s00253-009-2003-y

Nissilä ME, Lay C-H, Puhakka JA (2014) Dark fermentative hydrogen production from lignocellulosic hydrolyzates – a review. Biomass Bioenergy 67:145–159. doi:10.1016/j.biombioe.2014.04.035

Okamoto M, Miyahara T, Mizuno O, Noike T (2000) Biological hydrogen potential of materials characteristic of the organic fraction of municipal solid wastes. Water Sci Technol 41:25–32

Ozkan L, Erguder TH, Demirer GN (2011) Effects of pretreatment methods on solubilization of beet-pulp and bio-hydrogen production yield. Int J Hydrog Energy 36:382–389. doi:10.1016/j.ijhydene.2010.10.006

Oztekin R, Kapdan IK, Kargi F, Argun H (2008) Optimization of media composition for hydrogen gas production from hydrolyzed wheat starch by dark fermentation. Int J Hydrog Energy 33:4083–4090, doi: http://dx.doi.org/10.1016/j.ijhydene.2008.05.052

Pakarinen O (2008) Batch dark fermentative hydrogen production from grass silage: the effect of inoculum, pH, temperature and VS ratio. Int J Hydrog Energy 33:594–601. doi:10.1016/j.ijhydene.2007.10.008

Pakarinen OM, Tähti HP, Rintala JA (2009) One-stage H2 and CH4 and two-stage H2+CH4 production from grass silage and from solid and liquid fractions of NaOH pre-treated grass silage. Biomass Bioenergy 33:1419–1427. doi:10.1016/j.biombioe.2009.06.006

Pan C, Fan Y, Hou H (2008) Fermentative production of hydrogen from wheat bran by mixed anaerobic cultures. Ind Eng Chem Res 47:5812–5818. doi:10.1021/ie701789c

Perera KRJ, Nirmalakhandan N (2010) Enhancing fermentative hydrogen production from sucrose. Bioresour Technol 101:9137–9143. doi:10.1016/j.biortech.2010.06.145

Perera KRJ, Nirmalakhandan N (2011) Evaluation of dairy cattle manure as a supplement to improve net energy gain in fermentative hydrogen production from sucrose. Bioresour Technol 102:8688–8695. doi:10.1016/j.biortech.2011.02.044

Phowan P, Danvirutai P (2014) Hydrogen production from cassava pulp hydrolysate by mixed seed cultures: effects of initial pH, substrate and biomass concentrations. Biomass Bioenergy 64:1–10. doi:10.1016/j.biombioe.2014.03.057

Prakasham RS, Sathish T, Brahmaiah P, Subba Rao C, Sreenivas Rao R, Hobbs PJ (2009) Biohydrogen production from renewable agri-waste blend: optimization using mixer design. Int J Hydrog Energy 34:6143–6148. doi:10.1016/j.ijhydene.2009.06.016

Prakasham RS, Brahmaiah P, Nagaiah D, Rao PS, Reddy BVS, Rao RS, Hobbs PJ (2012) Impact of low lignin containing brown midrib sorghum mutants to harness biohydrogen production using mixed anaerobic consortia. Int J Hydrog Energy 37:3186–3190. doi:10.1016/j.ijhydene.2011.11.082

Quéméneur M, Bittel M, Trably E, Dumas C, Fourage L, Ravot G, Steyer J-P, Carrère H (2012) Effect of enzyme addition on fermentative hydrogen production from wheat straw. Int J Hydrog Energy 37:10639–10647. doi:10.1016/j.ijhydene.2012.04.083

Rabelo SC, Maciel Filho R, Costa AC (2009) Lime pretreatment of sugarcane bagasse for bioethanol production. Appl Biochem Biotechnol 153:139–150. doi:10.1007/s12010-008-8433-7

Ren N, Wang A, Cao G, Xu J, Gao L (2009) Bioconversion of lignocellulosic biomass to hydrogen: potential and challenges. Biotechnol Adv 27:1051–1060. doi:10.1016/j.biotechadv.2009.05.007

Rozendal R, Hamelers H, Euverink G, Metz S, Buisman C (2006) Principle and perspectives of hydrogen production through biocatalyzed electrolysis. Int J Hydrog Energy 31:1632–1640. doi:10.1016/j.ijhydene.2005.12.006

Saraphirom P, Reungsang A (2011) Biological hydrogen production from sweet sorghum syrup by mixed cultures using an anaerobic sequencing batch reactor (ASBR). Int J Hydrog Energy 36:8765–8773. doi:10.1016/j.ijhydene.2010.08.058

Sarma SJ, Brar SK, Le Bihan Y, Buelna G (2013) Bio-hydrogen production by biodiesel-derived crude glycerol bioconversion: a techno-economic evaluation. Bioprocess Biosyst Eng 36:1–10. doi:10.1007/s00449-012-0755-8

Shanmugam SR, Chaganti SR, Lalman JA, Heath D (2014) Using a statistical approach to model hydrogen production from a steam exploded corn stalk hydrolysate fed to mixed anaerobic cultures in an ASBR. Int J Hydrog Energy 39:10003–10015. doi:10.1016/j.ijhydene.2014.04.115

Shi X, Song H, Wang C, Tang R, Huang Z, Gao T, Xie J (2010) Enhanced bio-hydrogen production from sweet sorghum stalk with alkalization pretreatment by mixed anaerobic cultures. Int J Energy Res 34:662–672. doi:10.1002/er1570

Su H, Cheng J, Zhou J, Song W, Cen K (2009) Improving hydrogen production from cassava starch by combination of dark and photo fermentation. Int J Hydrog Energy 34:1780–1786. doi:10.1016/j.ijhydene.2008.12.045

Tenca A, Schievano A, Perazzolo F, Adani F, Oberti R (2011) Biohydrogen from thermophilic co-fermentation of swine manure with fruit and vegetable waste: maximizing stable production without pH control. Bioresour Technol 102:8582–8588. doi:10.1016/j.biortech.2011.03.102

Ullery ML, Logan BE (2015) Anode acclimation methods and their impact on microbial electrolysis cells treating fermentation effluent. Int J Hydrog Energy 40:6782–6791. doi:10.1016/j.ijhydene.2015.03.101

Urbaniec K, Markowski M, Budek A (2014) Studies on the energy demand of two-stage fermentative hydrogen production from biomass in a factory equipped with fuel-cell based power plant. Int J Hydrog Energy 39:7468–7475. doi:10.1016/j.ijhydene.2014.01.096

Varrone C, Giussani B, Izzo G, Massini G, Marone A, Signorini A, Wang A (2012) Statistical optimization of biohydrogen and ethanol production from crude glycerol by microbial mixed culture. Int J Hydrog Energy 37:16479–16488. doi:10.1016/j.ijhydene.2012.02.106

Venkata Mohan S, Mohanakrishna G, Goud RK, Sarma PN (2009) Acidogenic fermentation of vegetable based market waste to harness biohydrogen with simultaneous stabilization. Bioresour Technol 100:3061–3068. doi:10.1016/j.biortech.2008.12.059

Vollmer H, Scholz W (1985) Two-step biological treatment of slaughterhouse effluents. Flaischwirtschaft 65:1310, 1312–1316, 1364

Wang A, Sun D, Cao G, Wang H, Ren N, Wu W-M, Logan BE (2011) Integrated hydrogen production process from cellulose by combining dark fermentation, microbial fuel cells, and a microbial electrolysis cell. Bioresour Technol 102:4137–4143. doi:10.1016/j.biortech.2010.10.137

Wang K-S, Chen J-H, Huang Y-H, Huang S-L (2013) Integrated Taguchi method and response surface methodology to confirm hydrogen production by anaerobic fermentation of cow manure. Int J Hydrog Energy 38:45–53. doi:10.1016/j.ijhydene.2012.03.155

Wieczorek N, Kucuker MA, Kuchta K (2014) Fermentative hydrogen and methane production from microalgal biomass (Chlorella vulgaris) in a two-stage combined process. Appl Energy 132:108–117. doi:10.1016/j.apenergy.2014.07.003

Wu X, Yao W, Zhu J (2010) Effect of pH on continuous biohydrogen production from liquid swine manure with glucose supplement using an anaerobic sequencing batch reactor. Int J Hydrog Energy 35:6592–6599. doi:10.1016/j.ijhydene.2010.03.097

Wu J, Ein-Mozaffari F, Upreti S (2013a) Effect of ozone pretreatment on hydrogen production from barley straw. Bioresour Technol 144:344–349. doi:10.1016/j.biortech.2013.07.001

Wu J, Upreti S, Ein-Mozaffari F (2013b) Ozone pretreatment of wheat straw for enhanced biohydrogen production. Int J Hydrog Energy 38:10270–10276, doi: http://dx.doi.org/10.1016/j.ijhydene.2013.06.063

Wu X, Lin H, Zhu J (2013c) Optimization of continuous hydrogen production from co-fermenting molasses with liquid swine manure in an anaerobic sequencing batch reactor. Bioresour Technol 136:351–359. doi:10.1016/j.biortech.2013.02.109

Xing Y, Li Z, Fan Y, Hou H (2010) Biohydrogen production from dairy manures with acidification pretreatment by anaerobic fermentation. Environ Sci Pollut Res Int 17:392–399. doi:10.1007/s11356-009-0187-4

Yang H, Guo L, Liu F (2010) Enhanced bio-hydrogen production from corncob by a two-step process: dark- and photo-fermentation. Bioresour Technol 101:2049–2052. doi:10.1016/j.biortech.2009.10.078

Yang Z, Guo R, Xu X, Fan X, Luo S (2011) Hydrogen and methane production from lipid-extracted microalgal biomass residues. Int J Hydrog Energy 36:3465–3470. doi:10.1016/j.ijhydene.2010.12.018

Yokoyama H, Waki M, Moriya N, Yasuda T, Tanaka Y, Haga K (2007) Effect of fermentation temperature on hydrogen production from cow waste slurry by using anaerobic microflora within the slurry. Appl Microbiol Biotechnol 74:474–483. doi:10.1007/s00253-006-0647-4

Zhu J, Li Y, Wu X, Miller C, Chen P, Ruan R (2009) Swine manure fermentation for hydrogen production. Bioresour Technol 100:5472–5477. doi:10.1016/j.biortech.2008.11.045

Zhu Z, Shi J, Zhou Z, Hu F, Bao J (2010) Photo-fermentation of Rhodobacter sphaeroides for hydrogen production using lignocellulose-derived organic acids. Process Biochem 45:1894–1898. doi:10.1016/j.procbio.2010.08.017

Zong W, Yu R, Zhang P, Fan M, Zhou Z (2009) Efficient hydrogen gas production from cassava and food waste by a two-step process of dark fermentation and photo-fermentation. Biomass Bioenergy 33:1458–1463, doi: http://dx.doi.org/10.1016/j.biombioe.2009.06.008

5 Exploiting Biohydrogen Pathways of Cyanobacteria and Green Algae: An Industrial Production Approach

Anubha Kaushik and Mona Sharma

Abstract

Hydrogen is viewed as a clean and sustainable energy alternative of future that may change the present carbon-based economy to hydrogen-based economy in the years to come. Biohydrogen production by various microorganisms has emerged as a new area in energy generation that is moving ahead for industrial application. Cyanobacteria and green algae are photoautotrophic microbes that are capable of hydrogen generation by direct or indirect biophotolysis and photofermentation. Hydrogen production is mediated by hydrogenase and nitrogenase enzyme, both of which are oxygen sensitive. Various pathways and strategies of hydrogen production by these photoautotrophic microorganisms have been discussed in this chapter along with approaches to enhance hydrogen yields for prolonged duration using different photobioreactor designs. The possibility of using cyanobacteria and green algae for integrating hydrogen production with wastewater treatment and environmental implications thereof has also been discussed. In biological hydrogen production, there are no greenhouse gas emissions; thus, switching over to hydrogen as a future energy fuel would also help mitigate the global climate change problem.

5.1 Introduction

Global economy is largely dependent on energy derived from fossil fuels that is about 80 % of the total energy demands. The carbon-based economy has led to several alarming problems due to emission of various pollutants arising out of combustion of the fossil fuels. The most serious issue associated with fossil fuel consumption, however, is the global climate change problem, resulting from emission of carbon dioxide.

A. Kaushik (✉)
University School of Environment Management, GGS Indraprastha University, New Delhi 110078, India
e-mail: aks.es.10@gmail.com

M. Sharma
Department of Environmental Sciences, School of Earth, Environment and Space Studies, Central University of Haryana, Mahendergarh, India
e-mail: monasharma140@gmail.com

A. Singh, D. Rathore (eds.), *Biohydrogen Production: Sustainability of Current Technology and Future Perspective*, DOI 10.1007/978-81-322-3577-4_5

Since magnitude of energy consumption of a nation is generally taken as an index of its development, hence, energy use pattern and the type of energy used is a very important parameter of growth. Considering the finite nature of fossil fuel reserves and the environmental problems associated with their combustion, it has become extremely important to look for new renewable energy resources that would be cleaner alternatives to the fossil fuel energy and help to achieve the sustainability goals.

Hydrogen is considered to have tremendous potential as a clean and sustainable energy carrier that could serve as the fuel of the future leading to a shift from the present carbon-based economy to hydrogen-based economy in the years to come (Reith et al. 2003; Levin et al. 2004). Hydrogen is abundantly present on the earth in the form of water covering about three fourth of the earth's surface. On combustion, hydrogen produces water as the main product; hence, it is a clean fuel. It has a high calorific value of 3042 cal/m^3 with highest gravimetric energy density amongst all fuels known till now. Yet, substituting carbon by hydrogen as a fuel faces several technical challenges of efficient production, safe storage, transmission and distribution (Ren et al. 2011; Ramachandran and Menon 1998). Production of hydrogen at present is mostly by electrolysis of water or by steam reformation of methane.

In the past decade, biohydrogen production has emerged as a new and exciting area of research that is moving ahead for industrial application (Das and Veziroglu 2001) exploring the potential of various microorganisms (Mona and Kaushik 2014). Though there are diverse microorganisms capable of hydrogen production, green algae and cyanobacteria are the only organisms capable of both hydrogen production and photosynthesis. Due to their simple nutritional requirements and low cost of production in diverse climatic conditions, it seems promising to utilise the metabolic pathways of these photoautotrophic microbes for biohydrogen production. Another added advantage is that can be cultivated in open barren areas in tanks or ponds and would not compete with crop plants that need fertile lands to grow.

5.1.1 Biological Hydrogen Production

Biohydrogen can be produced by a variety of microorganisms including bacteria, green algae and cyanobacteria. All such processes producing biohydrogen depend fundamentally on the presence of the enzyme hydrogenase. However, the quantity of this enzyme is not the limiting factor for the amount of hydrogen being produced by any of the systems studied till now (Hallenbeck and Benemann 2002). Basically, hydrogenase enzyme catalyses the simple reaction:

$$2H^+ + 2e^- \rightarrow H_2 \tag{5.1}$$

Presently, three enzymes, namely, nitrogenase, Fe-hydrogenase and NiFe-hydrogenase, are recognised as the enzymes involved in biohydrogen production.

5.2 Biological Hydrogen Production Pathways

The basic biochemical pathways by which hydrogen production takes place are summarised in Table 5.1.

5.2.1 Biophotolysis

Photoautotrophic cyanobacteria and green algae, in the presence of light, break down water into hydrogen and oxygen. During the process of photosynthesis, these photoautotrophs convert the solar energy into biochemical energy. Various processes through which photobiological hydrogen can be produced are depicted in Table 5.1 indicating the basic biological reactions and the microorganisms involved therein.

5.2.1.1 Direct Biophotolysis

Prospects of developing various biological H_2 production processes seem attractive because of the renewable energy option and, particularly, because the processes operate under ambient temperature and pressure (Sinha and Pandey 2011). Amongst the various photobiological

Table 5.1 A summary of the biological hydrogen production processes (Das and Veziroglu 2001; Hawkes et al. 2002)

Production process	Reactions	Useful microorganism
Direct biophotolysis	$2H_2O + \text{light} \rightarrow 2H_2 + O_2$	Microalgae
Indirect biophotolysis	(a) $6H_2O + 6CO_2 + \text{light} \rightarrow C_6H_{12}O_6 + 6O_2$	Microalgae, cyanobacteria
	(b) $C_6H_{12}O_6 + 2H_2O \rightarrow 4H_2 + 2CH_3COOH + 2CO_2$	
	(c) $2CH_3COOH + 4H_2O + \text{light} \rightarrow 8H_2 + 4CO_2$	
	Overall reaction : $12H_2O + \text{light} \rightarrow 12H_2 + 6O_2$	
Photofermentation	$CH_3COOH + 2H_2O + \text{light} \rightarrow 4H_2 + 2CO_2$	Purple bacteria, microalgae

pathways, hydrogen production by direct biophotolysis is theoretically the simplest as it involves simple splitting of the water molecule by various cyanobacteria and green algae, producing H_2.

Cyanobacteria and microalgae possess the photosynthetic pigment chlorophyll that has its absorption maxima in the blue and red region and action spectrum in the visible range of electromagnetic spectrum of sun light. Direct biophotolysis appears to be an attractive process since generation of H_2 takes place in this method using abundantly available resources like water and sunlight. However, a major drawback of the process is that along with hydrogen, there is simultaneous production of oxygen. The key enzyme hydrogenase catalysing hydrogen production reaction of the algae is sensitive to oxygen; hence, the oxygen evolved has an inhibitory action on biological hydrogen production, resulting in generally less than 1.5 % conversion efficiency from solar light energy to hydrogen energy. The efficiency can, however, be increased to 3–10 %, if the oxygen produced is immediately removed. Genetic engineering methods may also prove useful to overcome oxygen inhibition of hydrogenase. It is also possible to improve hydrogen production by manipulating the concentrations of certain metals since the same may significantly affect the activity of hydrogenase. Metal ions can significantly influence enzyme activities associated with hydrogen production (Zhao et al. 2012).

5.2.1.2 Indirect Biophotolysis

In order to address the problem of oxygen inhibition of hydrogen production, a system is needed, wherein the two phases involving biological production of hydrogen and oxygen are separated in space or time, a process designated as indirect biophotolysis.

In spatial separation method, the oxygen and hydrogen producing processes of microalgae are spatially separated as demonstrated by Benemann (1997). In the first stage, open ponds were used, in which the microalgae fixed atmospheric CO_2 through photosynthesis into carbohydrates and evolved oxygen in the process. In a second phase, dark and anaerobic conditions were maintained using closed bioreactors, where the carbohydrates were degraded into acetic acid and subsequently into hydrogen in the presence of sunlight. It was proposed that the algae could be reused after allowing them to grow again by transferring them to open ponds. The reactions in the various steps are depicted in Table 5.1.

The temporal separation method of indirect biophotolysis was developed using the green microalga *Chlamydomonas reinhardtii*. Temporal separation of oxygen producing and hydrogen producing processes was achieved by altering the two phases over time. The first phase of photosynthetic O_2 evolution along with carbohydrate storage was followed by a second phase in which the growth medium was deprived of sulphur that led to blocking of photosynthesis and thus oxygen evolution (Melis et al. 2000). This facilitated hydrogenase activity in the alga cells and favoured production of hydrogen.

5.2.2 Photofermentation

Photofermentation is the process in which there is conversion of organic compounds like acetic acid into hydrogen and CO_2 by bacteria in the presence of sunlight and anaerobic conditions. This process may be coupled to dark hydrogen fermentation in which acetic acid is one of the end products (Yokoi et al. 1998; Hawkes et al. 2002). While this process is more common in bacteria, in the past some researchers have reported photofermentation in cyanobacteria and green algae under conditions of sulphur starvation. Enhanced hydrogen production by *Chlamydomonas reinhardtii* has been reported under conditions of sulphur deprivation. The sulphur-starved *C. reinhardtii* cells showed decline in oxygen evolution due to blocking of photosystem II, and anaerobic conditions are developed in the closed culture tubes. Even in the anaerobic conditions, the algae show survival as they shift to photofermentation (Winkler et al. 2002).

Photofermentation has been integrated with dark fermentation for cultivation of certain microalgae like *Chlorella vulgaris* using different substrates to achieve high biohydrogen yield up to 11.61 mol H_2/mol sucrose and production rate of 673.93 ml/h/l during an 80-day cycle of dark and light sequence. Since there was complete consumption of carbon, hence hydrogen produced did not show traces of carbon dioxide as impurity (Lo et al. 2010).

5.2.3 Dark Fermentation

Biomass is being used as both carbon and energy source by various microorganisms under anaerobic conditions leading to generation of biohydrogen by dark fermentation (Cardoso et al. 2014). Various bacteria in swamps, marshes, hot springs and sewage have been found to degrade the available organic matter to produce hydrogen. Some important metabolites like acetic acid, lactic acid, ethanol and amino acids are also produced in the process, which are utilised by methanogenic bacteria coexisting in these habitats. In natural habitats both hydrogen producers and consumers exist together; hence, net hydrogen production is not detected.

For designing a bioprocess to produce hydrogen from biomass, the basic requirement is to separate the processes of hydrogen generation and its consumption. For this we need to design a dual-step bioprocess in which first there is dark fermentation producing hydrogen, while utilisation of this hydrogen for metabolite synthesis is inhibited. The second step involves further transformation of organic metabolites produced in the first fermentation process into hydrogen through photobiological pathway of fermentation.

Approximate H_2 yield of the obligate anaerobic bacterium *Clostridium* is known to be 2 mol per mol glucose, whereas the yield from facultative anaerobes is even less than that. On the other hand, extreme thermophilic anaerobic bacteria have high hydrogen yields that may be very close to the maximum theoretical value of 4 mol H_2 mol^{-1} glucose and are hence found useful for dark fermenters. However, besides optimal molar yields of hydrogen, high rate of hydrogen production is also needed, which needs high cell density. Thermophilic bacteria generally have lower cell densities. Improved cell densities and proper biomass use are important factors for efficient working of dark fermenters (Reith et al. 2003).

A major limitation and challenge of the dark fermentation process is that as the pressure of hydrogen produced in the reactor increases, it starts acting as an inhibitor for the production process, and it becomes absolutely necessary to reduce the pressure by continuously removing the hydrogen at an average of 10–20 k Pa (Reith et al. 2003).

Pretreatment and hydrolysis of biomass is recommended both from a physiological and technological point of view before dark fermentation. While it is the prime aim to convert different types of biomass as a fermentation substrate for hydrogen production, it is also important to look out for methods to utilise the unfermented residual components.

In order to achieve economic viability of the biohydrogen production process, it is important that the other metabolites produced during dark fermentation are also optimally utilised, which is achieved by coupling the process with photofermentation. This ensures complete conversion of the substrate into hydrogen and carbon dioxide, resulting in enhanced hydrogen yield.

Rieth et al. (2003) provided the estimated cost of biohydrogen production as 2.74 Euro $kg^{-1}H_2$ or 19.2 Euro GJ^{-1}, which is almost at par with other carbon neutral technologies known for production of hydrogen. Since several types of wastes can be used as the substrates for biohydrogen production, therefore it may prove to be useful for commercialisation at local and small level. Various small-scale industries having organically rich wastewaters may develop their own waste-to-energy systems that could possibly meet the energy demands of the unit. Further, there are more than one energy options available including production of hydrogen that could be used directly as a source of energy or electricity generation through fuel cell technology, where hydrogen is fed into the cell. One of the countries leading in biohydrogen production is the Netherlands, where thermophilic bacteria are used in the fermentation process. This is especially useful considering the high temperatures of many wastewaters. However, more R&D is needed to transform different waste biomass into usable feedstock with good fermentative characteristics.

5.3 Physiology of Biohydrogen Production

Biological hydrogen production is mediated by hydrogenase and nitrogenase enzymes (Levin et al. 2004) which are discussed here.

5.3.1 Nitrogenase-Catalysed Hydrogen Formation

In the absence of nitrogen in the medium, many filamentous cyanobacteria show nitrogenase activity in some specialised cells known as heterocysts. Nitrogenase catalyses the reaction that converts dinitrogen to ammonia through the process of nitrogen fixation, and hydrogen is also produced in the process. The reaction mediated through use of ATP has been represented as follows (Moezelaar et al. 1996):

$$6ATP + 16H_2O + N_2 + 10H^+ + 8e^- \xrightarrow{\text{Nitrogenase}} 16ADP + 16Pi + 2NH_3 + H_2 \quad (5.2)$$

The enzyme nitrogenase comprises of two units – the dinitrogenase (220–240 kDa) having Mo-Fe protein moiety, responsible for breaking the triple bond between the two nitrogen atoms, and dinitrogenase reductase (60–70 kDa) with only Fe protein that has a key role in transferring electrons from the donor (ferredoxin or flavodoxin). These units are encoded by specific genes, namely, the nifD and nifK genes and nifH genes, respectively (Orme-Johnson 1992; Flores and Herrero 1994; Masepohl et al. 1997). The dinitrogenases may contain Mo, V or Fe as reported by different researchers (Lambert and Smith 1977; Philips and Mitsui 1983; Heyer et al. 1989; Lindberg et al. 2004).

While nitrogen-containing substrates like CN^- and N_2O suppress hydrogen production, dinitrogen (N_2) does not inhibit hydrogen production even at 50 atm pressure (Rivera-Ortiz and Burris 1975; Simpson and Burris 1984; Jensen and Burris 1986; Liang and Burris 1988).

It has been found that under a high partial pressure of N_2O, production of hydrogen becomes N_2 dependent (Liang and Burris 1988). Hence, there is always production of hydrogen when nitrogen fixation (reduction) reaction takes place. Interestingly, the H_2 produced inhibits reduction of N_2, but does not inhibit reduction of protons or other substrates (Wilson and Umbreit 1937).

5.3.2 Hydrogenase-Catalysed Hydrogen Production and Uptake

Hydrogenase catalyses the reversible oxidation of molecular hydrogen (H_2), as in Eqs. 5.3 and 5.4:

$$H_2 + A_{OX} \rightarrow 2H^+ + A_{red} \quad (5.3)$$

$$2H^+ + D_{red} \rightarrow H_2 + D_{OX} \quad (5.4)$$

Two forms of hydrogenase enzyme are known to occur in microalgae and cyanobacteria. These are uptake hydrogenase and bidirectional hydrogenase, which mediate oxidation of hydrogen and production/oxidation of hydrogen, respectively (Tamagnini et al. 2002). The uptake hydrogenases, encoded by hupSL and present in the thylakoid membranes of heterocysts are involved in electron transfer from hydrogen to produce the reducing power (NADH) of the cells. The bidirectional hydrogenases, on the other hand, which are encoded by hox FUYH, are attached to cytoplasmic membrane and mediate acceptance of electrons both from NADH and hydrogen, thereby involved in both hydrogen production and its consumption (Miyake et al. 1989). Thus, uptake hydrogenase uses up the hydrogen produced, resulting in no net hydrogen yield by the microorganism.

Uptake hydrogenase is a dimeric enzyme, while the reversible or bidirectional hydrogenase is multimeric (Flores and Herrero 1994).

Both nitrogenase and hydrogenase are extremely sensitive to oxygen, and, therefore, oxygen-free or low-oxygen-tension environment is a prerequisite for biological hydrogen production. Microalgae and cyanobacteria are both photoautotrophic in nature, evolving oxygen during the process of photosynthesis. Hence, there is a need to protect the nitrogenase enzyme from oxygen. The cyanobacteria have evolved the following strategies to deal with the problem:

(a) Two types of cells are developed in heterocystous cyanobacteria. The vegetative cells carry out photosynthesis and produce oxygen. The thick-walled heterocysts are the zones of low oxygen tension and contain nitrogenase, which mediate nitrogen fixation and production of hydrogen (Orme-Johnson 1992). This is a spatial separation strategy.
(b) The non-heterocystous cyanobacteria, on the other hand, have developed temporal separation strategy by performing photosynthesis during light hours and enzyme-mediated hydrogen production during night-time when there is dark (Thiel 1993).

The hydrogen uptake reaction (Eq. 5.3) involves production of reductants through transfer of electrons from H_2 to some electron acceptors (O_2, SO_4^{2-} ,NO_3^-, CO_2, etc.), whereas the reverse reaction (Eq. 5.4) involves transfer of electrons (from Fd) to H^+ to produce H_2, mediated via electron transport chain (Vignais et al. 2001).

Depending upon the metal component, Vignais et al. (2001) classified the hydrogenases into three types, namely, [NiFe]-, [FeFe]- and [Fe]-only hydrogenase. As shown in Fig. 5.1, these three types of hydrogenases share certain common features like possessing an active site with metal cluster, where actual catalysis occurs and a few more Fe-S clusters. Each metal forms coordination bonds with carbon monoxide (CO) and cyanide (CN) ligands (Fontecilla-Camps et al. 2007).

Different functions are performed by different hydrogenases. While both production and uptake of hydrogen are facilitated by [NiFe] hydrogenases, they are more actively involved in the latter process. Greening et al. (2014) observed wide variations in hydrogen affinity of H_2-oxidising hydrogenases.

Burgdorf et al. (2005) have discovered a novel hydrogenase from *Ralstonia eutropha*, which is oxygen tolerant, thus increasing the possibility of efficient utilisation of hydrogenases in photosynthetic production of hydrogen through photolysis of water.

In [FeFe] hydrogenases, a bridging dithiolate cofactor is present, and this type has exhibited a

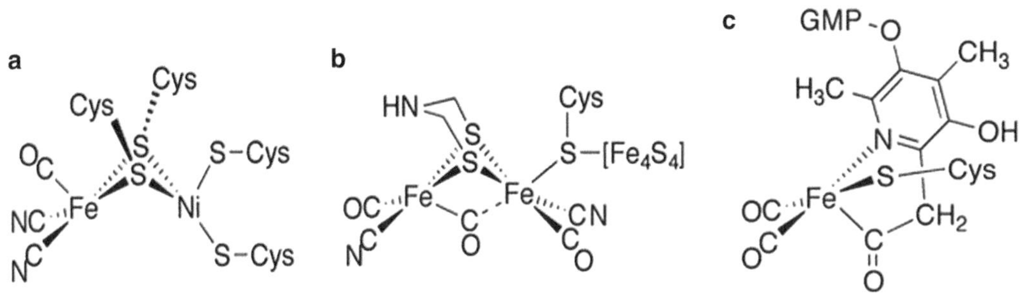

Fig. 5.1 Structure of three types of hydrogenase enzymes: (**a**) [NiFe] hydrogenase, (**b**) [FeFe] hydrogenase, (**c**) [Fe] hydrogenase (Adopted from Vignais et al. 2001)

high turnover rate (10,000 s^{-1}) in the case of *Clostridium* (Madden et al. 2012). Further three subtypes of [FeFe] hydrogenases have been recognised (Berggren et al. 2013), which are (i) monomeric hydrogenases present in the cytoplasm (*Clostridium pasteurianum*) that catalyse both H_2 production and uptake; (ii) periplasmic heterodimeric hydrogenases, catalysing mainly H_2 oxidation (*Desulfovibrio*); and (iii) monomeric hydrogenases, associated with chloroplasts which catalyses H_2 evolution (*Scenedesmus obliquus*). Due to high hydrogen yield facilitated by [FeFe] hydrogenase, more emphasis is now given to this type for sustainable production of H_2 (Smith et al. 2012).

Methanogenic bacteria possess [Fe] hydrogenase containing Fe instead of Ni or Fe-S clusters (Shima et al. 2008). The mode of catalysis is yet to be worked out in details. It is probably broken by Fe(II) and then hydride is transferred to the acceptor (Hiromoto et al. 2009).

Hydrogenases have a relatively low over potential and show even more efficient catalytic production of hydrogen than that mediated by platinum (Hinnemann et al. 2005).

[FeFe] hydrogenase is considered most suitable as a part of the solar H_2 production system. It is, however, necessary to engineer the hydrogenase enzyme for O_2 tolerance so that solar H_2 production becomes practically feasible. Efforts are underway to understand the mechanisms involved in O_2 inactivation of hydrogenases (Liebgott et al. 2010; Goris et al. 2011). Oxygen is reported to transform into a reactive species as it diffuses into [FeFe] hydrogenase, and then the Fe-S cluster at the active site is damaged (Stripp et al. 2009; Goris et al. 2011). In past few years, certain [NiFe] and [FeFe] hydrogenases have been discovered/developed that can tolerate oxygen, but these are the uptake hydrogenases, thus limiting their use in hydrogen production (Stripp et al. 2009; Bingham et al. 2012).

Since the atmosphere of earth initially had a high concentration of hydrogen, hence, hydrogenases seem to have evolved quite early to produce energy from H_2 and favour the growth of microorganisms (Vignais and Billoud 2007). The primary role of hydrogenases thus seems to be energy generation that could be used by the microbes themselves or by other organisms around. Evolution of reversible hydrogenases seems to have helped in regulating the levels of reducing equivalents by acting as "valves" in the photoautotrophs. This type of the function carried out by reverse hydrogenases has been important in anaerobic metabolism and hydrogen production by the microbes (Adams and Stiefel 1998).

In cyanobacteria and green algae, hydrogen production is mediated mainly by nitrogenase and hydrogenase enzyme, respectively (Figs. 5.2 and 5.3).

5.3.3 Environmental Factors Influencing H_2 Production

Biological hydrogen production has been reported to be largely affected by environmental factors like light intensity, light/dark period, temperature, pH, micronutrients, glucose

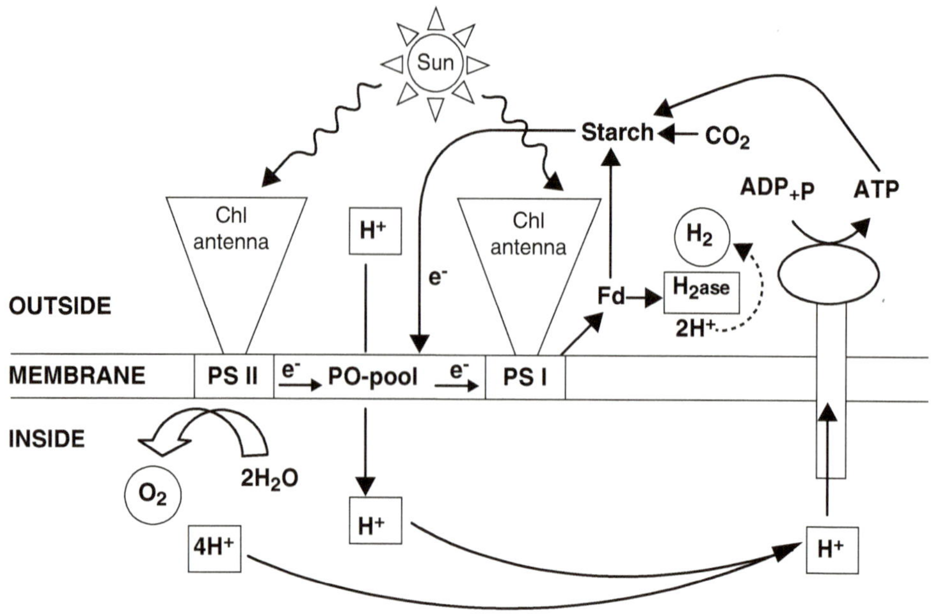

Fig. 5.2 Hydrogen production pathway in microalgae mediated by hydrogenase (Source: Tiwari and Pandey 2012)

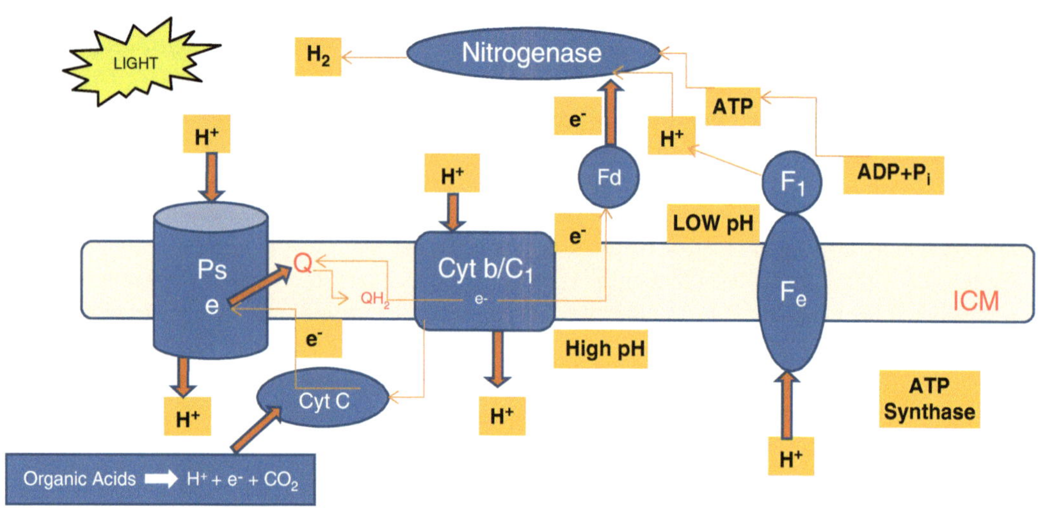

Fig. 5.3 Hydrogen production pathway in cyanobacteria mediated by nitrogenase

concentration, anaerobic gas mixture, carbon source, nitrogen source, temperature, salinity, etc. Studies reporting the effects of such factors on hydrogen production rates by microalgae and cyanobacteria are briefly discussed below:

5.3.3.1 Light

Light intensity and duration are important factors influencing algal hydrogen production. For optimal hydrogen production, microalgae have to be cultivated under moderately low light intensity, because as light intensity increases, there is evolution of f oxygen, which nullifies the anaerobic conditions produced by sulphur deprivation. Hydrogen production is significantly influenced by alternate light/dark cycle as photosynthesis takes place in the presence of light producing starch, which is then degraded into acetate, alcohol and hydrogen by fermentation during dark period (Miura et al. 1992). Different light (L) and dark (D) cycles (24D, 21L/3D and 18L/6D) have been reported to favour greater hydrogen production in different green algae and cyanobacteria (Mona et al. 2011), which is an important consideration for prolonged hydrogen production in bioreactors. Practical feasibility of such conditions is generally viewed with respect to the local conditions and availability of sunshine hours and natural light so as to cut on the additional costs on artificial illumination. In short-term experiments, 24 h dark conditions seem favourable for high yield of hydrogen, but such conditions would not give sustained yield because photosynthesis cannot take place in dark, and thus biomass and photosynthetic assimilates would diminish after some time (Kaushik and Anjana 2011).

5.3.3.2 Temperature

Temperature is an important factor that may shift the metabolic pathways in favour of enhanced hydrogen production. Temperature was reported to have a strong influence on hydrogen production by several cyanobacteria and microalgae (Litcht et al. 1997; Kaushik and Anjana 2011). Temperature optima vary with species as evident from various reports (Dutta et al. 2000; Shah 2000; Kaushik and Anjana 2011; Mona et al. 2011). Maximum rate of hydrogen production for *Nostoc linckia* isolated from the textile mill pond was at a temperature of 31 °C (Mona et al. 2011), whereas the temperature optima reported for *Nostoc muscorum* is 40 °C (Dutta et al. 2000; Shah 2000). Maximum hydrogen production rates reported for *Platymonas subcordiformis* is 27 °C, and that for marine green microalgae *Tetraselmis subcordiformis* is 55 °C (Guan et al. 2004).

5.3.3.3 Carbon Source

Carbon sources used in feedstock are important as they influence nitrogenase activity, which, in turn, affect hydrogen production by the cyanobacteria. Different concentrations of carbon sources cause variation in electron donation capabilities of nitrogenase and thus influence hydrogen production. The initial dose of glucose in the medium/feedstock is found to influence the hydrogen yield during photosynthesis/fermentation (Mona et al. 2011). The production decreases with increasing glucose concentration beyond 10 g L^{-1} (Alalayah et al. 2014).

5.3.3.4 pH

Hydrogen production has been reported to be influenced by the pH of the culture medium used for algal growth, and the pH requirement is species specific. It is favoured by slightly alkaline pH of 7.7 in microalga *Chlamydomonas* and 8.0 in *Phormidium* and *Lyngbya* (Kosourov et al. 2003; Kaushik and Anjana 2011) but decreases as pH becomes higher in the case of *Nostoc* (Guan et al. 2004; Mona et al. 2011).

5.3.3.5 Salinity

Dissolved salts have an effect on hydrogen production by cyanobacteria (Boison et al. 1999). The fresh water cyanobacteria show a decline in hydrogen production when dissolved salts in the medium increase. This seems to be due to the fact that some energy is diverted for the efflux and further prevention of influx of sodium ions (Asada and Kawamura 1984).

5.3.3.6 Gaseous Environment

Because of oxygen-sensitive nature of the enzymes involved in hydrogen production, it is important to have an anaerobic atmosphere. Biohydrogen production has been reported to be influenced by gaseous environment (Kaushik and Anjana 2011; Mona et al. 2011). In the presence of different concentrations of inert or anoxic gases like argon, carbon dioxide and methane, production of hydrogen is found to be significantly influenced. Yoon et al. (2002) reported increased hydrogen production by

Anabaena variabilis when injected repeatedly with carbon dioxide during growth. When CO_2 is administered, production of reductants increases, which give enhanced hydrogen yield on fermentation (Miura et al. 1995). Using suitable combinations of methane and carbon dioxide, hydrogen production by *Nostoc linckia* was significantly improved (Mona et al. 2011).

5.4 Industrial Approach to Biohydrogen Production by Cyanobacteria and Green Algae

5.4.1 Photofermentation Process Scaleup

In the two-stage biohydrogen reactors, photofermentation is carried out to convert the organic substrate completely into H_2 and CO_2 (Das and Veziroglu 2008). A basic requirement for scaleup of the system is to acquire a large surface area/volume ratio, which raises the cost of the photobioreactor. Solar collectors may provide an alternative to this; however, their production cost with the currently available technology is still quite high (Tsygankov et al. 1994; Bolton 1996; Fedorov et al. 1998).

Although different types of photobioreactors have been designed, at present the tubular photobioreactor fabricated by the IGV (Institut für Getreideverarbeitung GmbH, Germany) is being used commercially on a limited scale (Das 2009). The reactor has a surface area of 1.2 ha, with a cost of 660 Euro m^{-2}, which is slightly more than that of a system with solar panels on a 3 ha roof structure that costs around 580 Euro m^{-2} (Reith et al. 2003).

Energy potential of a photobioreactor is influenced by the available solar irradiance in that area (Ren et al. 2011). A thousand hectare photobioreactor system in the Netherlands or in the desert of Australia would perform quite differently producing hydrogen energy equivalent to 3 PJ and 5.3 PJ, respectively, on yearly basis, under conditions of maximum photo-conversion efficiency.

A photobiological system covering 1 ha ground area is found to possess ten times greater energy production potential than that by energy crops occupying same expanse of area. Besides this, hydrogen produced by the photobioreactor system is clean energy that is easy to transport and can be used directly in fuel cells. Reith et al. (2003) have calculated the rough cost of hydrogen production in a photobiological system over a hundred hectare area to be 10–15 Euro/GJ.

5.4.2 Photobioreactor Designs for Growth of Microalgae and Production of Hydrogen

Bioreactors are required for large-scale and prolonged production of hydrogen. Light being an essential factor determining algal and cyanobacterial growth, the photobioreactors have to be transparent (Bishop and Premakumar 1992; Dutta et al. 2000). All photobioreactors require adequate access to light, which is preferably the sunlight but may use controlled artificial light. The local sunshine hours are important in this context. Inside the photobioreactor, there is the illuminated zone, where photosynthesis occurs and the dark zone, where hydrogen is produced.

The hydrogen yield of a photobioreactor is limited by light, and it tends to decrease at higher intensities due to oxygen evolution leading to inactivation of hydrogenase enzyme. As the light intensity is reduced and diffused, production of hydrogen improves. Green algae and cyanobacteria absorb red light in the wavelength 680 nm. Hence, special panels of red light are used in the bioreactors. There is circulation of the algae between the illuminated and the dark zone of the photobioreactor at regular intervals, depending upon the reactor design. Growth of the algae and biological hydrogen generation are largely influenced by light source as well as gas liquid hydrodynamics effects (Ernst et al. 1979).

A photobioreactor is designed based on the following requisitions:

(i) It should have an enclosed system to ensure proper collection of the produced hydrogen.
(ii) There should be maximum surface area for efficient light capture required both for the growth of the algae and for hydrogen production.
(iii) The reactor should have provision for easy sterilization.
(iv) Surface to volume ratio of the reactor should be large.

The photobioreactors (PBRs) are generally categorised into three types: the vertical column type, tubular type and the flat-panel photobioreactor.

5.4.3 Vertical Column Reactor (Airlift Loop Reactor and Bubble Column)

These have columns made of transparent high-grade glass having a surrounding water jacket to regulate temperature and light. On the top as well as at the bottom, there is a port for the medium and inlet/outlet for gases (like Ar and H_2). Inoculum of desired microbes is added through septum to minimise chances of contamination (Lambert et al. 1979; Rawson 1985). In bubble columns, internal irradiance is increased at sunset and sunrise due to the bubbles. There are seasonal variations in biomass over the year showing peak during summer (Lambert et al. 1979; Rawson 1985). It is important to determine the dimensions of the vertical columns that depend on wind speed and strength of the glass or thermoplastics used in the reactor body. Generally, a diameter of 0.2 m and height of 4 m are considered suitable.

5.4.4 Flat-Panel Photobioreactor

Flat-panel PBR is made of a SS frame and three polycarbonate panels (Weissman and Benemann 1977). The reactor has two units, the front one of which has the culture. Required temperature of the culture is maintained by circulating water through a temperature-controlled water bath.

The source of artificial light is generally from tungsten-halogen lamps (500 W) with average light intensity of 175 W/m^2 or from red light emitting diodes (LEDs) put at one end of the reactor. Circulation of the gas is brought out using a pump and pressure is regulated by a pressure valve. Water vapours formed are condensed using a condenser to prevent their entry into the gas circulating system. Prior to use the reactor and the culture medium are autoclaved (Tao et al. 2007).

5.4.5 Tubular Photobioreactors

Tubular PBRs are made of long (10–100 m) transparent tubes (3–6 cm dia), into which the liquid culture is introduced using mechanical or airlift pumps (Molina et al. 2001). The tubes are placed horizontally, vertically or in a fence-like structure. The tubes may be straight or with U-shaped bends, coiled or parallel with multiple connections. Different designs influence the light regime by changing the photon flux density that falls upon the surface of the reactor (Tredici and Zittelli 1998).

Another special type of tubular PBR is the Tredici PBR made of flexible tubular plastic tubes filled with water and has an internal gas exchange system (Tredici et al. 1998). A corrugated plastic roofing sheet houses the tubes and keeps them in straight position. Inclination of the PBR at an angle facilitates free rise of gas bubbles, and a footer with compressed air line allows supply of air at the bottom of the reactor into selected tubes. A degasser at the top allows containment of the fluid displaced during aeration. In between, there are tubes that are not gassed to allow the return of the fluid, which helps in recirculation by airlift. Cooling of the system is through water spraying. The mass transfer characteristics of the tubular photobioreactor depend upon the shape of the reactor and type of mixing used (Ugwu et al. 2002). Different types of hydrogen photobioreactors and their salient features are briefly included in Table 5.2.

Table 5.2 Photobioreactor (PBR) types and properties

PBR type	Cyanobacteria used	Merits	Demerits	References
Vertical column	*Spirulina platensis*	(a) Simple and cost-effective design	(a) No control on incident light	Arik et al. (1996), Miro'n et al. (1999) and Miyamoto and Benemann (1988)
		(b) Better mass transfer in bubble columns	(b) Fluctuating productivity	
Flat panel	*Spirulina platensis*	(a) Better regulation of irradiant light	(a) Higher production costs	Hoekema et al. (2002) and Tredici and Zittelli (1998)
		(b) Gas pressure is controlled	(b) Complex design and greater maintenance required	
Tubular	*Arthrospira platensis*, *Anabaena variabilis PK84*, *Anabaena variabilis ATCC 29413*, *Anabaena variabilis PK84*	(a) Volume: surface area ratio is flexible	(a) Flexibility to irradiant light but mass transfer poor at places	Molina et al. (2001), Tredici and Zittelli (1998), Tredici et al. (1998), Ugwu et al. (2002), Carlozzi et al. (2005) and Borodin et al. (2000)
		(b) Light receiving surface keeps changing	(b) Time required for mixing more in internal static mixture	
		(c) Higher biomass with internal static mixture		

5.4.6 Approaches for Enhanced and Prolonged Hydrogen Production

One important method for increasing hydrogen production that has emerged in the last decade is by sulphur deprivation of the medium (Melis et al. 2000). Sulphur deprivation leads to selective and partial inactivation of the photosystem II of the algae, which induces anaerobiosis and hydrogenase expression. Hydrogen photo production through this technique has been demonstrated in *Chlamydomonas reinhardtii* for several days (Ghirardi et al. 2000).

Optimisation of various physicochemical parameters is also found effective in increasing hydrogen production. It is possible to increase hydrogen yield by manipulating various physicochemical conditions. Using appropriate combination in anoxic gaseous mixture in the bioreactor is found to increase hydrogen production many folds. Methane in headspace helps in scavenging the oxygen produced during photosynthesis and CH_4:Ar (11:2 v/v) results in up to eight times more yield as compared to Ar being used as an anoxic sparging gas in the headspace for *Lyngbya perelegans* (Anjana and Kaushik 2014) and 65 % increase in hydrogen when CO_2:N_2 (1:1 v/v) were used for *Anabaena* N7363 (Laurinavichene et al. 2008).

Another approach for increasing hydrogen production involves cell immobilisation in some suitable matrix, which is found to be very useful both for unicellular and filamentous algae and cyanobacteria (Kosourov and Seibert 2009; Anjana and Kaushik 2014). Cell immobilisation offers protection to various enzymes mediating metabolic activities against stress factors and the cells can be used for much longer period.

5.5 Coupling Biological H_2 Production with Wastewater Treatment Process

For making commercial application of biological hydrogen production a cost-effective reality, Benemann (2000) proposed that wastewater treatment and hydrogen production by microalgae should be integrated. Biosorption of toxic pollutants like heavy metals and dyes using algae and cyanobacteria has been extensively studied.

However, there is a need to systematically study the potential of integrating hydrogen production potential and pollutant bioremoval capacity of algae in combination.

An integrated system of sustainable hydrogen production and textile wastewater treatment technique was developed at lab scale using cyanobacterial biomass of *Nostoc linckia* by the authors (Kaushik et al. 2011). The cyanobacterium produced hydrogen for 3–4 weeks in a photobioreactor at sustained rates, and the spent biomass was then successfully used for biosorption of some toxic dyes (Reactive Red 198 and Crystal Violet) and heavy metals (chromium and cobalt) from simulated textile effluent in dynamic mode. Spent biomass from the hydrogen bioreactors taken out at different stages of reactor operation was successfully used for removal of desired metals and dyes from wastewater in packed bed biosorption columns (Mona et al. 2013).

In this context, exploitation of cyanobacteria isolated from polluted industrial sites may prove useful. *Nostoc linckia* isolated from within textile mill sites was reported to produce hydrogen at higher rate even in the presence of dyes and metals (Mona et al. 2011). Several strains of microalgae and cyanobacteria isolated from industrial sites were explored for hydrogen production and pollutant tolerance (Mona and Kaushik 2014). There is a need to look for more such strains for the dual purpose by developing energy self-dependent wastewater treatment techniques.

5.6 Implications of Biohydrogen in Climate Change Mitigation

Capability of algae and cyanobacteria to sequester carbon dioxide during photosynthesis, reduce GHG emissions and produce clean energy in the form of hydrogen is quite attractive. Carbon dioxide, the most important greenhouse gas produced during combustion of fuels, has become a cause of global concern, and the concept of carbon credits has emerged as a result of rising awareness about reducing these emissions. In the Kyoto Protocol, an international agreement was signed by more than 170 countries, and the market mechanisms were evolved through Marrakesh Accords that followed (Hoppe-Seyler 1987).

Farming of algae has come up as an easy and promising GHG reduction technology since the algae can consume tons of CO_2 for growth, produce hydrogen, thus reduce harmful greenhouse gases and earn carbon credit. The merits of CO_2 sequestration using algae lie in the fact that there is no need for high-purity CO_2 gas for cultivation and growth of the algae and, therefore, it is possible to feed the CO_2 containing flue gas to the photobioreactor directly and it is a renewable process. Algae have a very wide range of temperature tolerance (<0–70 °C) for their growth and, therefore, have a wide application range. An acre of algae consumes about 60 tons of CO_2 in a year (Singh et al. 2014). This suggests that using algae for energy production can go a long way in reducing emissions and the problem of global warming.

5.7 Conclusions and Future Prospects

Hydrogen production by cyanobacteria and microalgae is poised to be a useful, clean and sustainable process that would develop as a commercially usable technology in the years to come. Hydrogen can be used as a substitute for conventional fuel, may be used in fuel cells to generate electricity and can also be utilised for hydrogenation in food and chemical industries.

These photoautotrophic microorganisms are easy to grow with simple nutrient requirements; do not compete with crop plants that need to be grown in fertile lands; can sequester carbon dioxide and grow in the presence of flue gases, withstand high temperatures; and can also help in efficient bioremoval of various pollutants from wastewaters. All these features put them at an advantage over other organisms producing hydrogen.

Different designs of photobioreactors have been developed to achieve sustained hydrogen production, and enhanced hydrogen production rates have been achieved by optimising various physicochemical conditions and immobilisation of the algae in some suitable matrix. However, there is a need to reduce the cost of such algal hydrogen reactors and make the process more efficient so that all their microculture benefits are put to maximum use. Small-cell size and low-cell biomass levels in microalgal cultures often become limiting. Hence, there is a need to achieve high cell densities for various microalgal species.

Though the biological systems have relatively lower conversion efficiencies, their energy requirements are less, and the initial investment costs are quite low. The best aspect of the biological hydrogen production system is that it is produced from water that is abundantly available on the earth. To be economically competitive with various other methods of hydrogen production, intensive research is needed to integrate various benefits of these organisms, including flue gas utilisation for CO_2 sequestration, wastewater treatment and biomass production along with hydrogen production. From an environmental point of view, systems of microalgae and cyanobacterial cultures should be explored to capture CO_2 from industrial emissions consuming the nutrients present in the wastewaters and simultaneously produce hydrogen. From an engineering point of view, the costs associated with various energy intensive processes like harvesting and dewatering need to be reduced.

References

Adams MWW, Stiefel EI (1998) Biological hydrogen production: not so elementary. Science 282(5395):1842–1843

Alalayah WM, Alhamed YA, Al-Zahrani A, Edris G (2014) Experimental investigation parameters of hydrogen production by algae *Chlorella Vulgaris*. International conference on Chemical, Environment & Biological Sciences (CEBS-2014) Sept. 17–18, 2014 Kuala Lumpur (Malaysia) doi:10.1074/jbc.M111.302125 http://dx.doi.org/10.15242/IICBE.C914010

Anjana K, Kaushik A (2014) Enhanced hydrogen production by immobilized cyanobacterium *Lyngbya perelegans* under varying anaerobic conditions. Biomass Bioenergy 63:54–57

Arik T, Gunduz U, Yucel M, Turker L, Sediroglu V, Eroglu I (1996) Photoproduction of hydrogen by Rhodobacter sphaeroides O.U.001. In: Proceedings of the 11th world hydrogen energy conference, Stuttgart, Germany, 3:2417–2424

Asada Y, Kawamura S (1984) Hydrogen evolution by *Microcystis aeruginosa* in darkness. Agric Biol Chem 48:2595–2596

Benemann JR (1997) Feasibility analysis of photobiological hydrogen production. Int J Hydrog Energy 22:979–987

Benemann JR (2000) Hydrogen production by microalgae. J Appl Phycol 12:291–300

Berggren G, Adamska A, Lambertz C, Simmons TR, Esselborn J, Atta A, Gambarelli S, Mouesca JM, Reijerse E, Lubitz W, Happe T, Artero V, Fontecave M (2013) Biomimetic assembly and activation of [FeFe]-hydrogenases. Nature 499:66–69

Bingham AS, Smith PR, Swartz JR (2012) Evolution of an [FeFe] hydrogenase with decreased oxygen sensitivity. Int J Hydrog Energy 37:2965–2976

Bishop PE, Premakumar R (1992) Alternative nitrogen fixation systems. In: Stacey G, Burris RH, Evans HJ (eds) Biological nitrogen fixation. Chapman & Hall, New York, pp 736–762

Boison G, Bothe H, Hansel A, Lindblad P (1999) Evidence against a common use of the diaphorase subunits by the bidirectional hydrogenase and by the respiratory complex I in cyanobacteria. FEMS Microbiol Lett 174:159–165

Bolton JR (1996) Solar photoproduction of hydrogen. Sol Energy 57:37–50

Borodin VB, Tsygankov AA, Rao KK, Hall DO (2000) Hydrogen production by *Anabaena variabilis* PK84 under simulated outdoor conditions. Biotechnol Bioeng 69:478–485

Burgdorf T, Buhrke T, van-der Linden E, Jones A, Albracht S, Friedrich B (2005) [NiFe]-hydrogenases of ralstonia eutropha H16: modular enzymes for oxygen-tolerant biological hydrogen oxidation. J Mol Microbiol Biotechnol 10:181–196

Cardoso VL, Romao BB, Felipe TM, Silva, Julia G, Santos, Fabiana RX, Batista, Ferreira JS (2014) Hydrogen production by dark fermentation. Chem Engg Trans 38:481–486

Carlozzi P, Ena A, Carnevale S (2005) Hydrodynamic alterations during cyanobacteria (*Arthrospira platensis*) growth from low to high biomass concentration inside tubular photobioreactors. Biotechnol Prog 21:416–422

Das D (2009) Advances in biohydrogen production processes: an approach towards commercialization. Int J Hydrog Energy 34:7349–7357

Das D, Veziroglu TN (2001) Hydrogen production by biological processes: a survey of literature. Int J Hydrog Energy 26:13–28

Das D, Veziroglu TN (2008) Advances in biological hydrogen production processes. Int J Hydrog Energy 33:6046–6057

Dutta M, Nikki G, Shah V (2000) Cyanobacterial hydrogen production. World J Microbiol Biotechnol 16:8–9

Ernst A, Kerfi W, Spiller H, Boger P (1979) External factors influencing light-induced H_2 evolution by the blue-green algae, *Nostoc muscorum*. Z Naturforsch 34:820–825

Fedorov AS, Tsygankov AA, Rao KK, Hall DO (1998) Hydrogen photoproduction by *Rhodobacter sphaeroides* immobilised on polyurethane foam. Biotechnol Lett 20:1007–1009

Flores E, Herrero A (1994) Assimilatory nitrogen metabolism and its regulation. In: Bryant DA (ed) The molecular biology of cyanobacteria. Kluwer Academic Publishers, Dordrecht, pp 487–517

Fontecilla-Camps JC, Volbeda A, Cavazza C, Nicolet Y (2007) Structure/function relationships of [NiFe]- and [FeFe]-hydrogenases. Chem Rev 107(10):4273–4303

Ghirardi ML, Zhang L, Lee JW, Flynn T, Seibert M, Greenbaum E, Melis A (2000) Microalgae: a green source of renewable hydrogen. Trends Biotechnol 18:506–511

Goris T, Wait AF, Saggu M, Fritsch J, Heidary N, Stein M, Zebger I, Lendzian F, Armstrong FA, Friedrich B, Lenz O (2011) A unique iron-sulfur cluster is crucial for oxygen tolerance of a [NiFe]-hydrogenase. Nat Chem Biol 7:310–318

Greening C, Berney M, Hards K, Cook GM, Conrad R (2014) A soil actinobacterium scavenges atmospheric H_2 using two membrane-associated, oxygen-dependent hydrogenases. Proc Acad Natl Sci USA 111(11):4257–4261

Guan Y, Deng M, Yu X, Zhang W (2004) Two stage photo-biological production of hydrogen by marine green alga *Platymonas subcordiformis*. Biochem Eng J 19:69–73

Hallenbeck PC, Benemann JR (2002) Biological hydrogen production: fundamentals and limiting processes. Int J Hydrog Energy 27:1185–1193

Hawkes FR, Dinsdale R, Hawkes DL, Hussy I (2002) Sustainable fermentative hydrogen production: challenges for process optimisation. Int J Hydrog Energy 27:1339–1347

Heyer H, Stal LJ, Krumbein WE (1989) Simultaneous heterolactic and acetate fermentation in the marine cyanobacterium *Oscillatoria limosa* incubated anaerobically in the dark. Arch Microbiol 151:558–564

Hinnemann B, Moses PG, Bonde J, Jørgensen KP, Nielsen JH, Horch S, Chorkendorff I, Nørskov JK (2005) Biomimetic hydrogen evolution: MoS_2 nanoparticles as catalyst for hydrogen evolution. J Am Chem Soc 127:5308–5309

Hiromoto T, Warkentin E, Moll J, Ermler U, Shima S (2009) Iron-chromophore circular dichroism of [Fe]-hydrogenase: the conformational change required for H2 activation. Angew Chem Int 48:6457–6460

Hoekema S, Bijmans M, Janssen M, Tramper J, Wijffels RH (2002) A pneumatically agitated flat-panel photobioreactor with gas re-circulation: anaerobic photoheterotrophic cultivation of a purple non-sulfur bacterium. Int J Hydrog Energy 27:1331–1338

Hoppe-Seyler F (1987) Die Methanga¨rung der Essigsaure. Z Phys Chem 11:561–568

Jensen BB, Burris RH (1986) Nitrous oxide as a substrate and as a competitive inhibitor of nitrogenase. Biochemistry 25:1083–1088

Kaushik A, Anjana K (2011) Biohydrogen production by *Lyngbya perelegans*: influence of physico-chemical environment. Biomass Bioenergy 35(3):1041–1045

Kaushik A, Mona S, Kaushik CP (2011) Integrating photobiological hydrogen production with dye-metal bioremoval from simulated textile wastewater. Biores Technol 102:9957–9964

Kosourov NS, Seibert M (2009) Hydrogen photoproduction by nutrient-deprived *Chlamydomonas reinhardtii* cells immobilized within thin alginate films under aerobic and anaerobic conditions. Biotechnol Bioeng 102(1):50–58

Kosourov S, Seibert M, Ghirardi ML (2003) Effects of extracellular pH on the metabolic pathways in sulfur deprived, H_2 producing *Chlamydomonas reinhardtii* cultures. Plant Cell Physiol 44(2):146–155

Lambert GR, Smith GD (1977) Hydrogen formation by marine blue-green algae. FEBS Lett 83:159–162

Lambert GR, Daday A, Smith GD (1979) Hydrogen evolution from immobilized cultures of cyanobacterium *Anabena cylindrica*. FEBS Lett 101:125–128

Laurinavichene TV, Kosourov SN, Ghirardi ML, Seibert M, Tsygankov AA (2008) Prolongation of H_2 photoproduction by immobilized, sulfur-limited *Chlamydomonas reinhardtii* cultures. J Biotechnol 134:275–277

Levin DB, Lawrence P, Murray L (2004) Biohydrogen production: prospects and limitations to practical application. Int J Hydrog Energy 29:173–185

Liang J-H, Burris RH (1988) Interaction among nitrogen, nitrous oxide and acetylene as substrates and inhibitors of nitrogenase from *Azotobacter vinelandii*. Biochemistry 27:6726–6732

Liebgott PP, Leroux F, Burlat B, Dementin S, Baffert C, Lautier T, Fourmond V, Ceccaldi P, Cavazza C, Meynial-Salles I, Soucaille P, Fontecilla-Camps JC, Guigliarelli B, Bertrand P, Rousset M, Léger C (2010) Relating diffusion along the substrate tunnel and oxygen sensitivity in hydrogenase. Nat Chem Biol 6(1):63–70

Lindberg P, Lindblad P, Cournac L (2004) Gas exchange in the filamentous cyanobacterium *Nostoc punctiforme* Strain ATCC 29133 and its hydrogenase-deficient mutant Strain NHM5. Appl Environ Microbiol 70:2137–2145

Litcht R, Basin J, Hall DO (1997) The biotechnology of hydrogen production by Nostoc flagelliforme grown under chemostat conditions. Appl Microbiol Biotechnol 47:701–707

Lo YC, Chen CY, Lee CM, Chang JS (2010) Sequential dark–photo fermentation and autotrophic microalgal growth for high-yield and CO_2-free biohydrogen production. Int J Hydrog Energy 35:10944–10953

Madden C, Vaughn MD, Díez-Pérez I, Brown KA, King PW, Gust D, Moore AL, Moore TA (2012)

Catalytic turnover of [FeFe]-hydrogenase based on single-molecule imaging. J Am Chem Soc 134:1577–1582

Masepohl B, Schoelisch K, Goerlitz K, Kutzki C, Bohme H (1997) The heterocyst-specific fdxH gene product of the cyanobacterium *Anabaena* sp. PCC 7120 is important but not essential for nitrogen fixation. Mol Gen Genet 253:770–776

Melis A, Zhang L, Forestier M, Ghirardi ML, Seibert M (2000) Sustained photobiological hydrogen gas production upon reversible inactivation of oxygen evolution in the green alga *Chlamydomonas reinhardtii*. Plant Physiol 122:127–135

Miro'n AS, Go'mez AC, Camacho FG, Grima EM, Chisti Y (1999) Comparative evaluation of compact photobioreactors for large-scale monoculture of microalgae. J Biotechnol 70:249–270

Miura Y, Matsuoka S, Miyamoto K, Saltoh C (1992) Stably sustained hydrogen production with high molar yield through a combination of a marine green alga and a photosynthetic bacterium. Biosci Biotechnol Biochem 56:751–754

Miura Y, Akano T, Fukatsu K, Miyasaka H, Mizoguchi T, Yagi K, Maeda I, Ikuta Y, Matsumoto H (1995) Hydrogen production by photosynthetic microorganisms. Energy Convers Manag 36:903–906

Miyake J, Asada Y, Kawamura S (1989) Nitrogenase. In: Kitani O, Hall CW (eds) Biomass handbook. Gorton and Breach Science Publishers, New York, pp 362–370

Miyamoto K, Benemann JR (1988) Vertical tubular photobioreactor: design and operation. Biotechnol Lett 10:703–710

Moezelaar R, Bijvank SM, Stal LJ (1996) Fermentation and sulphur reduction in the mat-building cyanobacterium *Microcoleus chtonoplastes*. Appl Environ Microbiol 62:1752–1758

Molina E, Fernández J, Acién FG, Chisti Y (2001) Tubular photobioreactor design for algal cultures. J Biotechnol 92:113–131

Mona S, Kaushik A (2014) Screening metal-dye-tolerant photoautotrophic microbes from textile wastewaters for biohydrogen production. J Appl Phycol 27(3):1185–1194

Mona S, Kaushik A, Kaushik CP (2011) Hydrogen production and metal-dye bioremoval by a *Nostoc linckia* strain isolated from textile mill oxidation pond. Biores Technol 102:3200–3205

Mona S, Kaushik A, Kaushik CP (2013) Prolonged hydrogen production by *Nostoc* in photobioreactor and multi-stage use of the biological waste for column biosorption of some dyes and metals. Biomass Bioenergy 50:27–35

Orme-Johnson WH (1992) Nitrogenase structure: where to now? Science 257:1639–1640

Phlips EJ, Mitsui A (1983) Role of light intensity and temperature in the regulation of hydrogen photoproduction by the marine cyanobacterium *Oscillatoria* sp. Strain Miami BG7. Appl Environ Microbiol 45:1212–1220

Ramachandran R, Menon RK (1998) An overview of industrial uses of hydrogen. Int J Hydrog Energy 23:593–598

Rawson DM (1985) The effects of exogenous amino acids on growth and nitrogenase activity in the cyanobacterium *Anabena cylindrica* PCC 7122. J Gen Microbiol 134:2544–2549

Reith JH, Wijffels RH, Barten H (2003) Status and perspectives of biological methane and hydrogen production. Dutch Biological Hydrogen Foundation, The Netherlands ECN, The Hague

Ren N, Guo W, Liu B, Cao G, Ding J (2011) Biological hydrogen production by dark fermentation: challenges and prospects towards scaled-up production. Curr Opin Biotechnol 22:365–370

Rivera-Ortiz JM, Burris RH (1975) Interactions among substrates and inhibitors of nitrogenase. J Bacteriol 123:537–545

Shah V (2000) Exploitation of cyanobacteria for photohydrogen production and wastewater treatment. Ph.D. thesis, Sardar Patel University, VallabhVidyanogar, India

Shima S, Pilak O, Vogt S, Schick M, Stagni MS, Meyer-Klaucke W, Warkentin E, Thauer RK, Ermler U (2008) The crystal structure of [Fe]-hydrogenase reveals the geometry of the active site. Science 321(5888):572–575

Simpson FB, Burris RH (1984) A nitrogen pressure of 50 atmospheres does not prevent evolution of hydrogen by nitrogenase. Science 224(4653):1095–1097

Singh SK, Kishor K, Sundaram S (2014) Photosynthetic microorganisms mechanism for carbon concentration. Springer Cham Heidelberg, New York, pp 1–113

Sinha P, Pandey A (2011) An evaluative report and challenges for fermentative biohydrogen production. Int J Hydrog Energy 36:7460–7478

Smith PR, Bingham AS, Swartz JR (2012) Generation of hydrogen from NADPH using an [Fe Fe] hydrogenase. Int J Hydrog Energy 37:2977–2983

Stripp ST, Goldet G, Brandmayr C, Sanganas O, Vincent KA, Haumann M, Armstrong FA, Happe T (2009) How oxygen attacks [FeFe] hydrogenases from photosynthetic organisms. Proc Natl Acad Sci 106:17331–17336

Tamagnini P, Axelsson R, Lindberg P, Oxelfelt F, Wunschiers R, Lindblad P (2002) Hydrogenases and hydrogen metabolism of cyanobacteria. Microbiol Mol Biol Rev 66:1–20

Tao Y, Chen Y, Wu Y, He Y, Zhou Z (2007) High hydrogen yield from a two-step process of dark- and photo-fermentation of sucrose. Int J Hydrog Energy 32:200–206

Thiel T (1993) Characterization of genes for an alternative nitrogenase in the cyanobacterium *Anabaena variabilis*. J Bacteriol 175:6276–6286

Tiwari A, Pandey A (2012) Cyanobacterial hydrogen production – a step towards clean environment. Int J Hydrog Energy 37:139–150

Tredici MR, Zittelli GC (1998) Efficiency of sunlight utilization: tubular versus flat photobioreactors. Biotechnol Bioeng 57:187–197

Tredici MR, Zittelli GC, Benemann JR (1998) A tubular internal gas exchange photobioreactor for biological hydrogen production: preliminary cost analysis. In: Zaborsky O et al (eds) Bio hydrogen. Plenum Press, New York, pp 391–402

Tsygankov AA, Hirata Y, Miyake M, Asada Y, Miyake J (1994) Photobioreactor with photosynthetic bacteria immobilized on porous glass for hydrogen photoproduction. J Ferment Bioeng 77:575–578

Ugwu CU, Ogbonna JC, Tanaka H (2002) Improvement of mass transfer characteristics and productivities of inclined tubular photobioreactors by installation of internal static mixers. Appl Microbiol Biotechnol 58:600–607

Vignais PM, Billoud B (2007) Occurrence, classification, and biological function of hydrogenases: an overview. Chem Rev 107:4206–4272

Vignais PM, Billoud B, Meyer J (2001) Classification and phylogeny of hydrogenases. FEMS Microbiol Rev 25(4):455–501

Weissman JC, Benemann JR (1977) Hydrogen production by nitrogen starved cultures of *Anabaena cylindrica*. Appl Environ Microbiol 33:123–131

Wilson PW, Umbreit WW (1937) Mechanism of symbiotic nitrogen fixation. III. Hydrogen as a specific inhibitor. Arch Microbiol 8:440–457

Winkler M, Hemschemeier A, Gotor C, Melis A, Happe T (2002) [Fe]-hydrogenases in green algae: photo-fermentation and hydrogen evolution under sulfur deprivation. Int J Hydrog Energy 27:1431–1439

Yokoi H, Mori M, Hirose J, Hayashi S, Takasaki Y (1998) H_2 production from starch by a mixed culture of *Clostridium butyricum* and *Rhodobacter* sp. M-19. Biotechnol Lett 20:895–899

Yoon JH, Sim SJ, Kim MS, Park TH (2002) High cell density culture of *Anabaena variabilis* using repeated injections of carbon dioxide for the production of hydrogen. Int J Hydrog Energy 27:1265–1270

Zhao X, Xing D, Liu B, Lu L, Zhao J, Ren N (2012) The effects of metal ions and L-cysteine on hydA gene expression and hydrogen production by *Clostridium beijerinckii* RZF-1108. Int J Hydrog Energy 37:1711–1717

Characterization and Screening of Algal Strains for Sustainable Biohydrogen Production: Primary Constraints

6

Ramkrishna Ghosh, Punyasloke Bhadury, and Manojit Debnath

Abstract

Algae have emerged as one of the most promising sources for biofuel production. In particular, microalgae can provide several different types of renewable biofuels like biodiesel, ethanol, and biohydrogen. Oxygenic photosynthesis splits water to release oxygen gas and uses the hydrogen atoms to drive the reduction of carbon dioxide to sugars. Under some circumstances, cyanobacteria are able to release the reductant as hydrogen gas. Hydrogen is an excellent source for fuel cells and has some attractive features such as three times more potentiality than ethanol. Algal communities including cyanobacteria can produce H_2 through three main routes: (1) H_2 production directly from native bidirectional hydrogenase, (2) H_2 production from a native nitrogenase, and (3) H_2 production from an introduced hydrogenase. Over the last decade or so, several new algal hydrogenases have been reported in literature, and efforts have been undertaken by manipulation of genetic pathways and metabolic engineering approaches. However, such approaches have shown constraints in terms of scale-up at the industrial level. This chapter highlights the aspect of metabolic engineering approaches and underlying constraints for biohydrogen production from algae. This chapter mainly discusses biohydrogen production potential of algae with a focus on understanding of biomass production, optimization of H_2 production in response to strength of selected solution, and pH of the culture medium.

R. Ghosh
Systematics and Applied Phycology Laboratory, Postgraduate Department of Botany, Hooghly Mohsin College, Chinsurah, 712101 Hooghly, West Bengal, India

P. Bhadury
Integrative Taxonomy and Microbial Ecology Research Group, Department of Biological Sciences, Indian Institute of Science Education and Research Kolkata, Mohanpur, 741246 Nadia, West Bengal, India

M. Debnath (✉)
Systematics and Applied Phycology Laboratory, Postgraduate Department of Botany, Hooghly Mohsin College, Chinsurah, 712101 Hooghly, West Bengal, India

Integrative Taxonomy and Microbial Ecology Research Group, Department of Biological Sciences, Indian Institute of Science Education and Research Kolkata, Mohanpur, 741246 Nadia, West Bengal, India
e-mail: mandeb@rediffmail.com

A. Singh, D. Rathore (eds.), *Biohydrogen Production: Sustainability of Current Technology and Future Perspective*, DOI 10.1007/978-81-322-3577-4_6

6.1 Introduction

The progress of human civilization over centuries has been always dependent on energy. Supply of clean and sustainable energy is arguably the most important scientific and technical challenge facing humanity in this century (Lewis and Nocera 2006). Global energy consumption is projected to increase by at least twofold in the middle of this century due to population outburst and rapid economic growth. Presently, global energy consumption is approximately 15 TW, while it is estimated to increase to 27 TW in the year 2050 and may surpass 40 TW by 2100 (Lewis and Nocera 2006). The high energy demand could be met, in principle, from fossil energy resources, primarily derived from plant biomass. However, resulting carbon dioxide (CO_2) emissions due to burning of fossil fuels will severely affect global climate along with rise in air temperature (IPCC 2014). Therefore, it is important to develop technologies for carbon-neutral energy production which can meet global energy demand and at the same time safeguard environment. Except nuclear energy, other promising alternative renewable energy sources are mainly ethanol and hydrogen gas, which are derived from biological organisms. Organisms that produce hydrogen photobiologically can be broadly divided into two groups: photosynthetic bacteria and oxygenic photosynthetic organisms. Photosynthetic bacteria require organic compounds or reduced sulfur compounds as electron donors which are generally limited. On the other hand, oxygenic photosynthetic organisms use water as an electron donor, which is available in plentiful. Among oxygenic photosynthetic organisms, prokaryotic and eukaryotic microalgae have emerged as one of the most promising sources for biofuel production. In particular, microalga can be excellent source for various renewable biofuels such as biodiesel, ethanol, and biohydrogen (Darzin et al. 2010; Lakaniemi et al. 2011; Nayak et al. 2014). During oxygenic photosynthesis, water is split to oxygen gas (O_2), and hydrogen atoms are used to reduce CO_2 to sugars. Under certain conditions, microalgae are able to release the reductant as hydrogen gas (Melis 2002). Hydrogen (H_2) is an excellent fuel due to its highest energy efficiency (143GJ tonne^{-1}) (Nayak et al. 2014).

Algal communities including cyanobacteria can produce H_2 through three main routes: (i) H_2 production mediated by native bidirectional hydrogenase enzyme, (ii) H_2 production by native nitrogenase enzyme, and (iii) H_2 production from an introduced hydrogenase enzyme. Hydrogen production in prokaryotic microalgae (cyanobacteria) was first reported more than a century ago (Jackson and Ellms 1896). In cyanobacteria, an alternate nitrogenase-based H_2 production pathway is present, thereby catalyzing unidirectional production of hydrogen. In 1942, Hans Gaffron and co-workers demonstrated hydrogenase activity in the eukaryotic green microalga *Scenedesmus obliquus* (Gaffron and Rubin 1942; Homann 2003; Melis and Happe 2004). Hydrogenase activity is not observed in all eukaryotic microalgal groups which also include green algae (Brand et al. 1989; Boichenko et al. 2004; Melis and Happe 2004). It has been reported that some eukaryotic algae can produce H_2 using low-potential electrons derived either from light-driven photosynthetic pathways or during organic substrate fermentation (Brand et al. 1989; Happe and Kaminski 2002; Boichenko et al. 2004; Ghirardi et al. 2007). Over the last decade or so, several new algal hydrogenases have been reported in literature (e.g., Leino et al. 2014), and efforts involving manipulation of genetic pathways coupled with metabolic engineering approaches have been attempted for large-scale production of biohydrogen. However, such approaches have shown considerable constraints and challenges in terms of scale-up at the industrial level (Levin et al. 2004). Hydrogen production in microalgae is currently limited by technological constraints (Dubini et al. 2014). Specific limitations include:

(i) Extreme sensitivity of hydrogenases to molecular oxygen
(ii) Low reductant availability for hydrogenase activity due to the existence of competing metabolic pathways that converge at the level of ferredoxin (FD)

(iii) Downregulation of photosynthetic electron transport and establishment of cyclic electron transfer around photosystem I (PSI) under anaerobic H_2-producing conditions
(iv) Low level at which light saturation occurs during photosynthesis
(v) Reversible nature of hydrogenases that results in consumption of H_2 under high H_2 partial pressure
(vi) Low level of hydrogenase expression

The main emphasis of this chapter has been focused on elucidating biohydrogen production potential in microalgae such as prokaryotic cyanobacteria and eukaryotic green algae. Pertinent issues such as types of enzyme that catalyze biohydrogen production, microalgal biomass optimization, and the role of culture conditions from the context of pH and carbon source, in addition to selection of strains for H_2 production, have been thoroughly discussed. In the last part of this chapter, application of metabolic engineering approaches toward large-scale production of biohydrogen from microalgae and associated underlying constraints have been also highlighted.

6.2 Types of Biofuels from Algae

Algal biomass can be converted into biofuel by several processes including thermochemical and biological processes (Bridgwater 2003). Thermochemical processes such as gasification, pyrolysis, liquefaction, or even direct combustion can convert stored energy into gases like hydrogen (Bridgwater 2003; Murphy et al. 2013). Biological processes such as fermentation of biomass produce energy carriers, namely, bioethanol, biomethane, and biohydrogen (Hu and Gao 2003; Hu et al. 2008; Mata et al. 2010; Lakaniemi et al. 2011; Nayak et al. 2014) (Fig. 6.1). Optimization of microalgal biomass production is dependent on several biotic and abiotic factors. The selection of algal strain with most efficient biohydrogen potential is of utmost importance during scaling up of biomass.

6.3 Fundamentals of Biohydrogen Gas Production by Algae

6.3.1 Types of Algal Group

Algae represent a heterogeneous group of chlorophyll a containing oxygenic photosynthetic organisms (Falkowski et al. 2004). They are capable of fixing large amounts of carbon dioxide (approximately 50 Gt carbon fixed per year) while contributing approximately 40–50 % oxygen in the atmospheric pool (Field et al. 1998; Uitz et al. 2010; Giering et al. 2014). Algal com munities can be found in terrestrial and aquatic habitats and adapt to varying temperature, pH

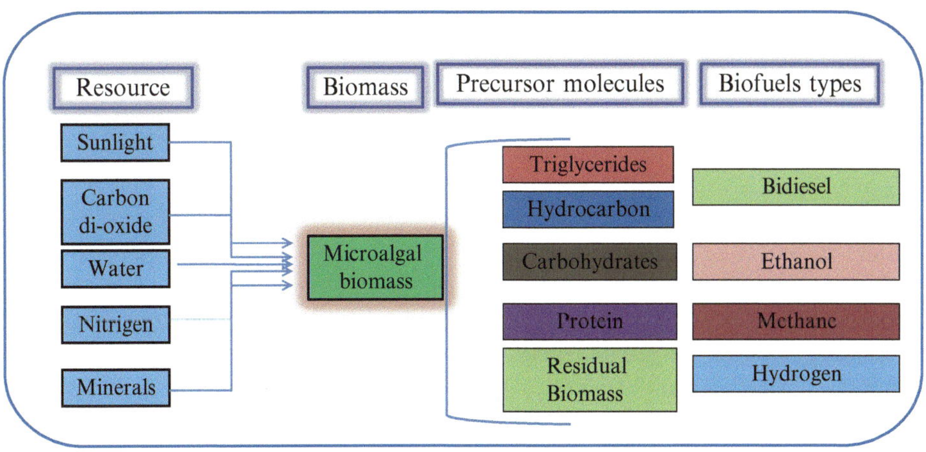

Fig. 6.1 Types of biofuel derived from microalgae grown in the presence of natural resources and inorganic minerals (Modified from Darzins et al. 2010)

and salinity conditions (Uitz et al. 2010). Microalgae are highly productive, represent 0.2 % of the world's total photosynthetic biomass inventory on a global scale, with rapid turnover of biomass (cell doublings of 1–4 per day) (Gallagher 2011; Slade and Bauen 2013). In microalgae, photosynthetic efficiency is higher by almost 20 % under available photosynthetically active radiation (PAR). In terrestrial crops, generally lower photosynthetic conversion efficiencies have been observed (Dismukes et al. 2008). For example, sugarcane, one of the most productive of all terrestrial crops, has a photosynthetic efficiency of 3.5–4 % (Odum 1971; Dismukes et al. 2008). The sugars formed during photosynthesis are converted to cellular molecules such as carbohydrates, proteins, and lipids that ultimately make up the biomass. Therefore, sustainable and carbon-free renewable energy sources such as biohydrogen can be effectively obtained from microalgal biomass while safeguarding the environment. In this context, microalgae have become a target for bioenergy production on an industrial scale.

Over 40,000 species of algae have been identified, and that number almost certainly represents a small fraction of the true unexplored diversity (Hu et al. 2008; Falkowski and Raven 2013). Algal classification is primarily dependent upon morphological and physiological attributes such as whole organismal morphology, cellular anatomy and ultrastructure, photosynthetic pigments, and metabolism (Lee 2009). The biological divisions that encompass various groups of algae are Cyanophyta (cyanobacteria); Prochlorphyta, Glaucophyta, and Rhodophyta (red algae); Cryptophyta (cryptomonads); Chlorophyta (green algae); Euglenophyta, Chloroarachniophyta, and Pyrrophyta (dinoflagellates); and Chromophyta (heterokonts) (Graham et al. 2000; Lee 2009). Generally macroalgae are not considered as potential candidates for commercial scale biohydrogen production due to slower growth rate and specific nutritional requirements (Pedersen and Borum 1996). According to Lee (2009), algae can be divided into four major evolutionary groups consisting of ten division including cyanobacteria and green microalgae (Chlorophyta) which are of global interest as major biohydrogen feedstock. There are other systems of algal classification where groupings are different under Chlorophyta and Chromophyta (Van den Hoek et al. 1995; Graham et al. 2008).

6.3.2 Microalgal Biohydrogen Gas Production

Microalgae use solar energy to convert water into hydrogen gas. Inside a microalgal cell, water is split into proton (H^+) and oxygen (O_2) in the presence of light. The H^+ gets converted into molecular hydrogen in the presence of hydrogenase enzyme through a process known as direct photolysis (Ayhan 2009). Hydrogen production in this process is low because (i) immediately after formation, H_2 and O_2 are converted into water and (ii) sensitivity of the enzyme hydrogenase to oxygen (Nath and Das 2004). Advantages and disadvantages of different light-dependent hydrogen production process have been summarized in Table 6.1.

The inhibitory effect can be overcome by adopting indirect biophotolysis during large-scale biohydrogen production. Indirect biophotolysis consists of two stages: in stage I, the cells synthesize organic compounds (mostly glucose) by photosynthesis, and during the process, oxygen is released. In stage II, algal cells degrade stored organic compounds under anaerobic condition (Melis and Melnicki 2006). During stage II, oxygen and hydrogen are evolved separately. Stage II can happen in the presence or absence of light, also known as photofermentation and dark fermentation, respectively (Guan et al. 2004). The concept of two-staged hydrogen production by microalgae and factors that affect hydrogen yield in stage I and stage II has been illustrated in Fig. 6.2. It has been also shown that growing algal cells in stage I produce optimum level of hydrogen (Rashid et al. 2013). Photosynthetic cyanobacteria and green algae provide a more promising pathway for generation of hydrogen on a large scale compared to hydrogen produced during non-photosynthetic fermentation (Roy

Table 6.1 Summary of advantages and disadvantages of light-dependent hydrogen production processes in cyanobacteria and eukaryotic microalgal biomass

Microalgal groups	Preferable light-dependent metabolic pathway for H_2 production	Advantages	Disadvantages
Cyanobacteria	Indirect biophotolysis through nitrogenase	H_2 evolution is separated from O_2 evolution	High-energy-dependent process
		Spatial separation in heterocystous N_2-fixing cyanobacteria	Biosynthesis and maintenance of heterocysts
		Temporal separation (light/dark) in nonheterocystous cyanobacteria	Significant ATP requirement for nitrogenase. The presence of uptake hydrogenase
			Reoxidize produced molecular hydrogen
Green microalgae	Direct and indirect biophotolysis through bidirectional hydrogenase	Hydrogen-economy strategy based on a virtually limitless and renewable source	Production of O_2 and H_2 simultaneous
		Energy cycle is carbon-free	Inhibition of hydrogenase by O_2
		Theoretical energy efficiency is much higher for hydrogen production from biophotolysis (40 %) compared to hydrogen production from biomass (1 %)	
		One of the most promising processes due to separate production of O_2 and H_2	

Modified from Dasgupta et al. 2010a, b

et al. 2014). Cyanobacteria and green microalgae absorb light through pigments that are associated with two photosystems, photosystem I (PSI) and photosystem II (PSII) (Fig. 6.3).

The absorbed light energy is transferred from antenna pigments to chlorophyll reaction center molecules where charge separation occurs, yielding oxidants and reductants (Hallenbeck 2012). The strong oxidant generated by PSII extracts electrons from water while releasing oxygen and protons as by-products (Fig. 6.2). The generated electrons reduce a series of membrane-bound and membrane-soluble carriers, ultimately reducing oxidant generated by PSI (Falkowski and Raven 2013). Photosystem I generates a reductant that eventually reduces the iron-sulfur protein ferredoxin, which plays a vital role in several metabolic processes such as cyclic and noncyclic photophosphorylation and nitrogen fixation (Bothe et al. 2010). The main function of PSI is to provide electrons to generate NADPH via ferredoxin-NADP oxidoreductase (FNR). The NADPH molecule, along with ATP, is needed for fixing carbon dioxide via Calvin-Benson-Bassham cycle ultimately resulting in the production of carbohydrate molecules. However, under anaerobic conditions in the absence of carbon dioxide, NADPH or reduced ferredoxin reduces protons to yield hydrogen gas, a reaction catalyzed by hydrogenase. In cyanobacteria, NADPH is the likely electron donor to hydrogenase, whereas ferredoxin links photosynthetic electron transport directly to hydrogen production in case of green microalgae (Hallenbeck 2012).

Fig. 6.2 Schematic representation of two-staged hydrogen production within the chloroplast or thylakoid of an oxygenic photosynthetic eukaryotic/prokaryotic cell (Modified from Rashid et al. 2013)

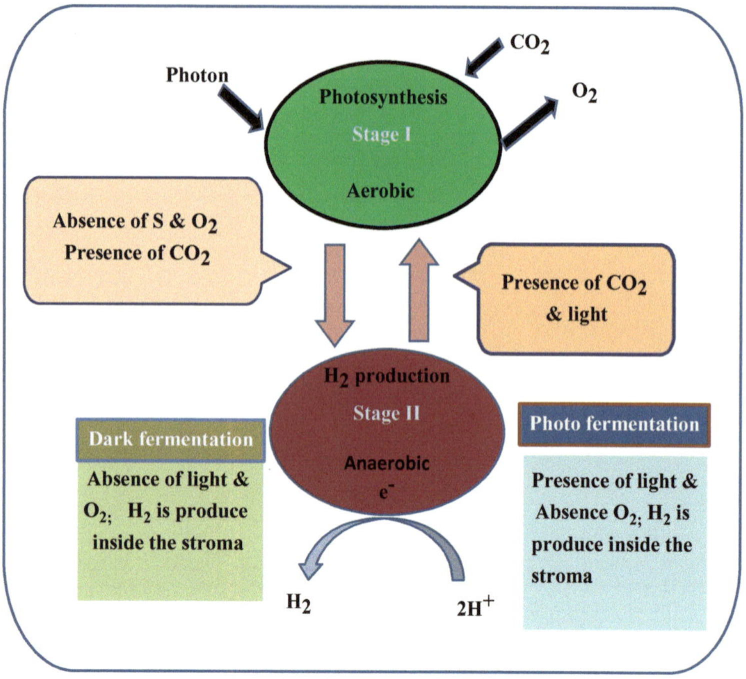

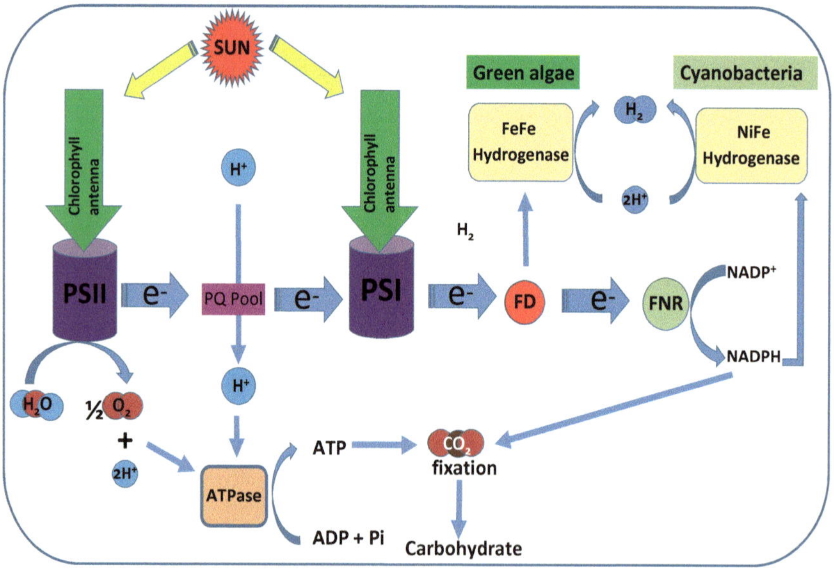

Fig. 6.3 Hydrogen production in cyanobacteria and green algae through photosynthesis-based pathways using different hydrogenases. *PS* photosystem, *PQ* plastoquinone, *FD* ferredoxin, *FNR* ferredoxin-NADP oxidoreductase (Modified from Hallenbeck 2012)

Immobilization and sulfur (S) starvation are the key intermediate steps of stage I and stage II, respectively, during indirect photobiolysis. For immobilization, microalgal cells are suspended in a solidifying material and cut into small pieces. Immobilized cells have higher stability and produce more hydrogen than free cells. Sulfur-deprived cells yield more hydrogen than sulfur-provided cells. In the presence of sulfur, the cell synthesizes proteins which suppress hydrogen production. In cyanobacteria, key enzymes involved in hydrogen production are hydrogenase and nitrogenase (Hallenbeck 2012; Nayak et al. 2014). Under in vivo condition, the enzyme nitrogenase is activated in the presence of light and absence of nitrogen (Bothe et al. 2010), whereas hydrogenase is activated at high light intensity and pH (Tamburic et al. 2011; Rashid et al. 2013). The mechanism during stage II as part of photobiological H_2 production is relatively well understood compared to stage I (Kosourov et al. 2002; Das and Veziroglu 2008; Rashid et al. 2011). An in-depth study of both stages is essential for optimization of hydrogen production in algal cells. Characteristics of photobiological hydrogen production processes along with mediating enzymes in cyanobacteria and green microalgae have been summarized in Table 6.2.

6.3.3 Types of Enzymes for Biohydrogen Production

Three enzymes, namely, hydrogenase (Hox), uptake hydrogenase (Hup), and nitrogenase (nif), are known to be involved in hydrogen generation in cyanobacteria and green microalgae (Tamagnini et al. 2007; Bothe et al. 2010; Berggren et al. 2013) (Fig. 6.4). In this subsection, characteristic features, mode of action, and H_2 production of these three enzymes have been detailed.

6.3.3.1 Hydrogenase

Hydrogenase is a key enzyme for biological hydrogen production which is present across all domains of life including bacteria and plant kingdom (Tamagnini et al. 2007).

Cyanobacteria and green algae also contain this particular class of enzyme. In a cyanobacterial cell, hydrogenase is present in the cytoplasm (Mathews and Wang 2009), whereas in a green algal cell, it is found in the chloroplast (Dubini et al. 2014). In cyanobacteria and green algae, protons and oxygen are produced by splitting of water, and generated protons are converted into hydrogen in both photosystems (PSI and PSII) and mediated by enzymes such as ferredoxin and hydrogenase (Fig. 6.3). The enzyme hydrogenase is sensitive to oxygen and gets deactivated at 2 % O_2 partial pressure (Ghirardi 1997). Hydrogenase oxidizes the low redox electron carrier ferredoxin during reversible reaction. In direct biophotolysis, light drives simultaneous O_2 evolution on the oxidizing side of PSII and H_2 production on the reducing side of PSI, with a maximum H_2:O_2 (mol/mol) ratio of 2:1 (Melis et al. 2000). Under anaerobic conditions, the activity of this enzyme is known to increase significantly (Melis et al. 2000; Bothe et al. 2010). Three phylogenetically distinct classes of this enzyme are known based on the composition of its metal center: [NiFe], [FeFe], or [Fe] (Berggren et al. 2013). The first two classes contain binuclear metal cores with unusual ligand spheres as catalytic centers, whereas the third class contains a mononuclear iron next to a special organic cofactor. Two types of hydrogenases, [FeFe] hydrogenase and [NiFe] hydrogenase, are known to be present in microalgae. The [FeFe] hydrogenase is 10–100 times more efficient than [NiFe] hydrogenase (Madden et al. 2011). The [FeFe] hydrogenase has protein containing [FeFe] catalytic core, while [NiFe] hydrogenase has selenium also in the form of selenocysteine (Volgusheva et al. 2013). The [FeFe] hydrogenases thermodynamically favor hydrogen production relative to [NiFe] hydrogenases, which are frequently regarded as predominantly H_2 uptake enzymes (Ducat et al. 2011). This has led to intense research focusing on the application of [FeFe] hydrogenase for sustainable production of H2 based on metabolic engineering approaches.

In all hydrogenases, including those found in microalgae, the active site has been found to be buried within the proteins. The hydrogenases contain three types of channels or pathways for

Table 6.2 Characteristic features of photobiological hydrogen production strategy and types of enzyme involved in hydrogen production across different microalgae

H_2 production strategies	H_2-evolving enzymes	Characteristic features	Requirements/comments	Microalgal groups
Photobiological hydrogen production utilizing oxygen-evolving photosynthesis				
1. Single stage: direct production of hydrogen				
A. Truly single stage	Hydrogenase	Simultaneous production of H_2 and O_2 in a single cell	Quick separation of H_2 from O_2	Cyanobacteria and green microalgae
B. Operationally single stage		Apparently continuous production of H_2	Operationally simple	
a. Spatial separation	Nitrogenase	Different H_2-producing cells and O_2-producing cells	Subsequent separation of H_2 from O_2	Heterocystous cyanobacteria
b. Temporal separation	Nitrogenase	Cyanobacteria and O_2-producing stages in the same cells	Subsequent separation of H_2 from O_2	Nonheterocystous cyanobacteria
2. Two stage				
Stage I: Accumulation of organic compounds by oxygenic photosynthesis				
Stage II				
A. Anaerobic light-driven H_2 production				
a. Linked to PSI	Hydrogenase	The presence of air	Change of gas phase	Cyanobacteria and green microalgae
b. Linked to PSI	Hydrogenase	PSII inactivation by sulfate depletion	Change of gas phase	Green microalgae
B. Dark fermentation	A variety of pathways			Bacteria

Modified from Bothe et al. (2010), Rashid et al. (2013)

gas access, proton as well as electron transfer. Such channels facilitate the movement of substrate and educts to move between active site and molecular surface (Fontecilla-Camps et al. 2007).

A soluble or loosely membrane associated [NiFe] hydrogenase present in some cyanobacteria. This [NiFe] hydrogenase can produce and take up hydrogen, known as bidirectional hydrogenase (Schmitz et al. 2002) (Fig. 6.4c). The bidirectional hydrogenase, purified from the cyanobacterium *Synechocystis* sp. PCC 6803 and *Synechococcus* sp. PCC 6301, is a complex dimer of proteins (HoxF, HoxU, HoxY, HoxH, HoxE) with a molecular weight of 375 kDa (Schmitz et al. 2002; Schwarz et al. 2010; Hallenbeck 2012). It has been also shown that some of the [NiFe] hydrogenase shows tolerance to oxygen (e.g., Ghirardi et al. 2007). Green microalgal cells are known to encode two distinct hydrogenases, namely, HYDA1 and HYDA2, with recent RNA interference-based approach showing that HYDA1 is the predominant H_2-producing enzyme (Godman et al. 2010). It has been shown that HYDA homologs are similar across members of green algae including *Scenedesmus obliquus* and *Chlamydomonas reinhardtii*, and they encode only H-cluster-active site domain and lack additional N-terminal iron-sulfur (FeS) cluster-binding domains (Fe-cluster) (Florin et al. 2001; Happe and Kaminski 2002). Fe-clusters are thought to mediate electron transfer between physiological donors/acceptors

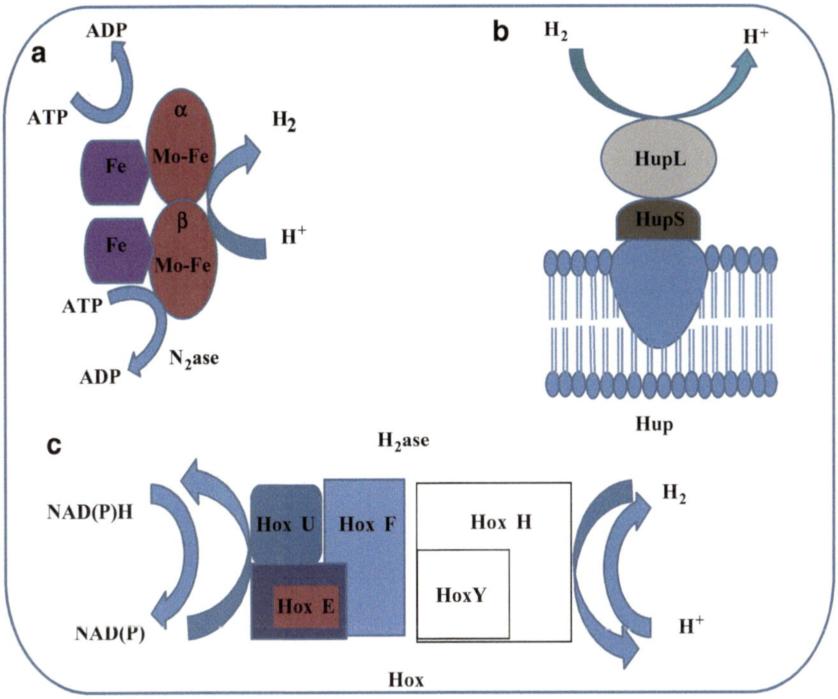

Fig. 6.4 Enzymes involved in hydrogen metabolism inside a cyanobacterial cell (Modified from Bothe et al. 2010; Hallenbeck 2012)

(Vignais and Billoud 2007). It has been also shown that *Chlorella variabilis* NC64A, a green algae, contain *HYDA* genes which code for accessory FeS cluster-binding domains (F-cluster), in contrast to other members of this group. In general, the two hydrogenases coded by *HYDA1* and *HYDA2* genes have different promoter regions and get transcribed and regulated differently in response to varying environmental conditions (Forestier et al. 2003). The Fe hydrogenases from green algae are monomeric proteins of about 45–50 kD in size (Roessler and Lien 1984; Happe and Naber 1993). The nucleus-encoded polypeptides are synthesized in the cytosol as precursor proteins, but the mature protein is localized in the chloroplast stroma of green alga *Chlamydomonas reinhardtii* (Happe et al. 1994).

6.3.3.2 Uptake Hydrogenase

Heterocystous cyanobacterial cells that fix nitrogen appear to have an uptake hydrogenase, whose function is to recover electrons lost during hydrogen production by nitrogenase (Tamagnini et al. 2000). The uptake hydrogenase comprises of two subunits encoded by *hupS* and *hupL* (Schmitz and Bothe 1996). In cyanobacteria, the *hupSL* genes appear to be transcribed predominantly or exclusively in heterocysts such as in *Anabaena* and *Nostoc*, consistent with their role in nitrogen fixation. However, there is one report that indicates that *hupL* is expressed in vegetative cells of *Anabaena variabilis* (Boison et al. 2000). The uptake hydrogenase is known to be resistant to oxygen. Two different mechanisms of oxygen tolerance have been noted in microalgal uptake hydrogenases. First, an additional CN^- ligand bound to nickel of [NiFe] site contributes to oxygen tolerance in this enzyme. Mutant proteins devoid of the nickel-bound CN^- ligand turned out to be oxygen sensitive (Burgdorf et al. 2005). Second, X-ray absorption spectroscopy revealed that the active site of this hydrogenase is coordinated by more oxygen ligands and less sulfur ligands. The different coordination of [NiFe]-active site may also contribute to oxygen tolerance in this enzyme (Burgdorf et al. 2005). For increased yield of hydrogen, uptake hydrogenase has been modified making them more oxygen tolerant based on protein engineering approaches in cyanobacteria and green algae (Das and Veziroglu 2001; Dasgupta et al. 2010a); however, this enzyme require high redox potential. Because of the limitation of this enzyme, it is not useful for maintaining a stable hydrogen production in microalgae (Lindblad et al. 2001; Rashid et al. 2013).

6.3.3.3 Nitrogenase

Nitrogenase converts N_2 into ammonia and produces hydrogen. Among oxygenic photosynthetic microalgae, only cyanobacteria contain this enzyme. The enzyme nitrogenase can be categorized based on metal, molybdenum (Mo), vanadium (V), and iron (Fe) present in the prosthetic group, namely, Mo-nitrogenase, V-nitrogenase, and Fe-nitrogenase (Tamagnini et al. 2002; Bothe et al. 2010). The reaction of nitrogenase enzyme is energetically inefficient due to irreversible reaction coupled by hydrolysis. Cyanobacteria can be broadly divided into nitrogen-fixing and nonnitrogen-fixing groups. Nitrogen fixation requires ATP (10–16 molecules for per molecule of dinitrogen) and reductants such as NADP or NADPH in cyanobacterial cell (Bothe et al. 2010). Majority of cyanobacterial cells show nitrogenase activity under anaerobic or micro-aerobic conditions (Bothe et al. 2010).

6.3.4 Biohydrogen Production in Cyanobacteria

Hydrogen production in cyanobacteria is potentially feasible using either reversible hydrogenase or nitrogenase (Bothe et al. 2010). The uptake hydrogenase recycles hydrogen produced by nitrogenase; thus, high levels of hydrogen production require inactivation of *Hup SL* (Fig. 6.5). In a study conducted by Weissman and Benemann (1977), *Anabaena cylindrica* was shown to produce hydrogen continuously for 7–19 days under an argon-CO_2 atmosphere in the presence of nitrogenase; however, the highest rates of production declined to half of the maximum within 5–7 days (in 30 ml per liter culture per hour). In *Anabaena* sp. CA and *Anabaena* sp. IF, hydrogen was produced at a rate of about 20–40 μl mg dry weight^{-1} hr^{-1} in air with higher yields at higher

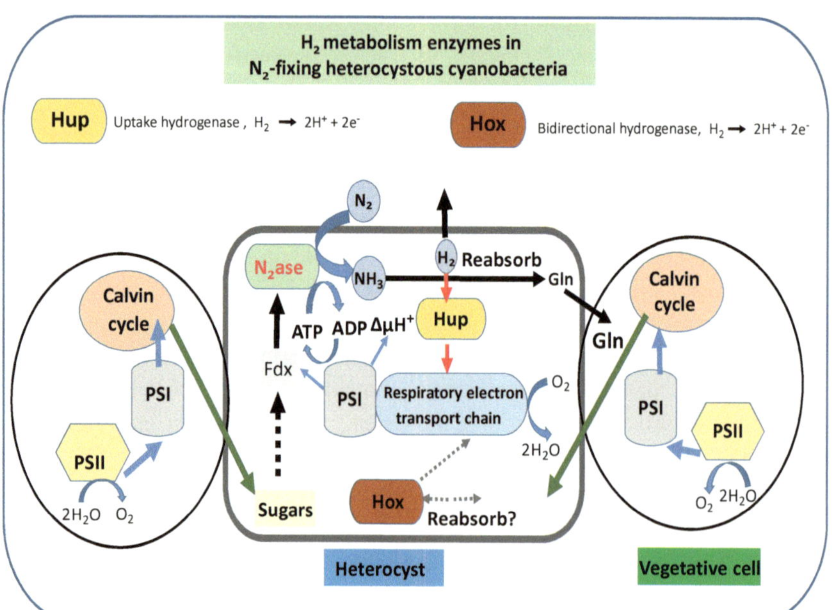

Fig. 6.5 Hydrogen metabolism in heterocystous cyanobacterial filament under active nitrogen-fixing condition (Modified from Schutz et al. 2004)

light intensities (≤180 μEm^{-2}) (Zhang et al. 1983). In *Anabaena* sp. CA, hydrogen production by uptake hydrogenase is dependent on nickel concentration present in the growth medium. Uptake hydrogenase in this train becomes inactivated in the presence of 100 nM Ni^{2+} (Smith et al. 1985). In most nitrogen-fixing cyanobacteria, net hydrogen production by nitrogenase is virtually nonexistent unless the uptake hydrogenase is eliminated (Happe et al. 2000; Masukawa et al. 2002; Tsygankov et al. 2002; Lindberg et al. 2004). A cluster of accessory proteins for [NiFe] hydrogenases, including *hypAB*, *hypCD*, and *hypEF*, is located about 4 kb upstream from *hupSL* in *Nostoc punctiforme*, *Anabaena variabilis*, and *Anabaena* sp. PCC 7120 (Table 6.3).

The functions of Hyp protein and *hyp* gene have not been thoroughly studied in cyanobacteria. In *E. coli*, Fe binds to *HypCD*, which then binds CO and CN^- ligands with the help of *HypEF* in an ATP-dependent reaction (Vignais and Colbeau 2004). Different types of hydrogenase and nitrogenase genes present in members representing cyanobacteria have been summarized in Tables 6.3 and 6.4.

In a comparison of hydrogen production among several heterocystous strains, higher levels of H_2 production have been noted in *Anabaena* sp. strain PCC 7120 compared to *Nostoc* sp. strains ATCC 73102, ATCC 38901, and ATCC 91911 (Schutz et al. 2004). Experiments with *A. variabilis* PK84 were also conducted outdoor in a 4.35 L bioreactor in summer months (United Kingdom) with the longest experiment lasting for 40 days (Tsygankov et al. 2002). The authors found that hydrogen production was highest at outdoor temperature of 30 °C, and they concluded that higher temperature can provide even higher yields of hydrogen (Tsygankov et al. 2002). The amount of hydrogen produced by various cyanobacteria (*Nostoc* spp., *Anabaena* spp., and *Anabaena variabilis* PK84) lacking the uptake hydrogenase in an aerobic environment is in the range of 50–150 nmol H_2 μg Chl*a*$^{-1}$h^{-1} (Borodin et al. 2000; Happe et al. 2000; Masukawa et al. 2002; Tsygankov et al. 2002; Lindblad et al. 2002; Schutz et al. 2004). Major finding in terms of comparative biohydrogen production has been summarized in Table 6.5.

The uptake hydrogenase may perform differently in different strains of cyanobacteria. A strain of *Cyanothece* sp. PCC 51142 when grown aerobically and incubated under argon produced higher amount of H_2. The H_2 production rate reached 373 mol mg chlorophyll *a* (Chl a)$^{-1}$h^{-1} under aerobic conditions and over 400 mol mg Chl a^{-1}h^{-1} when grown in the presence of 50 mM glycerol (Bandyopadhyay et al. 2010). The H_2 production rate in a *hupL* mutant strain of *Nostoc* sp. PCC 7422 increased by threefold compared to wild type (80 mol mg Chl a^{-1}h^{-1}) (Yoshino et al. 2007). In another *hupL* mutant strain of *Nostoc punctiforme* 29133, hydrogen production increased to 9 mol mg Chl a^{-1}h^{-1} in contrast to trace amounts produced in wild type (Ekman et al. 2011). Increased hydrogen production has been also reported in several other mutant strains of *Anabaena* sp. The uptake hydrogenase-deficient mutant of *Anabaena variabilis* PK84 strain produces more hydrogen (1670.6 mol/mg Chl a/h) compared to wild-type *A. variabilis* (39.4 mol mg Chl a^{-1}h^{-1}) when grown anaerobically with 25 % N_2, 2 % CO_2, and 73 % Ar (Sveshnikov et al. 1997; Dutta et al. 2005). Transcription regulation of uptake hydrogenase plays an important role in hydrogen production under nitrogen-fixing condition in cyanobacterial cells (Happe et al. 2000). In N_2-fixing cyanobacteria, *hupSL* transcription is coordinately regulated with nitrogenase along with the differentiation of heterocyst (Happe et al. 2000). In *Anabaena variabilis* ATCC 29413, *Anabaena* sp. PCC 7120, *Nostoc punctiforme* ATCC73102, and *Gloeothece* sp. strain ATCC 27152, the transcription of *hupSL* increases with the decrease in nitrogen fixation (Happe et al. 2000; Lindberg et al. 2000; Oliveira et al. 2004).

6.3.5 Biohydrogen Production in Eukaryotic Microalgae

In eukaryotic microalgal domains, *Chlamydomonas reinhardtii* has emerged as a model organism for studying H_2 metabolism,

Table 6.3 Different type of hydrogenase genes present in cyanobacteria

Strain	Hup	Hox	Hyp	References
Lyngbya majuscula	+	+	+	Tamagnini et al. (2000) and Schutz et al. (2004)
Nostoc sp. Mitsui 38901	+	–	+	Tamagnini et al. (2000) and Schutz et al. (2004)
Nostoc sp. Mitsui 56111	+	+	+	Tamagnini et al. (2000) and Schutz et al. (2004)
Anabaena sp. PCC 7120	+	+	+	Kaneko and Tabata (1997)
Unpublished data	Web site			
Synechococcus sp. PCC 7942	–	+	+	http://genome.jgipsf.org
Prochlorococcus marinus MIT9312	–	–	–	http://genome.jgipsf.org
Prochlorococcus marinus NATL 2A	–	–	–	http://genome.jgipsf.org
Crocosphaera watsonii WH8501	+	–	+	http://genome.jgipsf.org
Trichodesmium erythraeum IMS101	+	–	+	http://genome.jgipsf.org
Nostoc punctiforme ATCC 29133	+	–	+	http://genome.jgipsf.org
Anabaena variabilis ATCC 29413	+	+	+	http://genome.jgipsf.org
Unpublished data	GenBank accession number			
Synechococcus sp. PCC 6301	–	+	+	AP008231GR (Genome)
Synechococcus sp. PCC 7002	–	+	+	AAN03569.;AAN03573
Prochlorococcus marinus MIT9313	–	–	–	BX548175GR (Genome)
Prochlorococcus marinus CCMP1378	–	–	–	BX548174GR (Genome)
Prochlorococcus marinus CCMP1375	–	–	–	AE017126GR (Genome)
Prochlorothrix hollandica	NA	+	NA	AAB53705
Gloeobacter violaceus PCC7421	–	–	–	BA000045GR (Genome)
Arthrospira (spirulina)	NA	+	NA	AAQ63961
Spirulina subsalsa	NA	+	NA	AAO962
Anabaena siamensis TISTR8012	+	+	NA	AAN65267, AAN65266

and substantial progress has been made in recent years to understand mechanisms of H_2 production (Melis et al. 2000; Kruse et al. 2005; Kosourov and Seibert 2009). Detailed genetic and physiological studies describing several aspects of H_2 production in *C. reinhardtii* are available, and several of its mutants related to H_2 metabolism have been made and studied. It has been shown that *C. reinhardtii* has an unprecedented repertoire of metabolic capabilities that allow it to adapt to rapidly changing environmental conditions, including anoxia, which is important for H_2 production. The metabolic flexibility of *C. reinhardtii* and other phototrophic microorganisms (*Chlorella* sp., *Scenedesmus* sp., and *Nannochloropsis* sp.) likely facilitates acclimation to natural energetic fluxes arising from environmental conditions and diurnal light/dark cycles, allowing these organisms to readily adjust their metabolic flux in response to diverse

Table 6.4 Uptake and bidirectional hydrogenases as well as conventional and alternative nitrogenases in top 10 H_2-producing cyanobacterial strains

	Nitrogenase	Alternative nitrogenases	Uptake hydrogenase	Bidirectional hydrogenase	
Cyanobacterial strain	nifHDK1	nifHDK2	vnfDGK	hupLS	hoxY
Synechocystis sp. PCC 6803	–	–	–	–	+
Nostoc sp. Becid 19	+	–	–	+	+
Nostoc sp. XHIID A6	+	–	–	+	+
Anabaena sp. XSPORK 7B	NA	NA	NA	+	+
A. variabilis ATCC 29413	+	+	+	+	+
Anabaena sp. PCC 7120	+	–	–	NA	NA
Nodularia sp. AV33	+	–	–	+	+
Nodularia sp. TRO31	+	–	–	+	+
Calothrix sp. 336/3	+	–	–	+	+
Calothrix sp. XPORK 5E	+	–	–	+	+
Calothrix sp. XSPORK 11A	NA	NA	NA	–	–
Calothrix sp. XSPORK 11A	+	–	–	–	–
Calothrix sp. Becid 33	+	–	–	+	+

Strains source: University of Helsinki Cyanobacteria Collection, UHCC) (Modified from Leino et al. 2014

Table 6.5 Hydrogen production in heterocystous cyanobacteria (wild type/mutant strain) grown under varying growth conditions

Strain	Characteristic features	Growth conditions	H_2production nmol μg $Chla^{-1}h^{-1}$(maximum)	References
Nostoc punctiforme	$hupSL^-$	Air	6	Lindberg et al. (2004)
N. muscorum	Wild type	Air	4	Scherer et al. (1980)
A. variabilis AVM13	$hupSL^-$	Air	135	Happe et al. (2000)
A. variabilis PK84	$hupSL^-$ – Mo + V grown cell	Air	106	Borodin et al. (2000)
Anabaena sp. PCC7120	Wild type	Air	10	Masukawa et al. (2002)
Anabaena sp. PCC7120	$hupSL^-$	Air	52	Masukawa et al. (2002)
Anabaena sp. PCC7120	$hupSL^-$ $hoxH^-$	Air	50	Masukawa et al. (2002)
Anabaena cylindrica	Wild type	Air + 0.2 % CO + 10 % C_2H_2 + 3 % CO_2	66	Lambert et al. (1979)
Anabaena sp. CA	$hupSL^-$+ Ni and Ni-free	Air	0.4	Smith et al. (1985)
A. variabilis	Wild type	Air	10	Happe et al. (2000)
Anabaena sp. TU37-1	Wild type	Air	3	Kumazawa (2003)

Table 6.6 Hydrogen production by green microalgae in laboratory photobioreactor through direct biophotolysis process

Organism	Maximum evolution rate (mmolg^{-1} hr^{-1})	Maximum productivity (mmolL^{-1}hr^{-1})	Gas for growth; light intensity (wm^{-2})	Gas for H evolution; light intensity (wm^{-2})	References
Chlamydomonas reinhardtii CC124	5.94	0.094	97% air 3% CO_2; acetate (17 mM)	Argon; S-free acetate (17 mM)	Kosourov et al. (2002)
Chlamydomonas reinhardtii CC1036	5.91	0.48	Air; acetate (17 mM)	Argon; S-free acetate (17 mM)	Laurinavichene et al. (2006)
Platymonas subcordiformis	0.001	0.002	Air; seawater nutrients; (L/D)	N_2; S-free seawater	Guan et al. (2004)

Table 6.7 Hydrogen production by green microalgae grown in different culture medium

Green microalgae	Growth medium	Maximum rate of H_2 production (ml L^{-1} h^{-1})	References
Chlamydomonas reinhardtii	Acetate(TAP-S medium)	4.3	Laurinavichene et al. (2006)
Chlamydomonas reinhardtii	TAP-S medium	4.5	Winkler et al. (2002a)
Chlamydomonas reinhardtii	TAP(acetate) medium	80–140	Skjånes et al. (2010)
Chlamydomonas moewusii	TAP-S medium	10.0	Winkler et al. (2002a, b)
Chlamydomonas noctigama	TAP(acetate) medium	30–80	Skjånes et al. (2010)
Chlamydomonas euryale	TAP (acetate) medium	22	Skjånes et al. (2010)
Chlorella sorokiniana	TAP (acetate) medium	148	Kumar et al. (2013)
Chlorella pyrenoidosa	TAP (acetate) medium	10	Skjånes et al. (2010)
Scenedesmus obliquus	TAP-S medium	3.6	Winkler et al. (2002a, b)

challenges (Tables 6.6 and 6.7). Several proteins, including hydrogenases, are typically found in strictly anaerobic organisms and are used as part of anoxic metabolism. Although the presence of fermentation pathways in an oxygenic phototroph was initially considered somewhat paradoxical since these pathways are found in anaerobic chemotrophs, however it is now apparent that photosynthetic microbes frequently experience extended periods of limited O_2 availability (Quinn et al. 2002; Steunou et al. 2006; Mus et al. 2007). In eukaryotic microalgae, hydrogenases are solely responsible for H_2 production. The [FeFe] hydrogenase gene sequences derived from green algae indicate that majority of microalgal [FeFe]-hydrogenase genes encode small, monomeric proteins (approximately 45–50 kDa) containing only H-cluster-binding domain (Florin et al. 2001; Wunschiers et al. 2001; Happe and Kaminski 2002; Winkler et al. 2002a, b, 2004; Forestier et al. 2003). However, it was observed that a strain of *Chlorella* sp. NC64A possess an [FeFe] hydrogenase with F-cluster domains and exhibit both fermentative and H_2 photoproduction activities (Das and Veziroglu 2008). These additional FeS F-clusters are found in most bacterial [FeFe] hydrogenase enzymes and are putatively required for electron

Table 6.8 Genes coding for hydrogenase in green microalgae belonging to the division Chlorophyta

Green microalgae	Genes				References
	HYDA	HYDA1	HYDA2	HYDA3	
Chlamydomonas noctigama	–	+	–	+	Skjånes et al. (2010)
Chlamydomonas moewusii	–	–	+	–	Skjånes et al. (2010)
Chlamydomonas reinhardtii	–	+	+	–	Happe et al. (2002)
Chlorella vulgaris	–	+	–	–	Hwang et al. (2014)
Chlorella fusca	+				Hwang et al. (2014)
Scenedesmus obliquus	+	–	–	–	Florin et al. (2001)

+ and – denote the presence and absence

transport from/to electron mediators. Majority of algal [FeFe] hydrogenases lack additional F-clusters. This type of truncated [FeFe] hydrogenases are found only in green microalgae (Dubini et al. 2014). Eukaryotic microalgal hydrogenase genes have been detailed in Table 6.8. A third protein with [FeFe] hydrogenase homology is also present in the genome of *C. reinhardtii* (Accession number EDP03395). However, this protein has similarity with Narf-like protein family that are thought to play a role in FeS cluster assembly in some eukaryotes and possibly do not exhibit activities typical to that of hydrogenase (Balk et al. 2004). The microalgal [FeFe] hydrogenase contains transit peptides of variable length in N-terminus, which are required for translocation to the chloroplast. Majority of microalgal hydrogenase enzymes sequenced to date also contain an insertion of 15–54 amino acids in C-terminus and a smaller insertion, approximately nine amino acids to the N-terminal side of L1 motif. The physiological significance of these insertions is currently unknown; however, they may have roles in regulating enzyme activity, protein interactions, and cellular localization. Anaerobiosis is required to induce hydrogenase activity in *C. reinhardtii*, which is achieved in the laboratory in a variety of ways including (i) purging with inert gas, (ii) providing exogenous reductant, and (iii) allowing cellular respiration to metabolize dissolved O_2. Following the establishment of anaerobiosis, cultures are sealed to prevent introduction of O_2 from the atmosphere. Cultures grown in nutrient-replete media must be maintained under dark condition or at very low light levels to prevent O_2 accumulation from endogenous photosynthetic activity. Hydrogen photoproduction is observed at high initial rates immediately after illumination of dark in anaerobically adapted cells. However, in cultures grown in a nutrient-replete medium, these initial rates of H_2 photoproduction rapidly diminish as O_2 levels from photosynthesis increase and cells adapt to an aerobic metabolism along with fixation of CO_2. In Tables 6.6, 6.7, and 6.8, the rate of hydrogen production by green microalgae under direct biophotolysis using different growth medium in the presence of a variety of hydrogenase genes as reported in the literature has been summarized.

6.4 Primary Constraints of Algal Hydrogen Gas Production

6.4.1 Limiting Factors for Growth and Hydrogen Gas Production

In microalgae, hydrogen production efficiency is determined by physical and chemical factors such as light intensity, temperature, pH, carbon source, and composition of growth medium. Among these factors, we will discuss about the role pH and carbon sources for hydrogen production in microalgae. Some of the above factors have been summarized in Table 6.9.

Table 6.9 Rate of hydrogen production in cyanobacteria and green microalgae under optimal physicochemical condition

		Optimum physicochemical conditions				Hydrogen production rate			
Microalgal groups	Strains	pH	T °C	Light intensity	Carbon source	(ml L^{-1} culture h^{-1})	μmol mg dw^{-1} h^{-1}	μmol mg chl a^{-1} (protein) h^{-1}	References
Heterocystous									
Cyanobacteria	*Aphanocapsa montana*	NA	NA	4–6 Wm^{-2}	Air	NA	0.4	NA	Howarth and Codd (1985)
	Gloeobacter PCC 7421	NA	NA	20 $\mu Em^{-2}s^{-1}$	CO	NA	NA	1.38	Moezelaar et al. (1996)
	Anabaena cylindrica B 629	NA	NA	7000 lux	5 % CO_2	NA	0.103	NA	Lambert and Smith (1977)
	Anabaena variabilis	NA	30	150 $\mu Em^{-2}s^{-1}$	1 % CO_2	NA	0.25	NA	Berberoglu et al. (2008)
	Anabaena azollae	NA	NA	140 $\mu Em^{-2}s^{-1}$	2 % CO_2	13	NA	38	Tsygankov et al. (1998)
Nonheterocystous									
Cyanobacteria	*Synechococcus* PCC 602	NA	NA	20 $\mu Em^{-2}s^{-1}$	CO	NA	0.66	NA	Howarth and Codd (1985)
	Synechocystis sp. PCC 6803	NA	NA	50 $\mu Em^{-2}s^{-1}$	$NaHCO_3$	NA	NA	0.081	Burrows et al. (2008)
	Chroococcidiopsis thermalis CALU 758	NA	NA	70 $\mu Em^{-2}s^{-1}$	1 % CO_2	NA	NA	0.7	Serebryakova et al. (2000)
	Microcystis sp. PCC 7820	NA	NA	20 $\mu Em^{-2}s^{-1}$	CO	NA	NA	0.16	Moezelaar et al. (1996)
	Spirulina platensis	NA	NA	8 Wm^{-2}	NA	4.032	NA	NA	Aoyama et al. (1997)
	Oscillatoria limosa	NA	20	1200 lux	CO_2	NA	NA	0.83	Heyer et al. (1989)
	Oscillatoria miami BG7	7.75	35	100 $\mu Em^{-2}s^{-1}$	CO_2	NA	NA	0.25	Phlips and Mitsui (1983)
Green microalgae	*Chlamydomonas reinhardtii*	7	NA	100 $\mu Em^{-2}s^{-1}$	Acetate (17 mM); S deprived	2.1	NA	NA	Laurinavichene et al. (2006)
	Chlamydomonas sp. MGA 161	8	30	25 Wm^{-2}	5 % CO_2	4.48	NA	NA	Ohta et al. (1987)
	Platymonas subcordiformisa	NA	NA	22 Wm^{-2}	NA	0.048	NA	NA	Guan et al. (2004)
	Chlorella sorokiniana Ce	NA	NA	120 $\mu Em^{-2}s^{-1}$	Acetate;S deprived	1.35	NA	NA	Chader et al. (2009)

6.4.1.1 pH

In microalgae, cellular processes are dependent on intracellular pH (close to neutral), and most algae have limited abilities to tolerate variable pH conditions (Andersen 2005). In growth medium, hydrogen production process significantly depends upon pH. A subtle change in pH can change the end products (CO_2, acetate) of anaerobic process (Khanal et al. 2004; Dsagupta et al. 2010). During photosynthesis, initial pH decreases due to formation of carbonic acid as a result of a chemical reaction between CO_2 and water. After certain period, pH increases due to evolution of oxygen via photosynthesis. Microalgae usually grow at a pH ranging from 5.0 to 9.0 (Song et al. 2011). High pH in culture medium shortens lag time of hydrogen production and increases its rate of production (Khanal et al. 2004). Any changes in pH alter metabolic pathways that mediate hydrogen production. In mixed microbial flora, sucrose degradation increased with pH and maximum efficiency (95 %) was found at pH 9.0 (Lee et al. 2002). This fact can be explained in terms of enzyme activity; hydrogen-producing enzymes (hydrogenase and nitrogenase) are sensitive to pH. Initially, protons, generated by the degradation of endogenic or exogenic carbon source and by the splitting of water, are converted into hydrogen. Later, proton concentration increases; a few of them are entrapped by hydrogenase or nitrogenase (depending upon the light condition applied) and get converted into hydrogen; the rest of them remain unutilized. At low pH value (5.0), hydrogen-producing enzyme inactivates, thereby reducing hydrogen production rate. During sulfur (S) deprivation, pH of the culture medium fluctuates which may cause metabolic disturbance resulting in production of low amount of hydrogen in green microalgae (Khanal et al. 2004). According to Kosourov et al. (Kosourov et al. 2003), hydrogen production rate in S-deprived *Chlamydomonas reinhardtii* was high at pH 7.7 but decreased at pH 6.5. In cyanobacteria and green microalgae, pH requirement for photosynthesis and fermentation vary from species to species. Marine microalgae require different pH compared to freshwater microalgae due to low nitrate requirement in former case (Andersen 2005).

Some algal groups have adapted themselves in highly acidic environments by pumping protons out of the cell using efficient ATP-driven H^+ pumps, one example being the acidophilic *Chlamydomonas acidophila* (Gerloff-Elias et al. 2006). Under very low pH, as much as 50 % of the synthesized ATP inside an algal cell has been observed to be consumed by proton pumps (Bethmann and Schönknecht 2009). Cellular metabolic processes such as increase fatty acid saturation, production of acid-tolerant cell wall proteins, reduction of cell volume, reduction of starch reserves, and production of antioxidants in microalgal cell depend on low pH. Nitrate uptake along with carbon fixation can influence alteration of pH (Bothe et al. 2010, Rashid et al. 2013). It has been shown that in *Gloeocapsa alpicola*, optimal pH for H_2 production has been found to be ranging from 6.0 to 7.0 (Antal and Lindblad 2005). In *C. reinhardtii*, hydrogen production depends on pH of the medium (Antal et al. 2003). Optimization of pH is essential in aerobic and anaerobic phases of photobiological hydrogen production. The production of undesirable intermediate metabolic products can be controlled by developing a correlation between photosynthetic by-products in stage I and intermediate by-products during stage II with pH. Extensive research is available on pH optimization in stage I, but only few studies deal with the effect of pH on stage II (Das and Veziroglu 2008; Rashid et al. 2011, 2013; Dubini et al. 2014). Table 6.9 shows the pH tolerance for different microalgae species under anaerobic (stage II) conditions. Therefore, maintenance of pH is extremely difficult as well as costly when microalgal biomass are grown on a large scale with the objective to produce hydrogen at the industrial scale.

6.4.1.2 Carbon Sources

Carbon sources are also known to influence hydrogen production significantly while having an effect on nitrogenase activity (Mata et al. 2009). The flow of electron from carbon source to the nitrogenase may vary and thus influence hydrogen production in microalgae. Some strains of green microalgae are able to grow under both very high CO_2 concentrations (20–100 % bubbling of cultures) as well as high temperatures (49–56 °C) (Wang et al. 2008). One example is a strain of *Chlorella sorokiniana* isolated from a hot spring (Sakai et al. 1995); other known examples are strains of *Scenedesmus* sp. (de Morais and Costa 2007; Hanagata et al. 1992) and *Chlorococcum littorale* (Satoh et al. 2002). It is hypothesized that tolerance toward high CO_2 is connected to state transition in favor of PSI (Miyachi et al. 2003; Satoh et al. 2002). Microalgae store carbon in the form of starch or glycogen during photosynthesis and use them under anaerobic condition. These cells can accumulate limited amount of glycogen and starch, and ultimately there is a low yield of hydrogen during anaerobic phases. A significant increase in hydrogen yield is possible by introducing exogenic carbon source in early phase of anaerobiosis (Nayak et al. 2014). A wide variety of exogenic carbon sources are known to be used by algal cells for hydrogen production, namely, glucose, fructose, sucrose, malt extract, malic acid, acetate, and organic wastewater (Rashid et al. 2011). The yield of hydrogen varies according to the source of carbon as well as the cultured microalgal strains. Therefore, selection of carbon source is a prerequisite for establishment of large-scale cultures of microalgae with the objective to produce biohydrogen. Microalgae can use inorganic carbon (CO_2) as well as organic carbon sources (glucose, mannitol, acetate, sucrose) (Hu et al. 2003). Microalgae grown under heterotrophic condition (using organic carbon) could have more potential to produce hydrogen compared to autotrophic condition (inorganic carbon). In heterotrophic condition, high biomass is achieved. Unlike autotrophic condition, light is also not a prerequisite in heterotrophic cultivation. Therefore, heterotrophic cultivation can be cheaper than autotrophic cultivation (Nayak et al. 2014). However, bacterial contamination can be a serious concern in heterotrophic microalgal growth system (Das and Veziroglu 2008; Rashid et al. 2013). As soon as organic carbon is introduced in the medium, bacterial growth starts which outperforms the microalgae and consumes all the nutrients unless necessary steps are taken. Antibiotics are therefore needed to control the contamination at this stage (Das and Veziroglu 2008; Rashid et al. 2013). In biohydrogen production process, the effect of carbon source on microalgae cultivation is not fully explored. Wei et al. used glucose, sucrose, fructose, and malt extract as substrates for growing cells of *Microcystis aeruginosa* (cyanobacterium) and *Chlorella vulgaris* (green microalgae). Malt extract turned the maximum hydrogen yield of 1300 ml L^{-1} (of microalgae medium) in *Chlorella vulgaris*. Chen et al. (2008) used glucose, fructose, galactose, and sucrose as substrates with a concentration of 200 mg L^{-1} for growing cells of *Anabaena* sp. CH3 during hydrogen production. The authors found that preferred substrate in case of *Anabaena* sp. CH3 was fructose and glucose, producing 0.0016 and 0.004 mol of hydrogen, respectively. In *C. reinhardtii*, acetate was the most effective substrate producing 1.7 mol of hydrogen (Rashid et al. 2013). Table 6.9 shows the carbon sources used for growing different microalgae species during hydrogen production.

6.4.2 Photobioreactor and Its Utility for Hydrogen Gas Production

Bioreactors facilitated with illumination are essential for production of hydrogen and hence are called photobioreactors. In microalgae, H_2 production through a photobioreactor depends on two important steps: (i) culture specification for biomass (aerobic, stage I) production and (ii) culture conditions for H_2 production (Cuaresma et al. 2011). All photobioreactors require entry of light, which usually is sunlight, but in some photobioreactors, other artificial sources are also used for providing controlled light. Inside a photobioreactor, there should be a photic zone, close to the illuminated surface and a dark zone, further away from this surface. The dark zone is due to light absorption by algal cells and mutual shading. Hydrogen production inside a photobioreactor is light limited and tends to decrease at higher light intensities (photosynthesis diverts hydrogen production pathway); hence, light regime is determined by light gradient (must be diluted and distributed as much as possible; absolute dark condition responsible for highest production) (Oncel and Sabankay 2012). Liquid circulation time or aeration (enzymes for H_2 production are oxygen labile; anaerobic condition or inert gas environment is required) rate has a limiting effect on hydrogen production by microalgal biomass. The position of light source as well as gas liquid hydrodynamics also affects microalgal growth and resulting hydrogen production. To achieve maximum biomass from microalgae in photobioreactor, light panels are constructed along with speed controlled agitator (Menetrez 2012; Oncel and Kose 2014). As a result of agitation, microalgal cells will circulate between light and dark zone of the photobioreactor at a certain frequency and regular intervals, based on photobioreactor design and gas input (Oncel and Kose 2014). Photobioreactors used for hydrogen production can be broadly divided into three categories: vertical column photobioreactor, tubular-type photobioreactor, and flat panel photobioreactor (Table 6.10). Comparison of performance of reactor with respect to hydrogen production is given in Table 6.11. A photobioreactor for microalgal hydrogen production should meet following conditions:

1. Photobioreactor should be an enclosed system so that the produced hydrogen may be collected without any loss.
2. The photobioreactor design must allow sterilization with convenience and ease.
3. To maximize the area of incident light (thus allowing high growth and hydrogen production), photobioreactor design should provide high surface to volume ratio.

6.5 Progress in Metabolic Engineering for Hydrogen Gas Production

In case of eukaryotic microalgae, *Chlamydomonas reinhardtii* is an attractive candidate for hydrogen production due to its relatively high hydrogenase activity, through the [FeFe] hydrogenase HYDA1 (Meuser et al. 2012). Although hydrogen is naturally produced by *C. reinhardtii* under sulfur starvation and hydrogenase activity can be externally induced by adding 3-(3,4-dichlorophenyl)-1,1-dimethylurea (DCMU, a PSII electron chain uncoupler), the production cannot be sustained while photosynthesis is actively occurring since oxygen inactivates hydrogenase (Esquivel et al. 2011). In 2000, Melis and co-worker used sulfur deprivation strategy for lowering partial pressure of oxygen in *C. reinhardtii*. Oxygen sensitivity is a proton gradient-related problem which can be avoided by genetic insertion of a hydrogenase promoter programmed polypeptide proton channel into the algal thylakoid membranes. This necessitates a biphasic production strategy in which cells grow photosynthetically to accumulate biomass, which is then exploited for H_2 production under anoxic conditions. Several approaches to overcome this limitation have recently shown promising results (Oncel and Kose 2014; Dubini et al. 2014). Incorporation of leghemoglobin proteins (oxygen sequester in the nitrogen-fixing root nodules of legumes) within *Chlamydomonas* sp. cell can facilitate a fourfold increase of H_2 production (Wu et al. 2011). It is well known that in green microalgae about 90 % of the photons being captured by antenna systems are not being utilized (Hallenbeck and Benemann 2002). In a truncated antenna mutant of *C. reinhardtii* CC-4169, initially engineered for increased photosynthetic performance, an eightfold increase in H_2 production was observed under sulfur deprivation in presence of high light (350 $\mu Em^{-2}s^{-1}$) (Kosourov et al. 2011). In *C. reinhardtii*, photosynthetic efficiency improved by truncating chlorophyll antenna size of PSII (Perrine et al. 2012). A PSII protein D1 mutant of *C. reinhardtii* exhibit increased carbohydrate storage and H_2 production (Scoma et al. 2012); however, the highest reported yields for this alga (*C. reinhardtii* D1 mutant) are still three to five times below light to H_2 conversion than through direct photolysis (Esquivel et al. 2011). The H_2 production pathways predicted from in silico reconstruction suggest that increased production can occur under conditions of inhibited cyclic

Table 6.10 Type of photobioreactor with their optimal features

Type of photo bioreactor	S/V ratio	Agitation system	Temperature control	Gas exchange	Advantages	Disadvantages	References
Tubular reactors							
Vertical tubular	Small	Airlift, bubble column	NA	Open gas exchange at headspace	Good mixing, efficient CO_2 supply and O_2 removal	Scale-up is limited, major light is reflected due to angle	Martnez-Jeronimo and Espinosa-Chavez (1994) and Tamagnini et al. (2002)
Horizontal tubular	Large	Recirculation with diaphragm/mechanical pumps	Shading, overlapping, water spraying	Injection into feed, and dedicated degassing units	Adequate angle toward sunlight,	High shear due to pumps, risks of O_2 buildup, biofouling, separate gas exchange unit required	Iqbal et al. (1993) and Tredici et al. (1998)
Helical tubular	Large	Centrifugal pumps	Heat exchanger	-do-	High S/V, easy scale-up by increasing the number of units	O_2 buildup, separate gas exchange, pumps exert more shear, cell debris accumulate inside	Tsygankov et al. (1998) and Morita et al. (2000)
Table 10 (continued)							
Alpha shape reactor	Large	Airlift	-do-	Injection in the vertical units and degassed at top	High unidirectional flow rate with low air flow rate, high S/V	Foam formation due to high cell density	Lee et al. (1995)
Flat plate reactors							
Flat panel bubbled at bottom	Medium	Bubbling at bottom or from sides, recirculation	Heat exchange coils	Bubbling	Open gas transfer avoids O_2 buildup	Shear due to entrainment of cells till bubbles burst	Hu and Richmond (1996) and Tredici and Zittelli (1998)
Flat panel pivoted at center	Medium	Pulsating motion	Heat exchange coils	Degasser	Good mixing, low shear	Scale-up is difficult	Dasgupta et al. (2010a, b)

Panel with V-shape	Medium	Bubbling	-do-	-do-	Very high mixing rate, low shear	Agitation system can dilute H_2 formed	Iqbal et al. (1993)
Alveolar panel	Large	Bubbling	water circulation in lower row	-do-	High S/V due to alveolar panels, uniform distribution of light	O_2 buildup, high air flow rates required to move across the channels	Tredici et al. (1998) , Tredici et al. (1991), and Tredici et al. (1993)
Floating type bioreactor	Medium	Sea saw motion	No cooling required	-do-	Low energy for operation, good agitation, can be installed on lakes and sea floor	NA	Otsuki et al. (1998)
Fermentor type with internal/external lighting	Small	Impellers	Heat exchange coils	By sparger	High degree of control of various parameters	Light conversion efficiency is less	Pohl et al. (1988)
Torus shaped reactor	Medium	Marine impeller	Cooling fans	CO_2 inlet after impeller, outlet at top	Good mixing conditions owing to shape avoiding dead zones	NA	Pottier et al. (2005) and Fouchard et al. (2008)
Annular triple jacketed with lighting from innermost chamber	Medium	Magnetic stirrer	Outer water jacket	Open gas exchange	Good S/V and temperature control, open gas exchange	Scaling up is difficult, biofouling	Basak and Das (2009)

Adapted from Dasgupta et al. 2010a, b
NA not available

Table 6.11 Performance of different photobioreactor in terms of hydrogen production in microalgae

Reactor type	Organism	Surface to volume ratio	Volumetric productivity (g L^{-1} d^{-1})	Areal productivity (g m^{-2} d^{-1})	Percentage of photosynthetic efficiencies	References
Flat plate reactor	*Arthrospira platensis*	40	1.09	NA	4.84	Tredici and Zittelli (1998)
Helical tubular	*Arthrospira platensis*	53	0.9	NA	0.66	Tredici and Zittelli (1998)
Near horizontal tubular	*Arthrospira platensis*	70	1.4	28	5.6	Tredici and Zittelli (1998)
Panel type	*Chlamydomonas reinhardtii* CC124	NA	1.3 ± 0.05 ml L^{-1} h^{-1}	NA	NA	Oncel and Kose (2014)
Conical tubular	*Chlorella* sp.	NA	1.01	28.3	6.84	Morita et al. (2000)
Airlift	*Phaeodactylum tricornutum* UTEX 640	NA	1.2	20	NA	Fernandez et al. (2001)
Alveolar	*Nannochloropsis* sp.	80	1.45	NA	0.48	Tredici et al. (1991)

NA not available

electron flow (Dal'Molin et al. 2011), which is indeed observed in high H_2-producing *C. reinhardtii* mutant *Stm6Glc4* (Kruse et al. 2005). RNA interference (RNAi) has also been recently used to downregulate the entire family of light-harvesting complexes (LHC) in *C. reinhardtii*. The simultaneous knockdown of three LHC proteins (LHCMB 1, 2, and 3) was undertaken in the high H_2-producing *C. reinhardtii* mutant *Stm6Glc4* using an RNAi triple knockdown strategy (Oey et al. 2013), and this tool may prove to be extremely useful as part of metabolic engineering approaches.

In order to use cyanobacteria for biological production of hydrogen, it is important to thoroughly understand the regulation of hydrogen production machine and identification of bottlenecks that limit H2 production. The bidirectional hydrogenase from cyanobacteria does not require ATP to function and can suffer from a buildup of ATP which then inhibits electron flow (Lubitz et al. 2008). A variety of genetic tools also exist to express the bidirectional, oxygen-tolerant [NiFe] hydrogenase genes (such as *hydS* and *hydL*) in cyanobacterial species (Lukey et al. 2011). Recent studies have shown that incorporation of a heterologous [FeFe] hydrogenase (from *Shewanella oneidensis* MR-1) into the heterocysts of *Anabaena* sp. PCC 7210 could potentially provide a way to increase hydrogen production in this organism (Gartner et al. 2012). Major strategies developed for higher rate of H_2 production in cyanobacteria are detailed in Table 6.12:

- *Inactivation of uptake hydrogenase:* Preparation of single and double mutant of *Hup* and *Hox*, respectively, showed enhanced production of hydrogen in several *Hup* disrupted mutant strains of *Anabaena* sp., especially in strain PCC7120 (Masukawa et al. 2012). Thus, parental strain with high nitrogenase activity can perform high hydrogen production after *Hup* inactivation.
- *Modification of the catalytic activity center of nitrogenase*: In the presence of N_2, *Anabaena* wild strain PCC7120 effectively produces H_2 when *nif*V1 was inactivated along with high heterocyst frequency (Kufryk 2013).

Table 6.12 Major limitation and remedies for biohydrogen production in cyanobacteria and green microalgae

Microalgal groups	Challenges	Strategies	References
Cyanobacteria	Low H_2 production rate of [NiFe] hydrogenases	Requires constant sparging of inert gas	Silbert et al. (2001)
		Expressing the bidirectional, oxygen-tolerant [NiFe] hydrogenase genes, hudS and hydL in cyanobacteria	Lukey et al. (2011)
		Incorporation of [FeFe] hydrogenase into heterocysts	Gartner et al. (2012)
Green microalgae	Light conversion efficiency to H_2 (theoretically about 10 %)	Truncating the chlorophyll antenna size of PSII using RNAi method	Oey et al. (2013)
	Inhibition of [FeFe] hydrogenases by oxygen production through PSII	Sulfur deprivation	He et al. (2012)
		Genetic insertion of a hydrogenase promoter proton channel into the thylakoid membranes	

- *Overexpression of bidirectional hydrogenase*: In *Synechocystis* sp. PCC6803, increase in H_2 production was noted after overexpression of bidirectional hydrogenase and deletion of NDH-1 respiratory complex (Kufryk 2013).
- *Expression of heterologous hydrogenase*: Low specific activity of [NiFe] hydrogenase was 500 times more active in *Synechococcus elongatus* 7942, after expression of *Clostridium acetobutylicum* [FeFe] hydrogenase (*Hyd A*) (Weyman et al. 2011).
- *Inactivation of competitive biochemical pathways*: In *Synechococcus* sp. PCC 7002, genetic manipulation led to increased hydrogen production by increase in the ratio of $NADPH/NADP^+$ through bidirectional hydrogenase (Kufryk 2013).

Major limitation and strategies including metabolic and genetic engineering processes for overcoming such bottlenecks during biohydrogen production by microalgae are summarized in Tables 6.12 and 6.13.

6.6 Economic Viability

The research attention on biological hydrogen production has substantially increased over the last 10 years. However, only a limited number of studies have looked into the economic viability of biohydrogen production on a commercial scale. Reported analysis suggests that the cost of photobiologically produced hydrogen is much lower (\$25 m^{-3}) compared to that produced by photovoltaic process (\$170 m^{-3}) (Dutta et al. 2005). The experimental studies have shown that dark fermentation is a cheap method for generation of biohydrogen; however, yields are usually lower. On the contrary, photofermentation is a more efficient method, but it is relatively more expensive. The application of indirect photolysis methods of hydrogen production is predicted to cost around 1220\$ per GJ/year, while the capital cost is predicted to be 2.4\$/gigajoule/year (Resnick 2004; Menetrez 2012). For this reason, a hybrid production system has been proposed which integrates both light and dark fermentation

Table 6.13 Metabolic and genetic engineering strategies for overcoming bottlenecks during hydrogen production in cyanobacteria and green microalgae

Strategies	Advantages	Microalgae	References
Pigment reduction	Increasing the photosynthetic efficiency	*Chlamydomonas reinhardtii*, *Dunaliella salina*	Polle et al. (2003) and Mussgnug et al. (2007)
		Synechocystis sp. PCC 6803	Bernat et al. (2009)
Generating anaerobic environment	Activating hydrogen-producing enzyme	*C. reinhardtii* (arp mutant)	Melis (2007) and Ruhle et al. (2008)
		C. reinhardtii (D1 protein mutant)	Torzillo et al. (2009)
Oxygen-tolerant enzyme	Producing hydrogen in the presence of oxygen	*Chlamydomonas* sp.	Chen et al. (2003)
		Synechococcus sp. PCC7942	Chen et al. (2005)
Eliminating competitive inhibition by other e⁻ acceptor	Redirecting the e⁻ flux toward the hydrogen-producing enzyme	*C. reinhardtii* 137c	Lee and Greenbaum (2003)
Introducing foreign efficient hydrogen-producing enzyme	Enhancing hydrogen production in that particular microorganism, that may be efficient in other criteria	*Synechococcus elongatus*	Miyake and Asada (1997) and Asada et al. (2000)
Enhancing the capacity of deriving e⁻ from carbohydrates	Contributing in hydrogen production in anoxic dark condition	*C. reinhardtii* (Stm6 strain)	Kruse et al. (2005) and Doebbe et al. (2007)
Inhibiting or overexpressing the crucial metabolic enzymes	Redirecting the e⁻ flux toward hydrogen-producing enzyme	*Synechocystis* sp. (mutant M55)	Vignais et al. (2006)

processes for maximum biohydrogen yield from algal systems (Nayak et al. 2014). The use of external enzymes (e.g., amylase) for the breakdown of microalgal cell wall under dark fermentation is effective and can ultimately maximize H_2 production (Nayak et al. 2014). Hydrogen production using photobiological systems has the potential to become the most effective method for large-scale requirements (Dasgupta et al. 2010a, b). Hence, high setup cost is one of the major constraints that limit commercial scaling up of biohydrogen to meet large-scale energy demand in the present time. Summary of benefits of biohydrogen over fossil fuels through various routes of production and possibilities is represented in Fig. 6.6.

6.7 Conclusions

In the beginning of this chapter, we argued about sustainable and environmentally friendly carbon-free green energy for mankind and thus discussed mainly on the metabolism and mechanism of biological hydrogen production including scale-up and associated constraints. Biological hydrogen production has several advantages over conventional hydrogen production processes. Microalgae are sustainable and low-cost renewable source for biohydrogen production. However, biohydrogen production faces two major problems: (i) low hydrogen yield in dark fermentation (in microalgae, stored carbohydrates are converted into H_2 via pyruvate in dark) and (ii) high energy cost in photofermentation (in photosynthetic bacteria, stored organic compound produce H_2 under light). Therefore, hybrid production system is proposed to maximum biohydrogen yield from algal systems. Hydrogen production using photobiological systems has the potential to become the most effective method for large-scale requirements. However, innovative research approaches need to be formulated so as to improve the efficiency of microalgae for H_2 production including discovery and identification of new strains, species consortium, novel enzymes, and

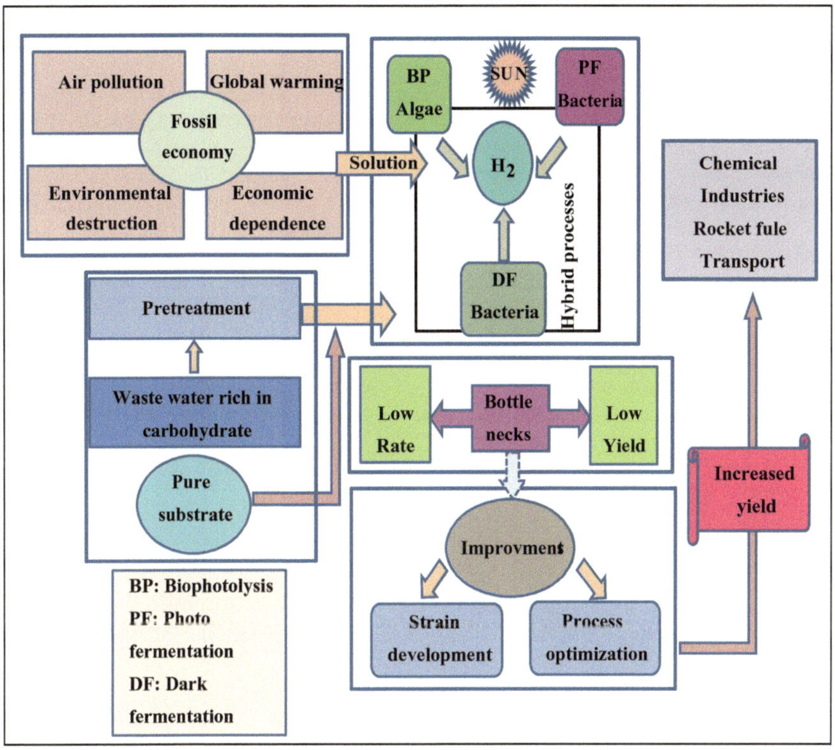

Fig. 6.6 Schematic summary of benefit of biohydrogen over fossil-based fuels and constraints in terms of production along with underlying challenges

manipulation of culture conditions. Development of low-cost photobioreactors (with maximum utilization of PAR) and fermentors is another challenge; when addressed, it can lead to commercial scale production of H_2. Detailed understanding of system biology including supply of carbon, reducing power, oxygen inactivation, metabolic engineering, and genetic manipulation can ultimately help toward cost-effective production of biohydrogen from microalgal biomass.

References

Andersen RA (ed) (2005) Algal cultural techniques. Phycological Society of America, Elsevier, p 556

Antal TK, Lindblad P (2005) Production of H_2 by sulphur-deprived cells of the unicellular cyanobacteria *Gloeocapsa alpicola* and *Synechocystis* sp. PCC 6803 during dark incubation with methane or at various extracellular pH. J Appl Microbiol 98:114–120. doi:10.1111/j.1365-2672.2004.02431.x

Antal TK, Krendeleva TE, Laurinavichene TV, Makarova VV, Ghirardi ML, Rubin AB (2003) The dependence of algal H_2 production on photosystem II and O_2 consumption activities in sulfur-deprived *Chlamydomonas reinhardtii* cells. Biochim Biophys Acta 1607(2–3):153–160. doi:10.1016/j.bbabio.2003.09.008

Aoyama K, Uemura I, Miyake J, Asada Y (1997) Fermentative metabolism to produce hydrogen gas and organic compounds in a cyanobacterium, *Spirulina platensis*. J Ferm Bioeng 83(1):17–20. doi:10.1016/S0922-338X(97)87320-5

Asada Y, Koike Y, Schnackenberg J, Miyake M, Uemura I, Miyake J (2000) Heterologous expression of clostridial hydrogenase in the cyanobacterium *Synechococcus* PCC7942. Biochim Biophys Acta 1490(3):269–278. doi:10.1016/S0167-4781(00)00010-5

Ayhan D (2009) Progress and recent trends in biodiesel fuels. Energy Convers Manag 50(1):14–34. doi:10.1016/j.enconman.2008.09.001

Balk J, Pierik AJ, Netz DJA, Mu-hlenhoff U, Lill R (2004) The hydrogenase-like Nar1p is essential for maturation of cytosolic and nuclear iron–sulphur proteins. The EMBO J 23(10):2105–2115. doi:10.1038/sj.emboj.7600216

Bandyopadhyay A, Jana Stöckel J, Hongtao Min H, Sherman LA, Himadri B, Pakrasi HB (2010) High rates of photobiological H_2 production by a cyanobacterium under aerobic condition. Nat Commun 1:139. doi:10.1038/ncomms1139

Basak N, Das D (2009) Photofermentative hydrogen production using purple non-sulfur bacteria Rhodobacter sphaeroides O.U.001 in an annular photobioreactor: a case study. Biomass Bioenergy 33(6):911–919. doi:10.1016/j.biombioe.2009.02.007

Berberoglu H, Jenny J, Laurent P (2008) Effect of nutrient media on photobiological hydrogen production by *Anabaena variabilis* ATCC 29413. Int J Hydrog Energy 33(4):1172–1184. doi:10.1016/j.ijhydene.2007.12.036'

Berggren G, Adamska A, Lambertz C, Simmons TR, Esselborn J, Atta M, Gambarelli S, Mouesca JM, Reijerse E, Lubitz W (2013) Biomimetic assembly and activation of [FeFe]-hydrogenases. Nature 499(7456):66–69. doi:10.1038/nature12239

Bernat G, Waschewski N, Rogner M (2009) Towards efficient hydrogen production: the impact of antenna size and external factors on electron transport dynamics in *Synechocystis* PCC 6803. Photosynth Res 99(3):205–216. doi:10.1007/s11120-008-9398-7

Bethmann B, Schönknecht G (2009) P[H] regulation in an acidophilic green alga-a quantitative analysis. New Phytol 183(2):327–339. doi:10.1111/j.1469-8137.2009.02862.x

Boichenko VA, Greenbaum E, Seibert M (2004) Hydrogen producton by photosynthetic microorganisms. In: Archer MD, Barber J (ed) Molecular and global photosynthesis. IC Press, London. Chapter 8, pp 397–455

Boison G, Bothe H, Schmitz O (2000) Transcriptional analysis of hydrogenase genes in the cyanobacteria *Anacystis nidulans* and *Anabaena variabilis* monitored by RT-PCR. Curr Microbiol 40(5):315–321. doi:10.1007/s002849910063

Borodin VB, Tsygankov AA, Rao KK, Hall DO (2000) Hydrogen production by *Anabaena variabilis* PK84 under simulated outdoor conditions. Biotechnol Bioeng 69(5):478–485. doi:10.1002/1097-0290(20000905)69:5<478

Bothe H, Schmitz O, Yates MG, Newton WE (2010) Nitrogen fixation and hydrogen metabolism in cyanobacteria. Microbiol Mol Biol Rev 74(4):529–551. doi:10.1128/MMBR.00033-10

Brand JJ, Wright JN, Lien S (1989) Hydrogen production by eukaryotic algae. Biotechnol Bioeng 33(11):1482–1488. doi:10.1002/bit.260331116

Bridgwater AV (2003) Renewable fuels and chemicals by thermal processing of biomass. Chem Eng J 91(2–3):87–102. doi:10.1016/S1385-8947(02)00142-0

Burgdorf T, Lo¨scher S, Liebisch P, Van der Linden E, Galander M, Lendzian F, Meyer-Klaucke W, Albracht SP, Friedrich B, Dau H, Haumann M (2005) J Am Chem Soc 127:576–592

Burrows EH, Chaplen FWR, Ely RL (2008) Optimization of media nutrient composition for increased photofermentative hydrogen production by *Synechocystis* sp. PCC 6803. Int J Hydrog Energy 33(21):6092–6099. doi:10.1016/j.ijhydene.2008.07.102

Chader S, Haceneb H, Agathos SN (2009) Study of hydrogen production by three strains of *Chlorella* isolated from the soil in the Algerian Sahara. Int J Hydrog Energy 34:4941–4946

Chen HC, Yokthongwattana K, Newton AJ, Melis A (2003) SulP, a nuclear gene encoding a putative chloroplast-targeted sulfate permease in *Chlamydomonas reinhardtii*. Planta 218(1):98–106. doi:10.1007/s00425-003-1076-6

Chen H, Newton A, Melis A (2005) Role of SulP, a nuclear-encoded chloroplast sulfate permease, in sulfate transport and H-2 evolution in *Chlamydomonas reinhardtii*. Photosyn Res 84(1–3):289–296. doi:10.1007/s11120-004-7157-y

Chen PPC, Fan SH, Chiang CL, Lee CM (2008) Effect of growth conditions on the hydrogen production with cyanobacterium *Anabaena* sp. strain CH3. Int J Hydrog Energy 33:1460–1464

Cuaresma M, Marcel Janssen M, Vílchez C, Wijffels RH (2011) Horizontal or vertical photobioreactors? How to improve microalgae photosynthetic efficiency. Bioresour Technol 102(8):5129–5137. doi:10.1016/j.biortech.2011.01.078

Dal'Molin CG, Quek LE, Palfreyman RW, Nielsen LK (2011) AlgaGEM; a genome-scale metabolic reconstruction of algae based on the *Chlamydomonas reinhardtii* genome. BMC Genomics 12(Suppl 4):S5. doi:10.1186/1471-2164-12-S4-S5

Darzins Al (NREL), Philip Pienkos (NREL), Les Edye (BioIndustry Partners) (2010) Current status and potential for algal biofuels production executive summary report T39-T2. August 6. Cited 14 Jan 2015

Das D, Veziroglu TN (2001) Hydrogen production by biological processes a survey of literature. Int J Hydrog Energy 26(1):13–28. doi:10.1016/S0360-3199(00)00058-6

Das D, Veziroglu TN (2008) Advances in biological hydrogen production processes. Int J Hydrog Energy 33(21):6047–6057. doi:10.1016/j.ijhydene.2008.07.098

Dasgupta CN, Gilbert JJ, Lindblad P, Heidorn T, Borgvang SA, Skjanes K, Das D (2010a) Current status, barriers and developments in biohydrogen production by microalgae. Int J Hydrog Energy 33:1–21

Dasgupta CN, Gilbert JJ, Lindblad P, Heidorn T, Borgvang SA, Skjanes K, Das D (2010b) Recent trends on the development of photobiological processes and photo bioreactors for the improvement of hydrogen production. Int J Hydrog Energy 35:10218–10238

de Morais MG, Costa JAV (2007) Isolation and selection of microalgae from coal fired thermoelectric power plant for biofixation of carbon dioxide. Energy Conv Manag 48(7):2169–2173. doi:10.1016/j.enconman.2006.12.011

Dismukes GC, Carrieri D, Bennette N, Ananyev GM, Posewitz MC (2008) Aquatic phototrophs: efficient alternatives to land based crops for biofuels. Curr Opin Biotechnol 19(3):235–240. doi:10.1016/j.copbio.2008.05.007

Doebbe A, Rupprecht J, Beckmann J, Mussgnug JH, Hallmann A, Hankamer B (2007) Functional integration of the HUP1 hexose symporter gene into the genome of *C. reinhardtii* impacts on biological H_2 production. J Biotechnol 131:27–33. doi:10.1016/j.jbiotec.2007.05.017

Dubini A, Maria L, Ghirardi ML (2014) Engineering photosynthetic organisms for the production of biohydrogen. Photosynth Res Rev 84(1–3):289–296. doi:10.1007/s11120-014-9991-x

Ducat DC, Sachdeva G, Silver PA (2011) Rewiring hydrogenase-dependent redox circuits in cyanobacteria. Proc Natl Acad Sci U S A 108(10):3941–3946. doi:10.1073/pnas.1016026108

Dutta D, Debojyoti D, Chaudhuri S, Bhattacharya S (2005) Hydrogen production by cyanobacteria. Microb Cell Fact 4:36. doi:10.1186/1475-2859-4-36

Ekman M, Ow SY, Holmqvist M, Zang X, van Wagenen J, Wright PC, Stensjo K (2011) Metabolic adaptation in a H_2 producing heterocyst-forming cyanobacterium: potentials and implication for biological engineering. J Proteomics Res 10(4):1772–1784. doi:10.1021/pr101055v

Esquivel MG, Amaro HM, Pinto TS, Fevereiro PS, Malcata FX (2011) Efficient H_2 production via *Chlamydomonas reinhardtii*. Trends Biotechnol 29(12):595–600. doi:10.1016/j.tibtech.2011.06.008

Falkowski PG, Raven JA (2013) An introduction to photosynthesis in aquatic system, (Chapter 1). Aquatic photosynthesis. Princeton University Press, USA, pp 1–43

Falkowski PG, Katz M, Knoll AH, Quigg A, Raven JA, Schofield OM, Taylor FJR (2004) The evolution of modern eukaryotic phytoplankton. Science 305:354–360. doi:10.1126/science.1095964

Fernandez FGA, Sevilla JMF, Perez JAS, Molina GE, Christi Y (2001) Airlift-driven external-loop tubular photobioreactors for outdoor production of microalgae: assessment of design and performance. Chem Eng Sci 56:2721–2732

Field CB, Behrenfeld MJ, Randerson JT, Falkowski P (1998) Primary production of the biosphere: integrating terrestrial and oceanic component. Science 281(5374):237 240. doi:10.1126/science.281.5374.237

Florin L, Tsokoglou A, Happe T (2001) A novel type of Fe-hydrogenase in the green alga *Scenedesmus obliquus* is linked to the photosynthetical electron transport chain. J Biol Chem 276:6125–6132. doi:10.1074/jbc.M008470200

Fontecilla-Camps JC, Volbeda A, Cavazza C, Nicolet Y (2007) Structure/function relationships of [NiFe]- and [FeFe]-hydrogenases. Chem Rev 107:4273–4303

Forestier M, King P, Posewitz M, Schwarzer S, Happe T, Zhang L, Ghirardi ML, Seibert M (2003) Expression of two [Fe]-hydrogenases in *Chlamydomonas reinhardtii* under anaerobic conditions. Eur J Biochem 270:2750–2758. doi:10.1046/j.1432-1033.2003.03656

Fouchard S, Pruvost J, Legrand J (2008) Investigation of H_2 production by microalgae in a fully-controlled photobioreactor. Int J Hydrog Energy 33:3302–3310

Gaffron H, Rubin J (1942) Fermentative and photochemical production of hydrogen in algae. J Gen Physiol 26:219–240. doi:10.1085/jgp.26.2.219

Gallagher BJ (2011) The economics of producing biodiesel from algae. Renew Energy 36(1):158–162. doi:10.1016/j.renene.2010.06.016

Gartner K, Lechno-Yossef S, Cornish AJ, Wolk CP, Hegg EL (2012) Expression of *Shewanella oneidensis* MR-1 [FeFe]-hydrogenase genes in *Anabaena* sp. strain PCC7120. Appl Environ Microbiol 78(24):8579–8586. doi:10.1128/AEM.01959-12

Gerloff-Elias A, Barua D, Mölich A, Spijkerman E (2006) Temperature and pH-dependent accumulation of heat-shock proteins in the acidophilic green alga *Chlamydomonas acidophila*. FEMS Microbiol Ecol 56(3):345–354. doi:10.1111/j.1574-6941.2006.00078.x

Ghirardi ML (1997) Oxygen sensitivity of algal H_2-production. Appl Biochem Biotechnol 63:141–151. doi:10.1007/978-1-4612-2312-2-14

Ghirardi ML, Posewitz MC, Maness PC, Dubini A, Yo J, Seibert M (2007) Hydrogen and hydrogen photoproduction in oxygenic photosynthetic organisms. Annu Rev Plant Biol 58:71–91. doi:10.1146/annurev.arplant.58.032806.103848

Giering SLC, Sanders R, Lampitt RS, Anderson TR, Tamburini C, Boutrif M, Zubkov MV, Marsay CM, Henson SA, Saw K, Cook K, Mayor DJ (2014) Reconciliation of the carbon budget in the ocean's twilight zone. Nature 507:480–483. doi:10.1038/nature13123

Godman JE, Molna'r A, Baulcombe DC Balk J (2010) RNA silencing of hydrogenase (-like) genes and investigation of their physiological roles in the green alga *Chlamydomonas reinhardtii*. Biochem J 431:345–351

Graham JE, Graham LE, Wilcox LW (2008) Algae, 2nd edn. Benjamin Cummings, USA

Graham JE, Wilcox LW, Graham LE (2000) Algae, 2nd edn. Prentice Hall, Upper Saddle River

Guan YF, Deng MC, Yu XJ, Zhang W (2004) Two-stage photobiological production of hydrogen by marine green alga *Platymonas subcordiformis*. Biochem Eng J 19:69–73

Hallenbeck PC (2012) Hydrogen production by cyanobacteria. In: Hallenbeck PC (ed) Microbial technologies in advanced biofuels production, part 2. Springer, New York, pp 15–28. doi:10.1007/978-1-4614-1208-3-2

Hallenbeck PC, Benemann JR (2002) Biological hydrogen production; fundamentals and limiting processes. Int J Hydrog Energy 27(11–12):1185–1193. doi:10.1016/S0360-3199(02)00131-3

Hanagata N, Takeuchi T, Fukuju Y, Barnes DJ, Karube I (1992) Tolerance of microalgae to high CO_2 and high temperature. Phys Chem Chem Phys 31(10):3345–3348. doi:10.1016/0031-9422(92)83682-O

Happe T, Kaminski A (2002) Differential regulation of the Fe-hydrogenase during anaerobic adaptation in the green alga *Chlamydomonas reinhardtii*. Eur J Biochem 269(3):1022–1032. doi:10.1046/j.0014-2956.2001.02743.x

Happe T, Naber JD (1993) Isolation, characterization and N-terminal amino acid sequence of hydrogenase from the green alga *Chlamydomonas reinhardtii*. Eur JBiochem214:475–481.doi:10.1111/j.1432-1033.1993.tb17944.x

Happe T, Mosler B, Naber JD (1994) Induction, localization and metal content of hydrogenase in the green alga

Chlamydomonas reinhardtii. Eur J Biochem 222:769–774. doi:10.1111/j.1432-1033.1994.tb18923.x

Happe T, Schutz K, Bohme H (2000) Transcriptional and mutational analysis of the uptake hydrogenase of the filamentous cyanobacteria *Anabaena variabilis* ATCC29413. J Bacteriol 182(6):1624–1631. doi:10.1128/JB.182.6.1624-1631.2000

He M, Li L, Zhang L, Liu J (2012) The enhancement of hydrogen photoproduction in *Chlorella protothecoides* exposed to nitrogen limitation and sulfur deprivation. Int J Hydrog Energy 37(22):16903–16915

Heyer H, Stal LJ, Krumbein WE (1989) Simultaneous heterolactic and acetate fermentation in the marine cyanobacterium *Oscillatoria limosa* incubated anaerobically in the dark. Arch Microbiol 151(6):558–564. doi:10.1007/BF00454875

Homann PH (2003) Hydrogen metabolism of green algae. Discovery and early research-a tribute to Hans Gaffron and his coworkers. Photosynth Res 76(1–3):93–103. doi:10.1023/A:1024935223225

Howarth DC, Codd GA (1985) The uptake and production of molecular hydrogen by unicellular cyanobacteria. J Gen Microbiol 131(7):1561–1569. doi:10.1099/00221287-131-7-1561

Hu H, Gao K (2003) Optimization of growth and fatty acid composition of a unicellular marine picoplankton, *Nannochloropsis* sp., with enriched carbon sources. Biotechnol Lett 25(5):421–425. doi:10.1023/A:1022489108980

Hu Q, Richmond A (1996) Productivity and photosynthetic efficiency of *Spirulina platensis* as affected by light intensity, algal density and rate of mixing in a flat plate photobioreactor. J Appl Phycol 8(2):139–145. doi:10.1007/BF02186317

Hu Q, Sommerfeld M, Jarvis E, Ghirardi M, Posewitz M, Seibert M, Darzins A (2008) Microalgal triacylglycerols as feedstocks for biofuel production: perspectives and advances. Plant J 54(4):621–639. doi:10.1111/j.1365-313X.2008.03492.x

Hwang JH, Kim HC, Choi JA, Abou-Shanab RAI, Dempsey BA, Regan JM, Kim JR, Song H, Nam IH, Su-Nam Kim SN, Lee W, Park D, Kim Y, Choi J, Ji MK, Jung W, Jeon BH (2014) Photoautotrophic hydrogen production by eukaryotic microalgae under aerobic conditions. Nat Commun 5:3234. doi:10.1038/ncomms4234

IPCC (2014) Summary for policymakers. In: Field CB, Barros VR, Dokken DJ, Mach KJ, Mastrandrea MD, Bilir TE, Chatterjee M, Ebi KL, Estrada YO, Genova RC, Girma B, Kissel ES, Levy AN, MacCracken S, Mastrandrea PR, White LL (eds) Climate change 2014: impacts, adaptation, and vulnerability. Part A: global and sectoral aspects. Contribution of working group II to the fifth assessment report of the intergovernmental panel on climate change. Cambridge University Press, Cambridge, pp 1–32

Iqbal M, Grey D, Stepan-Sarkissian F, Fowler MW (1993) A flat-sided photobioreactor for continuous culturing microalgae. Aquacult Eng 12:183–190

Jackson DD, Ellms JW (1896) On odors and surface waters with special reference to *Anabaena*, a microscopical organism found in certain water supplies of Massachusetts. Rep Mass State Board Health 410–420

Kaneko T, Tabata S (1997) Complete genome structure of the unicellular cyanobacterium *Synechocystis* sp. PCC6803. Plant Cell Physiol 38(11):1171–1176

Khanal SSK, Chen WH, Li L, Sung S (2004) Biological hydrogen production: effects of pH and intermediate products. Int J Hydrog Energy 29:1123–1131

Kosourov SN, Seibert M (2009) Hydrogen photoproduction by nutrient-deprived chlamydomonas reinhardtii cells immobilized within thin alginate films under aerobic and anaerobic conditions. Biotechnol Bioeng 102(1):50–58. doi:10.1002/bit.22050

Kosourov S, Tsygankov A, Seibert M, Ghirardi ML (2002) Sustained hydrogen photoproduction by *Chlamydomonas reinhardtii*: effects of culture parameters. Biotechnol Bioeng 78(7):731–740. doi:10.1002/bit.10254

Kosourov S, Seibert M, Ghirardi ML (2003) Effects of extracellular pH on the metabolic pathways in sulfur-deprived, H_2-producing *Chlamydomonas reinhardtii* cultures. Plant Cell Physiol 44(2):146–155. doi:10.1093/pcp/pcg020

Kosourov SN, Ghirardi ML, Seibert M (2011) A truncated antenna mutant of *Chlamydomonas reinhardtii* can produce more hydrogen than the parental strain. Int J Hydrog Energy 36(3):2044–2048. doi:10.1016/j.ijhydene.2010.10.041

Kruse O, Rupprecht J, Bader K, Thomas-Hall S, Schenk PM, Finazzi G, Hnkamer B (2005) Improved photobiological H_2 production in engineered green algal cells. J Biol Chem 280(40):34170–34177. doi:10.1074/jbc.M503840200

Kufryk G (2013) Advances in utilizing cyanobacteria for hydrogen production. Adv Microbiol 3:60–68. doi:10.4236/aim.2013.36A008

Kumar K, Roy S, Das D (2013) Continuous mode of carbon dioxide sequestration by *C. sorokiniana* and subsequent use of its biomass for hydrogen production by *E. cloacae* IIT-BT 08. Bioresour Technol 145:116–122. doi:10.1016/j.biortech.2013.01.137

Kumazawa S (2003) Photoproduction of hydrogen by the marine heterocystous cyanobacterium *Anabaena* species TU37-1 under a nitrogen atmosphere. Mar Biotechnol, NY 5(3):222–226. doi:10.1007/s10126-002-0106-x

Lakaniemi AM, Hulatt CJ, Thomas DN, Tuovinen OH, Puhakka JA (2011) Biogenic hydrogen and methane production from *Chlorella vulgaris* and *Dunaliella tertiolecta* biomass. Biotechnol Biofuels 4:34. doi:10.1186/1754-6834-4-34

Lambert GR, Smith GD (1977) Hydrogen formation by marine blue-green algae. FEBS Lett 83(1):159–162. doi:10.1016/0014-5793(77)80664-9

Lambert GR, Daday A, Smith GD (1979) Hydrogen evolution from immobilized cultures of the cyanobacterium *Anabaena cylindrica* B629. FEBS Lett 101:125–128. doi:10.1016/0014-5793(79)81309-5

Laurinavichene TV, Fedorov AS, Ghirardi ML, Seibert M, Tsygankov AA (2006) Demonstration of hydrogen photoproduction by immobilized sulfur-deprived Chlamydomonas reinhardtii cells. Int J Hydrog Energy 31(5):659–667. doi:10.1016/j.ijhydene.2005.05.002

Lee RE (2009) Phycology. Cambridge University Press, Cambridge

Lee JW, Greenbaum E (2003) A new oxygen sensitivity and its potential application in photosynthetic H_2 production. Appl Biochem Biotechnol 106(1–2–3):303–313

Lee YK, Ding SY, Low CS, Chang YC (1995) Design and performance of an a-type tubular photobioreactor for mass cultivation of microalgae. J Appl Phycol 7:47–51

Lee YYJ, Miyahara T, Noike T (2002) Effect of p^H on microbial hydrogen fermentation. J Chem Technol Biotechnol 77(6):694–698. doi:10.1002/jctb.623

Leino H, Shunmugam S, Isoja¨rvi J, Oliveira P, Mulo P, Saari L, Battchikova N, Sivonen K, Lindblad P, Aro EV, Allahverdiyev Y (2014) Characterization of ten H_2 producing cyanobacteria isolated from the baltic sea and finnish lakes. Int J Hydrog Energy 39(17):8983–8991. doi:10.1016/j.ijhydene.2014.03.171

Levin DB, Pitt L, Love M (2004) Biohydrogen production: prospects and limitations to practical application. Int J Hydrog Energy 29:173–185. doi:10.1016/S0360-3199(03)00094-6

Lewis NS, Nocera DG (2006) Powering the planet: chemical challenges in solar energy utilization. Proc Natl Acad Sci U S A 103(43):15729–15735. doi:10.1073/pnas.0603395103

Lindberg P, Hansel A, Lindblad P (2000) hupS and hupL constitute a transcription unit in the cyanobacterium *Nostoc* sp. PCC73102. Arch Microbiol 174(1–2):129–133. doi:10.1007/s002030000186

Lindberg P, Lindblad P, Cournac L (2004) Gas exchange in the filamentous cyanobacterium *Nostoc punctiforme* strain ATCC29133 and Its hydrogenase-deficient mutant strain NHM5. Appl Environ Microbiol 70(4):2137–2145. doi:10.1128/AEM.70.4.2137-2145.2004

Lindblad P, Christensson K, Lindberg P, Fedorov A, Pinto F, Tsygankov A (2002) Photoproduction of H_2 by wild type *Anabaena* PCC7120 and a hydrogen uptake deficient mutant: from laboratory experiments to outdoor culture. Int J Hydrog Energy 27(11):1271–1281. doi:10.1016/S0360-3199(02)00111-8

Lubitz W, Reijerse EJ, Messinger J (2008) Solar water-splitting into H_2 and O_2: design principles of photosystem II and hydrogenases. Energy Environ Sci 1(1):15–31. doi:10.1039/b808792j

Lukey MJ, Roessler MM, Parkin A, Evans RM, Davies RA, Lenz O, Friedrich B, Sargent F, Armstrong FA (2011) Oxygen-tolerant [NiFe]- hydrogenases: the individual and collective importance of supernumerary cysteines at the proximal Fe-S cluster. J Am Chem Soc 133(42):16881–16892. doi:10.1021/ja205393w

Madden C, Vaughn MD, Daez-Parez I, Brown KA, King PW, Gust D, Moore AL, Moore TA (2011) Catalytic turnover of [FeFe]-hydrogenase based on single-molecule imaging. J Am Chem Soc 134(3):1577–1582. doi:10.1021/ja207461t

Martnez-Jeronimo F, Espinosa-Chavez FA (1994) Laboratory-scale system for mass culture of freshwater microalgae in polyethylene bags. J Appl Phycol 6:423–425

Masukawa H, Mochimaru M, Sakurai H (2002) Disruption of the uptake hydrogenase gene, but not of the bidirectional hydrogenase gene, leads to enhanced photobiological hydrogen production by the nitrogen-fixing cyanobacterium *Anabaena* sp. PCC 7120. Appl Microbiol Biotechnol 58(5):618–624. doi:10.1007/s00253-002-0934-7

Masukawa H, Kitashima M, Kazuhito Inoue K, Hidehiro Sakurai H, Hausinger RP (2012) Genetic engineering of cyanobacteria to enhance biohydrogen production from sunlight and water. AMBIO 41(2):169–173. doi:10.1007/s13280-012-0275-4

Mata TM, Martins AA, Caetano NS (2009) Microalgae for biodiesel production and other applications: a review. Renew Sust Energy Rev 14(1):217–232. doi:10.1016/j.rser.2009.07.020

Mata TA, Nidia AA, Caetano S (2010) Microalgae for biodiesel production and other applications: a review. Renew Sustain Energy Rev 14(1):217–232

Mathews J, Wang G (2009) Metabolic pathway engineering for enhanced biohydrogen production. Int J Hydrog Energy 34:7404–7416. doi:10.1016/j.ijhydene.2009.05.078

Melis A (2002) Green alga hydrogen production: progress, challenges and prospects. Int J Hydrog Energy 27:1217–1228

Melis A (2007) Photosynthetic H_2 metabolism in *Chlamydomonas reinhardtii* (unicellular green algae). Planta 226(5):1075–1086. doi:10.1007/s00425-007-0609-9

Melis A, Happe T (2004) Trails of green alga hydrogen research; from Hans Gaffron to new frontiers. Photosynth Res 80(1–3):401–409. doi:10.1023/B:PRES.0000030421.31730

Melis A, Melnicki MR (2006) Integrated biological hydrogen production. Int J Hydrog Energy 31(11):1563–1573. doi:10.1016/j.ijhydene.2006.06.038

Melis A, Zhang L, Forestier M, Ghirardi ML, Seibert M (2000) Sustained photobiological hydrogen gas production upon reversible inactivation of oxygen evolution in the green alga *Chlamydomonas reinhardtii*. Plant Physiol 122:127–136

Menetrez MY (2012) An overview of algae biofuel production and potential environmental impact. A critical review. Environ Sci Technol 46(13):7073–7085. doi:10.1021/es300917r

Meuser JE, D'Adamo S, Jinkerson RE, Mus F, Yang W, Ghirardi ML, Seibert M, Grossman AR, Posewitz MC (2012) Genetic disruption of both *Chlamydomonas reinhardtii* [FeFe]- hydrogenases: insight into the role of HYDA2 in H_2 production. Biochem Biophys Res Commun 417(2):704–709. doi:10.1016/j.bbrc.2011.12.002

Miyachi S, Iwasaki I, Shiraiwa Y (2003) Historical perspective on microalgal and cyanobacterial acclimation to low- and extremely high-CO_2 conditions. Photosynth Res 77(2–3):139–153

Miyake M, Asada Y (1997) Direct electroporation of clostridial hydrogenase into cyanobacterial cells. Biotechnol Tech 11(11):787–790. doi:10.1023/A:1018417023074

Moezelaar R, Bijvank SM, Stal LJ (1996) Fermentation and sulfur reduction in the mat-building cyanobacterium *Microcoleus chtonoplastes*. Appl Environ Microbiol 62:1752–1758

Morita M, Watanable Y, Saiki H (2000) Investigation of photobioreactor design for enhancing the photosynthetic productivity of microalgae. Biotechnol Bioeng 69(6):693–698. doi:10.1002/1097-0290(20000920)69:6

Murphy M, Devlin G, Deverel R, McDonnel K (2013) Biofuel production in Ireland: an approach to 2020 targets with a focus on algal biomass. Energies 6(12):6391–6412. doi:10.3390/en6126391

Mus F, Dubini A, Seibert M, Posewitz MC, Grossman AR (2007) Anaerobic acclimation in *Chlamydomonas reinhardtii*: anoxic gene expression, hydrogenase induction, and metabolic pathways. J Biol Chem 282(35):25475–25486. doi:10.1074/jbc.M701415200

Mussgnug JH, Thomas-Hall S, Rupprecht J, Foo A, Klassen V (2007) Engineering photosynthetic light capture: impacts on improved solar energy to biomass conversion. Plant Biotechnol J 5(6):802–814. doi:10.1111/j.1467-7652.2007.00285.x

Nath K, Das D (2004) Improvement of fermentative hydrogen production: various approaches. Appl Microbiol Biotechnol 65:520–529. doi:10.1007/s00253-004-1644-0

Nayak BK, Roy S, Das D (2014) Biohydrogen production from algal biomass (Anabaena sp. PCC7120) cultivated in airlift photobioreactor. Int J Hydrog Energy 3(14):7553–7560. doi:10.1016/j.ijhydene.2013.07.120

Odum HT (1971) Environment, power and society. Wiley-Interscience, New York, USA, p 331

Oey M, Ross IL, Stephens E, Steinbeck J, Wolf J, Radzun KA, Kagler J, Ringsmuth AK, Kruse O, Hankamer B (2013) RNAi knock-down of LHCBM1, 2 and 3 increases photosynthetic H_2 production efficiency of the green alga *Chlamydomonas reinhardtii*. PLoS One 8(4):61375. doi:10.1371/journal.pone.0061375

Ohta S, Miyamoto K, Miura Y (1987) Hydrogen evolution as a consumption mode of reducing equivalents in green algal fermentation. Plant Physiol 83:1022–1026

Oliveira P, Leitao E, Tamagnini P, Moradas-Ferreira P, Oxelfelt F (2004) Characterization and transcriptional analysis of *hupSLW* in *Gloeothece* sp. ATCC27152: an uptake hydrogenase from a unicellular cyanobacterium. Microbiology 150(11):3647–3655. doi:10.1099/mic.0.27248-0

Oncel S, Kose A (2014) Comparison of tubular and panel type photobioreactors for biohydrogen production utilizing *Chlamydomonas reinhardtii* considering mixing time and light intensity. Bioresour Technol 151:265–270. doi:10.1016/j.biortech.2013.10.076

Oncel S, Sabankay M (2012) Microalgal biohydrogen production considering light energy and mixing time as the two key features for scale-up. Bioresour Technol 121:228–234. doi:10.1016/j.biortech.2012.06.079

Otsuki T, Uchiyama S, Fujiki K, Fukunaga S (1998) Hydrogen production by a floating-type photobioreactor. In: Zaborsky OR (ed) Biohydrogen. Plenum Press, London, pp 369–374. doi:10.1007/978-0-585-35132-2-45

Pedersen MF, Borum J (1996) Nutrient control of algal growth in estuarine waters. Nutrient limitation and the importance of nitrogen requirements and nitrogen storage among phytoplankton and species of macroalgae. Mar Ecol Prog Ser (MEPS) 142:261–272. doi:10.3354/meps142261

Perrine Z, Negi S, Sayre RT (2012) Optimization of photosynthetic light energy utilization by microalgae. Algal Res 1(2):134–142. doi:10.1016/j.algal.2012.07.002

Phlips EJ, Mitsui A (1983) Role of light intensity and temperature in the regulation of hydrogen photoproduction by the marine cyanobacterium *Oscillatoria* sp. strain Miami BG7. Appl Environ Microbiol 45:1212–1220

Pohl P, Kohlhase M, Martin M (1988) Photobioreactors for the axenic mass cultivation of microalgae. In: Stadler T, Mollion J, Verdus MC, Karamanos Y, Morvan H, Christiaen D (eds) Algal biotechnology. Elsevier, New York, pp 209–218

Polle J, Kanakagiri S, Melis A (2003) tla1, a DNA insertional transformant of the green alga *Chlamydomonas reinhardtii* with a truncated light-harvesting chlorophyll antenna size. Planta 271(1):49–59. doi:10.1007/s00425-002-0968-1

Pottier L, Pruvost J, Deremetz J, Cornet JF, Legrand J, Dussap CG (2005) A fully predictive model for one-dimensional light attenuation by *Chlamydomonas reinhardtii* in a torus photobioreactor. Biotechnol Bioeng 91(5):569–582. doi:10.1002/bit.20475

Quinn JM, Eriksson M, Moseley JL, Merchant S (2002) Oxygen deficiency responsive gene expression in Chlamydomonas reinhardtii through a copper-sensing signal transduction pathway. Plant Physiol 128(2):463–471. http://dx.doi.org/10.1104/pp.010694

Rashid N, Lee K, Mahmood Q (2011) Bio-hydrogen production by *Chlorella vulgaris* under diverse photoperiods. Bioresour Technol 102(2):2101–2104. doi:10.1016/j.biortech.2010.08.032

Rashid N, Rehman MSU, Memonb S, Rahman ZU, Lee K, Han J (2013) Current status, barriers and developments in biohydrogen production by microalgae. Renew Sustain Energy Rev 22:571–579. doi:10.1016/j.rser.2013.01.051

Resnick RJ (2004) The economics of biological methods of hydrogen production. Massachusetts Institute of Technology. (Thesis (S.M.M.O.T.) Massachusetts Institute of Technology, Sloan School of Management, Management of Technology Program, 2004. Includes bibliographical references, pp 98–108. URI: http://hdl.handle.net/1721.1/17885. Cited 14th Jan 2015

Roessler PG, Lien S (1984) Activation and de novo synthesis of hydrogenase in *Chlamydomonas*. Plant Physiol 76:1086–1089

Roy S, Kumar K, Ghosh S, Das D (2014) Thermophilic biohydrogen production using pre-treated algal biomass as substrate. Biomass Bioenergy 61:157–166. doi:10.1016/j.biombioe.2013.12.006

Ruhle T, Hemschemeier A, Melis A, Happe T (2008) A novel screening protocol for the isolation of hydrogen producing Chlamydomonas reinhardtii strains. BMC Plant Biol 8(107):1471–2229. doi:10.1186/1471-2229-8-107

Sakai N, Sakamoto Y, Kishimoto N, Chihara M, Karube I (1995) *Chlorella* strains from hot springs tolerant to high temperature and high CO_2. Energy Convers Manag 36(6–9):693–696. doi:10.1016/0196-8904(95)00100-R

Satoh A, Kurano N, Senger H, Miyachi S (2002) Regulation of energy balance in photosystems in response to changes in CO_2 concentrations and light intensities during growth in extremely-high-CO_2-tolerant green microalgae. Plant Cell Physiol 43(4):440–451. doi:10.1093/pcp/pcf054

Scherer S, Kerfin W, Boger P (1980) Increase of nitrogenase activity in the blue-green alga *Nostoc muscorum* (Cyanobacterium). J Bacteriol 144:1017–1023

Schmitz O, Bothe H (1996) The diaphorase subunit HoxU of the bidirectional hydrogenase as electron transferring protein in cyanobacterial respiration? Naturwissenschaften 83(11):525–527. doi:10.1007/BF01141957

Schmitz O, Boison G, Salzman H, Bothe H, Schutz K, Wang S, Happe T (2002) HoxE- a subunit specific for the pentameric bidirectional hydrogenase complex (HoxEFUYH) of cyanobacteria. Biochim Biophys (BBA) Bioenergetics 1554(1–2):66–74. doi:10.1016/S0005-2728(02)00214-1

Schutz K, Happe T, Troshina O, Lindblad P, Leitao E, Oliveira P, Tamagnini P (2004) Cyanobacterial H_2 production; a comparative analysis. Planta 218:350–359

Schwarz S, Poss Z, Hoffmann D, Appel J (2010) Hydrogenases and hydrogen metabolism in photosynthetic prokaryotes. In: Hallenbeck PC (ed) Recent advances in phototrophic prokaryotes. Springer, New York, pp p305–p348

Scoma A, Krawietz D, Faraloni C, Giannelli L, Happe T, Torzillo G (2012) Sustained H_2 production in a *Chlamydomonas reinhardtii* D1 protein mutant. J Biotechnol 157(4):613–619. doi:10.1016/j.jbiotec.2011.06.019

Serebryakova LT, Sheremetieva ME, Lindblad P (2000) H_2-uptake and evolution in the unicellular cyanobacterium *Chroococcidiopsis thermalis* CALU 758. Plant Physiol Biochem 38:525–530

Skjånes K, Pinto FL, Lindblad P (2010) Evidence for transcription of three genes with characteristics of hydrogenases in the green alga *Chlamydomonas noctigama*. Int J Hydrog Energy 35:1074–1088

Slade R, Bauen A (2013) Micro-algae cultivation for biofuels: cost, energy balance, environmental impacts and future prospects. Biomass Bioenergy 53:29–38. doi:10.1016/j.biombioe.2012.12.019

Smith RL, Kumar D, Zhang XK, Tabita FR, Baalen VC (1985) H_2, N_2, and O_2 metabolism by isolated heterocysts from *Anabaena* sp. strain CA. J Bacteriol 162(2):565–570

Song W, Rashid N, Choi W, Lee (2011) Biohydrogen production by immobilized *Chlorella* sp. using cycles of oxygenic photosynthesis and anaerobiosis. Bioresour Technol 102(18):8676–8681. doi:10.1016/j.biortech.2011.02.082

Steunou AS, Bhaya D, Bateson MM, Melendrez MC, Ward DM, Brech E, Peters JW, Kuhl M, Grossman AR (2006) In situ analysis of nitrogen fixation and metabolic switching in unicellular thermophilic cyanobacteria inhabiting hot spring microbial mats. Proc Natl Acad Sci U S A 103(7):2398–2403. doi:10.1073/pnas.0507513103

Sveshnikov DA, Sveshnikova NV, Rao KK, Hall DO (1997) Hydrogen metabolism of mutant forms of *Anabaena variabilis* in continuous cultures and under nutritional stress. FEMS Microbiol Lett 147(2):297–301. doi:10.1111/j.1574-6968.1997.tb10257.x

Tamagnini P, Costa JL, Almeida L, Oliveira MJ, Salema R, Lindblad P (2000) Diversity of cyanobacterial hydrogenases, a molecular approach. Curr Microbiol 40(6):356–361. doi:10.1007/s002840010070

Tamagnini P, Axelsson R, Lindberg P, Oxelfelt F, Wunschiers R, Lindblad P (2002) Hydrogenases and hydrogen metabolism of cyanobacteria. Microbiol Mol Biol Rev 66(1):1–20. doi:10.1128/MMBR.66.1.1-20.2002

Tamagnini P, Leita˜o E, Oliveira P, Ferreira D, Pinto F, Harris D (2007) Cyanobacterial hydrogenases: diversity, regulation and applications. FEMS Microbiol Rev 31:692–720. doi:10.1111/j.1574-6976.2007.00085.x

Tamburic B, Zemichael FW, Maitland GC, Hellgardt K (2011) Parameters affecting the growth and hydrogen production of the green alga *Chlamydomonas reinhardtii*. Int J Hydrog Energy 36(13):7872–7876. doi:10.1016/j.ijhydene.2010.11.074

Torzillo G, Scoma A, Faraloni C, Ena A, Johanningmeier U (2009) Increased hydrogen photoproduction by means of a sulfur-deprived *Chlamydomonas reinhardtii* D1 protein mutant. Int J Hydrog Energy 34(10):4529–4536. doi:10.1016/j.ijhydene.2008.07.093

Tredici MR, Zittelli GC (1998) Efficiency of sunlight utilization: tubular versus flat photobioreactors. Biotechnol Bioeng 57(2):187–197. doi:10.1002/(SICI)1097-0290(19980120)5

Tredici MR, Carlozzi P, Zittelli GC, Materassi R (1991) A vertical alveolar panel (VAP) for outdoor mass cultivation of microalgae and cyanobacteria. Bioresour Technol 38(2):153–159. doi:10.1016/0960-8524(91)90147-C

Tredici MR, Zittelli GC, Biagiolini S, Materassi R (1993) Novel photobioreactor for the mass cultivation of *Spirulina* sp. Bull Inst Oceanogr 12:89–96

Tredici MR, Zittelli GC, Benemann JR (1998) A tubular integral gas exchange photobioreactor for biological hydrogen production. In: Zaborsky OR (ed) Biohydrogen. Plenum Press, London, pp 391–340. doi:10.1007/978-0-585-35132-2-48

Tsygankov AA, Hall DO, Liu J, Rao KK (1998) An automated helical photobioreactor incorporating cyanobacteria for continuous hydrogen production. In: Zaborsky OR (ed) Biohydrogen. Plenum Press, London, pp 431–440. doi:10.1007/978-0-585-35132-2-52

Tsygankov AA, Fedorov AS, Kosourov SN, Rao KK (2002) Hydrogen production by cyanobacteria in an automated outdoor photobioreactor under aerobic conditions. Biotechnol Bioeng 80(7):777–783. doi:10.1002/bit.10431

Uitz J, Claustre H, Gentili B, Stramski D (2010) Phytoplankton class-specific primary production in the world's oceans: seasonal and interannual variability from satellite observations. Global Biogeochem Cycles 24:GB3016. doi:10.1029/2009GB003680

Van den Hoek C, Mann DG, Jahns HM (1995) Algae: an introduction to phycology. Cambridge University Press, Cambridge, p 623

Vignais PM, Billoud B (2007) Occurrence, classification, and biological function of hydrogenases: an overview. Chem Rev 107:4206–4272

Vignais PM, Colbeau A (2004) Molecular biology of microbial hydrogenases. Curr Issues Mol Biol 6(2):159–188

Vignais PM, Magnin JP, Willison JC (2006) Increasing biohydrogen production by metabolic engineering. Int J Hydrog Energy 31:1478–1483. doi:10.1016/j.ijhydene.2006.06.013

Volgusheva A, Stenbjo¨rn S, Fikret M (2013) Increased photosystem II stability promotes H2 production in sulfur-deprived Chlamydomonas reinhardtii. Proc Natl Acad Sci U S A 110(18):7223–7228

Wang B, Li YQ, Wu N, Lan CQ (2008) CO_2 bio-mitigation using microalgae. Appl Microbiol Biotech 79(5):707–718

Weissman JC, Benemann JR (1977) Hydrogen production by nitrogen-starved cultures of *Anabaena cylindrica.* Appl Environ Microbiol 33:123–131

Weyman PD, Smith HO, Xu Q (2011) Genetic analysis of the Alteromonas macleodii [NiFe]-hydrogenase. FEMS Microbiol Lett 322(2):180–187. doi:10.1111/j.1574-6968.2011.02348.x

Winkler M, Heil B, Happe T (2002a) Isolation and molecular characterization of the [Fe]-hydrogenase from the unicellular green alga *Chlorella fusca.* Biochim Biophys Acta 1576:330–334

Winkler M, Heschemeirer A, Gotor C, Melis A, Happe T (2002b) [Fe]-hydrogenases in green algae: photo-fermentation and hydrogen evolution under sulfur deprivation. Int J Hydrog Energy 27:1431–1439

Winkler M, Macurer C, Heschemeirer A, Happe T (2004) The isolation of green algal strains with outstanding H_2 productivity. In: Miyake J, Igarshi Y, Roger M (eds) Biohydrogen III. Elsevier, New York, pp 103–115

Wu S, Xu L, Huang R, Wang Q (2011) Improved biohydrogen production with an expression of codon-optimized hemH and lba genes in the chloroplast of *Chlamydomonas reinhardtii.* Bioresour Technol 102(3):2610–2616. doi:10.1016/j.biortech.2010.09.123

Wunschiers R, Stangier K, Senger H, Schulz R (2001) Molecular evidence for a Fe-hydrogenase in the green alga *Scenedesmus obliquus.* Curr Microbiol 42:353–360. doi:10.1007/s002840010229

Yoshino F, Ikeda H, Masukawa H, Sakurai H (2007) High photobiological hydrogen production activity of a *Nostoc* sp. PCC7422 uptake hydrogenase-deficient mutant with high nitrogenase activity. Mar Biotechnol 9(1):101–112. doi:10.1007/s10126-006-6035-3

Zhang XK, Haskell JB, Tabita FR, Baalen VC (1983) Aerobic hydrogen production by the heterocystous cyanobacteria *Anabaena* spp. strains CA and 1F. J Bacteriol 156(3):1118–11

Challenges in the Design and Operation of an Efficient Photobioreactor for Microalgae Cultivation and Hydrogen Production

7

Surajbhan Sevda, Sourish Bhattacharya, Ibrahim M. Abu Reesh, S. Bhuvanesh, and T.R. Sreekrishnan

Abstract

The major challenge in the production of biofuels from microalgae is the need to generate sufficient quantities of microalgal biomass and an environmentally friendly and cost-effective method for extraction of oil from the biomass. Biomass can be generated by cultivating microalgae in open ponds or closed photobioreactor systems. When using a photobioreactor system, it is possible to have better control over parameters such as temperature, pH, light intensity, dissolved oxygen and dissolved carbon dioxide. However, they consume more energy and are expensive to operate. Cultivation of microalgae in open ponds is cheaper, and it utilises less energy as compared to closed photobioreactors. But, it is not possible to control physical parameters like temperature and light intensity as they depend on the environmental conditions. Also, contamination from other predators, parasites and weeds needs to be addressed. Considering, the overall cost-effectiveness, it may be possible to cultivate microalgae in open ponds under semi-continuous systems. Direct production of hydrogen using photosynthetic microorganisms such as microalgae may also be

S. Sevda (✉)
Department of Chemical Engineering, School of Engineering, Qatar University, 2713, Doha 2713, Qatar

Department of Biochemical Engineering and Biotechnology, Indian Institute of Technology Delhi, New Delhi 110016, India
e-mail: sevdasuraj@gmail.com

S. Bhattacharya
Process Design and Engineering Cell, Central Salt and Marine Chemicals Research Institute, Bhavnagar 364002, Gujarat, India

I.M.A. Reesh
Department of Chemical Engineering, School of Engineering, Qatar University, 2713, Doha 2713, Qatar

S. Bhuvanesh • T.R. Sreekrishnan
Department of Biochemical Engineering and Biotechnology, Indian Institute of Technology Delhi, New Delhi 110016, India

A. Singh, D. Rathore (eds.), *Biohydrogen Production: Sustainability of Current Technology and Future Perspective*, DOI 10.1007/978-81-322-3577-4_7

considered since it can be energetically more favourable than cultivating, harvesting and processing the biomass for biofuel production. In such cases, degradation of the hydrogen produced by the hydrogenase enzyme present in the system needs to be managed. Considering future energy demands, the possibility of CO_2 sequestration and bioenergy production from microalgae and the overall ease of cultivation, it may be possible to use semi-continuous cultivation in open ponds for generating microalgal biomass with better biomass yield.

7.1 Introduction

Combustion of fossil fuels is the main anthropogenic source of greenhouse gas (GHG) emissions. Carbon dioxide is one of the main GHG, which causes not only global warming, but, also other environmental problems (Boden et al. 2012). Agriculture and deforestation associated with it is the second major source of GHG emissions (Metz et al. 2007). In this scenario, the need to find clean and renewable energy with minimum footprint is one of the most challenging problems. There is a need to replace fossil fuels with biofuels to prevent further deterioration to the environment. Biofuels can replace currently used fossil fuels like diesel and gasoline with a little or no modifications to the currently used engines.

7.1.1 Microalgae

Microalgae are a group of prokaryotic and eukaryotic photosynthetic microorganisms present in a wide range of environmental conditions, both in aquatic and terrestrial ecosystem. Examples of eukaryotic microalgae are green algae (Chlorophyta) and diatoms (Bacillariophyta) and prokaryotic microorganisms is Cyanobacteria (Cyanophyceae) (Georgianna and Mayfield 2012; Li et al. 2008). Microalgae have the capability to live in harsh conditions and grow rapidly due to their unicellular or simple multicellular structure. They are photoautotrophs with a short doubling time. Being suited for growth under harsh conditions, they can grow almost anywhere with sunlight and a few nutrients.

Recently, it has been identified that it is possible to generate biofuels from microalgal biomass. Generation of biofuels using microalgae has various advantages. It is estimated that more than 50,000 species of microalgae exists, but only around 30,000 species have been studied and analysed (Richmond 2003). They are known to grow rapidly under a variety of environmental conditions and survive under harsh environment. Hence, they can be grown on land not suitable for agriculture.

7.1.2 Advantages of Using Microalgae for Biofuel Production

One of the major advantages is that barren or unused land can be utilised for microalgal cultivation (Mata et al. 2010). This has been possible because microalgae are easy to cultivate and can grow with little attention on a marginal land using wastewater or water unsuitable for human consumption.

Microalgae can potentially be used as a source of various renewable fuels (Kroger and Muller-Langer 2012). The microalgal biomass can be utilised for combustion as the crude oil present in the microalgae can be utilised for direct combustion for supplementing other transportation fuels such as diesel, gasoline and jet fuel (kerosene) (Lestari et al. 2009). Anaerobic digestion of the microalgal biomass can produce biogas and hydrogen (Markou et al. 2013; Zamalloa et al. 2011). Fermentation of the carbohydrates derived from microalgae can be used for ethanol production (Ho et al. 2013; Matsumoto et al. 2003). Bioethanol can also be produced directly through microalgal photosynthesis (et al. 2011; Williams 2009).

Microalgae as a potential source of biofuel production have many advantages over traditional biofuel crops. There are various reports describing the additional benefits of using microalgae over conventional feedstock for production of biofuels (Georgianna and Mayfield 2012; Li et al. 2008). As compared to terrestrial crops, which take a season to grow and contain a maximum of about 50% of dry oil content, microalgae grow quickly and contains higher oil content (Chisti 2008; Mata et al. 2010). Further, microalgae have much higher growth rates and productivity as compared to conventional biofuel crops (Table 7.1). Cultivation of microalgae for biofuel production has additional advantages such as (i) biological sequestration of greenhouse gases like CO_2 (various strains of *Chlorella* sp. have been studied for CO_2 sequestration in order to reduce the effects of global warming (Keffer and Kleinheinz 2002; Yanagi et al. 1995)); (ii) treatment of wastewater for removal of NH_4^+, NO_3^- and PO_4^{3-} by microalgal growth on wastewater (Wang et al. 2008); (iii) after extracting oil from the microalgae, the utilisation of the resulting biomass containing nitrogen, phosphorus and potassium as biofertilizer (Wang et al. 2008); and (iv) production of few microalgae species for valuable products like fine chemicals, antioxidants, sugars, oil, pigments, polyunsaturated fatty acids, natural dyes, high-value bioactive compounds and other fine chemicals (Li et al. 2008; Raja et al. 2008).

It is important to look for an alternate source of energy since our fossil fuels are fast depleting and causing various environmental issues. Source of sustainable energy such as the microalgae can be an attractive solution for biofuel production. This chapter aims to discuss the critical parameters essential for designing an efficient bioreactor in a cost-effective manner for cultivating microalgae for the production of biofuel and biohydrogen as a source of alternate energy.

7.2 Microalgal Cultivation

Microalgae are capable of metabolic shifts in response to the environmental conditions and can be cultivated under autotrophic, heterotrophic and mixotrophic growth conditions. Photoautotrophic growth is based on photosynthesis where light energy is used for conversion of CO_2 to biomass. Photoheterotrophic growth is based on the growth of microalgae utilising light and organic carbon source. Chemoheterotrophic growth is based on cellular consumption of organic compounds as a source of energy and carbon. Mixotrophic growth uses a combination of autotrophic and heterotrophic growth.

Along with carbon, in the form of carbon dioxide, microalgae requires nitrogen, which is required for their primary metabolism. Lipids are synthesised as a food reserve that may be required during periods of nutritional stress. Phosphorus, in the form of phosphates, is another major nutrient for growth of microalgae. In addition to these components, trace metals such as Mg, Ca, Mn, Zn, Cu and Fe and vitamins are needed for higher productivity (Lardon et al. 2009; Liu et al. 2011; Mata et al. 2010). Microalgae are known to thrive at higher temperatures, while lower temperatures inhibit their growth. The optimum temperature for most of the species of microalgae is in the range of 15–26 Â°C (Bacellar et al. 2013).

There are two important means of cultivating phototrophic microalgae, viz. open pond system and photobioreactor (PBR). Various designs have been reported for cultivation of microalgae in open ponds and PBR with varying yields according to the process (Georgianna and Mayfield, 2012). Microalgae can be cultivated in open ponds having 0.2–0.5 m depth by mixing it with

Table 7.1 Comparison of oil content and yield of microalgae with other conventional biodiesel feedstocks (Chisti 2008; Mata et al. 2010)

Source	Oil content (%)	Oil yield (L/ha.year)
Corn	44	172
Soybean	18	446
Canola	41	1190
Jatropha	28	1892
Oil palm	36	5950
Microalgae (low oil content)	30	58,700
Microalgae (high oil content)	70	136,900

paddle wheel in order to prevent settling. PBR system consists of transparent line up of tubes or plates and circulation of culture is from a central reservoir. Although PBR systems have better control over culture environment, it is more expensive as compared to open ponds. Microalgae can be cultivated in batch processes or continuous processes. Each process has its own advantages and disadvantages.

7.2.1 Batch Cultivation

Batch cultivation of microalgae is usually carried out in closed photobioreactors. These photobioreactors are made of a translucent container with the light source. Since, the system is closed, all essential nutrients must be introduced into the system at the beginning of cultivation. Components are neither added nor removed during the cultivation process. An open pond covered with a greenhouse could also be used for batch cultivation of microalgae. However, closed photobioreactor have some specific advantages over open systems (Posten and Schaub 2009; Pulz 2001). A closed batch photobioreactor is designed to provide optimal illumination, nutrients, CO_2, mass transfer and mixing to the phototrophic organisms in the liquid suspension. Hence, they have better control over the growth environment than open ponds. It also allows control over the type of species that needs to be grown in the reactor. In photobioreactors, light intensity available to individual cell is ten times higher than any open system which increases cell density of culture. This reduces the amount of water that must be processed for harvesting the cells and thus the overall cost of operation (Chisti 2008). Further, the evaporation loss in a photobioreactor is much lower than in an open pond. Also, cultivation of microalgae can be carried out in closed photobioreactor in uncultivable land. The limiting factors of using a photobioreactor are cost of construction of the reactor and energy requirements for the reactor operation. In a batch cultivation process, light and nutrient levels cannot be maintained uniformly, due to continuously increasing cell density. Therefore, steady-state continuous processes are required with which cell density can be controlled.

7.2.2 Continuous Cultivation

Continuous and semi-continuous cultivation of microalgae is generally carried out in open ponds. The most preferred continuous system for mass cultivation of microalgae includes shallow ponds, tanks, circular ponds and raceway ponds (Oron et al. 1979; Seshadri and Thomas 1979; Vonshak et al. 1985). In a semi-continuous system, microalgae are periodically harvested followed by addition of nutrient media for further growth. Semi-continuous processes can be carried out in open ponds or closed bioreactors.

For mass cultivation of microalgae in open ponds/raceway ponds, commercial grade chemicals can be used. Freshwater or seawater can be used whenever necessary. Inoculum for the large-scale open ponds may be generated in pilot photobioreactors. Table 7.2 summarises the microalgae production from continuous cultivation in photobioreactors.

Microalgal biomass may be produced in open ponds in a cost-effective manner. However, biomass productivity depends on weather, solar intensity, nutrient composition and the type of water. Other limitations of open ponds include

Table 7.2 Biomass production in continuous cultivation of microalgae

Strain	Biomass concentration (g/L)	PBR design	Reference
Chlorella vulgaris	1.72	Tubular	Feng et al. (2011)
Chlorella minutissima	0.726	CSTR	Tang et al. (2012)
Scenedesmus obliquus	2.5–3.5	Tubular	Hulatt and Thomas (2011)
Cyanobium sp.	1.2	Tubular	Wang et al. (2012)

poor light availability for the cells, evaporative water loss, requirement of huge areas of land and large quantities of water and low biomass productivity (Posten and Schaub 2009). However, it may not be right to compare biomass productivity of open systems to closed systems, since it varies with the geographical location.

7.2.3 Accelerated Stat (A-Stat) Cultivation

One of the challenges in a photobioreactor design is the uniform light availability. Steady-state cultivation is more often used for increasing the biomass productivity using chemostat cultivations. However, for light limited growth kinetics of microalgae, A-stat cultivation can be used, which is a continuous culture with smooth controlled change. In this type of cultivation, the dilution rate is increased smoothly at the best estimated rate. This acceleration rate should be a balance between making the cultivation time short and at the same time adapt the metabolism of the microorganism to the changes in the environment. A-stat cultivations are usually performed in bubble column reactors (Barbosa et al. 2003).

The advantages and disadvantages of the different cultivation methods are mentioned in Table 7.3.

7.2.4 Issues with Microalgal Cultivation

Although greater algal biomass concentrations can be generated in PBR systems, it consumes more energy at the cultivation stage as compared to open ponds. Other issue in terms of energy is the energy involved in pumping the culture around the PBR (energy fraction for tubular PBRs is 86 % and 92 %; the energy fraction for flat plat PBRs is 22 %) (Watanabe et al. 1995).

The main energy consumption in mass cultivation of microalgae in open ponds is the electricity desired for circulating the culture which accounts for 22–79 % of the total energy required in mass cultivation which may be reduced by wind energy to rotate the paddle (Watanabe et al. 1995). Another major drawback of open systems is poor light utilisation by the cells, losses due to evaporation and larger area for cultivation. However, insufficient stirring mechanisms in open ponds may lead to less biomass productivity due to indigent mass transfer rates.

7.3 Photobioreactor: System Concept and Design

The major bottleneck for the production of biofuel from microalgae at a commercial scale is the generation of biomass. This is necessary for the

Table 7.3 Advantages and disadvantages of various microalgal cultivation techniques

Cultivation method	Advantages	Disadvantages
Batch cultivation	Larger H/D ratio for increasing volumetric cell densities	Requirement of periodic cleaning due to biofilm formation
	Superior long-term culture maintenance	Problems with reactor scale-up
	Lower loss of water than open ponds	Issue of continuous availability of cost-effective carbon substrates such as lignocellulosic sugars
Continuous cultivation	Lower capital costs and water evaporation helps in maintaining temperature	Open pond continuous cultures are subjected to daily and seasonal changes in temperature and humidity
		Inherent difficultly in the maintenance of monocultures in open ponds
A-stat cultivation	Higher biomass productivity	Difficult to scale up in other PBRs other than bubble column PBR

process to be economically feasible and to ensure uninterrupted supply for the consumer markets (Brennan and Owende 2010). The basic necessities for the cultivation of microalgae are light, CO_2, minerals and water. However, the specific requirements are dependent on the species or strain of microalgae that are being cultivated. Microalgae could be mass cultivated in open ponds or closed PBRs (Oswald 1992; Tredici 2003). Since, the 1950s, industrial-scale cultivation of microalgae is being carried out in open systems (Oswald 1992). Large-scale open ponds can be constructed with plastic, concrete and bricks for cultivation (Oswald 1992; Tredici 2003). The most commonly used open system is the raceway pond, an oval form resembling a car-racing circuit (Chisti 2008; Lee and Lee 2001; Pulz 2001). Basically these cultivation systems require relatively low construction and operating cost, require lesser maintenance and energy to operate and utilise solar energy for growth, and pilot systems can be made on lands that have degraded and are unfit for agriculture (Brennan and Owende 2010; Chen and Johns 1996a; Tredici 2003). However, open systems have the problems of contamination from other microorganisms such as protozoa, viruses and other species of algae and fungi. Further, it is not possible to control the temperature of water and the amount of light, and they depend on the environmental conditions. The main source of energy consumption in mass cultivation of microalgae is the electricity required for recirculating the culture which accounts to 22–79 % of the total energy utilisation (Brennan and Owende 2010).

Photobioreactors are systems that are generally used for growing photoautotrophic microorganisms such as cyanobacteria and microalgae, which use light energy to convert CO_2 to organic carbon. In the context of production of biofuel from microalgae and cyanobacteria, photobioreactor is an important equipment. Unlike open ponds, photobioreactor permits culturing single species of microalgae under controlled conditions for prolonged durations, making it suitable for commercial-scale applications. The major problem of using a photobioreactor is the difficulty in design to evenly distribute light for microalgal growth (Chen and Johns 1996b; Pulz 2001).

Benefits of PBR as compared to open cultivation are as follows:

- Prevents or minimises contamination during cultivation
- Has better control over cultivation conditions such as pH, pCO_2, pO_2, temperature, nutrient supply, etc.
- Reduces water usage by minimising evaporation loss
- Lower CO_2 loss
- Possibility of achieving higher cell concentration and thus higher volumetric productivity

A well-designed photobioreactor generally provides better growth conditions through its innovative design such as efficient distribution of light, recirculation and proper aeration and agitation and online measurements of vital parameters (Table 7.4). Because of this, they have the capability of higher biomass productivity. This is desirable for substituting fossil fuel with biofuels with minimal use of land.

Table 7.4 The various components of a PBR

S. no.	Systems of PBR	Key components of PBR
1	Light and optical transmission system	Light sensor
2	Air handling system	Air sparging enabling oxygen availability for culture
3	Mixing system	Water inlet valve and purge valve, feed tank
4	Pumps	Pumps for recirculating culture, harvesting and pH maintenance
5	Filtration system	Purge valve, collection filters, water inlet valve
6	Various sensor systems	Oxygen, CO_2, temperature, conductivity and pH sensors
7	Instrumentation systems	PLC control panel

Various types of photobioreactor have been designed and are being used for the cultivation of different algae.

Prior to designing a PBR, certain critical parameters needs to be considered. They are:

- Possibility of easy cleaning and use
- Maintenance of proper temperature and removal of heat
- Proper sparger design for maintaining bubble size/number, mass transfer and feed gas pressure
- Sufficient mixing to avoid oxygen holdup and biofouling
- Desired light intensity for light penetration during photoperiod
- Material of construction with the ability to withstand extreme pH, temperature and salinity

The different types of photobioreactors used for microalgal cultivation are tubular PBR, helical PBR, vertical PBR and flat panel PBR. Flat panel PBRs are further classified into vertical column PBR, bubble column PBR, air lift PBR, stirred tank PBR and immobilised PBR based on the agitation and aeration systems in these reactors. The different types of photobioreactor are:

7.3.1 Tubular Photobioreactor

Tubular PBRs are suitable for mass cultivation of microalgae outdoors. In tubular PBR, the culture is recirculated through pumps in order to achieve the required dissolved oxygen concentration (Molina et al. 2001; Richmond et al. 1993; Torzillo et al. 1986; Ugwu et al. 2008). However, when large-scale tubular PBRs are used, they have poor mass transfer properties as compared to other PBRs. Further, they have other disadvantages, such as photoinhibition and difficulty in maintaining reactor temperature when grown outdoors.

The basic diagram of a tubular photobioreactor is shown in Fig. 7.1. The bioreactor comprises of tubular sections of different lengths made of acrylic glass. The height and diameter of the bioreactor depends on the scale of operation and the type of microalgae.

Different designs of tubular photobioreactor have been reported. The commonly used designs are:

- Airlift and bubble column: It is basically made of a vertical cylinder that is transparent enough for light penetration. Carbon dioxide is supplied by bubbling the gas through the bottom of the reactor.
- Horizontal tubular bioreactor: It comprises of a horizontal transparent cylinder, containing gas transfer systems.
- Helical tubular bioreactor: It is made of a coiled plastic tube over an annular framework (Table 7.5).

7.3.2 Helical Photobioreactor

Helical photobioreactor is made of illuminated coiled tubes having a small diameter along with a degassing unit. The culture is circulated from the tube to the degassing unit with the help of a pump. These systems provide better CO_2 availability for the culture due to higher mass transfer from the gas phase to liquid phase (Watanabe et al. 1995). However, such systems are prone to fouling. Various researchers have suggested many designs of helical photobioreactors for increasing microalgal biomass productivity. One such design uses a UV resistance translucent PVC tubing mounted in a helical fashion inside an acrylic glass box with illumination from the outer side using warm circular fluorescent lamp (Oswald 1992). Briassoulis et al. (2010) suggested another design for growing *Nannochloropsis* sp. wherein large ratio of culture volume to surface area and optimised light path were considered. Another design suggested by Travieso et al. (2001) comprises of a cylindrical tube (0.9 m high) with a 0:25 m^2 basal area and a photostage comprising of a 60 m transparent PVC tubing with an inner diameter of 1.6 cm. The growth medium was circulated with 4% CO_2 at a flow rate of 1 vvm.

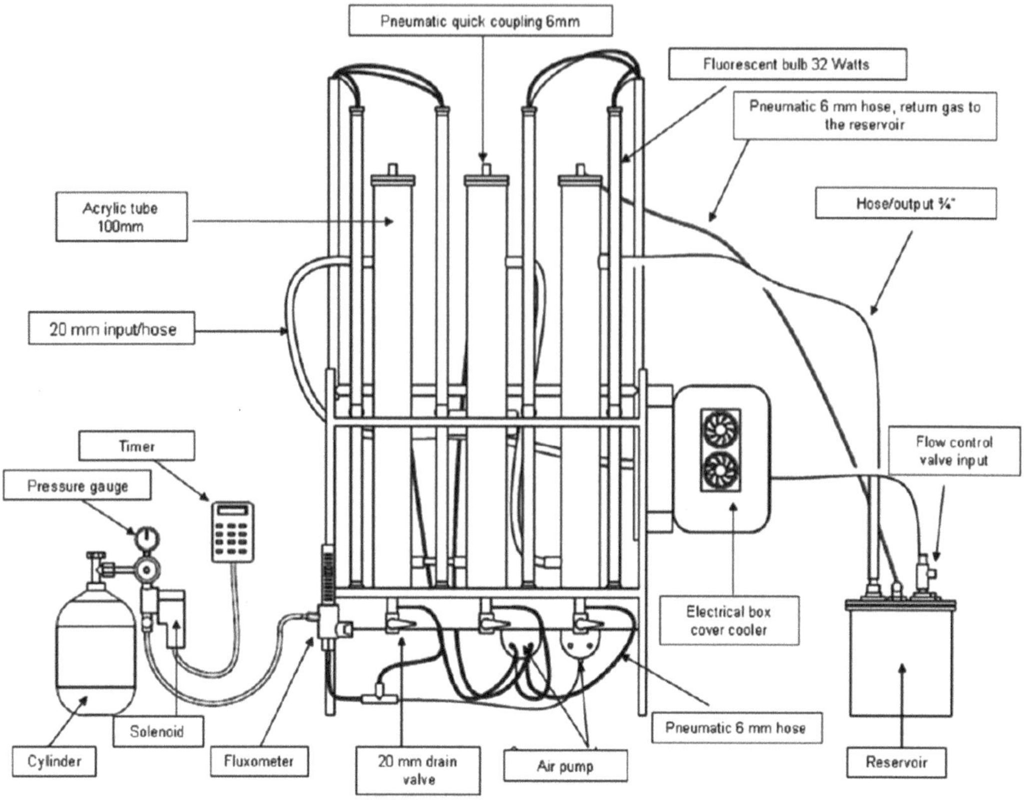

Fig. 7.1 Design of tubular photobioreactor (Gressler et al. 2014)

Table 7.5 Advantages and disadvantages of a tubular photobioreactor

Advantages	Disadvantages
The reactor can be easily scaled up	Productivity is lower as compared to stirred tank bioreactor
Faster growth rate of microalgae due to higher CO_2 transfer rate	Energy intensive process
Circulation is achieved without moving essential parts and this reduces the potential for contamination	Continuous removal of oxygen is a problem
Absence of mechanical pumping reduces cell damage	

7.3.3 Vertical Photobioreactor

Vertical column PBRs are compact and economical and can easily be operated under axenic conditions. Further, they are suitable for large-scale cultivation of microalgae. Biomass productivity attained in airlift and bubble column PBR has been comparable to the values reported in a tubular PBR, the reactor which has been reported to be most suitable for mass cultivation of microalgae (Miron et al. 2002). The disadvantages of a vertical photobioreactor include small surface area available for illumination, requirement of sophisticated materials for construction, the problem of shear over the microalgal culture imparted due to the reactor design and difficulty of illumination on scale-up.

Airlift and bubble column reactors are commonly used vertical photobioreactors, comprising of polyethylene bags/glass tubes. These systems are transparent enough to permit good light penetration. Sparging of air from the bottom of the reactor provides adequate mixing, sufficient supply of CO_2 and efficient removal of O_2 (Carvalho et al. 2006). Various researchers have suggested various designs for growing microalgae with increased biomass yields (Fig. 7.2; Tables 7.6 and 7.7).

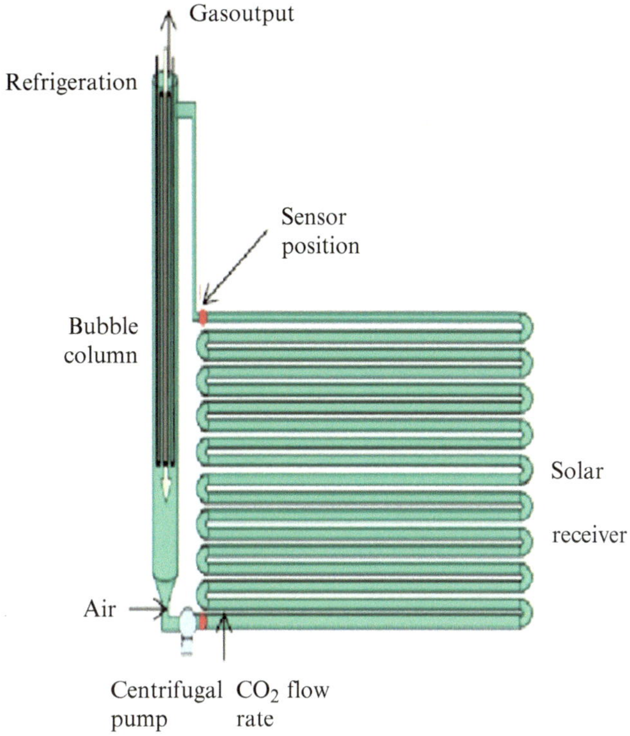

Fig. 7.2 Schematic diagram of helical photobioreactor (Dormido et al. 2014)

Table 7.6 Advantages and disadvantages of helical photobioreactor

Advantages	Disadvantages
Higher volumetric productivity lower biofouling in the system	Lower productivity as compared to open systems
Less oxygen build-up	The light regime and the narrow light path cannot be effectively used by slow growing microalgae
Large culture volume to surface area ratio with optimal light penetration depth	Energy intensive and expensive
Easy control over temperature and contaminants and effective spatial	
Proper distribution of fresh air and CO_2	

Torzillo et al. (1986) suggested long tubular reactor made of acrylic glass, placed horizontally along with a diaphragm pump for proper mixing of cultures and allowing adequate gas transfer within the tube. Chae et al. (2006) demonstrated a design comprising of vertical and horizontal tube wherein sunlight and flue gas was used for growing microalgae.

7.3.4 Flat Panel Photobioreactor

Flat panel photobioreactors are made of a flat plate and have a large illumination surface area. The material of construction for flat panel photobioreactor is transparent glass enabling solar light to pass inside the cells (Ugwu et al. 2008). Also, accumulation of dissolved oxygen in these reactors is relatively low compared to horizontal tubular photobioreactor. Guan et al. (2014) designed a flat panel membrane bioreactor with composite membrane consisting of a dense layer and support layer. Another design was suggested by Li et al. (2014) for growing *Chlorella sorokiniana*, and *Scenedesmus obliquus* consists of a

Table 7.7 Advantage and disadvantage of vertical column photobioreactor

Advantages	Disadvantages
Compact, low cost and easy to operate axenically	Lower volumetric productivity as compared to tubular PBR
Low power consumption and desired mass transfer coefficient are easily attainable	Cell damaging hydrodynamic due to shear stress
Liquid circulation velocity at a relatively low power input	Possibility of cell sedimentation
Requirement of less land for commercial production	

flat panel constructed of transparent polycarbonate (PC) divided into nine parallel flow chambers along with airlift module prepared with tubes and plates (Fig. 7.3). The airlift module comprises of a riser as well as a down comer adhered to gas liquid separator desired for adequate mixing of culture. To enable proper mixing, air is passed through a perforated sparger. Sometimes, CO_2 is also mixed with air for supplying carbon source (Li et al. 2014).

Another design has been suggested by Bergmann et al. (2013) wherein aeration is provided by introducing a membrane in order to supply sterile air and CO_2 for growing the desired microalgae. Generated foam and exhaust air are dissipated on continuously (Bergmann et al. 2013) (Table 7.8).

7.4 Critical Design Elements of PBR for Microalgal Cultivation

The critical elements required for the design of a PBR should focus on achieving efficient utilisation of light and overcoming operational challenges such as overheating, oxygen build-up and biofouling. The elemental design criteria for photobioreactor consist of reactor configuration with respect to light gradient and light/dark cycles, surface-to-volume ratio and mixing and degassing devices (Enzing et al. 2014).

7.4.1 Light Intensity

Light intensity plays an important role in microalgae productivity. Hence, a PBR designed with higher light intensity/penetration will perform better in terms of the biomass productivity. Light at the proper intensity is required since at saturated levels of light, the excess energy is dissipated as heat.

7.4.2 Surface-to-Volume Ratio

Systematic exposure of light to the cells is among the major issues in a PBR design. The light exposure depends on the surface-to-volume (S/V) ratio. The surface-to-volume ratio estimates the concentration of light penetration per unit volume. Operating the reactor at a high S/V ratio will result in a higher cell concentration and volumetric productivity of the culture. The hydrodynamics in the reactor and the effect of that on the culture is also dependent on the S/V ratio.

7.4.3 Agitation

The type of device used for mixing and circulation of the culture suspension is another important parameter that has to be considered while designing a PBR. The capital and operational cost and the performance of the reactor in terms of productivity will depend on the type of mixing device. Mixing and aeration is important for:

- Preventing auto settling of microalgal cells
- Avoiding temperature and pH gradients along the vessel
- Homogeneous mixing of nutrients for its uniform distribution
- Removal of oxygen generated from the process of photosynthesis
- Supplying CO_2 to the culture

Fig. 7.3 Schematic diagram of flat panel photobioreactor (Li et al. 2014)

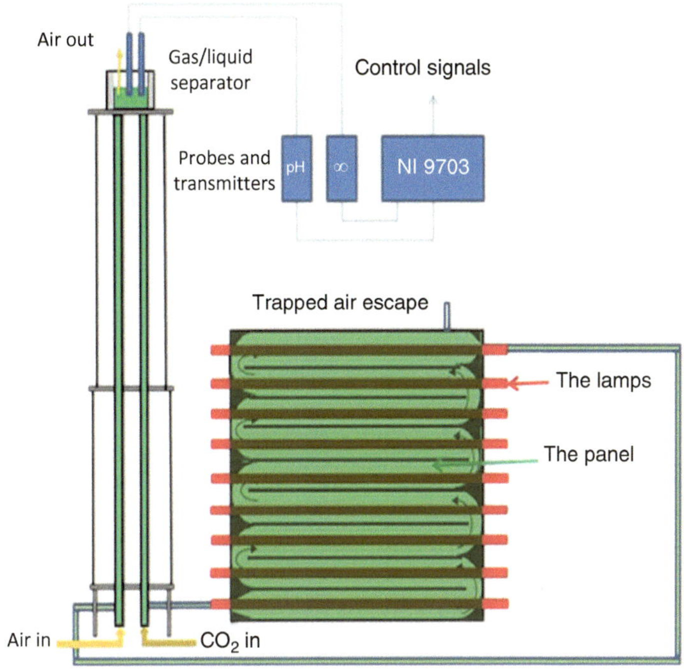

Table 7.8 Advantage and disadvantage of flat panel photobioreactor

Advantages	Disadvantages
Optimal light distribution to all cells	Material and operating cost can be higher per unit area
Minimal shear stress on microalgae	No liquid bulk flow in the PBR
Higher cell density compared to other systems	Biofouling occurs frequently
Reduction in aeration rate at low light intensities	Issues with scaling up

7.4.4 Nutrient Requirements

The two major nutrients required for controlling the growth rate and lipid production of microalgae are nitrogen and phosphorus. Other essential nutrients are hydrogen, oxygen, carbon, sulphur, magnesium, calcium, sodium, chlorine and potassium. Micronutrients that are required in trace quantities include copper, manganese, iron, boron, molybdenum, selenium, nickel, vanadium, silicon and cobalt (Suh and Lee 2003). Subjecting the organism to stress in the form of nutrient limitation (particularly P or N) can increase lipid concentrations inside the biomass. Microalgae can be cultivated under three different types of nutritional conditions, viz. nutrient deficient, nutrient limited and nutrient sufficient. Nutrient-sufficient condition is when excess nutrients are added to the system such that enough nutrients are available for the growth of the microalgae during all phases of growth. Nutrient-limited condition is when cells are grown in an environment with lower concentration of the limiting nutrients. However, under such conditions, the cells are generally known to adapt themselves for growth. Nutrient-deficient conditions are when the microalgal culture is dependent on the endogenous reserves since there are no readily available nutrients in the environment.

7.4.5 Temperature

Temperature is an important parameter for growth of microalgae. Though microalgae can grow in a wide range of temperature, optimal growth is constrained to a specific temperature range, and it depends on the strain of microalgae cultivated. The strain of microalgae grown outdoors varies depends on the outdoor temperature, and it is representative to the geographic region of cultivation. Routine seasonal variations in temperature can interfere with the biomass productivity. Sometimes, temperature reaches as high as 30 Â°C greater than ambient temperature in open systems as well as closed PBRs without temperature control equipment (Suh and Lee 2003). In order to reduce the evaporation rate, various cooling and shading techniques are employed (Lee 2001).

7.4.6 Water

Water consumption is another important parameter to be considered for the growth of microalgae especially while cultivating microalgae for production of biofuel. However, it may be noted that cultivation of various strains of microalgae for fuel production does not require freshwater. Many studies have shown that microalgae can be grown in brackish water, fresh drinking water, saline water and wastewater effluent (Yun et al. 1997).

7.4.7 CO_2 and pH

Microalgae convert CO_2 to biomass through photosynthesis and hence CO_2 is one of the limiting factors for biomass growth. CO_2 concentration in the range of 1–5 % (by volume) is required for efficient biomass productivity. Higher yields of biomass can be achieved at higher CO_2 concentration due to the improvement in the photosynthetic efficiency of the microalgae (Zhu et al. 2013). Sodium bicarbonate is added to the medium to help maintain pH in the medium and also serve as a source of CO_2, improving biomass productivity (Zhu et al. 2013). The photosynthetic efficiency of photosystem II upswings with growing CO_2 levels and captured light energy gets transformed into chemical energy (Zhu et al. 2013). Flue gases produced in the cultivation system are rich in oxygen and it may be utilised to reduce greenhouse gas emission.

The cultivation pH should be controlled during microalgal growth to enhance absorption and consumption of CO_2 by microalgae. The pH can distinctly affect the permeability of the cell and influence CO_2 absorption. CO_2 in the culture (either from the environment or from the bicarbonate in the medium) is consumed by the microalgae during photosynthesis causing an increase in the pH of the system. Therefore, the pH needs to be controlled from increasing beyond the tolerance range of the microalgae. For adjusting pH, compared to mineral acids, organic acids, like acetic acid, and bicarbonate salts have the advantage of serving as a source of carbon as well as maintaining pH enhancing growth rate of microalgae (Zhu et al. 2013).

7.5 Photobiological Hydrogen Production: Current Challenges and Perspectives

Photobiological hydrogen production is the process of biological photolysis of water molecule into oxygen and hydrogen. Some photosynthetic microbes produce hydrogen through photodecomposition of organic materials. Production of hydrogen directly from water using photosynthetic microorganisms such as the microalgae which are energetically more favourable than growing, harvesting and processing microalgae for biofuels. Hydrogen yield varies from organism to organism, which depends on the type of fermentative products generated (Yu and Takahashi 2007).

Photobiological hydrogen production is the process of biological photolysis of water molecule into oxygen and hydrogen. It is usually carried out by microorganisms such as microalgae and cyanobacteria (Markov 2012). Generally biophotolysis by prokaryotic microalgae, cyanobacteria and other photosynthetic bacteria is possible due to the presence of hydrogenase and nitrogenase enzymes in the cells (Maness et al. 2005; Kaidi et al. 2012). Most of the hydrogenases catalyse the conversion of hydrogen into electron and protons. However, a reversible or bidirectional hydrogenase accepts electron directly from reduced ferredoxin for generation of hydrogen. It

$$H_2O \rightarrow PS\ II \rightarrow PS\ I \rightarrow Fd \rightarrow Hydrogenase \rightarrow H_2$$
$$\downarrow$$
$$O_2$$

Fig 7.4 Direct photoproduction of hydrogen (photosynthetic systems-PS I & PS-II)

should be noted that hydrogenases and nitrogenases are sensitive to oxygen, and hydrogen production takes place only in the absence of oxygen. Photobiological hydrogen production from water can take place through two major routes, viz. indirect and direct biophotolysis.

7.5.1 Direct Conversion of Water to Hydrogen

Direct biophotolysis involves simultaneous production of oxygen and hydrogen in a single step. This is shown schematically in Fig. 7.4.

Direct biophotolysis can be used in a system where the oxygen and hydrogen are flushed out of the system. It should be noted that mixture of hydrogen and oxygen are potentially flammable and the gases needs to be handled with care.

7.5.2 Indirect Biophotolysis

In indirect biophotolysis, oxygen evolves separately with the reducing equivalents stored in the organism in the first stage. In the second stage, these protons are reduced to hydrogen with the use of light energy. Since, the oxygen is removed in a separate stage, it may be possible to overcome the problem of oxygen sensitivity of hydrogenase enzyme by cultivating the microalgae in separate aerobic and anaerobic stages. The basic reaction for hydrogen production through the indirect biophotolysis method may be represented as:

$$12H_2O + 6CO_2 + \text{light energy} \rightarrow C_6H_{12}O_6 + 6O_2$$

and

$$C_6H_{12}O_6 + 12H_2O + \text{light energy} \rightarrow 12\,H_2 + 6O_2$$

The comparison of the advantages and disadvantages of direct and indirect biophotolysis is shown in Table 7.9.

Table 7.9 Advantages and disadvantages of various processes for H_2 production using microalgae

	Advantages	Disadvantages
Direct photolysis	Hydrogen production directly from water and sunlight	Requires high intensity of light
	Ten times higher solar conversion efficiency with respect to lignocellulosic biomass	Lower photochemical efficiency
		O_2 can be dangerous for the system
Indirect photolysis	Hydrogen production from water using cyanobacteria	Hydrogenase enzymes produced during the process needs to be removed in order to stop degradation of H_2
	Nitrogen fixation from atmosphere	Gas mixture consists of 30 % O_2

7.6 Conclusions and Future Prospects

In order to meet the current fuel demands, biofuel based on microalgae is an attractive solution. Since, energy and fuel requirements are on the rise and expected to rise in the future too, it is important to look for sustainable ways of generating energy. Microalgae as a source of sustainable energy will help us not only with biofuel production but also help us in sequestering the CO_2 present in the environment. Microalgae are capable of fixing severalfold more carbon dioxide per unit area than agricultural crops or trees. Cultivation of microalgae for biofuel production and various other commercially important compounds is an attractive option since microalgal fuels are a better alternative to the existing fossil fuels. Focus on biofuel production from microalgae needs to be given for reduction in processing cost, CO_2 sequestration along with energy production, design of low cost PBRs which are energy efficient and direct hydrogen production. At present, globally, more than 100 firms are focusing on the production of biofuels from microalgae and more firms expected to join the race in the near future.

References

Bacellar Mendes L, Vermelho A (2013) Allelopathy as a potential strategy to improve microalgae cultivation. Biotechnol Biofuels 6(1):152. doi:10.1186/1754-6834-6-152

Barbosa MJ, Hoogakker J, Wijffels RH (2003) Optimisation of cultivation parameters in photobioreactors for microalgae cultivation using the A-stat technique. Biomol Eng 20(4):115–123. doi:10.1016/S1389-0344(03)00033-9

Bergmann P, Ripplinger P, Beyer L, Trosch W (2013) Disposable flat panel airlift photobioreactors. Chemie Ingenieur Technik 85(1–2):202–205. doi:10.1002/cite.201200132

Boden TA, Marland G, Andres RJ (2012) Global, regional, and national fossil-fuel CO_2 emissions. Information Analysis Center, Oak Ridge National Laboratory, U.S. Department of Energy, Oak Ridge, Tenn., U.S.A. doi:10.3334/CDIAC/00001_V2012

Brennan L, Owende P (2010) Biofuels from microalgae-a review of technologies for production, processing, and extractions of biofuels and co-products. Renew Sustain Energy Rev 14:557–577. doi:10.1016/j.rser.2009.10.009

Briassoulis D, Panagakis P, Chionidis M, Tzenos D, Lalos A, Tsinos C, Berberidis K, Jacobsen A (2010) An experimental helical-tubular photobioreactor for continuous production of *Nannochloropsis* sp. Bioresource Technol 101(17):6768–6777. doi:10.1016/j.biortech.2010.03.103

Carvalho AP, Meireles LA, Malcata FX (2006) Microalgal reactors: a review of enclosed system designs and performances. Biotechnol Prog 22(6):1490–1506. doi:10.1021/bp060065r

Chae SR, Hwang EJ, Shin HS (2006) Single cell protein production of Euglena gracilis and carbon dioxide fixation in an innovative photo-bioreactor. Bioresour Technol 97(2):322–329. doi:10.1016/j.biortech.2005.02.037

Chen F, Johns MR (1996a) Heterotrophic growth of Chlamydomonas reinhardtii on acetate in chemostat culture. Process Biochem 31:601–604. doi:10.1016/S0032-9592(96)00006-4

Chen F, Johns MR (1996b) Relationship between substrate inhibition and maintenance energy of *Chlamydomonas reinhardtii* in heterotrophic culture. J Appl Phycol 8:15–19. doi:10.1007/BF02186216

Chisti Y (2008) Biodiesel from microalgae beats bioethanol. Trends Biotechnol 26:126–131. doi:10.1016/j.tibtech.2007.12.002

Dormido R, Sanchez J, Duro N, Dormido-Canto S, Guinaldo M, Dormido S (2014) An interactive tool for outdoor computer controlled cultivation of microalgae in a tubular photobioreactor system. Sensors (Basel, Switzerland) 14(3):4466–4483. doi:10.3390/s140304466

Enzing C, Ploeg M, Barbosa M, Sijtsma L (2014) Microalgae-based products for the food and feed sector: an outlook for Europe. doi:10.2791/3339

Feng Y, Li C, Zhang D (2011) Lipid production of *Chlorella vulgaris* cultured in artificial wastewater medium. Bioresour Technol 102(1):101–105. doi:10.1016/j.biortech.2010.06.016

Georgianna DR, Mayfield SP (2012) Exploiting diversity and synthetic biology for the production of algal biofuels. Nature 488(7411):329–335. doi:10.1038/nature11479

Gressler P, Bjerk T, Schneider R, Souza M, Lobo E, Zappe A, Corbellini VA, Moraes M (2014) Cultivation of *Desmodesmus subspicatus* in a tubular photobioreactor for bioremediation and microalgae oil production. Environ Technol 35(2):209–219. doi:10.1080/09593330.2013.822523

Guan J, Xiao Y, Song J, Miao J (2014) A typical flat-panel membrane bioreactor with a composite membrane for sulfur removal. Front Earth Sci 8(1):142–149. doi:10.1007/s11707-013-0370-2

Ho SH, Kondo A, Hasunuma T, Chang JS (2013) Engineering strategies for improving the CO_2 fixation and carbohydrate productivity of Scenedesmus obliquus CNW-N used for bioethanol fermentation. Bioresour Technol 143:163–171. doi:10.1016/j.biortech.2013.05.043

Hulatt CJ, Thomas DN (2011) Energy efficiency of an outdoor microalgal photobioreactor sited at mid-temperate latitude. Bioresour Technol 102(12):6687–6695. doi:10.1016/j.biortech.2011.03.098

Kaidi F, Rihani R, Ounnar A, Benhabyles L, Naceur MW (2012) Photobioreactor design for hydrogen production. Procedia Eng 33:492–498. doi:10.1016/j.proeng.2012.01.1229

Keffer JE, Kleinheinz GT (2002) Use of *Chlorella vulgaris* for CO_2 mitigation in a photobioreactor. J Industr Microbiol Biotechnol 29:275–280. doi:10.1038/sj.jim.7000313

Kroger M, Muller-Langer F (2012) Review on possible algal-biofuel production processes. Biofuels 3(3):333–349. doi:10.4155/bfs.12.14

Lardon L, Helias A, Sialve B, Steyer JP, Bernard O (2009) Life-cycle assessment of biodiesel production from microalgae. Environ Sci Tech 43:6475–6481. doi:10.1021/es900705j

Lee YK (2001) Microalgal mass culture systems and methods: their limitation and potential. J Appl Phycol 13:307–315. doi:10.1023/A:1017560006941

Lee K, Lee C-G (2001) Effect of light/dark cycles on wastewater treatments by microalgae. Biotechnol Bioproc Eng 6:194–199. doi:10.1007/BF02932550

Lestari S, Maki-Arvela P, Beltramini J, Lu GQM, Murzin DY (2009) Transforming triglycerides and fatty acids into biofuels. ChemSusChem 2:1109–1119. doi:10.1002/cssc.200900107

Li Y, Horsman M, Wu N, Lan CQ, Dubois-Calero N (2008) Biofuels from microalgae. Biotechnol Prog 24(1):815–820. doi:10.1021/bp.070371k

Li J, Stamato M, Velliou E, Jeffryes C, Agathos SN (2014) Design and characterization of a scalable airlift flat panel photobioreactor for microalgae cultivation. J Appl Phycol 27(1):75–86. doi:10.1007/s10811-014-0335-1

Liu J, Huang J, Chen F (2011) Microalgae as feedstocks for biodiesel production. In: Biodiesel feedstocks and processing technologies. InTech, Rijeka, pp 133–160

Lu J, Sheahan C, Fu P (2011) Metabolic engineering of algae for fourth generation biofuels production. Energy Environ Sci 4(7):2451–2466. doi:10.1039/c0ee00593b

Maness PC, Weaver PF (2002) Hydrogen production from a carbon-monoxide oxidation pathway in *Rubrivivax gelatinosus*. Int J Hydr Energy 27(11–12):1407–1411. doi:10.1016/S0360-3199(02)00107-6

Maness PC, Huang J, Smolinski S, Tek V, Vanzin G (2005) Energy generation from the CO oxidation-hydrogen production pathway in *Rubrivivax gelatinosus*. Appl Environ Microbiol 71(6):2870–2874. doi:10.1128/AEM.71.6.2870-2874.2005

Markou G, Angelidaki I, Georgakakis D (2013) Carbohydrate-enriched cyanobacterial biomass as feedstock for bio-methane production through anaerobic digestion. Fuel 111:872–879. doi:10.1016/j.fuel.2013.04.013

Markov SA (2012) Hydrogen production in bioreactors: current trends. Energy Procedia 29:394–400. doi:10.1016/j.egypro.2012.09.046

Mata TM, Martins AA, Caetano NS (2010) Microalgae for biodiesel production and other applications: a review. Renew Sustain Energy Rev 14:217–232. doi:10.1016/j.rser.2009.07.020

Matsumoto M, Yokouchi H, Suzuki N, Ohata H, Matsunaga T (2003) Saccharification of marine microalgae using marine bacteria for ethanol production. Appl Biochem Biotechnol 105–108:247–254. doi:10.1385/ABAB:105:1-3:247

Metz B, Davidson OR, Bosch PR, Dave LA (eds) (2007) Climate change 2007: mitigation of climate change contribution of working group III to the fourth assessment report of the intergovernmental panel on climate change (IPCC). Cambridge University Press, Cambridge, Retrieved from http://www.mnp.nl/ipcc/

Miron AS, Garcia MCC, Camacho F, Molina Grima E, Chisti Y (2002) Growth and biochemical characterization of microalgal biomass produced in bubble column and airlift photobioreactors: studies in fed-batch culture. Enzyme Microb Technol 31:1015–1023. doi:10.1016/S0141-0229(02)00229-6

Molina E, Chisti Y (2001) Tubular photobioreactor design for algal cultures. J Biotechnol 92:113–131. doi:10.1016/S0168-1656(01)00353-4

Oron G, Shelef G, Levi A (1979) Growth of Spirulina maxima on cow-manure wastes. Biotechnol Bioeng 21:2169–2173. doi:10.1002/bit.260211203

Oswald WJ (1992) Microalgae and wastewater treatment. In: Borowitzka A, Borowitzka LJ (eds) Microalgal biotechnology. Cambridge University Press, Cambridge, pp 305–328

Posten C, Schaub G (2009) Microalgae and terrestrial biomass as source for fuels-a process view. J Biotechnol 142:64–69. doi:10.1016/j.jbiotec.2009.03.015

Pulz O (2001) Photobioreactors: production systems for phototrophic microorganisms. Appl Microbiol Biotechnol 57:287–293. doi:10.1007/s002530100702

Raja R, Hemaiswarya S, Kumar NA, Sridhar S, Rengasamy R (2008) A perspective on the biotechnological potential of microalgae. Crit Rev Microbiol 34:77–88. doi:10.1016/0167-7799(92)90282-Z

Richmond A (2003) Handbook of microalgal culture. In: Richmond A (ed) Biotechnology and applied phycology. Blackwell Publishing Ltd, Oxford. doi:10.1002/9780470995280

Richmond A, Boussiba S, Vonshak A, Kopel R (1993) A new tubular reactor for mass production of microalgae outdoors. J Appl Phycol 5:327–332. doi:10.1007/BF02186235

Seshadri CV, Thomas S (1979) Mass culture of spirulina using low-cost nutrients. Biotechnol Lett 1:287–291. doi:10.1007/BF01386766

Suh IS, Lee C-G (2003) Photobioreactor engineering: design and performance. Biotechnol Bioproc Eng 8:313–321. doi:10.1007/BF02949274

Tang H, Chen M, Simon Ng KY, Salley SO (2012) Continuous microalgae cultivation in a photobioreactor. Biotechnol Bioeng 109(10):2468–2474. doi:10.1002/bit.24516

Torzillo G, Pushparaj B, Bocci F, Balloni W, Materassi R, Florenzano G (1986) Production of spirulina biomass in closed photobioreactors. Biomass 11:61–74. doi:10.1016/0144-4565(86)90021-1

Travieso L, Hall DO, Rao KK, Benitez F, Sanchez E, Borja R (2001) A helical tubular photobioreactor producing spirulina in a semicontinuous mode. Int Biodeterioration Biodegradation 47(3):151–155. doi:10.1016/S0964-8305(01)00043-9

Tredici MR (2003) In: Richmond A (ed) Handbook of microalgal culture. Blackwell Publishing Ltd, Oxford. doi:10.1002/9780470995280

Ugwu CU, Aoyagi H, Uchiyama H (2008) Photobioreactors for mass cultivation of algae. Bioresour Technol 99:4021–4028. doi:10.1016/j.biortech.2007.01.046

Vonshak A, Cohen Z, Richmond A (1985) The feasibility of mass cultivation of porphyridium. Biomass 8:13–25. doi:10.1016/0144-4565(85)90032-0

Wang B, Li Y, Wu N, Lan CQ (2008) CO2 bio-mitigation using microalgae. Appl Microbiol Biotechnol 79:707–718. doi:10.1007/s00253-008-1518-y

Wang B, Yang X, Lu J, Zhou Y, Su J, Tian Y, Zhang J, Wang G, Zheng T (2012) A marine bacterium producing protein with algicidal activity against Alexandrium tamarense. Harmful Algae 13:83–88. doi:10.1016/j.hal.2011.10.006

Watanabe Y, De la Noue J, Hall DO (1995) Photosynthetic performance of a helical tubular photobioreactor incorporating the cyanobacterium Spirulina platensis. Biotechnol Bioeng 47:261–269. doi:10.1002/bit.260470218

Williams D (2009) Algenol biofuels announces plan to build and operate a pilot-scale algae-based integrated biorefinery. J Can Petroleum Technol 48:6–8

Yanagi M, Watanabe Y, Saiki H (1995) CO_2 fixation by *Chlorella* sp. HA-1 and its utilization. Energy Conver Manag 36:713–716. doi:10.1016/0196-8904(95)00104-L

Yu J, Takahashi P (2007) Biophotolysis-based hydrogen production by cyanobacteria and green microalgae. Commun Curr Res Educ Topics Trends Appl Microbiol 1:79–89

Yun YS, Lee SB, Park JM, Lee CI, Yang JW (1997) Carbon dioxide fixation by algal cultivation using wastewater nutrients. J Chem Technol Biotechnol 69:451–455. doi:10.1002/(SICI)1097-4660(199708)69:4<451::AID-JCTB733>3.0.CO;2-M

Zamalloa C, Vulsteke E, Albrecht J, Verstraete W (2011) The techno-economic potential of renewable energy through the anaerobic digestion of microalgae. Bioresour Technol 102:1149–1158. doi:10.1016/j.biortech.2010.09.017

Zhu J, Rong J, Zong B (2013) Factors in mass cultivation of microalgae for biodiesel. Chin J Catalys 34:80–100. doi:10.1016/S1872-2067(11)60497-X

8 Sustainability of Biohydrogen Production Using Engineered Algae as a Source

Khorcheska Batyrova and Patrick C. Hallenbeck

Abstract

The increased scarcity of fossil fuels and concerns over climate change have focused attention on alternative energy and the production of renewable fuels. Hydrogen is a promising substitute as an energy carrier to the fossil fuels, since it has high conversion efficiency and high energy content and is environmentally friendly. Nowadays, hydrogen is mostly produced via chemical reformation of fossil fuels; therefore, the biological production of hydrogen is seen as very attractive. Biological hydrogen production represents a renewable means of generating this biofuel and can be performed by a wide range of microorganisms from strict anaerobic bacteria to eukaryotic green algae. Different strains of photoautotrophic green algae have the remarkable ability to reduce protons to H_2 using light energy. Photobiological hydrogen production by green algae is particularly attractive due to the fact that water and solar energy as main inputs for the process are plentiful on our planet. In this chapter we are focusing on the recent developments in photobiological H_2 production by green algae with highlights on the barriers that prevent H_2 production and how those limitations can be addressed, through genetic and metabolic engineering.

K. Batyrova
Département de Microbiologie, infectiologie et immunologie, Université de Montréal, CP 6128 Succ. Centre-ville, Montréal, Québec H3C 9J7, Canada
e-mail: horcheska@gmail.com

P.C. Hallenbeck (✉)
Département de Microbiologie, infectiologie et immunologie, Université de Montréal, CP 6128 Succ. Centre-ville, Montréal, Québec H3C 9J7, Canada

Life Sciences Research Center, Department of Biology, United States Air Force Academy, 2355 Faculty Drive, USAF Academy, Colorado Springs, CO 80840, USA
e-mail: patrick.hallenbeck@umontreal.ca

A. Singh, D. Rathore (eds.), *Biohydrogen Production: Sustainability of Current Technology and Future Perspective*, DOI 10.1007/978-81-322-3577-4_8

8.1 Introduction

The ability of green algae to produce hydrogen under illumination was discovered more than 65 years ago by Gaffron and Rubin (1942). This discovery was an important evidence supporting van Niel's general scheme relating plant photosynthesis and bacterial photosynthesis:

$$CO_2 + 2H_2A = [CH_2O] + 2A + H_2O$$

Subsequently, several diverse phototrophic microorganisms, including numerous species of green algae, in particular strains of *Chlamydomonas*, have been shown to produce H_2 (Melis and Happe 2001; Melis 2002). *Chlamydomonas* is a genus of green algae that has served as a significant model for studies of a variety of fundamental processes such as photosynthesis and the responses to light (Harris and Witman 2008). In green algae the formation of hydrogen gas is catalyzed by a ferredoxin-dependent [FeFe] hydrogenase (Forestier et al. 2003; Happe and Kaminski 2002). Although hydrogenase activity is not observed in all genera of green algae, at present at least 30 genera of green algae are known to have H_2 metabolism (Brand et al. 1989; Boichenko et al. 2004; Melis and Happe 2004). *Chlamydomonas reinhardtii* is currently the main model organism for investigating H_2 metabolism in water-oxidizing, photosynthetic eukaryotes, and significant progress has been made in understanding the mechanisms of H_2 production, as well as in improving H_2 photoproduction yields (Weaver et al. 1980; Ghirardi et al. 2000, 2005; Kosourov et al. 2002, 2012; Fedorov et al. 2005; Kruse et al. 2005). At present, all participants of the electron transport chain leading from light absorption to H_2 formation are known. Detailed studies are available that describe the structure and maturation of [FeFe] hydrogenases (Ghirardi et al. 2007; Bock et al. 2006; Meyer 2007). Also, much is known about the biophysics and genomics of green algae H_2 production (Boichenko et al. 2004; Antal et al. 2003; Melis and Happe 2004). These research achievements, in combination with the complete *C. reinhardtii* genome sequence, have dramatically enhanced our understanding of algal H_2 production.

8.2 Physiology of H_2 Production in Green Algae

8.2.1 Limits to Photosynthesis

Oxygenic or anoxygenic photosynthesis is a process that can be described as the conversion of light energy into chemical energy through a series of redox reactions. Pigment molecules of the photosystem protein complex first absorb light, leading to the production of high-energy electrons. Through the oxidation of electron donors these electrons then are shuttled through an electron transport chain, generating energy-rich intermediate compounds (ATP and NAD(P)H). These energy carriers are then consumed to drive a variety of reductions; the most important of which is carbon fixation whereby CO_2 is reduced to synthesize fructose which then feeds into central metabolism.

In oxygenic photosynthesis that occurs in plants, cyanobacteria, and algae, water is used as an electron donor leading to the generation of protons and molecular oxygen (biophotolysis).

Although all plants use this energy for synthesizing sugar via CO_2 fixation, some species of water-oxidizing, phototrophic algae such as *Chlamydomonas* can utilize low-potential electrons from the photosynthetic electron transport chain to reduce protons to H_2. Consequently, renewable H_2 production can be achieved in these microorganisms using the energy provided by solar irradiation and electrons derived from water.

Water as a substrate is naturally abundant on our planet, and since for the H_2 production clean water is not required, some types of wastewater that contain necessary growth nutrients are more preferable. However, there are still many serious technical limitations that should be addressed before the practical application of this technology (Beer et al. 2009; Ghirardi and Mohanty 2010). Some of these limitations are inefficient use of light due to uneven light distribution, limiting effective light penetration into cultures, and the dissipation of excess absorbed light as heat; thus, overall photosynthetic efficiencies are no more than 1 % often closer to 0.1 %. In various studies, it was estimated that the efficiency of light utilization has to be at least 10 % in order to have an economically viable process (Hallenbeck 2011, 2012; Tsygankov 2012).

Secondly, although green algal hydrogen production depends upon biophotolysis, oxygenic photosynthesis and photobiological hydrogen production are antagonistic processes due to the fact that evolved in photosynthesis molecular oxygen quickly destroys the O_2-sensitive [FeFe] hydrogenase, stopping H_2 production after only a few minutes (Seibert 2007). This is one of the major fundamental obstacles to this approach.

Thirdly, hydrogen-evolving enzymes compete for the electrons with carbon fixation pathway which is the major sink of electrons in cells. Therefore, in order for biophotolysis to become a sustainable platform for the biological hydrogen production and compete with current methods in petrochemical industry, such challenges must be resolved through metabolic and genetic engineering. Strategies for improving H_2 production have been described in multiple reviews (Hallenbeck et al. 2012; Srirangan et al. 2011; Esper et al. 2006; Ghirardi et al. 2007).

8.2.2 Pathways of Electron Flow to Hydrogen Production

Green algae of the genera *Chlamydomonas*, *Scenedesmus*, *Lobochlamys*, and *Chlorella* are known to exhibit hydrogenase activity, derived predominantly from the [FeFe] hydrogenase, and therefore can reduce protons to produce molecular hydrogen (Forestier et al. 2003; Happe and Kaminski 2002; Meuser et al. 2009, 2012). *C. reinhardtii* contains two highly similar, differentially regulated, oxygen-sensitive [FeFe] hydrogenases, HydA1 and HydA2 (Forestier et al. 2003), of which only HydA1 is thought to play a role in H_2 production (Godman et al. 2010). In fact, the normal physiological function of hydrogen production as a secondary metabolite is thought to be to poise the photosynthetic electron transport chain for light-driven activity after a period of darkness. In *Chlamydomonas*, cellular redox conditions regulate the complex interactions of glycolysis, fermentation, chlororespiration, mitochondrial respiration, and photosynthesis, all of which can ultimately influence hydrogenase activity.

In green algae, H_2 is known to be produced through two light-dependent pathways and one light-independent fermentative pathway (Fig. 8.1). The first of the photoproduction pathways is a consequence of direct biophotolysis; in this process, photosystem II (PSII) catalyzes the photolysis of water, producing O_2 and releasing electrons into the photosynthetic electron transport (PET) chain, ultimately generating reduced ferredoxin (Fd) (gene designation PetF), that can be used as an electron donor by hydrogenase to reduce protons to hydrogen (Greenbaum et al. 1983). But as soon as cells begin to produce oxygen at high rates, the hydrogen evolution rapidly stops, due to the oxygen sensitivity of [FeFe] hydrogenases (Stripp et al. 2009b; Ghirardi et al. 1997).

The second hydrogen photoproduction pathway, known as indirect biophotolysis, involves non-photochemical reduction of plastoquinone pool by electrons from nicotinamide adenine dinucleotide phosphate (NADPH). Electrons in this pathway are derived from catabolism of endogenous carbohydrates (glycolytic pathway and citric acid cycle) and from other organic molecules such as lipids, followed by light dependent ferredoxin reduction by photosystem I (PSI), with subsequent transfer of electrons to hydrogenases (Kosourov et al. 2003). This pathway is independent of PSII; however, it is dependent on NADPH-plastoquinone oxidoreductase (NPQR) activity (Mus et al. 2005; Cournac et al. 2000).

The third pathway for hydrogen production in algae cultures occurs under dark anaerobic conditions. In order to sustain basal level of metabolism under dark anaerobic condition, algae cells degrade endogenous reservoirs of starch hydrolyzing them to sugars, and the sugars then are converted to pyruvate by glycolysis (Mus et al. 2007). In the fermentation processes, pyruvate is the primary metabolite, and it serves as substrate for pathways generating organic acids, acetyl-CoA, ethanol, CO_2, and H_2. In cells of *C. reinhardtii*, there are two key enzymes that catalyze

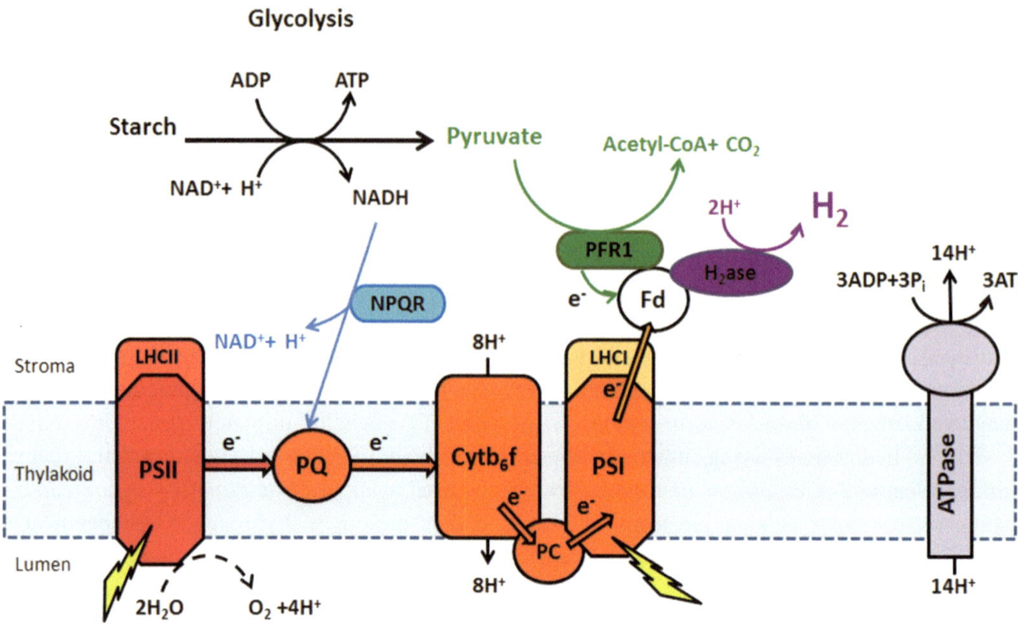

Fig. 8.1 H_2 production pathways in *C. reinhardtii*. Solar energy is being captured first by light-harvesting complexes (*LHCI*/*LHCII*) of photosystems I and II (*PSI*/*PSII*). There are two pathways associated with the photoproduction of hydrogen: PSII-dependent (*red*) and PSII-independent. PSII-independent pathway is known as NADPH-plastoquinone oxidoreductase (*NPQR*)-dependent (*blue*). Both the PSII-dependent (*red*) and NPQR-dependent (*blue*) pathways require reduction of the PQ pool and photosystem I (*PSI*) activity (*orange*). Electrons from PSI are able to reduce ferredoxin (*Fd*), the physiological electron donor to hydrogenase (*H_2ase*). Under *dark* anaerobic conditions, H_2 produced through the third pathway (*green*); in this case, pyruvate oxidation occurs with simultaneous reduction of Fd by enzyme pyruvate:Fd oxidoreductase 1 (*Pfr1*); electrons from reduced Fd are transferred then to hydrogenase. PSII-independent and dark H_2 production pathways use electrons generated by oxidation of organic compounds that are predominantly derived from starch degradation (*black*)

pyruvate: pyruvate formate lyase 1 (Pfl1) and pyruvate:ferredoxin oxidoreductase 1 (Pfr1, in some reports PFOR). Pfl1 present in both the mitochondria and the chloroplasts catalyzes the nonoxidative conversion of pyruvate to acetyl-CoA and formate (Kreuzberg et al. 1987; Atteia et al. 2006; Wagner et al. 1992). The chloroplast-located Pfr1 oxidizes pyruvate to acetyl-CoA, CO_2, and two molecules of reduced Fd. Electrons from Fd can thenbe used by hydrogenases to produce H_2 (Müller 2003).

8.2.3 Interplay of Light and Dark Reaction Pathways

The light reactions of photosynthesis in algae are organized according to the so-called Z-scheme (Govindjee 2008). Light energy allows green algae to extract electrons from high redox potential compounds such as water (in oxygenic photosynthesis) or organic acids (in nonoxygenic photosynthesis), producing energy in the form of ATP and low redox potential reductants. The electrons and protons derived from water are normally used to fix CO_2 and produce sugars using the Calvin-Benson-Bassham (CBB) cycle. However, in nature green algae can experience dark anaerobic conditions on a daily basis since when in the dark (at night), green algae and other organisms consume oxygen via respiration bringing about anaerobiosis. Under dark, anaerobic conditions, the CO_2 fixation pathway is inactivated and a new enzymatic system consisting of a [FeFe] hydrogenase enzyme and supporting pathways is expressed.

When dark-adapted *C. reinhardtii* cells are suddenly illuminated, the photosynthetic apparatus activates very quickly, and as a result, Fd is reduced to a higher extent than under darkness alone. A high level of reduced NADPH in cells prohibits further reduction of $NADP^+$, and when active hydrogenase is present, cells evolve significant amounts of hydrogen for a short period of time. In the beginning of illumination, green algae produce a high amount of hydrogen gas, and in some strains of green algae, it can be as high as light-dependent oxygen evolution ~100–300 μmol mg^{-1} Chl h^{-1} (Boicheo and Hoffmann 1994; Boichenko et al. 2004). The rate of H_2 production in anaerobically adapted *Chlorella vulgaris* cultures upon illumination (0.03 W m^{-2}) reaches a maximum after 2.5 s, and with subsequent kinetics which are linear for at least 1 min, but at higher light intensities (2 W m^{-2}), cells show a maximum rate after 0.6 s that starts to decline after 1 s of illumination (Boichenko et al. 1983). Other researchers demonstrated a decline in the hydrogen production rate within minutes of illumination (Greenbaum 1980; Yanyushin 1982). The decline in H_2 production rate is a result of photosynthetic O_2 evolution by PSII, with consequent inhibition of hydrogenase (Ghirardi et al. 1997), and the CBB cycle takes over its function as the major sink of electrons (Stuart and Gaffron 1972; Cinco et al. 1993). Interestingly, initial exposure to very high light levels allows subsequent H_2 evolution under low light, presumably through a photoinhibition of O_2 evolution by PSII (Markov et al. 2006). Some studies have shown that PSII participation is not absolutely necessary for the initial H_2 production. The addition of DCMU (3-(3,4-dichlorophenyl)-1,1-dimethylurea), which is an inhibitor of electron flow from PSII, has no effect on H_2 production during the first second but it starts to decline after 1 s of illumination (Boichenko et al. 1983). When DCMU is present, endogenous reduced compounds serve as the electron donors for PSI.

8.3 H_2 Production by Nutrient-Deprived Green Algae

A number of different strategies might be applied in attempts to increase hydrogen yields and rates (Table 8.1), including the highly studied use of nutrient deprivation. These are discussed in what follows.

8.3.1 Different Nutrient Deprivation Regimes

High initial rate of hydrogen production by green algae appeared to be very promising for practical application in case if it would be maintained for a longer period of time. In order to sustain photohydrogen production by green algae, cultures have to be kept free of oxygen that can be achieved by a variety of methods, including addition of O_2 scavengers (chromous chloride, dithionite) or purging the cultures with inert gases such as nitrogen or helium (Healey 1970; Randt and Senger 1985; Gfeller and Gibbs 1984; Graves et al. 1990). It has been proposed by Benemann in 1996 that reversible inactivation of PSII-dependent O_2 evolution activity might lead to temporal separation of H_2 and O_2 evolution in green algae ("indirect biophotolysis"); according to his model in the absence of PSII activity, endogenous starch reserves would serve as electron donors for H_2 production. Since then, and consistent with this proposal, it has been shown that the absence of essential macronutrients in the culture environment, such as sulfur, nitrogen, or phosphorus, leads to reversible inactivation of PSII (Melis et al. 2000; Batyrova et al. 2012; Philipps et al. 2012; He et al. 2012), with the subsequent temporal separation of O_2 and H_2 production.

Acclimation of algae cultures to the nutrient deficiencies usually accompanied with significant changes in cell physiology, including changes in protein, starch, and lipid biosynthesis/degradation, leads to decrease in photosynthetic activity alone with the reduction of linear photosynthetic electron transport from PSII and causes an increase in the rate of cellular respiration (Wykoff et al. 1998; Antal et al. 2003). Previous reports demonstrated that sulfur deprivation significantly inhibits PSII-dependent O_2-evolving activity in algal cultures (Wykoff et al. 1998) which, under certain conditions , results in the establishment of anaerobiosis and leads to H_2 photoproduction (Melis et al. 2000). The reduc-

Table 8.1 Possibilities for increasing hydrogen production by green algae

Strategy	Positive	Negative
Metabolic engineering		
Increase O_2 tolerance of hydrogenase	Increased yields and longevity	No clear engineering principles
Create antenna mutants	Increased photosynthetic efficiencies at high light	Overall fitness reduced
Eliminate competing pathways	Increased yields and longevity	Overall fitness reduced
		Inability to build up reserves
Create PSII mutants with lowered activity	Increased longevity	Overall fitness reduced
	Possible elimination of need for nutrient deprivation	Possible reduction in photosynthetic efficiencies
Manipulation for increased starch synthesis and mobilization	Increased overall yields and longevity of production	Possible unknown metabolic side effects
Physiological/bioprocess manipulation		
Nutrient deprivation	Facilitate anaerobiosis by decreasing PSII activity	Decreased photosynthetic efficiencies
		Requirement for centrifugation
Supply respiratory substrate (tris-acetate-phosphate (TAP))	Facilitate anaerobiosis by increasing respiration	Fixed carbon input requirement
Inert gas sparging	Reduces O_2 partial pressure	Dilution of product gas stream

tion of O_2-evolving activity was also shown under Fe, Mn (Ghirardi et al. 2000), phosphorus (Wykoff et al. 1998), and nitrogen (Philipps et al. 2012) deprivation, but bulk amounts of H_2 photoproduction have been demonstrated only under phosphorous-deprived (Batyrova et al. 2012) and nitrogen-deprived (Philipps et al. 2012; He et al. 2012) conditions. The most commonly used protocol for achieving sustained H_2 production under light involves starving cultures of sulfur (Melis et al. 2000). Hydrogen evolution in sulfur-deprived suspension cultures normally lasts for about 100 h and can be described as passing through five phases: aerobic, oxygen consumption, anaerobic, hydrogen production, and termination (Kosourov et al. 2002). Aerobic and O_2 consumption phases last between 18 and 40 h (Kosourov et al. 2002) but are strain- (Chochois et al. 2009, 2010) and condition-dependent (Kosourov et al. 2002). During the aerobic stage, cells continue to evolve oxygen (Ghirardi et al. 2000; Kosourov et al. 2002; Melis et al. 2000) and accumulate energy reserves in the form of starch to amounts 8–20 times higher than under non-stressed conditions (Tsygankov et al. 2002; Zhang et al. 2002), as well as accumulating triacylglycerides (TAGs) (Timmins et al. 2009b).

The accumulation of starch as a response to nutrient limitation can also be seen during nitrogen starvation (Ball et al. 1990; Philipps et al. 2012; He et al. 2012) and salt stress (Siaut et al. 2011). In the absence of sulfate, protein synthesis is geared toward those products that are necessary for viability under conditions of decreased metabolic activity (Wykoff et al. 1998). As a consequence, the D1 protein of PS II (which is essential for photosynthesis but turns over rapidly due to continuous photodamage) cannot be replaced quickly, leading to the gradual loss of O_2 evolution. Initially, respiratory activity in the mitochondrion remains largely unimpaired (Zhang et al. 2002) causing the rate of O_2 evolution in algae to drop below respiration. Thus, cultures respire all available O_2 and become anaerobic in the light if the photobioreactors are sealed (Melis et al. 2000). Reduction of PSII activity at the early stages of anaerobiosis is also associated with over-reduction of the PQ pool, as a consequence of the lack of electron sinks, since CBB cycle is almost completely inhibited due to the very low level of RuBisCO, and active hydrogenase has not been expressed yet (Antal et al. 2003; Zhang et al. 2002). Such conditions in cells lead to switch from linear to cyclic electron transport, when electrons are shuttled around PSI generating ATP. The lack of electron sinks together with anaerobiosis activates hydrogenase expression with subsequent hydrogen production (Zhang et al. 2002). Hydrogenase protein expression can be detected after about 3–4 h (Forestier et al. 2003); however, measurable amounts of H_2 can be detected only after another 5 h (Antal et al. 2003). Due to limited O_2 availability that causes reduced rates of oxidative phosphorylation in the mitochondria, glycolysis and anaerobic fermentation become major pathways for ATP formation (Timmins et al. 2009b). The products of anaerobic fermentation in algae cultures are ethanol, acetate, and formic acid, produced through a pyruvate formate lyase-dependent pathway; accumulation of such products in algal cells causes significant drop of pH during the H_2 photoproduction (Hemschemeier et al. 2008a; Matthew et al. 2009).

Presently, it is widely accepted that hydrogenase gets electrons for hydrogen production from a direct (PSII-dependent) and indirect (PSII-independent) pathway (Chochois et al. 2009; Fouchard et al. 2005). The direct pathway utilizes electrons derived from water-splitting reaction and relies on oxygen consumption through respiratory activities; the indirect pathway fed by electrons derived from the catabolism of endogenous carbohydrate stores and involves non-photochemical reduction of PQ pool (Fouchard et al. 2005). The indirect pathway was found to be absent in starch-deficient mutants, indicating that carbohydrate reserves are the main source of reductant for PSII-independent H_2 production and not protein or TAGs (Hemschemeier et al. 2008b). More evidence of this theory comes from mutants affected in starch catabolism that show a general decrease in the amount of PSII-independent H_2 production and those with a reduced rate of starch degradation which show delayed H_2 evolution (Chochois et al. 2010).

However, in the experiments with the ADP-glucose pyrophosphorylase-deficient mutant sta6, which is unable to accumulate starch, wild-type levels of H_2 production during sulfur deprivation were observed (Chochois et al. 2009). This result suggests that starch is dispensable for PSII-dependent hydrogen production, and probably, unlike wild-type cultures, sta6 consumes acetate during the hydrogen production phase, which could act as a replacement for carbohydrate in maintaining respiration (Chochois et al. 2009).

The precise contribution of each of these pathways in H_2 production has yet to be fully resolved. However, the low-level PSII or PSI activity is required to sustain hydrogen production, since it has been shown that sulfur-deprived cultures don't produce hydrogen in the absence of light (Melis et al. 2000). Direct measurements of PSII photochemical activity during the H_2 production phase and experiments with DCMU addition demonstrated that PSII is active and donates electrons via the cytochrome b_6f complex and PSI to ferredoxin (PetF) (Antal et al. 2003; Kosourov et al. 2003; Winkler et al. 2009), which transfers them to the hydrogenase enzyme (Hyd1). The importance of PSII activity for the process was first identified by the inability of mutants lacking PSII to evolve H_2 (Hemschemeier et al. 2008b; Zhang et al. 2002). In experiments with wild-type *C. reinhardtii*, addition of DCMU at the beginning of sulfur depletion gave a ~20 % inhibition of PSII activity at the beginning of H_2 evolution (Hemschemeier et al. 2008b). In another case it was suggested that up to 90 % of the electrons for H_2 production come from PSII (Chochois et al. 2009, 2010), but this value is likely to be strain- and condition-dependent (Chochois et al. 2009). Following approximately 72 h of S deprivation, the rate of fermentative starch catabolism starts to decrease, the pH partially stabilizes, and the rate of hydrogen production declines (Matthew et al. 2009). Hydrogen production eventually stops after 4 or 5 days of sulfur deprivation.

In addition to sulfur deprivation, phosphorus deficiency also inhibits PSII-dependent O_2-evolving activity in algae cells (Wykoff et al. 1998). But in contrast with S deprivation, the inhibition effect of phosphorus deprivation is slower and cultures establish anaerobiosis later (Wykoff et al. 1998). Phosphorus deprivation also causes a decline in the in vivo, light-saturated rate of photosynthesis in higher plants (Dietz and Heilos 1990; Jacob and Lawlor 1993); the decline occurs due to limitation in the rate of CO_2 fixation because of the depletion of the pool of phosphorylated intermediates of the reductive pentose phosphate cycle (Jacob and Lawlor 1993; Brooks 1985). Bulk amounts of H_2 gas can be obtained under P-deprived conditions with the green alga *C. reinhardtii*, but only after the intracellular pools of reserve phosphorus are depleted (Batyrova et al. 2012). Continuous H_2 production in P-deprived algae was demonstrated using a dilution approach that was firstly developed for sulfur deprivation (Laurinavichene et al. 2002). Cultures washed free of phosphorus and diluted to below 2 mg Chl l^{-1} in phosphate-free medium were able to establish anaerobiosis in the photobioreactor and to produce H_2 in the same quantities as sulfur-deprived cultures. Experiments using a microprocessor-controlled photobioreactor system demonstrated that phosphorus-deprived cultures pass through the same physiological stages as sulfur-deprived cultures: aerobic/photosynthetic, O_2 consumption, anaerobic, H_2 production, and termination stages. The major difference between the two culture types is that establishment of anaerobiosis in phosphorus-deprived cultures occurs 100 h after the beginning of phosphorus deprivation, significantly later than in the sulfur-deprived cultures obtained by the centrifugation method (18–40 h). On the other hand, like phosphorous-deprived cultures, sulfur-deprived cultures obtained by the dilution method start to produce H_2 gas after 75–100 h of growth in the sulfur-depleted medium (Laurinavichene et al. 2002). H_2 appears in the system after about 100 h of phosphorus deprivation and stopped after about 300 h.

An advantage to the phosphorous deprivation approach is that it can be used to induce H_2 production in marine strains of green algae, where S deprivation is impossible due to the high concentration of sulfate in seawater. Application of seawater for H_2 production presents a great deal of interest, since freshwater resources on our planet are limited. Recently, Batyrova and coworkers demonstrated the ability of the marine green microalgae *Chlorella* sp. to produce H_2 under phosphorous-deprived conditions (Batyrova et al. 2015). The dilution approach was applied in order to achieve H_2 photoproduction in P-deprived *Chlorella* sp. Cultures diluted to about 0.5–1.8 mg Chl l^{-1} at the beginning of the experiment were able to establish anaerobic conditions after the initial growth period, when cells utilize all intracellular reserves of phosphorus, with subsequent H_2 production. It appears that marine microalgae *Chlorella* sp. passed through the same physiological stages of adaptation to P deprivation as freshwater microalgae. The presence of inorganic carbon was essential for starch accumulation and subsequent hydrogen production by *Chlorella* sp. The final H_2 production yield was 40 ml H_2 gas per 1iter of the culture, which is comparable to that obtained with phosphorous-deprived *C. reinhardtii* cultures.

In 2012 Philipps and coworkers reported that *C. reinhardtii* wild-type strain CC124 is capable of producing considerable amounts of H_2 under nitrogen-deprived conditions, if incubated under anaerobic conditions in sealed flasks. However, N-deprived cultures start accumulating H_2 for about 48 h later then under S-deprived conditions and in a less amounts. However, the amount of accumulated starch during the N deprivation was about twofold higher than under S deprivation, despite that efficient starch degradation in N-deprived cultures was not observed. Analysis of the photosynthetic apparatus of N-deprived *C. reinhardtii* cultures revealed that the delay in H_2 production was due to prolonged PSII activity and that N deprivation causes degradation of cytochrome b_6f complex. H_2 photoproduction under N-deprived conditions was also reported in *Chlorella protothecoides*, where N limitation resulted in considerable starch accumulation, reduction in chlorophyll biosynthesis, blockage of photosynthetic electron transfer, and decline in PSII oxygenic activity (He et al. 2012). Also in the case of N-deprived *C. reinhardtii*, significant starch degradation in *Chlorella protothecoides* under N deprivation was not observed.

Fuller understanding of physiological differences related to H_2 production under nutrient deprivations will require studying of regulatory mutants impaired in the specific acclimation responses to S (sac1, sac3, snrk2.1) (Davies et al. 1996; Davies et al. 1999), P (psr1) (Wykoff et al. 1999) and N (nit2) (Camargo et al. 2007) deficiencies. For instance, sac1 and snrk2.1 mutants showed a light-dependent bleaching phenotype under S deprivation with a subsequent fast cell death (Gonzalez-Ballester et al. 2008; Davies et al. 2006), indicating the importance of S-deficient response mechanism for cell survival. The other research demonstrated that D1 degradation upon S deprivation is controlled by the FstH protease (Malnoe et al. 2014). Therefore, D1 degradation under S deprivation is not just a result of protein damage but also part of S-specific acclimation process.

It is known that modulation of the light-harvesting complex (LHC) proteins is crucial in regulating light absorption by photosynthetic organisms. Under nutrient-replete conditions, the level of *LHCBM9* transcript in *Chlamydomonas* cells is barely detectable; however, it increases by about 1000-fold under S deprivation and becomes one of the most abundant mRNA in S-starved cells (Gonzalez-Ballester et al. 2010). It has been demonstrated that mutants of green algae impaired in *LHCBM9* expression experience more pronounced oxidative damage caused by singlet oxygen and produce less H_2 under S-deprived conditions (Grewe et al. 2014).

8.3.2 The Price to Be Paid for Nutrient Deprivation

In the absence of any of these nutrients, cell division is arrested and cultures stop growing (Hase et al. 1958). Under phosphorous deprivation, cultures stop growing after 4–5 days, whereas H_2 production stops after 10–12 days. Nitrogen limitation causes cessation of cell growth after 2 days and H_2 evolution ceases after approximately 7 days. In both cases production of H_2 under either nitrogen or phosphorous deprivation eventually stops, despite the continuing presence of energy reserves in the form of starch, TAGs, and acetate, which could be due to the toxic nature of the accumulated metabolites or as a result of the long-term consequences of nutrient deprivation. Hydrogen production stops after about 4–5 days of sulfur deprivation, and the algal cells exhibit a spherical morphology and a significant reduction in cell mass (Zhang et al. 2002). Presently, it is not entirely known what events cause termination of hydrogen production. However, metabolomic analysis of algal cultures at the termination stage of hydrogen production has revealed significant starch reserves, as well as triacylglycerides, indicating that depletion of energy stores is not the cause of the cessation of hydrogen production (Matthew et al. 2009). One of the possible reasons for the termination of hydrogen production is accumulation of toxic fermentation end products such as ethanol and formate.

Although sulfur deficiency causes significant metabolic stress in the algal cell, H_2 production can be sustained for longer periods by operating algal cell suspensions in cycles of +S/−S (Ghirardi et al. 2000) or by physically separating the O_2 from the H_2 evolution stages in separate photobioreactors (Fedorov et al. 2005). The duration of hydrogen evolution has also been extended by immobilization of cells on inert matrices such as glass fibers (Laurinavichene et al. 2006) or alginate films (Kosourov and Seibert 2009). Immobilizing cells also provides a higher tolerance to O_2, greater achievable cell densities, and better light utilization, leading to maximum recorded rates of H_2 production at around 12.5 mmol/mgChl/h and an increase in light energy conversion efficiency from 0.24 % in liquid cultures to ~1 % in immobilized cells (Kosourov and Seibert 2009).

Determination of the precise mechanisms involved in H_2 production under different nutrient deficiencies as well as metabolic and genetic engineering might lead to processes where sustainable H_2 production is possible without the application of nutrient deprivation.

8.3.3 H_2 Without Nutrient Deprivation

There are a number of disadvantages of nutrient depletion, including of course that it ultimately results in cell death, is only suitable as a batch process, and requires continuous illumination over several days for maximum H_2 production (Oncel and Sukan 2011). Therefore, controllable expression of PSII could be used to reduce oxygen evolution to a rate below respiration and drive cultures into anaerobiosis. It has been shown that the copper-sensitive cytochrome c_6 promoter can be used to repress PSII assembly, thereby inducing anaerobiosis and hydrogen production when cells were transferred from copper-free to copper-replete media (Surzycki et al. 2007). However, it was also found that the promoter used was stimulated by anaerobiosis, even in the presence of copper, resulting in the reestablishment of aerobic conditions shortly after the initiation of H_2 evolution, bringing the process to a halt (Surzycki et al. 2007). Nevertheless, this approach demonstrates that controlling photosynthesis through inducible promoters is an effective method of stimulating H_2 evolution. Inactivation of PSII in *C. reinhardtii* cells has also been demonstrated by application of RNA antisense technique against a sulfate transporter gene (SulP); obtained mutants were able to produce H_2 even in the presence of 100 μM sulfate (Chen et al. 2005). Another possible way to decrease PSII activity is through mutations in the D1 protein, the key protein of PSII that is essential for photosynthesis, and it has been reported that S-deprived *C. reinhardtii* D1 mutant that carries a double amino acid substitution produces more H_2 than wild type. In this mutant the leucine residue L159 of D1 protein was replaced by isoleucine, and the asparagine N230 was replaced by tyrosine (L159I-N230Y). This mutant is very efficient in prolonged H_2 production, as well as having decreased chlorophyll content, and a higher respiration rate, both of which contribute to a higher H_2 production yield (Torzillo et al. 2009). The other attempt to decrease PSII activity was done in *Chlorella* sp. strain DT, where expression of PsbO protein of PSII (part of the oxygen evolution center) was knocked down through a short interference antisense RNA; as a result mutant exhibited a tenfold higher hydrogen evolution under low illumination and semi-aerobic conditions compared to wild type (Lin et al. 2013).

Another approach to avoiding nutrient deprivation is the identification of mutants with altered rates of photosynthesis to respiration which would automatically go anaerobic when placed in sealed containers. A forward genetic screen was used to identify such strains based on a colorimetric analysis of dissolved oxygen concentrations, allowing one to be found which was named apr1 for attenuated photosynthesis to respiration, showing dramatically reduced photosynthetic rates and a slight increase in respiration (Ruhle et al. 2008). However, despite going anaerobic when placed in sealed bioreactors, H_2 was only produced in the light after the CBB cycle was inhibited through the addition of glycolaldehyde (Ruhle et al. 2008). These results suggested that downregulation of the CBB cycle is a necessary step for stimulating H_2 production which otherwise acts as a preferential electron sink (Ruhle et al. 2008).

It was also demonstrated that the mutant CC-2803, which lacks RuBisCO (a key enzyme of the CBB cycle), has a light-sensitive phenotype and a dramatically reduced rate of photosynthesis. Consequently, cultures transit to anaerobiosis and produce H_2 in sealed bioreactors even in the presence of all required nutrients (Hemschemeier et al. 2008b). RuBisCO can also serve as a target to decrease the specificity of carboxylation over oxygenation reactions, which would result in a higher rate of oxygen consumption, or reducing flux through the CBB cycle (Chen et al. 1988; Genkov et al. 2006; Satagopan and Spreitzer 2004). Control of the CBB cycle or RuBisCO activity therefore represents a potentially novel method of inducing H_2 production (Marin-Navarro et al. 2010) by removing the major sink of electrons for reduced ferredoxin generated by the light reactions. This could be achieved through inducible control of RuBisCO or CBB cycle enzyme expression or control of CO_2 supply to carbon-concentrating mutants (Spalding 2008).

In addition to altering the specificity of RuBisCO to increase the oxygenation rate, direct reduction of internal O_2 levels without affecting PSII activity could be achieved by overexpressing O_2-consuming enzymes. For example, transformation in chloroplast of *C. reinhardtii* cells a codon-optimized leghemoglobin protein from soy, which serves as an oxygen sequester in the nitrogen-fixing root nodules of soy, along with a ferrochelatase from a nitrogen-fixing bacterium (for assembling the heme group) resulted in a fourfold increase in H_2 production (Wu et al. 2011). Likewise, pyruvate oxidase from *Escherichia coli* was expressed in *C. reinhardtii* in order to reduce intracellular oxygen concentration. Pyruvate oxidase decarboxylates pyruvate into acetyl phosphate while consuming one molecule of oxygen. Obtained transgenic algal strain was able to produce up to 2.5-fold more hydrogen than the parental strain under very low light (30 μE m^{-2} s^{-1}) and sulfur-replete conditions (Xu et al. 2011), although it remains to be seen whether the benefits of decreasing O_2 levels would outweigh the loss of electrons which could potentially be fed to hydrogenase. In this regard it would be very important to follow up on the recent report of two novel strains of *Chlorella vulgaris* which exhibited hydrogen production under atmospheric oxygen concentrations (Hwang et al. 2014).

8.4 Engineering for Sustainable Hydrogen Production

8.4.1 Genetic Engineering to Improve Hydrogen Production

Genetic engineering of green algae is in its infancy but holds a great promise and can help in improving H_2 production (Lee 2013). Metabolic and genetic engineering can help to address limitations associated with photohydrogen production specifically: hydrogenase activity and its O_2 sensitivity, anaerobiosis induction, and competition for electrons with other pathways (Melis et al. 2007; Beer et al. 2009; Esquível et al. 2011; Dubini and Ghirardi 2014).

Recent achievements in metabolic and genetic engineering have pushed forward the biohydrogen research and provided additional insights into complex interactions between different pathways involved in the photohydrogen production. Here, we discuss some of the genetic modification strategies which have led to improved hydrogen production.

Increasing hydrogenase tolerance to O_2 is one of the major challenges to commercial feasibility of microalgal H_2 production. Three main approaches can be taken to modify the O_2 sensitivity of the [FeFe] hydrogenases: (1) bioprospecting for enzymes from nature with increased resistance to O_2 inactivation, (2) use of random mutagenesis followed by high-throughput screening to identify enzymes that are more tolerant to O_2, and (3) molecular engineering of hydrogenases to decrease O_2 access to their catalytic site (Nagy et al. 2007; Boyer et al. 2007; Ghirardi et al. 2005; Melis et al. 2007; Seibert et al. 1998).

A variety of methods such as site-directed mutagenesis, molecular dynamics simulations, potential mean energy estimates, and solvent accessibility maps were used in order to identify gas diffusion pathways in [FeFe] hydrogenase (Chang et al. 2007; Long et al. 2009). Although the results obtained so far are encouraging from a scientific perspective in understanding enzyme structure and function, this approach was not successful, since results of observations demonstrated that the amino acid residues responsible for binding of the catalytic cluster are also involved in the formation of the gas channels. Therefore, mutants that were affected on these residues were unable to properly fold hydrogenase and as a result had lower activity and higher O_2 sensitivity (Mulder et al. 2010; Liebgott et al. 2010).

Many attempts have been made to generate O_2-tolerant hydrogenases using random mutagenesis in vivo and in vitro (Stapleton and Swartz 2010; Ghirardi et al. 1997). Unfortunately, these efforts were also not successful and resulted only to small changes in O_2 tolerance. Alternatively, the endogenous hydrogenase was overexpressed in *Chlorella* sp. strain DT that lead to seven- to tenfold increase in H_2 production under aerobic sulfur-replete conditions (Chien et al. 2012). Thus, this hydrogenase appears to be more oxygen-tolerant than the hydrogenase from *Chlamydomonas*.

It has been estimated that the vast diversity of hydrogenase sequences remains underrepresented in databases (Beer et al. 2009), and there may yet be a suitable sequence present in nature. The establishment of metagenomics, including analysis of termite gut microbiota and the Global Ocean Sampling program, has allowed the identification of novel hydrogenases from the environment (Boyd et al. 2009; Maroti et al. 2009; Rusch et al. 2007; Warnecke et al. 2007), but as yet the oxygen tolerance of these enzymes is unknown.

For optimal photosynthetic activity, photosynthetic organisms must balance the amount of light energy absorbed by the PSII and PSI reaction centers through transfer of mobile light-harvesting complexes in a process known as state transitions (Lemeille and Rochaix 2010). In *C. reinhardtii* this process allows the switch between linear and cyclic electron flow (CEF). CEF potentially occurs by two pathways in *C. reinhardtii* which can be distinguished by their sensitivity to antimycin A (Ravenel et al. 1994). The efficacy of inhibiting CEF for increasing H_2 yields was demonstrated in short-term inhibitor studies using antimycin A, which resulted in twofold increase of H_2 production as a result of antimycin A's activity in blocking CEF (Antal et al. 2009). It is possible to screen for mutants affected in CEF through chlorophyll fluorescence video imaging, which has allowed the identification of mutants unable to switch between linear and cyclic electron transport (Depege et al. 2003). This screen also led to the discovery a hydrogen-efficient *C. reinhardtii* strain 'stm6' (Kruse et al. 2005). In this strain, cyclic electron transfer is inhibited, and the algae exhibit a high rate of respiratory O_2 consumption (due to an upregulated alternative oxidase (AOX)) and an extra accumulation of starch, thus leading to about a ninefold increase in H_2 production relative to the parental strain, with a maximum production rate of 4 ml H_2/l culture/h.

The major pathway competing for electrons with hydrogenase is carbon fixation by RuBisCO (CBB cycle); in the presence of active CO_2 fixation, the reductant flux available for H_2 production is low; therefore, downregulation of carbon fixation can improve H_2 production. It was reported that expression of a mutated small subunit of RuBisCO (RBCS-Y67A) from the nucleus of an RBCS-deficient *C. reinhardtii* strain resulted in significant decrease of PSII activity and 10- to 15-fold increase in H_2 production under S-deprived conditions (Pinto et al. 2013). In the other research work, in order to increase the flux toward hydrogenase, a hexose uptake protein (HUP1) symporter from *Chlorella kessleri* was incorporated into the *Chlamydomonas* stm6 mutant strain; the newly obtained mutant stm6Glc4 can use externally supplied glucose for heterotrophic growth in the dark, and 1.5-fold increase in H_2 production was observed under the light (Doebbe et al. 2007).

Efficient starch metabolism is an important factor for hydrogen production, since starch reserves serve as a source of electrons for hydrogenase via PSII-independent H_2 production pathway. Presently, only two mutant strains affected in starch catabolism present an interest for improving H_2 production. In the first mutant (std3), the amount of accumulated starch under S deprivation was similar to the wild type; however, the amount of residual starch left in the termination stage of H_2 production was lower indicating that faster starch degradation correlated with higher H_2 production. The second mutant (sda6) exhibited slower rate of starch degradation accompanied with lower initial rate of H_2 production that was lower than in wild type; however, the final H_2 production yield of mutant appeared to be much higher than that in the wild type (Chochois et al. 2010).

One of the most important factors for low-cost generation of H_2 via microalgal-mediated processes is the photosynthetic productivity and light utilization efficiency that can be attained. It has been proposed in the past that decrease of the chlorophyll antenna size can improve light utilization efficiency in algae cultures (Melis et al. 2000). Decreasing the light-harvesting capacity of the photosystems reduces excess light absorption by individual cells, thereby increasing photon and energy conversion efficiencies as well as light penetration in a culture (Melis et al. 1989; Mussgnug et al. 2007; Polle et al. 2002). Mutants with a truncated light-harvesting antennae (such as the tla mutants) have been identified on account of their pale phenotype (Berberoglu et al. 2008; Polle et al. 2002) or through reverse genetic approaches, such as by downregulation of light-harvesting complexes through RNAi (Mussgnug et al. 2007) or overexpression of the RNA-binding protein, NAB1, known to block translation of light-harvesting subunits (Beckmann et al. 2009). Knockdown of the three major proteins (LHCMB1, 2, 3) of light-harvesting complex II resulted in a twofold increase in H_2 production under sulfur deprivation compared to the parental strain (Oey et al. 2013). Analysis of the *Chlamydomonas* tla1 truncated antenna mutant demonstrated an increase in H_2 productivity in long-term experiments under different light intensities ranging from 19 to 350 μE m^{-2} s^{-1} PAR; in these experiments, wild-type and tla1 S-deprived mutant cells were immobilized in alginate films and monitored for H_2 photoproduction (Kosourov et al. 2011). In this experiment mutant was able to produce H_2 gas for over 250 h under all light intensities that were tested and exhibited a four to eight times higher maximum specific rate of H_2 production in the range between 285 and 350 μE m^{-2} s^{-1} in comparison to wild-type cells.

8.4.2 Sustainability

One issue that is of great importance but which is only beginning to be addressed in the area of biofuel production is sustainability. On the one hand, many systems, such as hydrogen production by engineered green algae, are barely at the proof of principle stage and hence are not amenable to realistic well-to-wheel life cycle assessments (LCAs) or other measures of sustainability. On the other hand, present-day use of fossil fuels is so huge that any system that makes even a very small impact on this will nevertheless have an enormous environmental footprint. Thus, it is instructive to at least consider a few of what are likely to be sustainability issues with this process (Table 8.2).

Of course water is a very large issue with almost any biofuel production scheme, but especially for any system that relies on microalgae (Abdelaziz et al. 2013a, b; Leite et al. 2013). This potential problem could be possibly circumvented through the use of wastewater or saltwater. However, the challenge would be how to treat or otherwise use the spent algal medium after fuel production has ceased. Large-scale microalgal culture would demand enormous quantities of nutrients, placing any production system in direct competition with food production. In principle, some wastewaters could be used to support algal growth and fuel production, but this also brings its own challenge in terms of not being of the proper chemical composition to favor hydrogen production (-S, -N, etc.). Hydrogen is a gaseous product and hence the culturing must be done in large arrays consisting of enclosed photobioreactors. How these would be manufactured or disposed of once no longer serviceable is an open question but one which will also lead to concerns over their environmental footprints. Finally, most of the algal biomass is not consumed in the hydrogen production process and will have to be disposed of in some ecologically sound manner. Of course the best would be to develop some type of biorefinery where other products and/or energy sources could also be produced.

Table 8.2 Sustainability issues in hydrogen production by green algae

Issue	Strategy	Challenge
Water usage	Use wastewater or saltwater	Process water treatment
Nutrient requirements	Use wastewater	Obtaining wastewaters with suitable composition (-S, -N, -P, -Mg)
Hydrogen capture	Use transparent enclosed photobioreactors	High cost of fabrication
Excess (waste) biomass	Use in biorefinery to produce pigments and proteins	High cost of harvesting

8.5 Conclusion

Biological hydrogen production offers one possible alternative to fossil fuels, but due to the complexity of the metabolism involved in H_2 production, it requires precise control and regulation of the involved process pathways. Metabolic and genetic engineering was employed in order to overcome some of the limitations associated with biological hydrogen production, and most of the attempts were successful and improved H_2 production; however, much research is still required to improve yields and bring down costs.

References

Abdelaziz AEM, Leite GB, Hallenbeck PC (2013a) Addressing the challenges for sustainable production of algal biofuels: I. Algal strains and nutrient supply. Environ Technol 34(13–14):1783–1805

Abdelaziz AEM, Leite GB, Hallenbeck PC (2013b) Addressing the challenges for sustainable production of algal biofuels: II. Harvesting and conversion to biofuels. Environ Technol 34(13–14):1807–1836

Antal T, Krendeleva T, Laurinavichene T, Makarova V, Ghirardi ML, Rubin A, Tsygankov A, Seibert M (2003) The dependence of algal H_2 production on photosystem II and O 2 consumption activities in sulfur-deprived *Chlamydomonas reinhardtii* cells. Biochim Biophys Acta 1607:153–160

Antal TK, Volgusheva AA, Kukarskih GP, Krendeleva TE, Rubin AB (2009) Relationships between H_2 photoproduction and different electron transport pathways in sulfur-deprived Chlamydomonas reinhardtii. Int J Hydrog Energy 34:9087–9094

Atteia A, van Lis R, Gelius-Dietrich G, Adrait A, Garin J, Joyard J, Rolland N, Martin W (2006) Pyruvate formate-lyase and a novel route of eukaryotic ATP synthesis in *Chlamydomonas* mitochondria. J Biol Chem 281:9909–9918

Ball SG, Dirick L, Decq A, Martiat JC, Matagne R (1990) Physiology of starch storage in the monocellular alga *Chlamydomonas reinhardtii.* Plant Sci 66:1–9

Batyrova K, Tsygankov A, Kosourov S (2012) Sustained hydrogen photoproduction by phosphorus-deprived *Chlamydomonas reinhardtii* cultures. Int J Hydrog Energy 37:8834–8839

Batyrova K, Gavrisheva A, Ivanova E, Liu J, Tsygankov A (2015) Sustainable hydrogen photoproduction by phosphorus-deprived marine green microalgae *Chlorella* sp. Int J Mol Sci 16:2705–2716

Beckmann J et al (2009) Improvement of light to biomass conversion by de-regulation of light-harvesting protein translation in Chlamydomonas reinhardtii. J Biotechnol 142:70–77

Beer LL, Boyd ES, Peters JW, Posewitz MC (2009) Engineering algae for biohydrogen and biofuel production. Curr Opin Biotechnol 20:264–271

Benemann J (1996) Hydrogen biotechnology: progress and prospects. Nat Biotechnol 14:1101–1103

Berberoglu H, Pilon L, Melis A (2008) Radiation characteristics of Chlamydomonas reinhardtii CC125 and its truncated chlorophyll antenna transformants tla1, tlaX and tla1-CWþ. Int J Hydrog Energy 33:6467–6483

Bock A, King PW, Blokesch M, Posewitz MC (2006) Maturation of hydrogenases. Adv Microb Physiol 51:1–71

Boichenko V, Arkhipov V, Litvin F (1983) Simultaneous measurements of fluorescence induction and hydrogen evolution of *Chlorella* under anaerobic conditions. Biophysics (Russ) 28:976–979

Boichenko V, Greenbaum E, Seibert M (2004) Hydrogen production by photosynthetic microorganisms. In: Archer MD, Barber J (eds) Photoconversion of solar energy, molecular to global photosynthesis, 2nd edn. Imperial College Press, London, pp 397–452

Boicheo VA, Hoffmann P (1994) Photosynthetic hydrogen production in prokaryotes and eukaryotes: occurrence, mechanism, and functions. Photosynthetica 30:527–552

Boyd ES, Spear JR, Peters JW (2009) [FeFe] hydrogenase genetic diversity provides insight into molecular adaptation in a saline microbial mat community. Appl Environ Microbiol 75:4620–4623

Boyer M, Stapleton J, Kuchenreuther J et al (2007) Cell-free synthesis and maturation of [FeFe] hydrogenases. Biotechnol Bioeng 99:59–67

Brand JJ, Wright JN, Lien S (1989) Hydrogen production by eukaryotic algae. Biotechnol Bioeng 33:1482–1488

Brooks A (1985) Effects of phosphorus nutrition on ribulose-1,5-bisphosphate carboxylase activation, photosynthetic quantum yield and amounts of some Calvin-cycle metabolites in spinach leaves. Aust J Plant Physiol 13:221e37

Camargo A, Llamas A, Schnell RA, Higuera JJ, Gonzalez-Ballester D, Lefebvre PA, Fernandez E, Galvan A (2007) Nitrate signaling by the regulatory gene NIT2 in Chlamydomonas. Plant Cell 19:3491–3503

Chang C, King P, Ghirardi M, Kim K (2007) Atomic resolution modeling of the ferredoxin:[FeFe] hydrogenase complex from Chlamydomonas reinhardtii. Biophys J 93(9):3034–3045

Chen ZX, Chastain CJ, Al-Abed SR, Chollet R, Spreitzer RJ (1988) Reduced CO_2/O_2 specificity of ribulose-bisphosphate carboxylase/oxygenase in a temperature-sensitive chloroplast mutant of *Chlamydomonas*. Proc Natl Acad Sci U S A 85:4696–4699

Chen H-C, Newton AJ, Melis A (2005) Role of SulP, a nuclear-encoded chloroplast sulfate permease, in sulfate transport and H2evolution in Chlamydomonas reinhardtii. Photosynth Res 84:289–296

Chien L, Kuo T, Liu B, Lin H, Feng T, Huang C (2012) Solar-tobioH_2 production enhanced by homologous overexpression of hydrogenase in green alga Chlorella sp. DT. Int J Hydrog Energy 37(23):17738–17748

Chochois V, Dauvillee D, Beyly A, Tolleter D, Cuine S, Timpano H, Ball S, Cournac L, Peltier G (2009) Hydrogen production in *Chlamydomonas*: PSII Solar-dependent and independent pathways differ in their requirement on starch metabolism. Plant Physiol 151:613–640

Chochois V, Constans L, Dauville ED, Beyly A, Soliveres M, Ball S, Peltier G, Cournac L (2010) Relationships between PSII-independent hydrogen bioproduction and starch metabolism as evidenced from isolation of starch catabolism mutants in the green alga *Chlamydomonas reinhardtii*. Int J Hydrog Energy 35:10731–10740

Cinco R, Macinnis J, Greenbaum E (1993) The role of carbon-dioxide in light-activated hydrogen-production by *Chlamydomonas reinhardtii*. Photosynth Res 38:27–33

Cournac L, Redding K, Ravenel J, Rumeau D, Josse EM, Kuntz M, Peltier G (2000) Electron flow between photosystem II and oxygen in chloroplasts of photosystem I-deficient algae is mediated by a quinol oxidase involved in chlororespiration. J Biol Chem 275:17256–17262

Davies J, Yildiz F, Grossman AR (1996) Sac1, a putative regulator that is critical for survival of Chlamydomonas reinhardtii during sulfur deprivation. EMBO J 15:2150–2215

Davies JP, Yildiz FH, Grossman AR (1999) Sac3, an Snf1-like serine/threonine kinase that positively and negatively regulates the responses of Chlamydomonas to sulfur limitation. Plant Cell 11:1179–1190

Davies KM, Skamnaki V, Johnson LN, Venien-Bryan C (2006) Structural and functional studies of the response regulator HupR. J Mol Biol 359:276–288

Depege N, Bellafiore S, Rochaix J-D (2003) Role of chloroplast protein kinase Stt7 in LHCII phosphorylation and state transition in Chlamydomonas. Science 299:1572–1575

Dietz KJ, Heilos L (1990) Carbon metabolism in spinach leaves as affected by leaf age and phosphorus and sulfur nutrition. Plant Physiol 93:1219e25

Doebbe A, Rupprecht J, Beckmann J, Mussgnug J, Hallmann A, Hankamer B, Kruse O (2007) Functional integration of the HUP1 hexose symporter gene into the genome of C-reinhardtii impacts on biological H2 production. J Biotechnol 131(1):27–33

Dubini A, Ghirardi ML (2014) Engineering photosynthetic organisms for the production of biohydrogen. Photosynth Res 2:1–13

Esper B, Badura A, Rögner M (2006) Photosynthesis as a power supply for (bio-)hydrogen production. Trends Plant Sci 11(11):543–549

Esquível MG, Amaro HM, Pinto TS, Fevereiro PS, Malcata FX (2011) Efficient H2 production via Chlamydomonas reinhardtii. Trends Biotechnol 29:595–600

Fedorov A, Kosourov S, Ghirardi M, Seibert M (2005) Continuous hydrogen photoproduction by *Chlamydomonas reinhardtii*. Appl Biochem Biotechnol 121:403–412

Forestier M, King P, Zhang L, Posewitz M, Schwarzer S, Happe T, Ghirardi ML, Seibert M (2003) Expression of two [Fe]-hydrogenases in *Chlamydomonas reinhardtii* under anaerobic conditions. Eur J Biochem 270:2750–2758

Fouchard S, Hemschemeier A, Caruana A, Pruvost K, Legrand J, Happe T, Peltier G, Cournac LB (2005) Autotrophic and mixotrophic hydrogen photoproduction in sulfur-deprived *Chlamydomonas* cells. Appl Environ Microbiol 71:6199–6205

Gaffron H, Rubin J (1942) Fermentative and photochemical production of hydrogen in algae. J Gen Physiol 26:219–240

Genkov T, Du Y-C, Spreitzer RJ (2006) Small-subunit cysteine-65 substitutions can suppress or induce alterations in the large-subunit catalytic efficiency and holoenzyme thermal stability of ribulose-1,5-bisphosphate carboxylase/oxygenase. Arch Biochem Biophys 451:167–174

Gfeller RP, Gibbs M (1984) Fermentative metabolism of *Chlamydomonas reinhardtii*. 1. analysis of fermentative products from starch in dark and light. Plant Physiol 75:212–218

Ghirardi ML, Mohanty P (2010) Oxygenic hydrogen photoproduction – current status of the technology. Curr Sci 98:499–507

Ghirardi ML, Togasaki RK, Seibert M (1997) Oxygen sensitivity of algal H(2)-production. Appl Biochem Biotechnol 63–65:141–151

Ghirardi ML, Zhang L, Lee JW, Flynn T, Seibert M, Greenbaum E, Melis A (2000) Microalgae: a green source of renewable H_2. Trends Biotechnol 18:506–511

Ghirardi ML, King PW, Posewitz MC, Maness PC, Fedorov A, Kim K, Cohen J, Schulten K, Seibert M (2005) Approaches to developing biological H2-photoproducing organisms and processes. Biochem Soc Trans 33:70–72

Ghirardi ML, Posewitz MC, Maness PC, Dubini A, Yu J, Seibert M (2007) Hydrogenases and hydrogen photoproduction in oxygenic photosynthetic organisms. Annu Rev Plant Physiol 58:71–91

Godman JE, Molnár A, Baulcombe DC, Balk J (2010) RNA silencing of hydrogenase (–like) genes and investigation of their physiological roles in the green alga chlamydomonas reinhardtii. Biochem J 431:345–351

Gonzalez-Ballester D, Pollock SV, Pootakham W, Grossman AR (2008) The central role of a SNRK2 kinase in sulfur deprivation responses. Plant Physiol 147:216–227

Gonzalez-Ballester D, Casero D, Cokus S, Pellegrini M, Merchant SS, Grossman AR (2010) RNA-seq analysis of sulfur-deprived Chlamydomonas cells reveals aspects of acclimation critical for cell survival. Plant Cell 22:2058–2084

Govindjee (2008) In: Govindjee (ed) Advances in photosynthesis and respiration, vol 126. Springer, Dordrecht

Graves DA, Spradlin GM, Greenbaum E (1990) Effect of oxygen on photoautotrophic and heterotrophic growth of Chlamydomonas reinhardtii in an anoxic atmosphere. Photochem Photobiol 52(3):585–590

Greenbaum E (1980) Simultaneous photoproduction of hydrogen and oxygen by photosynthesis. Biotechnol Bioeng Symp 10:1–13

Greenbaum E, Guillard RRL, Sunda WG (1983) Hydrogen and oxygen photoproduction by marine algae. Photochem Photobiol 37:649–655

Grewe S, Ballottari M, Alcocer M, D'Andrea C, Blifernez-Klassen O, Hankamer B, Mussgnug JH, Bassi R, Kruse O (2014) Light-Harvesting complex protein LHCBM9 is critical for photosystem II activity and hydrogen production in *Chlamydomonas reinhardtii*. Plant Cell 26:1598–1611

Hallenbeck PC (2011) Microbial paths to renewable hydrogen production. Biofuels 2:285–302

Hallenbeck PC (2012) Hydrogen production by cyanobacteria. In: Hallenbeck PC (ed) Microbial technologies in advanced biofuels production. Springer, New York, pp 15–28

Hallenbeck P, Abo-Hashesh M, Ghoshet D (2012) Strategies for improving biological hydrogen production. Bioresour Technol 110:1–9

Happe T, Kaminski A (2002) Differential regulation of the Fe-hydrogenase during anaerobic adaptation in the green alga *Chlamydomonas reinh*ardtii. Eur J Biochem 269:1022–1032

Harris EH, Witman GB (2008) The Chlamydomonas sourcebook, 2nd edn. Elsevier, Boston

Hase E, Morimura Y, Mihara S, Tamiya H (1958) The role of sulfur in the cell division of *Chlorella*. Arch Mikrobiol 31:87–95

He M, Li L, Zhang L, Liu J (2012) The enhancement of hydrogen photoproduction in *Chlorella protothecoides* exposed to nitrogen limitation and sulfur deprivation. Int J Hydrog Energy 37:16903–16915

Healey FP (1970) The mechanism of hydrogen evolution by *Chlamydomonas moewusii*. Plant Physiol 45:153–159

Hemschemeier A, Fouchard S, Cournac L, Peltier G, Happe T (2008a) Hydrogen production by *Chlamydomonas reinhardtii*: an elaborate interplay of electron sources and sinks. Planta 227:397–407. doi:10.1007/s00425-007-0626-8

Hemschemeier A, Fouchard S, Cournac L, Peltier G, Happe T (2008b) Hydrogen production by *Chlamydomonas reinhardtii*: an elaborate interplay of electron sources and sinks. Planta 227:397–407

Hwang J-H, Kim H-C, Choi J-A, Abou-Shanab R, Dempsey BA, Regan JM, Kim JR, Song H, Nam I-H, Kim S-N (2014) Photoautotrophic hydrogen production by eukaryotic microalgae under aerobic conditions. Nat Commun 5:60–67

Jacob J, Lawlor DW (1993) In vivo photosynthetic electron transport does not limit photosynthetic capacity in phosphate deficient sunflower and maize leaves. Plant Cell Environ 16:785e95

Kosourov S, Seibert M (2009) Hydrogen photoproduction by nutrient-deprived *Chlamydomonas reinhardtii* cells immobilized within thin alginate films under aerobic and anaerobic conditions. Biotechnol Bioeng 102:50–58

Kosourov S, Tsygankov A, Seibert M, Ghirardi ML (2002) Sustained hydrogen photoproduction by *Chlamydomonas reinhardtii*: effects of culture parameters. Biotechnol Bioeng 78:731–740

Kosourov S, Seibert M, Ghirardi ML (2003) Effects of extracellular pH on the metabolic pathways in sulfur deprived, H2-producing *Chlamydomonas reinhardtii* cultures. Plant Cell Physiol 44:146–155

Kosourov S, Ghirardi M, Seibert M (2011) A truncated antenna mutant of Chlamydomonas reinhardtii can produce more hydrogen than the parental strain. Int J Hydrog Energy 36(3):2044–2048

Kosourov S, Batyrova K, Petushkova K, Tsygankov A, Ghirardi ML, Seibert M (2012) Maximizing the hydrogen photoproduction yields in chlamydomonas reinhardtii cultures: the effect of the H_2 partial pressure. Int J Hydrog Energy 37:8850–8858

Kreuzberg K, Klöck G, Grobheiser D (1987) Subcellular distribution of pyruvate-degrading enzymes in *Chlamydomonas reinhardtii* studied by an improved protoplast fractionation procedure. Physiol Plant 69:481–488

Kruse O, Rupprecht J, Bader KP, Thomas-Hall S, Schenk PM, Finazzi G, Hankamer B (2005) Improved photobiological H 2 production in engineered green algal cells. J Biol Chem 280:34170–34177

Laurinavichene TL, Tolstygina I, Galiulina RR, Ghirardi M, Seibert M, Tsygankov AA (2002) Dilution methods to deprive *Chlamydomonas reinhardtii* cultures of sulfur for subsequent hydrogen photoproduction. Int J Hydrog Energy 27:1245–1249

Laurinavichene T, Fedorov A, Ghirardi M, Seibert M, Tsygankov A (2006) Demonstration of sustained hydrogen photoproduction by immobilized, sulfur deprived *Chlamydomonas reinhardtii* cells. Int J Hydrobiol Energy 31:659–667

Lee JW (2013) Designer transgenic algae for photobiological production of hydrogen from water. In: Lee JW (ed) Advanced biofuels and bioproducts, I, Chapter 20. Springer, New York, pp 371–404

Leite GB, Abdelaziz AEM, Hallenbeck PC (2013) Algal biofuels: challenges and opportunities. Bioresour Technol 145:134–141

Lemeille S, Rochaix J-D (2010) State transitions at the crossroad of thylakoid signalling pathways. Photosynth Res 106:33–46

Liebgott P, Leroux F, Burlat B, Dementin S, Baffert C, Lautier T, Fourmond V, Ceccaldi P, Cavazza C, Meynial-Salles I, Soucaille P, Fontecilla-Camps J, Guigliarelli B, Bertrand P, Rousset M, Leger C (2010) Relating diffusion along the substrate tunnel and oxygen sensitivity in hydrogenase. Nat Chem Biol 6(1):63–70

Lin H-D, Liu B-H, Kuo T-T, Tsai H-C, Feng T-Y, Huang C-C, Chien L-F (2013) Knockdown of PsbO leads to induction of HydA and production of photobiological H 2 in the green alga Chlorella sp. DT. Bioresour Technol 143:154–162

Long H, King P, Ghirardi M, Kim K (2009) Hydrogenase/ferredoxin charge-transfer complexes: effect of hydrogenase mutations on the complex association. J Phys Chem A 113(16):4060–4067

Malnoe A, Wang F, Girard-Bascou J, Wollman F-A, de Vitry C (2014) Thylakoid FTSH protease contributes to photosystem ii and cytochrome b(6)f remodeling in Chlamydomonas reinhardtii under stress conditions. Plant Cell 26:373–390

Marin-Navarro J, Esquivel M, Moreno J (2010) Hydrogen production by *Chlamydomonas reinhardtii* revisited: rubisco as a biotechnological target. World J Microbiol Biotechnol 26:1785–1793

Markov SA, Eivazova ER, Greenwood J (2006) Photostimulation of H_2 production in the green alga Chlamydomonas reinhardtii upon photoinhibition of its O_2-evolving system. Int J Hydrog Energy 31:1314–1317

Maroti G, Tong Y, Yooseph S, Baden-Tillson H, Smith HO, Kovacs KL, Frazier M, Venter JC, Xu Q (2009) Discovery of [NiFe] hydrogenase genes in metagenomic DNA: cloning and heterologous expression in Thiocapsa roseopersicina. Appl Environ Microbiol 75:5821–5830

Matthew T, Zhou WX, Rupprecht J, Lim L, Thomas-Hall SR, Doebbe A, Kruse O, Hankamer B, Marx UC, Smith SM, Schenk PM (2009) The metabolome of *Chlamydomonas reinhardtii* following induction of anaerobic H_2 production by sulfur depletion. J Biol Chem 284:23415–23425

Melis A (2002) Green algal hydrogen production: progress, challenges and prospects. Int J Hydrog Energy 27:1217–1228

Melis A, Happe T (2001) Hydrogen production. Green algae as a source of energy. Plant Physiol 127:740–748

Melis A, Happe T (2004) Trails of green alga hydrogen research: from Hans Gaffron to new frontiers. Photosynth Res 80:401–409

Melis A, Mullineaux CW, Allen JF (1989) Acclimation of the photosynthetic apparatus to photosystem I or photosystem II light: evidence from quantum yield measurement and fluorescence spec- troscopy of cyanobacterial cells. Z Naturforsch Teil C 44:109–118

Melis A, Zhang L, Forestier M, Ghirardi ML, Seibert M (2000) Sustained photobiological hydrogen gas production upon reversible inactivation of oxygen evolution in the green alga *Chlamydomonas reinhardtii*. Plant Physiol 122:127–136

Melis A, Seibert M, Ghirardi ML (2007) Hydrogen fuel production by transgenic microalgae. Adv Exp Med Biol 616:108–121

Meuser JE, Ananyev G, Wittig LE, Kosourov S, Ghirardi ML, Seibert M, Dismukes GC, Posewitz MC (2009) Phenotypic diversity of hydrogen production in chlorophycean algae reflects distinct anaerobic metabolisms. J Biotechnol 142:21–30

Meuser JE, D'adamo S, Jinkerson RE, Mus F, Yang W, Ghirardi ML, Seibert M, Grossman AR, Posewitz MC (2012) Genetic disruption of both Chlamydomonas reinhardtii [FeFe]-hydrogenases: insight into the role of HYDA2 in H(2) production. Biochem Biophys Res Commun 417:704–709

Meyer J (2007) [FeFe] hydrogenases and their evolution: a genomic perspective. Cell Mol Life Sci 64:1063–1084

Mulder D, Boyd E, Sarma R, Lange R, Endrizzi J, Broderick J, Peters J (2010) Stepwise [FeFe]-hydrogenase H-cluster assembly revealed in the structure of HydA(DeltaEFG). Nature 465(7295):248–251

Müller M (2003) Energy metabolism. Part 1: anaerobic protozoa. In: Marr JJ, Nilsen TW, Komunieck R (eds) Molecular medical parasitology. Academic, Amsterdam, pp 125–139

Mus F, Cournac L, Cardettini V, Caruana A, Peltier G (2005) Inhibitor studies on non-photochemical plastoquinone reduction and H_2 photoproduction in *Chlamydomonas reinhardtii*. Biochim Biophys Acta 1708:322–332

Mus F, Dubini A, Seibert M, Posewitz MC, Grossman AR (2007) Anaerobic acclimation in *Chlamydomonas reinhardtii*: anoxic gene expression, hydrogenase induction, and metabolic pathways. J Biol Chem 282:25475–25486

Mussgnug JH, Thomas-Hall S, Rupprecht J, Foo A, Klassen V, McDowall A, Schenk PM, Kruse O, Hankamer B (2007) Engineering photosynthetic light capture: Impacts on improved solar energy to biomass conversion. Plant Biotechnol J 5:802–814

Nagy LE, Meuser JE, Plummer S et al (2007) Application of gene-shuffling for the rapid generation of novel [FeFe]-hydrogenase libraries. Biotechnol Lett 29:421–430

Oey M, Ross IL, Stephens E, Steinbeck J, Wolf J, Radzun KA, Kügler J, Ringsmuth AK, Kruse O, Hankamer B (2013) RNAi knock-down of LHCBM1, 2 and 3 increases photosynthetic H2 production efficiency of the green alga Chlamydomonas reinhardtii. PLoS One 8:e61375

Oncel S, Sukan FV (2011) Effect of light intensity and the light: dark cycles on the long term hydrogen production by batch cultures. Biomass Bioenergy 35:1066–1074

Philipps G, Happe T, Hemschemeier A (2012) Nitrogen deprivation results in photosynthetic hydrogen production in *Chlamydomonas reinhardtii*. Planta 235:729–745

Pinto T, Malcata F, Arrabaça J, Silva J, Spreitzer R, Esquível M (2013) Rubisco mutants of Chlamydomonas reinhardtii enhance photosynthetic hydrogen production. Appl Microbiol Biotechnol 97:5635–5643

Polle JEW, Kanakagiri S, Jin E, Masuda T, Melis A (2002) Truncated chlorophyll antenna size of the photosystems – a practical method to improve microalgal productivity and hydrogen production in mass culture. Int J Hydrog Energy 27:1257–1264

Randt C, Senger H (1985) Participation of the two photosystems in light dependent hydrogen evolution in *Scenedesmus obliquus*. Photochem Photobiol 42:553–557

Ravenel J, Peltier G, Havaux M (1994) The cyclic electron pathways around photosystem I in Chlamydomonas reinhardtii as determined in vivo by photoacoustic measurements of energy storage. Planta 193:251–259

Ruhle T, Hemschemeier A, Melis A, Happe T (2008) A novel screening protocol for the isolation of hydrogen producing *Chlamydomonas reinhardtii* strains. BMC Plant Biol 8:107

Rusch DB, Halpern AL, Sutton G, Heidelberg KB, Williamson S, Yooseph S, Wu D, Eisen JA, Hoffman JM, Remington K, Beeson K, Tran B et al (2007) The sorcerer II global ocean sampling expedition: northwest Atlantic through eastern tropical pacific. PLoS Biol 5:e77

Satagopan S, Spreitzer RJ (2004) Substitutions at the Asp-473 latch residue of *Chlamydomonas* ribulose bisphosphate carboxylase/oxygenase cause decreases in carboxylation efficiency and CO_2/O_2 specificity. J Biol Chem 279:14240–14244

Seibert M (2007) Hydrogenases and hydrogen photoproduction in oxygenic photosynthetic organisms. Annu Rev Plant Physiol 58:71–91

Seibert M, Flynn T, Benson D, Ghirardi M (1998) Development of selection and screening procedures for rapid identification of H2-producing algal mutants with increased O_2 tolerance. In: Zaborsky OR (ed) Biohydrogen. Plenum, New York, pp 227–234

Siaut M, Cuine S, Cagnon C, Fessler B, Nguyen M, Carrier P, Beyly A, Beisson F, Triantaphylides C, Li-Beisson Y, Peltier G (2011) Oil accumulation in the model green alga *Chlamydomonas reinhardtii*: Characterization, variability between common laboratory strains and relationship with starch reserves. BMC Biotechnol 11:7

Spalding MH (2008) Microalgal carbon-dioxide-concentrating mechanisms: *Chlamydomonas* inorganic carbon transporters. J Exp Bot 59:1463–1473

Srirangan K, Pyne ME, Chou CP (2011) Biochemical and genetic engineering strategies to enhance hydrogen production in photosynthetic algae and cyanobacteria. Bioresour Technol 102:8589–8604

Stapleton J, Swartz J (2010) Development of an in vitro compartmentalization screen for high-throughput directed evolution of [FeFe] hydrogenases. PLoS One 5(12):e15275

Stripp ST, Goldet G, Brandmayr C, Sanganas O, Vincent KA, Haumann M, Armstrong FA, Happe T (2009) How oxygen attacks [FeFe] hydrogenases from photosynthetic organisms. Proc Natl Acad Sci U S A 106:17331–17336

Stuart TS, Gaffron H (1972) The mechanism of hydrogen photoproduction by several algae. Planta 106:101–112

Surzycki R, Cournac L, Peltier G, Rochaix J-D (2007) Potential for hydrogen production with inducible chloroplast gene expression in *Chlamydomonas*. Proc Natl Acad Sci U S A 104:17548–17553

Timmins M, Zhou W, Rupprecht J, Lim L, Thomas-Hall SR, Doebbe A, Kruse O, Hankamer B, Marx UC, Smith SM, Schenk PM (2009) The metabolome of *Chlamydomonas reinhardtii* following induction of anaerobic H_2 production by sulfur depletion. J Biol Chem 284:35996

Torzillo G, Scoma A, Faraloni C, Ena A, Johanningmeier U (2009) Increased hydrogen photoproduction by means of a sulfur-deprived Chlamydomonas reinhardtii D1 protein mutant. Int J Hydrog Energy 34:4529–4536

Tsygankov AA (2012) Hydrogen production: light-driven processes – green algae. In: Hallenbeck PC (ed) Microbial technologies in advanced biofuels production. Springer, New York

Tsygankov A, Kosourov S, Seibert M, Ghirardi M (2002) Hydrogen photoproduction under continuous illumination by sulfur-deprived, synchronous Chlamydomonas reinhardtii cultures. Int J Hydrog Energy 27:1239–1244

Wagner AF, Frey M, Neugebauer FA, Schafer W, Knappe J (1992) The free radical in pyruvate formate-lyase is located on glycine-734. Proc Natl Acad Sci U S A 89:996–1000

Warnecke F, Luginbuhl P, Ivanova N, Ghassemian M, Richardson TH, Stege JT, Cayouette M, McHardy AC, Djordjevic G, Aboushadi N, Sorek R, Tringe SG et al (2007) Metagenomic and functional analysis of hindgut microbiota of a wood-feeding higher termite. Nature 450:560–565

Weaver PF, Lien S, Seibert M (1980) Photobiological production of hydrogen. Solar Energy 24:3–45

Winkler M, Kuhlgert S, Hippler M, Happe T (2009) Characterization of the key step for light-driven hydrogen evolution in green algae. J Biol Chem 284:36620–36627

Wu S, Xu L, Huang R, Wang Q (2011) Improved biohydrogen production with an expression of codon-optimized hemH and lba genes in the chloroplast of Chlamydomonas reinhardtii. Bioresour Technol 102:2610–2616

Wykoff DD, Davies JP, Melis A, Grossman AR (1998) The regulation of photosynthetic electron transport during nutrient deprivation in *Chlamydomonas reinhardtii*. Plant Physiol 117:129–139

Wykoff DD, Grossman AR, Weeks DP, Usuda H, Shimogawara K (1999) Psr1, a nuclear localized protein that regulates phosphorus metabolism in Chlamydomonas. Proc Natl Acad Sci U S A 96:15336–15341

Xu F-Q, Ma W-M, Zhu X-G (2011) Introducing pyruvate oxidase into the chloroplast of Chlamydomonas reinhardtii increases oxygen consumption and promotes hydrogen production. Int J Hydrog Energy 36:10648–10654

Yanyushin M (1982) Hydrogenase activation and hydrogen photoevolution in a synchronous culture of *Chlamydomonas reinhardtii* during anaerobic adaptation under light. Sov Plant Physiol 29:863–869

Zhang L, Happe T, Melis A (2002) Biochemical and morphological characterization of sulfur-deprived and H2 producing *Chlamydomonas reinhardtii* (green alga). Planta 214:552–561

Biohydrogen Production from Microalgae: An Enzyme Perspective

9

Ayse Kose and Suphi S. Oncel

Abstract

Biological hydrogen production from microbial origins especially from microalgal species has been an attractive source for the world to compensate the extreme fuel consumption of civilized population. The attribution of biohydrogen production is thought to be effective on macroscale considering global energy market; however there exist lots of biochemical reactions in a single cell to produce hydrogen. From this point of view, the aim of this chapter is to highlight the enzymes responsible for biohydrogen production in microalgae and to discuss enzymatic reactions focusing on cell dynamics, metabolism, structure, function, and challenges regarding sustainable biohydrogen production.

9.1 Introduction

Biologically produced hydrogen, biohydrogen, is actually a metabolic waste to balance redox potential of individual cells in order to sustain metabolic processes in which H^+ is utilized as final electron acceptor under anoxic conditions (Meyer 2007). At the end of a chain of serial reactions, hydrogen gas (H_2) is released from cytosol to extracellular environment rather than utilized in any metabolic reaction within the cell. The biotechnological importance of this so-called metabolic reaction in photosynthetic organisms as cyanobacteria and microalgae is recognized later after the first studies covering algal hydrogen production metabolism (Gaffron and Rubin 1942; Wykoff et al. 1998; Melis et al. 2000; Happe and Kaminski 2002; Posewitz et al. 2004; Melis 2007; Ghirardi and Mohanty 2010; Oncel 2013). Biohydrogen production is an evolutionary regulation of cellular metabolism to survive under anoxic conditions which can be seen in diverse groups of microorganisms such as bacteria, archaea, cyanobacteria, and microalgae (Melis et al. 2000; Ghirardi et al. 2006).

Reactions for the hydrogen gas production are catalyzed only with the activation of certain hydrogenase enzymes under anaerobic conditions. The existence of hydrogenase enzymes is thought to be an important asset for biotechnological studies, and there are diverse topics to

A. Kose • S.S. Oncel (✉)
Department of Bioengineering, Ege University, Izmir, Turkey
e-mail: suphi.oncel@ege.edu.tr

A. Singh, D. Rathore (eds.), *Biohydrogen Production: Sustainability of Current Technology and Future Perspective*, DOI 10.1007/978-81-322-3577-4_9

clarify the existence, evolutionary development, function, and structure of hydrogenase enzymes regarding the species, strains, and environmental conditions (Miyake et al. 1999; Nicolet et al. 2000, 2002; Marr et al. 2001; Cavazza et al. 2008; Sun et al. 2005; Meyer 2007; Rosenberg et al. 2008; Shafaat et al. 2013; Ginovska-Pangovska et al. 2014). The first hydrogen production studies from microalgae have been evaluated under dark conditions, based on the information that dark metabolism triggers hydrogen production (considering the bacterial biohydrogen production as a model) (Gaffron and Rubin 1942; Gibbs et al. 1986; Melis and Happe 2001); however under illuminated conditions the change in hydrogen production efficiency demonstrated that light is also an important parameter (Greenbaum 1982; Ghirardi et al. 2000; Melis et al. 2000; Happe and Kaminski 2002; Laurinavichene et al. 2004; Kima et al. 2006; Saleem et al. 2012; Oncel and Kose 2014; Oncel et al. 2014).

Algal hydrogen production is not a new concept; it has been the topic of interest since the pioneering studies of Gaffron and Rubin and the following researchers who provided a strong database and know-how for both microalgal biohydrogen production and hydrogenase enzymes (Gaffron and Rubin 1942; Melis et al. 2000; Lindblad et al. 2002; Kojima and Lin 2004; Guan et al. 2004; Posewitz et al. 2004; Oncel 2013; Lubitz et al. 2014; Zhang et al. 2014). However one of the most outstanding studies that has been presented by Melis et al. (2000) resulted in the sustained biohydrogen production using *Chlamydomonas reinhardtii* as model microalgae under sulfur-deprived conditions. Results encouraged the biotechnological point of view and algal hydrogen studies have been accelerated (Torzillo et al. 2014).

The discovery of hydrogenase enzymes has also a noteworthy importance in the hydrogen researches that was first reported by Stephenson and Stickland in 1931 using *E. coli* as the model organism (Stephenson and Stickland 1931). They claimed that these microorganisms can produce hydrogen under anaerobic condition. For a long time, bacteria have been thought to be the only source of hydrogenase enzymes; however algal hydrogen production proved that certain microalgal species also have active hydrogenase enzymes (Florin et al. 2001; Posewitz et al. 2004). Rapid developments in screening of the potential strains, understanding hydrogen metabolism, and focusing on hydrogenase enzymes as hydrogen-producing machines increased the attention to algal hydrogen production (Vignais et al. 2001; Nicolet et al. 2002; Winkler et al. 2002a, b; Forestier et al. 2003; Posewitz et al. 2004; Meyer 2007; Lubitz et al. 2014).

The aim of this chapter is to give information about microalgal hydrogen production and state-of-the-art, hydrogen-catalyzing enzymes found in microalgae (green algae and cyanobacteria) and model microalgal species regarding genetic engineering and transcriptomic and metabolomic studies, giving ideas about biotechnological applications and future studies for a sustainable biohydrogen production technology under the scope of enzymatic regulation at cellular basis.

9.2 State of the Art

The conceptual development of photobiological hydrogen production is first introduced by Gaffron and Rubin in 1942. After the study of Melis and coworkers which finally managed to sustain long-term biohydrogen production, several microalgal species have been identified with hydrogen production capabilities and hydrogenase activity (Gaffron 1944; Kessler 1962; Healey 1970; Melis et al. 2000; Shepard et al. 2011). Today it can clearly be stated that microalgal biohydrogen production is a relatively young area in comparison with other renewable energy sources (Hallenbeck 2011). Even if microalgal biohydrogen production is promising, there exist several bottlenecks and challenges from micro- to macroscale regarding yield, efficiency, scale-up, bioreactor/photobioreactor design, and most importantly finding appropriate species to sustain biohydrogen production considering oxygen sensitivity as a main challenge (Posewitz et al. 2004; Mathews and Wang 2009; Faraloni and Torzillo

2010; Srirangan et al. 2011; Oncel 2013; Oncel and Kose 2014).

The mechanism of algal hydrogen production was a mystery for a long time because the first studies are designed under dark anaerobic conditions which exogenous substrates are utilized for electron donors (Kessler 1974). The role of light supply for hydrogen production has been revealed later according to various studies which all resulted that light acts as a driving force (biophotolysis) for hydrogen production directly affecting the overall efficiency (Melis and Happe 2001; Melis 2007). Sustainable biohydrogen production can only be achieved under anaerobic conditions due to oxygen sensitivity of hydrogenase enzymes (Vignais et al. 2001; Posewitz et al. 2004; Meyer 2007; Heinekey 2009; Lubitz et al. 2014).

Sulfur, being one of the key nutrients in algal nutrition, has a major role in the regulation and operation of photosynthesis. When sulfur is depleted from nutrient medium, the photosynthetic activity of individual cells starts to decrease dramatically under aerobic conditions. Photosystem II (PSII) activity decreased but respiration still remains (Wykoff et al. 1998). These findings construct the basis to understand hydrogen production mechanism in algal cells. Sulfur deprivation is used by Melis and coworkers which finally achieved sustained long-long term biohydrogen production (Melis et al. 2000). They used "two-stage protocol," separation of aerobic phase (biomass production) from anaerobic phase (biohydrogen production) (Faraloni and Torzillo 2010; Scoma et al. 2012).

In this regard, biohydrogen production can be divided into six main steps: (i) utilization of existing oxygen within the sealed vessel, (ii) PSII activity decrease due to the sulfur starvation and adaptation of cells to anaerobic environment, (iii) induction of hydrogenase enzyme-encoding genes and synthesis of hydrogenase enzymes, (iv) logarithmic hydrogen production, (v) decrease in the hydrogen production rate due to inhibition of metabolic oxygen generation, and (vi) termination (Melis et al. 2000; Melis and Happe 2001; Melis 2007; Tsygankov et al. 2002; Kosourov et al. 2007).

Why oxygen containing environments are being a threat for biohydrogen production?: Oxygen sensitivity of hydrogenase enzymes

A rapid anaerobiosis triggers cells to synthesize hydrogenase enzymes ([FeFe] hydrogenases in microalgae; [NiFe] hydrogenases in cyanobacteria) using free protons as final electron acceptor (Posewitz et al. 2004; Meyer 2007). The basic reaction of hydrogen production ($2H^{+}+2e^{-} \rightarrow H_2$) occurs at the active site of hydrogenase enzyme which is buried deep inside the 3D mosaic (Lubitz et al. 2014). Nevertheless enzymatic regulation of biohydrogen production is limited even by trace amount of oxygen (Ghirardi et al. 2000). It is known that oxygen can travel through the micro-channels of enzyme to active site and blocks the activation of hydrogenase enzyme (Srirangan et al. 2011). Not only mature enzyme but also maturation proteins are inhibited by oxygen which is a transcriptional inhibitor in gene level (Happe et al. 1994; Happe and Kaminski 2002). The ultimate goal is to obtain oxygen-resistant species with high yield of single-step biohydrogen production.

9.3 Biophotolysis

Biophotolysis is defined as the light-driven splitting of water or organic compounds to its electrons and oxygen, which is found in photosynthetic microorganisms: as it is the topic of this chapter, microalgae and cyanobacteria (Melis et al. 2000; Oh et al. 2011; Srirangan et al. 2011; Oncel 2013). Biophotolysis is the initial step of photosynthesis resulting in the electron flow through the electron transport chain with the assistance of several accessory proteins in the chloroplast membrane or photosynthetic chain in the cytoplasm (Masojidek et al. 2013). The process is unique to microalgal species and thought to be only sustainable biohydrogen production metabolism regarding microbiological species (Carrieri et al. 2011).

The photosynthetic reactions in microalgae occur in the chloroplasts; however in cyanobacteria the reactions take place in the cytoplasm

(Srirangan et al. 2011). Biophotolysis also varies in between microalgae and cyanobacteria, and even the reaction steps are well conserved among species. Microalgae possess direct biophotolysis; meanwhile cyanobacteria conduct indirect pathway to shuttle electrons to hydrogenase enzymes (Benemann 1997).

9.3.1 Direct Biophotolysis

Direct biophotolysis is the source for electrons finally transferred to hydrogenase enzymes to be used for reduction of protons (H^+) to form H_2 in gaseous form (Melis and Happe 2001). The reaction is free from carbon metabolism; only water is involved in the process. Solar energy is used in natural systems; however artificial illumination is another alternative for bioprocessing of biohydrogen. Light-harvesting complex (LHC), PSII, and electron transport chain (ETC) are the main structural components to sustain direct biophotolysis (Winkler et al. 2009; Oh et al. 2011). Besides the role of PSII, photosynthetic ferredoxin (PetFd) and PQ pool are vital for hydrogen production (Melis et al. 2000; Melis 2007; Tóth et al. 2007; Faraloni and Torzillo 2010). It would be better to emphasize the importance of solar energy conversion in microalgae while discussing the role of direct biophotolysis in microalgal biohydrogen production. The electron flow through the hydrogenase enzymes is the indicator for the efficiency of hydrogen production activity under anaerobic conditions. With water photolysis a sink of protons (H^+) is occurred with a pool of free electrons. Benemann suggested that 22 % of light energy is converted into chemical energy; however the actual values are lower to 1–2 % (Benemann 1997; Torzillo and Seibert 2013). In direct biophotolysis , one of the main challenges is light conversion efficiency (solar energy to hydrogen). Light is absorbed from LHC and in PSII reaction center water is split into hydrogen and oxygen ($H_2O \rightarrow 2H^+ + O_2$). The activity of PSII reaction center is crucial to direct biophotolysis (Wykoff et al. 1998) which is decreased at sulfur-deprived conditions. When photosynthetic conversion efficiency which is related to the antenna size of the cells and the capacity of the chloroplasts (Melis and Happe 2001) is increased, the activity of PSII reaction center is also increased (Polle et al. 2002; Melis 2009).

9.3.2 Indirect Biophotolysis

Cyanobacteria are prokaryotic organisms which are capable of producing hydrogen via several diverse pathways using nitrogenases (N_2 fixation) and hydrogenases ([NiFe] type hydrogenase enzymes) combining with photosynthesis (CO_2 fixation accompanied with O_2 generation) (Benemann 1997; Carrieri et al. 2011; Lubitz et al. 2014). H_2O acts as an indirect electron donor in which photosynthesis reactions use H_2O-driven electrons to fix CO_2 into organic carbon bodies which is the main concept of photosynthesis. Rather than electron transfer, the fermentation efficiency is crucial for this process (Hallanbeck et al. 2012). The oxygen sensitivity is not a serious problem as it is in direct biophotolysis (Oh et al. 2011). However, because it uses endogenous carbon sources, the cell maintenance during anaerobiosis is crucial. Besides the anabolism of culture media nutrients into complex organic carbon sources should be mentioned. Benemann suggested a two-stage approach to indirect biophotolysis (1997) which is quite similar to the two-stage protocol presented by Melis et al. (2000). The suggested process covers an aerobic biomass production stages in open ponds under illuminated conditions to accumulate endogenous substrates later to be utilized via indirect biophotolysis and transfer of the cultures to a dark and anaerobic environment to activate hydrogenase/nitrogenase enzymes to increase the electron flow (Benemann 1997).

9.4 Photobiological Hydrogen Production

Microalgae use light energy either natural or artificial to fix CO_2 into complex carbon bodies to provide energy to sustain metabolic reactions; the overall process is called *photosynthesis*, a light-driven carbon fixation in the thylakoid membranes of the chloroplast (Masojidek et al. 2013). Photosynthesis is a series of redox reactions which shuttles electrons to the final electron acceptor, oxygen. Thus it is also called as oxygenic photosynthesis (McKinlay and Harwood 2010). Photosynthesis in prokaryotes occurs specialized structures in cytosol as a representative of photosynthetic membrane (Srirangan et al. 2011). Oxygenic photosynthesis has light and dark reactions. Light reactions take place in thylakoid membranes which is composed of light-harvesting antenna (LHA), PSII, cytochrome *b6/f* complex, PQ, photosynthetic ferredoxin (PetFd), and ATP synthase which are also components for photobiological hydrogen production except ATP synthase. Under oxygenic conditions, Fd transfers electrons to RUBISCO enzyme in Calvin-Benson cycle to synthesize complex organic molecules. However the scenario changes when metabolism is reorganized to generate hydrogen (Happe and Hemsheimer 2011).

Under anaerobic conditions, Calvin cycle of oxygenic photosynthesis is downregulated; Fd acts as redox partner of the active site of hydrogenase enzymes and donates electrons to the H-cluster (Florin et al. 2001; Melis 2007) rather than RUBISCO and cycles between reduced and oxidized form shuttling electrons to [FeFe] hydrogenases (Forestier et al. 2003; Sybirna et al. 2013). Simultaneous formation of hydrogen and oxygen under aerobic conditions results in short-term hydrogen sensing (max 2 min) because of the strong suppression effect of oxygen to hydrogenase enzyme-encoding genes (*hyd*A1 and *hyd*A2) (Greenbaum et al. 1983; Ghirardi et al. 1997).

Light-dependent water splitting in PSII reaction center is the first step of hydrogen production. Water is the main electron and proton donor for algal hydrogen evolution (Bishop and Gaffron 1963; Melis 2007). Released electrons are transferred to PSII and Fd, respectively (Doebbe et al. 2007).

The role of photosystems in algal hydrogen production reveals that PSII-deficient strains cannot synthesize hydrogen; however deficiencies in PSI do not affect the release of hydrogen in gas form (Hemschemeier et al. 2008). Even starch and other endogenous carbon sources drive energy and help cells to survive under anaerobic conditions; direct biophotolysis is proved to be essential to sustain hydrogen evolution (Chochois et al. 2009). These results show that algal hydrogen production is dependent on PSII which is a light-mediated reaction center consisted of D1 and D2 heterodimer proteins (Faraloni and Torzillo 2010).

Some arguments have been proposed about the relation between electron sources and hydrogen production regarding gene expression (Melis et al. 2000; Mus et al. 2005; Hemschemeier et al. 2008). However studies revealed that PSII-deficient mutant strains of *C. reinhardtii* cannot produce hydrogen despite *in vitro* hydrogenase activity being observed (Hemschemeier et al. 2008). Even if PSII reaction center is the main electron source for hydrogen production (Antal et al. 2003), under aerobic conditions the active PSI center regulates the action of Calvin cycle to produce starch (Zhang et al. 2002; Melis 2007).

While questioning the role of photosynthetic pathways on hydrogen production, the hydrogenase enzyme synthesis and activity should be taken into consideration too. The role of PSII has been displayed well, and when starch production and degradation are blocked, cells sustained hydrogen production; however hydrogen production yield of mutant strain with modified antenna size and lack of cyclic electron flow metabolism (*stm*6) was lower than wild-type strain (Kruse et al. 2005). As proposed by Melis et al., cells utilize respiration metabolism in order to survive under anaerobic and S-free media conditions. During this time a rapid starch accumulation is observed which is later used as energy source and an electron sink for hydrogen production also providing protons to active enzyme (Melis et al. 2000; Melis 2007; Hemschemeier and Happe 2011).

The accumulation and transfer of electrons to active hydrogenase enzyme is the vital part of the overall hydrogen production process (Antal et al. 2011, 2003; Doebbe et al. 2007). Cyclic electron transport around PSI and the size of the PQ pool which is a reservoir that accumulate and transport electrons to electron transport chains in the chloroplast (Finazzi et al. 1999) are the limiting steps of enzyme activity when anaerobiosis is well established and mature hydrogenase enzymes are observed (Winkler et al. 2013). During this process endogenous substrates and supportive medium components as acetate are the secondary sources regulating electron flow through ferredoxin (Melis 2007).

The case study of hydrogen metabolism: Sulfur starved microalgal metabolism during anaerobiosis and gene regulation of hydrogenase enzyme synthesis

Sulfur as one of the main components in algal metabolism has a crucial role in the regulation of PSII photodamage repair as mentioned above under the title of direct biophotolysis. However the role of S starvation and the relation with the hydrogenase enzyme activity is also another point to be addressed. Catabolic and anabolic regulation during anaerobic sulfur-starved cultures displays unique features of metabolic reorganization of microalgal hydrogen production. Responses of cells to sulfur starvation have a critical role to be used in genomic and metabolomic studies as well as artificial hydrogenase systems regarding cell-free generation of biohydrogen.

Protons as final electron acceptors regulate the redox balance of cells under anaerobic conditions which reflects as a stress to microalgal metabolism (Antal et al. 2011). Under anaerobic, sulfur (S)-depleted conditions, cells sustain their viability for a few days. Besides physical changes in cell morphology, cells start to get larger in volume during early hours of starvation. Under sustained anaerobic conditions, cells start to disrupt and lose viability which is also observed as a color change from green to yellow. Under S starvation, the PSII photorespiration is downregulated, and partially on account of blocking the S-containing amino acid synthesis, inhibition in the D1 protein coding occurs (Faraloni and Torzillo 2010). Reversible inhibition in PSII triggers the rapid decrease in the photosynthetic activity and with respiration, the oxygen is consumed in the culture chamber. During this short time interval (~24 h), cells started to accumulate starch as an energy reservoir for anaerobiosis (Zhang et al. 2002). Decrease in the photosynthetic efficiency, rapid accumulation of starch, and S starvation trigger the regulation of multiple gene activation, and mainly the hydrogenase maturation enzyme-encoding genes are activated and mature hydrogenase enzyme is catalyzed in the chloroplasts of microalgal cells (Posewitz et al. 2004; Melis 2007; Meyer 2007; Mathews and Wang 2009).

9.5 Hydrogen-Catalyzing Enzymes

Since 1931, the enzymatic studies deal with micro- and macroscale regulations, transcription aspects, and amino acid sequencing besides taxonomic variations which can be seen as the key point to understand how an enzyme works to catalyze hydrogen (Adams 1990; Happe et al. 1994; Meyer 2007; Winkler et al. 2009; Godman et al. 2010; Meuser et al. 2011; Yacoby et al. 2012; Sybirna et al. 2013). One of the main challenges regarding microalgal hydrogenase is the inadequate know-how on microalgal hydrogen production machinery. The structural studies are able to cover the bacterial hydrogenases well using different characterization techniques such as X-ray crystallography, FTIR spectroscopy, and computational modeling of the enzyme structure as well as environmental stress physiology, biochemical organization of enzyme machinery, and genetic engineering studies (Nicolet et al. 2001; Mayfield and Franklin 2005; Peters et al. 2006; Roseboom et al. 2006; Beer et al. 2009; Lubitz et al. 2014).

For a long time, bacterial hydrogenases have been the model for hydrogenase studies (Lubitz et al. 2014). Structure and protein mosaics have been highlighted and can be found in the databases (Adams 1990; Peters et al. 1998; Friedrich et al. 2005; Meyer 2007; Shafaat et al. 2013; Greening and Cook 2014). What we need to know is where to use hydrogenase enzymes and how to trigger them to circumvent main challenges regarding a sustained hydrogen production with *in vivo* or *in vitro* applications.

The hydrogenase enzymes are classified into three groups according to their active site organization and metallic atoms: [NiFe] hydrogenases, [FeFe] hydrogenases, and [Fe] hydrogenases (Kim and Kim 2011; Lubitz et al. 2014). [Fe] hydrogenases are synthesized by methanogenic archaea species. The active site confirmation, electron transport pathway, and molecular size differ from [NiFe] and [FeFe] hydrogenases. Because [Fe] hydrogenases are not found in microalgae and cyanobacteria, this enzyme is not mentioned further in this chapter.

[NiFe] hydrogenases are the most diversified cluster among them (Horch et al. 2012); however [FeFe] hydrogenases display the highest catalytic activity under certain conditions (Volbeda et al. 1995; Meyer 2007; Nicolet et al. 2010; Lindahl 2012). [NiFe] and [FeFe] hydrogenases have common building blocks in mature enzyme structure (CO and CN^- ligands, bimetallic active site, iron atom existence in the active site mosaic); however they exhibit distinct 3D mosaic (Heinekey 2009; Corr and Murphy 2011; Mulder et al. 2011; Lubitz et al. 2014). X-ray crystallography and FTIR studies display that in order to have an active hydrogen-catalyzing enzyme, these ligands are necessary accessories (Happe et al. 1994; Kamp et al. 2008; Lubitz et al. 2014).

9.5.1 Hydrogenases

The universal enzyme responsible for hydrogen catalysis in certain microbial sources is hydrogenase enzyme which catalyzes reversible generation of H_2, diversified according to microorganism, active site confirmation, enzyme structure, molecular weight, and specific enzyme activity (Lee et al. 2010). Hydrogenases are thought to be found only for bacteria and some anaerobic organisms (Horner et al. 2002), but later studies showed that eukaryotic microalgae and some protists also have hydrogenase enzymes (Skjanes et al. 2010). Cyanobacteria are capable of synthesizing both [FeFe] and [NiFe] hydrogenases; however eukaryotic species only have [FeFe] hydrogenases, while cyanobacteria have [NiFe] type (Ludwig et al. 2006). [NiFe] and [FeFe] hydrogenases are distinct enzymes even though they catalyze the same reaction. The structural studies present similarities especially regarding the structure and building of the active sites of the enzymes (Vignais et al. 2001). However the turnover rate and the specific activity of the enzymes vary in between the species (Oh et al. 2011). Crystal structure studies mostly conducted on [FeFe] hydrogenases because [NiFe] hydrogenases are more complex structure and bigger in shape (Casalot and Rousset 2001; Volbeda and Fontecilla-Camps 2005; Greening and Cook 2014)

9.5.1.1 [FeFe] Hydrogenases

The first isolated hydrogenase enzyme was iron-sulfur hydrogenase enzymes, [FeFe] hydrogenases, was found in eubacteria such as *Clostridium* and *Desulfovibrio* species and green microalgae (Peters et al. 1998; Nicolet et al. 1999; Posewitz et al. 2004). According to current databases known, sequenced algal hydrogenase enzymes are monomeric and around 48 kDa in molecular weight with 350 amino acid residues (Meyer 2007) known as the smallest hydrogenase enzyme compared to bacterial hydrogenases (Happe and kaminski 2002). When anaerobiosis is achieved, gene regulation regarding hydrogenase synthesis occurs. The genes are activated in ribosomes and mature enzyme is localized in chloroplasts. In order to contain active mature enzyme, the accessory maturation proteins are required. *hyd*A1 and *hyd*A2 are the genes having the role of controlling hydrogenase synthesis (Shepard et al. 2011). Even if the overall function of these genes is not known, it is found that *hyd*A1 is the main gene controlling the activity of the hydrogenase enzyme; however *hyd*A2 is still unclarified (Meuser et al. 2011). Isolated enzyme sequences also suggest that *hyd*A1 and *hyd*A2 genes are present when cells are under anoxic environment. Also aerated cells lack these genes which provide a reliable basis for the relation of gene regulation and environmental conditions. Anaerobiosis not only regulates *hyd*A1 and *hyd*A2 synthesis but also triggers the genes encoding maturation enzymes (Posewitz et al. 2004; Rubach et al. 2005; McGlynn et al. 2008; Nicolet et al. 2010). *hyd*EF and *hyd*G are the maturation enzyme-encoding genes found in algae (Nicolet et al. 2010). The cDNA sequence analysis highlighted a 1,494 base pair open reading frame and the protein encoded is calculated as 53.1 kDa.

The amino acid sequences of diverse microalgal hydrogenases are useful to understand the conserved sequences in between microalgal species (Schmitter et al. 1998; Happe and Naber 2003). Recent studies display that Lys^{396} are conserved among algal species (Winkler et al. 2009). Another study discussed Arginine at 171 residue which is also conserved among algal species, and the results showed dramatic decrease in the hydrogen production capacities of mutant cells in terms of site direction mutations regarding the amino acid sequences (Sybirna et al. 2013). It can be said that conserved sequences are responsible for the activation of enzyme; however site-directed arginine mutagenesis did not affect the active site conformation. It is thought to be related with the electrostatic interaction between enzyme and PSI (Sybirna et al. 2013).

The structural organization of [FeFe] hydrogenases has unique characteristics which has a harmony in its active site called as H-cluster (Lubitz et al. 2014). It is a monomeric enzyme relatively simpler in shape when compared to [NiFe] hydrogenases (Oh et al. 2011). The active site is embedded in deeper parts of the enzyme. Strict anaerobic conditions regulate the activity of the enzyme and affinity to the substrate (protons). The protons, electrons, and oxygen are thought to run through micro-channels to reach active site (Fig. 9.1). The location of active site makes it hard to reach the available structural data.

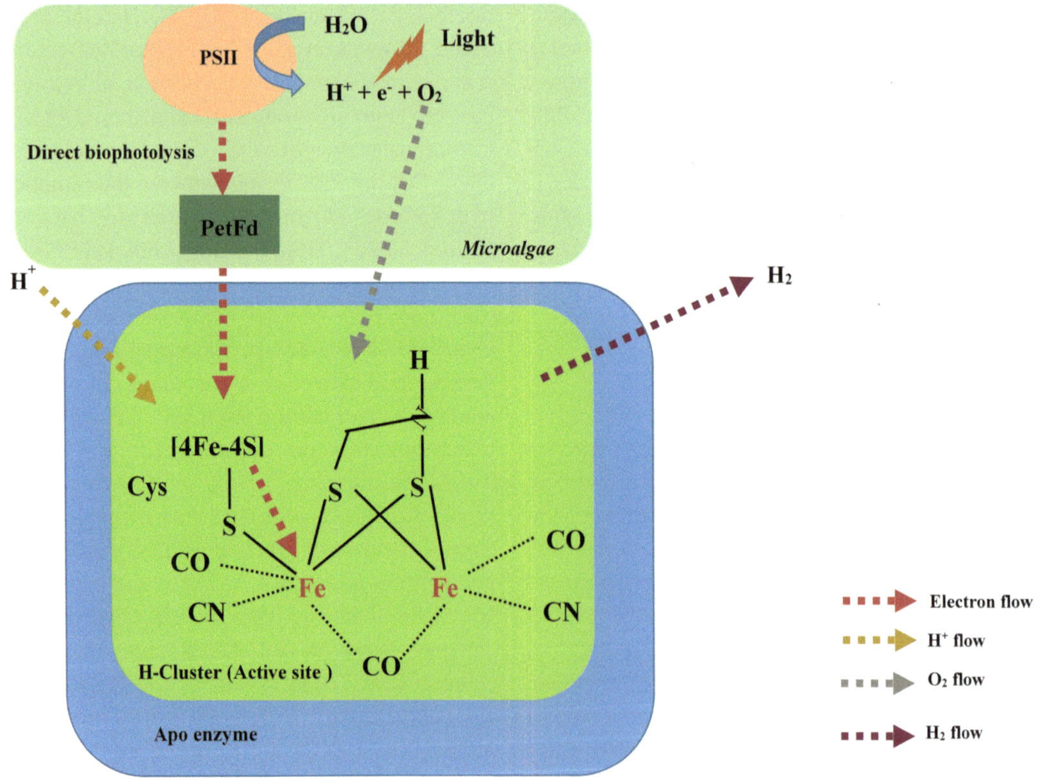

Fig. 9.1 Electron flow pathway through [FeFe] hydrogenases in microalgae (adopted from Kim and Kim 2011)

These enzymes have two iron atoms in their active site (Peters et al. 1998). Crystal structure studies have been done on bacterial hydrogenases, and *Desulfovibrio desulfuricans* and *Clostridium* species have been used as models (Lubitz et al. 2014). In these organisms a single hydrophobic channel through the embedded active site is found (Nicolet et al. 2002).

Because H-cluster only has Fe atoms, [FeFe] hydrogenases are also called as "iron only" hydrogenases (Nicolet et al. 2000). This di-iron complex has cyanide (CN) and carbon monoxide (CO) ligands attached. Active cluster of the enzyme only contains cysteine as organic residues. The metal center is ligated via sulfur bridges which hold the structure as a whole (Heinekey 2009). The ligating inorganic substances are first detected via FTIR spectroscopy in bacteria (Volbeda et al. 1995). Algal hydrogenases display differences in terms of the crystal mosaic of active site; H-cluster contains six iron complexes attached to the active site of the enzyme. The simpler structural developments of algal hydrogenases represent better structural data for crystallography and spectral studies (Lambertz et al. 2014). The conformational regulation of active site is affected by the existence of oxygen as an inhibitor agent, and protons as substrates besides catalyzed hydrogen gas are another factor (Lubitz et al. 2014; Lambertz et al. 2014). However the solubility of hydrogen in water is very low which does not affect the activity of enzyme until the partial pressure of hydrogen in sealed chamber altered and saturation point is reached (Kosourov et al. 2012).

The primary function of the hydrogenases in certain cells is either to provide electrons from H_2 or to balance cells overall redox potential (Lubitz et al. 2014). Especially in microalgae, light-driven water splitting generates electrons to flow through the transport chain which finally shuttles to hydrogenase enzyme under anoxic conditions (Melis 2009). The excess electrons can be reacted with protons within the cells. The specific activity of hydrogenase enzyme is an important parameter for biotechnological applications. [FeFe] hydrogenase shows the highest affinity to protons to generate hydrogen in gas form. It is reported that some of the hydrogenase enzymes can generate 10^4 molecule of approximately H_2 per second at room temperature (Volbeda et al. 1995; Lubitz et al. 2014).

Microalgal *hyd*A genes have been identified in *Chlamydomonas reinhardtii* and *Scenedesmus obliquus*. Two copies of *hyd*A genes in microalgae clearly state the diversity in microalgal expression system; however it has been found out that most of the structural and functional role of the enzyme is controlled by *hyd*A1 rather than *hyd*A2 (Meuser et al. 2011). Microalgal hydrogenases are lack of additional FeS cluster which represents a simpler motif (Florin et al. 2001; Happe and kaminski 2002). With this basic structure, algal hydrogenases represent a good model for understanding 3D structure, maturation, and evolution (Lambertz et al. 2014).

Electron transport in *hyd*A type hydrogenases is simpler in theory because the algal enzymes are monomers and only have H-clusters (L1, L2, and L3 motifs) (Vignais et al. 2001). Enzymes with unique CO, cysteine, and CN ligands are thought to be a regulation to function at low redox states. Centric Fe atoms are coordinated as distal and proximal iron centers Fe_D and Fe_P which represent the electron flow directions through the H cluster. X-ray studies display that the gas channel in [FeFe] hydrogenases is shorter than that in [NiFe] type, which is related to rapid inhibition of [FeFe]-type hydrogenases when oxygen exists (Lubitz et al. 2014).

9.5.1.2 [NiFe] Hydrogenases

NiFe hydrogenases are another major class of hydrogen-catalyzing enzymes which are found in methanogenic bacteria and later found to have many other classes of prokaryotic organisms including cyanobacteria (Heinekey 2009). The genetic and functional diversity of [NiFe] hydrogenases drives a classification to understand function and species relationships; there are four groups of [NiFe] hydrogenases where cyanobacterial hydrogenases are also taken into account (Kim and Kim 2011). Cyanobacteria have [NiFe] uptake hydrogenases (group II) which oxidize H_2 to protons and electrons to provide electron to cellular machinery, and the other type is [NiFe] bidirectional hydrogenases (group III, NAD/NADP reducing) responsible for H_2 production to balance the redox potential of cyanobacteria cells (Prince and Kheshgi 2005; Mathews and Wang 2009; Kim and Kim 2011; Srirangan et al. 2011; Sakurai et al. 2013). Uptake hydrogenase reactions are unwanted in terms of sustainable hydrogen production and suppression of *hup* genes encoding uptake hydrogenases, and the regulation of these gene sets is an important parameter to enhance hydrogen production from cyanobacteria. Uptake hydrogenases are encoded by *hup*S and *hup*L genes. The role of these enzymes in cyanobacteria is oxidizing the H_2 released from nitrogenase activity (Khetkorn et al. 2012). Because of relatively simpler structure compared to microalgae, genetic engineering tools are highly efficient to enhance hydrogen production activity; nevertheless the structural organization of [NiFe] hydrogenases is hard to understand that we need to use electrochemical, spectroscopic, and engineering tools.

[NiFe] hydrogenases are composed of two units, small (28 kDa) and large (60 kDa) (heterodimer), respectively (Volbeda et al. 1995). The active site has Ni and Fe bonds and 3[Fe-S] ligands attached to metallic ions (Lubitz et al. 2014). The active site of the enzyme is more tolerant to oxygen; however the specific activity of [NiFe] hydrogenases is 100-fold lower than [FeFe] hydrogenases (Kim and Kim 2011). The active site organization has some structural similarities with [FeFe] hydrogenases; CN and CO ligands are also found in [NiFe] type (Fig. 9.2). NiFe site has two cysteine bridges where Ni atom has also two more cysteine attachments (Volbeda and Fontecilla-Camps 2005). These ligands are thought to be a necessary ancestor for catalytic activation of mature enzyme. The enzyme has an approximately 100 kDa molecular weight and is found in the periplasmic space of the cell, where the photosynthetic apparatus of cells is located (Shafaat et al. 2013). Small subunit consisted of [4Fe4S] and [3Fe4S] clusters. [4Fe4S] is arranged as proximal and distal termination (Lubitz et al. 2014).

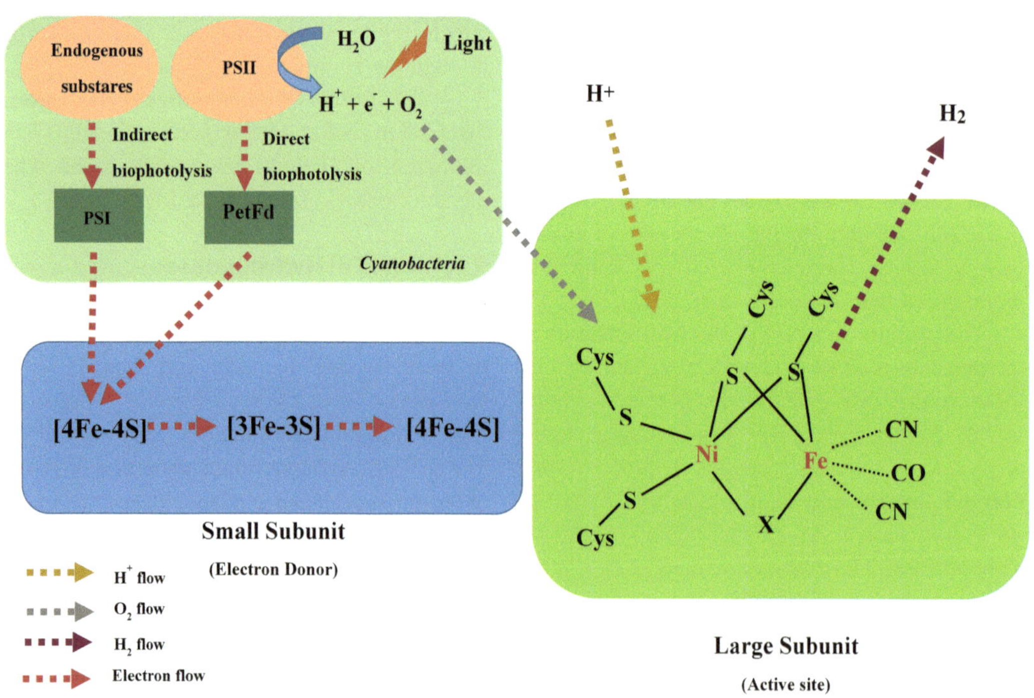

Fig. 9.2 Electron flow pathway through [NiFe] hydrogenases in cyanobacteria

The uptake hydrogenases are encoded by *hup* genes catalyzing the irreversible H_2 oxidation to protons and electrons. Meanwhile bidirectional hydrogenases are encoded by *hox* genes, using NAD/NADP as electron donors directed through the active site of the enzyme. As [FeFe] hydrogenases, *hox* hydrogenases are also sensitive to the oxygen and rapid inhibition of enzyme activity occurs when cells are exposed to oxygen. Small and large subunits are controlled by *hox*Y and *hox*H genes, respectively (Tamagnini et al. 2007). The electrons are shuttled from the Fe-S cluster of small subunit and transported to the Ni-Fe active site of larger subunit (Lubitz et al. 2014). Proton transfer is also an important issue to emphasize. The micro-hydrophobic channels of enzyme are thought to act as proton pathway; however the actual transport pathway of protons to reach the active site of the enzyme is still unclear (Lubitz et al. 2014).

The role of nitrate is essential in the activation of *hox* genes. When nitrogenase activity is suppressed (inhibition in the nitrate synthesis), *hox*-mediated H_2 evaluation is increased at *Synechocystis* sp. strain PCC6803 (Baebprasert et al. 2011). The role of inhibitors such as DCMU, KCN, and CCCP on electron flow to nitrogenase and bidirectional hydrogenases suggested that certain suppression in the *hup* genes has been an effective strategy to enhance cyanobacterial hydrogen production using *hox* hydrogenases (Khetkorn et al. 2012).

9.5.1.3 Nitrogenases

Nitrogenase enzymes catalyze nitrogen fixation reaction, an ability which is limited to a number of organisms (Peters et al. 2006; Tsygankov2007; Peters and Szilagyi 2006). Like hydrogenase enzymes nitrogenase enzymes are also sensitive to oxygen (Thompson and Zehr 2013). N_2 fixation is specific to prokaryotes. Theories about the origins of cyanobacterial N_2 fixation are thought to be either an evolutionary regulation or by certain genes that are transferred from N_2-fixing species to cyanobacteria (Thompson and Zehr 2013). With the scope of this chapter, only cyanobacterial nitrogenases will be emphasized. Atmospheric nitrogen or dinitrogen (N_2) is converted into ammonium and later used by higher plants or microorganisms (Reese and Howard 2000). Nitrogen in the form of ammonium is crucial for amino acid, protein, and nucleic acid synthesis, required for cell maintenance and growth. Nitrogenase enzymes are mostly found in purple non-sulfur bacteria (PNSB) and some species of cyanobacteria (Kim and Kim 2011). When N_2 exists in the atmosphere, cells fix N_2 to NH_3. Nitrogenase-catalyzed reactions require ATP which is not thought to be an energy-efficient process. Two ATP per electron transfer and in general 16 ATP are required to fix 1 mole of N_2 (Reese and Howard 2000; Zhao et al. 2006). When NH_3 is depleted in the culture environment, nitrogenases tend to catalyze H_2 using ATP (Kim and Kim 2011; Oh et al. 2011). Nitrogenase enzyme-encoding genes are dependent on the NH_4^+ existence in the culturing medium. Even at low concentrations of NH_4, enzyme synthesis is inhibited at posttranscriptional levels (Oh et al. 2011).

The first cyanobacterium with N_2-fixing characteristics found was *Trichodesmium* which is a free-living filamentous nonheterocystic form of marine cyanobacteria described by James Cook because of its eye-visible characteristics (Thompson and Zehr 2013). At first cyanobacterial N_2 fixation thought to be a symbiotic relation with N_2-fixing marine bacteria (Taylor et al. 1973; Thompson and Zehr 2013). However because of photosynthetic metabolism of cyanobacteria immunolabeling techniques, transmission electron microscopy and later mass spectrometry display active nitrogenase proteins in *Trichodesmium* which approves that cyanobacteria are also capable of N_2 fixation with their photosynthetic nature (Bergman and Carpenter 1991; Thompson and Zehr 2013). Active site models and arrangement of enzyme during H_2 formation process have been reviewed according to several aspects (Apte and Prabhavathi 1994; Zhao et al. 2006; Schawarz et al. 2009; Hoffman et al. 2013; Thompson and Zehr 2013).

Cyanobacteria with heterocyst formation display the spatial isolation of hydrogen-producing anaerobic environment from aerobic phase with their unique heterocyst structure (Kallas et al.

1983; Thompson and Zehr 2013). Heterocysts are thought to be a protection mechanism of nitrogenase enzymes from highly oxygenic environment; thus the thick wall organization of heterocysts only permits trace amount of oxygen to enter the cell (Leite and Hallenbeck 2014). Besides, heterocysts do not have active PSII reaction center where oxygen is generated. High respiration rates in cyanobacterial heterocysts, other than the rapid consumption of the oxygen in the environment provide ATP for nitrogenase enzymes serves another advantage for production. Non-heterocyst species requires more attention because of the oxygen sensitivity of the enzymes. Even if nitrogenases are sensitive to oxygen, their tolerance limit is higher than the hydrogenases, resulting in a decreased enzyme degradation due to oxygenic conditions (Kufryk 2013). The first nitrogenase enzyme is isolated in 1966 and the X-ray crystallography of the active site and enzyme mosaic has been presented at the early 1990s (Seefeldt et al. 2012; Hu and Ribbe 2013; Okhi 2014). Nitrogenase enzymes are diversified according to their metal-containing active sites (Kim and Kim 2011); however the most abundant one is molybdenum-iron (MoFe) nitrogenases that consisted of MoFe and Iron (Fe) subclusters. Iron (Fe) cluster is combined to a protein structure and is a nucleotide dependent (Rees and Howard 2000). Electrons are donated from Fe cluster and reduction reactions occur at MoFe site of the enzyme (Seefeldt et al. 2012). Fe protein binds cubane [4Fe4S] cluster and also has an MgATP site (Schawarz et al. 2009). MoFe, a heterodimer structure ($\alpha_2\beta_2$) active site, consisted of P-cluster [8Fe7S] ligation part and MoFe bimetallic cofactor site called Mo-Fe-Co (Okhi 2014). Mo atom is ligated with an organic homocitrate body (Dos Santos et al. 2005). Cofactor body is covalently linked with histidine imidazole group at the Mo site and Fe site is linked with cysteine thiolate. Fe protein is a ~60 kDa homodimer and MoFe protein is a ~230 kDa heterodimer. Structural organization of MoFe site is more complex than Fe protein. The enzyme is encoded by *nif*K and *nif*D genes (Kallas et al. 1983).

Electron flow is controlled and regulated by the oxidation and reduction of MgATP and [4Fe4S] site of Fe protein through MoFe protein (Dos Santos et al. 2005; Seefeldt et al. 2012). To be able to obtain an active MoFe site, electron accumulation in Fe protein is required (Fig. 9.3). Accumulated electrons are driven through the MoFe site where substrates are reduced to hydrogen (Hu and Ribbe 2013).

The complex organization of nitrogenase enzyme requires a more brief explanation of the encoding *nif* genes, a combination of Fe and MoFe protein, activation of a catalytic subunit, and conformational regulation of the active site through electron transport and release of NH_3 or H_2 from the active site; reaction rates and dynamics of the overall mechanism are of importance to imagine the 3D structure of the active enzyme and reaction steps (Hoffman et al. 2013). The catalytic modeling of the active site and spectroscopic techniques to identify structural organization of nitrogenase enzymes directed the research to a new area (Ohki 2014). Identifying active site of the enzyme using site-directed mutagenesis emerged to discover the role of amino acid sequences and relation with the catalytic and structural importance of the enzyme (Zehr et al. 2003).

Even if nitrogenase-encoding genes are found to be highly conserved, 16SrRNA analysis shows sequence similarities (Postgate and Eady 1988; Young 1992: Zehr et al. 2003); however the distribution of nitrogenase enzymes among species also points out the diversities of the organizational regulation at the genome level (Hoffman et al. 2013).

9.5.2 Maturation of Hydrogen-Catalyzing Enzymes and Evolutionary Studies

Maturation of hydrogenase enzymes is an important complex process, as well as the reactions catalyzed in the active site of the enzyme. Regulatory and structural genes and gene sets are involved in the maturation process. Cofactors and accessory proteins also have an active role in the

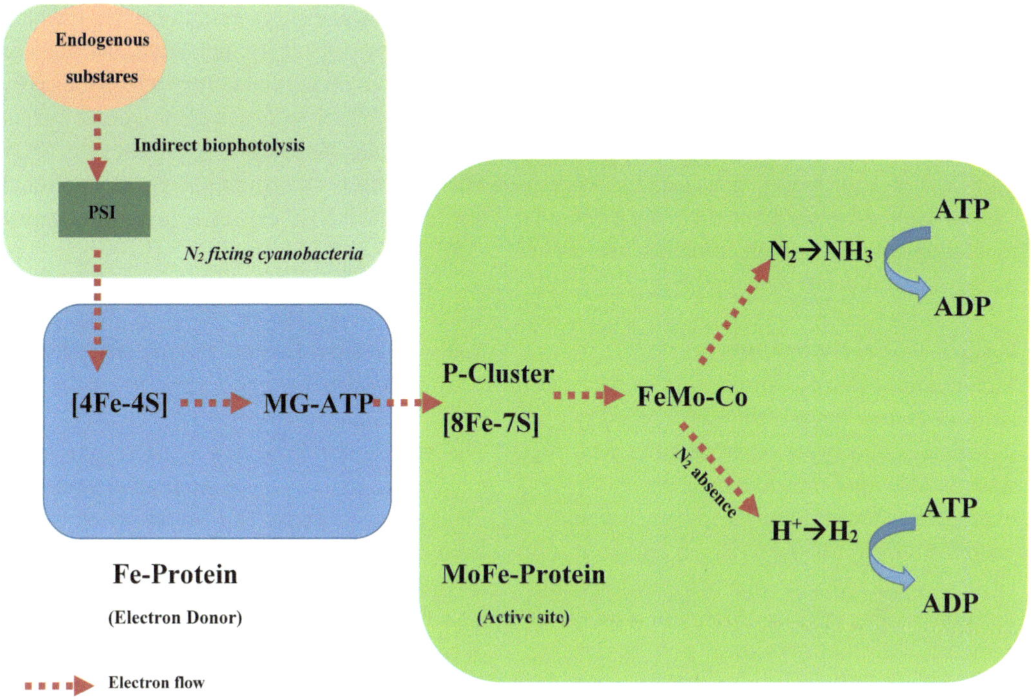

Fig. 9.3 Electron flow pathway through nitrogenases in N_2-fixing cyanobacteria

process; however some of them are not well defined yet (Pierik et al. 1998; Nicolet et al. 2010). Heterologous expression of maturation enzymes is thought to be a functional strategy, but because the maturation process is also sensitive to oxygen, understanding the concept of metabolism still seems challenging (Girbal et al. 2005). In this part, the fundamentals in the maturation process of hydrogenase enzymes are revealed.

9.5.2.1 [FeFe] Hydrogenases

Maturation of [FeFe] hydrogenase enzymes requires complex organization of several genes and their transcription products (Posewitz et al. 2004). In bacterial cells maturation proteins are called to be *hyd*E, *hyd*F, and *hyd*G which are encoded separately; however in *C. reinhardtii* these genes encode *hyd*E and *hyd*F as one protein, *hyd*EF (Ghirardi et al. 2006). The maturation process consisted of several steps: synthesis of CN, dithiolate, and CO ligands, assembly of di-iron active site, combination of H-cluster with cubane [4Fe4S] structure, and activation of the mature enzyme (Peters et al. 1998; Heinekey 2009). Heterologous expression of *hyd*A genes results as inactive hydrogenase enzyme which indicates that maturation proteins are required to obtain active enzyme (Rubach et al. 2005). Interspecies expression systems also show that maturation proteins are also provided via heterologous expression of active enzyme (hydA1 from green algae to *C. acetobutylicum*) (Nicolet et al. 2010).

*hyd*E and *hyd*F genes are encoded separately and *hyd*G is located at the gene-encoding *hyd*E protein (Rubach et al. 2005; Berggren et al. 2013). *hyd*E and *hyd*G proteins belong to radical S-adenosylmethionine (SAM) enzyme family and *hyd*G acts like GTPase (Posewitz et al. 2004). *hydF* is thought to be responsible from activation of H-cluster (McGlynn et al. 2008). The activation procedure and maturation machinery have not been fully revealed, but protein engineering techniques and heterologous expression models are promising to determine the overall frame of the process (Lubitz et al. 2014). *hyd*F is related to the C terminus 4Fe4S cluster and *hyd*EG is responsible for the N-terminus 4Fe4S and cofactor assembly (Rubach et al. 2005; Lubitz et al. 2014).

Expression models and structural analysis of maturation concept has been established in details (Nicolet et al. 2010; Lubitz et al. 2014). The lack of knowledge on maturation systems can be named as understanding organizational regulation of maturation proteins to trigger hydrogenase subunits to assembly.

9.5.2.2 [NiFe] Hydrogenases

Maturation of [NiFe] hydrogenase enzymes also requires maturation proteins and accessory units in order to obtain an active enzyme with high specific activity. Comparative analysis of [NiFe] hydrogenase enzymes revealed that they share a common ancestor with conserved sequence similarities including maturation and functional organization of the enzyme. Maturation genes are named as *hox* and *hyp*. In *E. coli* model, *hyp*ABCDEF, *hox*W, and *hox*EFUYH genes are required to express an active enzyme (Dos Santos et al. 2005; Tamagnini et al. 2007). In cyanobacteria *hox*H is thought to be important for the activity of the enzyme and *hup*L (Ludwig et al. 2006; Barz et al. 2010). *hox* genes can be activated both under oxic or anoxic conditions; however the oxygen controls the activity of the mature enzyme. The existence of nitrogen and illumination periods triggers or suppresses the transcription of bidirectional hydrogenases. Illumination triggers enzyme synthesis and under dark conditions enzyme synthesis rate is decreased. Maturation of [NiFe] hydrogenases is a complex organization of several gene sets and also there are some steps to be clarified.

9.5.2.3 Nitrogenases

The changes in the environmental conditions trigger gene activation called *nif* genes encoding nitrogenase enzymes (Howard and Rees 1996). MoFe protein is encoded by *nif*K and *nif*D; Fe protein is encoded by *nif*H genes (Rice et al. 1982). Transcriptional regulation and maturation of nitrogenase enzymes are thought to be very complicated; there are nine operons called *nif* E, N, B, F, Q, M, V, S, and U (Mevarech et al. 1980; Rubio and Ludden 2005). The transcriptional organization and posttranslational regulation of these genes still remain unknown, but there are strong observations about their role in the activation and assembly of mature nitrogenase enzyme (Mulligan and Haselkorn 1989: Zehr et al. 2003). The activation of maturation genes is strongly regulated with NH_4^+ existence in the environment (Mevarech et al. 1980).

9.6 Genomic and Transcriptomic Studies

The sophisticated metabolic characteristics of microalgae have been an emerging topic for a long time (Oliveira Dal'Molin et al. 2011). Biohydrogen production is a metabolic organization which is shared with bacteria and archaea as an ancient evolutionary regulation of cell metabolism when anoxic environment was prevalent (Ghirardi et al. 1997). Fundamental knowledge on regulation of algal metabolism during hydrogen production can be utilized to develop futuristic ideas and innovative approaches to algal biotechnology accompanied with renewable energy (Mayfield and Franklin 2005). Today we are at a state that microalgal biohydrogen production requires comprehensive evaluation of what is known and what can be done to achieve industrial production (Lopez et al. 2011). Nevertheless the general aspects and advanced studies that have been accomplished so far are a powerful tool for genomic and transcriptomic studies to pursue the ultimate goal, single-step hydrogen generation (Matthew et al. 2009; Grossman 2005; Nguyen et al. 2008; Mathews and Wang 2009; Rupprecht 2009).

The lack of knowledge regarding algal genomics is the main challenge to develop an efficient transformation and recombination. Today we only have full genome sequences of only a limited number of microalgae like *Chlamydomonas reinhardtii*, *Chlorella* sp., *Phaeodactylum tricornutum*, and *Thalassiosira pseudonana* (Grossman 2005; Merchant et al. 2007; Rosenberg et al. 2008; Weckwerth 2011; Blaby et al. 2014). However genome sequences, cDNA libraries, and gene mapping (including taxonomical relations) are required to display and visualize what is hidden inside the microscale individual algae cells. We can see the green, red, brown, or yellowish samples, and we can extract certain useful substances, but how are those substances produced, how do cells manipulate their metabolism, and which pathway is required to trigger certain reactions? These questions also refer to hydrogen production in microalgae in genomic and metabolomic aspects.

System biology approach is a convenient tool to understand hydrogen metabolism accompanied with high-throughput and omics technology (proteomics, genomics, transcriptomics, and metabolomics) targeting a model organism to represent other related species (Leon-Banares et al. 2004; Nguyen et al. 2008; Gimpel et al. 2013; Godaux et al. 2013). Computational designs assist in covering metabolic regulation besides enabling inter- and intra-species comparisons (May et al. 2009; Rupprecht 2009; Lopez et al. 2011).

Transcriptome analysis reveals the function of certain gene product within the cellular regulation (Nguyen et al. 2011). Transcriptomic tools are used to identify the expression level of certain genes, transcriptional units, and regulation of gene expression as a response to environmental conditions and metabolic changes, the expression scores of certain traits among other species, and the rate of expression (Hemaiswarya et al. 2013). Responses at molecular level are of importance to display the operation of cellular machinery (Rismandi Yazdi et al. 2011). Rather than sequencing whole genome, transcriptomic and metabolomic studies are more practical, easier, and applicable to particular studies, as photobiological hydrogen production which is discussed in here (Nguyen et al. 2008).

Together with transcriptomics, metabolomic and proteomic studies are also functional to discover structural, functional, developmental, and evolutionary aspects of cellular organization considering posttranslational regulation and metabolic flux (Grossman 2005; Bochenek et al. 2013). Today microalgal omic tools are used to highlight biofuels, nutraceuticals, and pharmaceuticals from microalgal species (Jamers et al. 2009). Together with genetic engineering tools, transcriptomic studies of algal metabolism during hydrogen production are beneficial to understand the role of gene activation , maturation machinery, and enzyme activity as it has been demonstrated by using *Chlamydomonas reinhardtii* (Florence et al. 2007; Nguyen et al. 2008; Chen et al. 2010), *Chlorella* (Meuser et al. 2011), *Dunaliella tertiolecta* (Rismandi Yazdi et al. 2011), and *Tetraselmis subcordiformis* species (Cao et al. 2008). Besides the role of sulfate deficiency has been demonstrated on *Emiliania huxleyi* which can be used as a beneficial literature to understand the relationship between sulfate and hydrogen (Bochenek et al. 2013). A detailed microalgal omic and genomic review has been done to understand the role of algae as a model for eukaryotic studies (Grossman 2005; Walker et al. 2005; Weckwerth 2011; Guaernieri and Pienkos 2013). Rather than discussing the algal omics, we will highlight the model systems regarding photobiological hydrogen production and hydrogen-catalyzing enzyme aspect.

9.6.1 Model Microorganisms

Microalgae are a diverse group of organisms mostly found in marine and freshwater environments (Kützing 1849). The photosynthetic nature is an important metabolic asset to discover fundamental regulations in photosynthesis studies. Microalgal species share an enormous diversity both on metabolic and structural meaning. Value-added compounds for nutraceuticals, pharmaceuticals, and cosmetics purposes encourages the enlargement commercial consideration and investments (Spolaore et al. 2006; Borowitzka 2013).

Considering hydrogen production, the most promising species is thought to be *Chlamydomonas* species as we know today; however other biohydrogen-producing species are *Anabaena* sp., *Synechococcus elongatus*, *Chlorella vulgaris*, *Chlorella zofingiensis*, and *Scenedesmus* sp. (Gaffron 1944; Kessler 1962; Healey 1970; Oncel 2013). However physiological responses of *Chlamydomonas* to anoxic conditions help to understand the hydrogenase activity, genome mapping, and characterization of metabolic responses to anaerobic conditions on the basis of hydrogenase enzymes. The enhanced genome sequencing and success on the genomic studies accompanied with transcriptomics and metabolomics enabled a better vision for further applications and research.

9.6.1.1 Green Algae

Green algae represent the eukaryotic model organism. The genome sequencing projects have been focused on mitochondrial, nuclear, and chloroplast genomic materials (Mayfield and Franklin 2005). Fully sequenced three genomes give an idea about cellular regulation, structural organization, responses to environmental stress conditions, metabolic repression regulation, and comparison studies.

Commercially important strains also represent the algal physiology and metabolism for food, fuel, and pharmaceutical industry (Mayfield and Franklin 2005; Rosenberg et al. 2008). Among those species *Chlorella*, *Chlamydomonas*, *Haematococcus*, *Dunaliella*, *Tetraselmis*, and *Phaeodactylum* species have been the model organisms (Lopez et al. 2011; Rosenberg et al. 2008). Manipulating microalgal metabolism within the cell has been one of the dream ideas to increase product yield, especially heterologous expression studies are still exciting and intriguing (Walker et al. 2005).

After the very first studies of microalgal biohydrogen production, *Chlamydomonas reinhardtii* species gained attraction when Melis et al. showed significant amount of hydrogen production under sulfur-deprived illuminated and anaerobic conditions. In 1996 the first eukaryotic hydrogenase is isolated from *Trichomonas vaginalis* (Bui and Johnson 1996). However most of the hydrogen-related studies have been done using *Chlamydomonas reinhardtii*, which has been a global model strain.

9.6.1.2 Cyanobacteria

The concept and regulation of photosynthesis and diverse hydrogen production pathways direct a new understanding of multienzyme hydrogen studies which is thought to be a sustainable method because some of the microalgae can generate hydrogen under aerobic conditions; however eukaryotes require strict anaerobiosis (Kufryk 2013). Besides the evolutionary regulation of heterocyst formation, nitrogenases provide databases for biomimetic studies (Wilmotte and Herdman 2001; Gugger and Hoffman 2004).

The biotechnology of cyanobacteria has a long history (Tamagnini et al. 2002); however as other algal biotechnology development, the biotechnological importance of cyanobacteria has been known during the 1950s (Belay 2013). The well-defined production systems for algae cultivation have been developed mostly using the model cyanobacteria *Spirulina platensis* (Switzer 1981). Today, *Spirulina*-based products are well marketed and physiological studies are also well established. However other cyanobacteria species are also promising for biotechnological approaches. Especially the lack of adequate information about algal genetics directed to map the whole cyanobacterial genomes. *Anabaena*, *Oscillatoria*, *Synechococcus*, *Calothrix*, *Cylindrospermum*, *Nostoc*, and *Microcystis* species are models for cyanobacteria research. However regarding biohydrogen production, *Gloeocapsa*, *Anabaena*, *Spirulina*, *Cyanothece*, and *Nostoc* species are promising for further development (Oncel 2013). Because [NiFe] hydrogenases are only found in cyanobacteria species, this type of enzyme regarding photobiological hydrogen production, enzyme structure, morphology, and regulation studies is required to be isolated from those species (Carrieri et al. 2011).

9.6.2 Hydrogen-Producing Machinery Concept: Protein Engineering

Hydrogen production in green algae is dependent on the enzymatic regulation during anaerobiosis as a response to changes into metabolic rearrangement (Melis et al. 2000). Environmental conditions are driving forces of hydrogen production (Grossman 2000). Changes from oxic conditions to anoxic metabolism require new structural and functional regulations within the cells. The absence of the oxygen drives cells to define a new electron acceptor, H^+ (Matthew et al. 2009).

Previous transcriptomic studies display that hydrogen production is highly related with photosynthesis, starch synthesis and degradation, Calvin-Benson cycle, and sulfur utilization (Surzycki et al. 2007; Nguyen et al. 2008; Beer et al. 2009). However the level of protein interaction is crucial for model system engineering considering commercial production as the ultimate goal. Light to biomass and light to fuel concept can be adaptable to photobiological hydrogen production systems (Oncel 2013). The definition of production strategies can be interchangeable according to the approach adopted. In this regard, considering protein engineering as a sustainable tool for hydrogenase synthesis, maturation, activation, and challenges involving in enzymatic machinery must be highlighted (Nguyen et al. 2008).

The organizational and functional regulation of hydrogenases provides reliable infrastructure to be able to conduct new ideas for further development (Lubitz et al. 2014). Protein engineering concept can be designed directly on hydrogenase enzymes or indirectly applied on the pathways controlling hydrogenase enzyme synthesis and specific activity (Posewitz et al. 2004). The direct methods are related to the maturation enzyme conformation, genetic regulation and distribution of hydrogenase genes among diverse species enabling heterologous expression, building up synthetic biology tools to bio-mimic the active site conformation to design cell-free hydrogenase-mediated hydrogen production, and enhancing the oxygen tolerance of active site structure regulating the micro-channel configuration of enzyme from the outer space to active site (Berggren et al. 2013). The amino acid sequences, gene mapping strategies, PCR-based studies, and conserved trait analysis can be reliable tools to understand intra- and interspecies regulation of hydrogenase activity and also residual behavior of enzymatic substructure and the role of enzyme assembly and activity as it has been shown previously via site-directed mutagenesis of certain conserved amino acid sequences among microalgal species (Happe et al. 1994; Chen and Melis 2004; Posewitz et al. 2004; Surzycki et al. 2007; Kamp et al. 2008; Rolland et al. 2009).

However indirect tools are designed to manipulate environmental or intracellular regulation of pathway (Melis 2007). Understanding the role of sulfur depletion and cellular regulation was a well-established method to sustain hydrogen providing suitable environment to activate hydrogen-encoding genes (Wykoff et al. 1998; Melis et al. 2000). Understanding the starch metabolism, discovering the D1 protein role and contribution to hydrogenase enzyme activation, and enhancing light utilization capacity via interfering the size of light-harvesting antenna can be considered as indirect methods to engineer hydrogen generation metabolism (Faraloni and Torzillo 2010; Oncel et al. 2014).

However with current perspective of hydrogenase enzymes either bacterial or microalgal, the lack of proposed production technology drives the attention to protein engineering to be able to use each individual cell as hydrogen-catalyzing factories considering microalgal cell factory concept.

9.7 Biotechnological Advances

Basic cultivation techniques have been provided a strong basis to keep organisms alive while manipulating hydrogen synthesis (Melis 2007). Meanwhile conventional biochemical techniques and basic sequencing studies also helped to understand metabolism of hydrogen production from various organisms (Beer et al. 2009). It should be noted that still bacteria and bacterial hydrogenases are highlighted well and algal hydrogenases are compared to known structures (Lubitz et al. 2014). Genetic engineering tools have been utilized to sequence unknown genomes and genes encoding the hydrogenase enzymes from maturation to active enzyme synthesis. Comparative studies also helped to display the conservations and differences among hydrogenase-containing species (Rosenberg et al. 2008; Nguyen et al. 2008; Lubitz et al. 2014).

We can state that biodiversity and searching new strains with upgraded tools may be beneficial to design a hydrogen economy (Faraloni and Torzillo 2010).

The main challenges regarding hydrogenase studies can be listed as oxygen sensitivity of maturation proteins and active enzymes, electron flow through active site (cyclic electron transfer around PSI), gene sequencing, and *in vitro* enzyme studies. However protein engineering, heterologous expression of enzymes, recombination, genetic engineering, and screening new strains offer more feasible platform for biotechnological studies.

Active site studies of hydrogenase enzymes ([NiFe] and [FeFe] hydrogenases) enable biomimetic tools to design artificial active sites or analog models of hydrogen-producing machinery which requires structural biology infrastructure. In this aspect artificial photosynthesis (Meyer 1989; Gust et al. 2009; Kalyanasundaram and Graetzel 2010) and artificial hydrogenase studies are thought to be promising for futuristic biotechnological studies (King et al. 2006; Hambourger et al. 2008).

Hydrogenases are not only enzymes capable to catalyze reactions *in vivo*. Yet so far there have been good examples to use *in vitro* hydrogenases to conduct certain biochemical reactions either to produce energy or to monitor certain reactions. Fuel cell development using hydrogenase enzymes as electrodes is an urgent topic to be used combined with artificial hydrogenase or artificial photosynthesis studies. Conventional fuel cells are dependent to platinum electrodes to generate electricity; however utilization of hydrogenase enzymes in fuel cell technology is thought to be more efficient than regular designs. Indeed [FeFe] hydrogenases isolated from *Clostridium* showed promising biocatalyst effect on fuel cell development, and [NiFe] hydrogenases with higher oxygen toleration are thought to be also promising. Considering hydrogenase enzymes as redox proteins, the cellular machinery is a whole challenge which can be divided into subclasses (Jones et al. 2003; Lamle et al. 2003; Martens and Liese 2004). The technology can be able to design a system to shuttle electrons to the enzyme to conduct the reactions is required, in which an active enzyme isolation and sustainable enzyme activity are necessary. Immobilization of isolated enzyme is a tool to overcome stability problems, but recombinant gene technology and protein engineering are thought to be more promising regarding the alterations in the active site structure configuration and also covering the micro-channel theory where oxygen, protons, and electrons travel through the active site (Chenevier et al. 2013).

In vitro hydrogenase studies are still at an early development stage which requires further data sets regarding the combination of relevant structural mosaics regarding isolation or mimicking. Considering isolation of hydrogenase heterologous expression and defining a sustainable and stable expression system are crucial for further development.

9.8 Recent Trends and Future Prospects

What do we know about hydrogen production from microalgae and what we should develop for a better hydrogen production? We should design our knowledge and technology around this question. The dependence of the world population on petroleum-derived fuel sources is increasing day by day with an estimation that it may reach 50 % at 2030 (Carrieri et al. 2011). Thus development of emerging fuel technologies from renewable sources is crucial. In that case hydrogen as being one of the energy-dense source and clean combustion characteristics seems to be promising for the future. Additional benefits of utilization biological sources such as microalgae and cyanobacteria with CO_2 sequestering capabilities seem to be greener.

Cultivation of certain algal species for sustainable hydrogen production and an efficient procedure to be able to sustain hydrogen production have been showed and clearly stated for a long time (Aparico et al. 1985; Fedorov et al. 2005; Melis 2007; Dickson et al. 2009; Antal et al. 2011;

Oncel 2013). Different cultivation procedures and the role of environmental conditions, such as nutrient distribution and light utilization, have been also stated with different perspectives (Laurinavichene et al. 2004; Kima et al. 2006; Kosourov et al. 2007; Giannelli and Torzillo 2012; Oncel and Kose 2014; Oncel and Kose 2014; Oncel et al. 2014). Immobilization as an efficient tool to preserve the cell viability for a longer period of time is one of the emerging techniques to control cyclic hydrogen production using the same culture for a longer period of time (Laurinavichene et al. 2004; Kosourov and Seibert 2009).

The microlevel changes regulate the hydrogen production efficiency in macrolevel. The design of microalgal hydrogen production scheme is dependent on understanding the metabolic regulations. Sustainable biohydrogen production in larger scale is dependent on suitable photobioreactor or fermenter designs and see the effects of scale-up criteria for sustained hydrogen production (Srirangan et al. 2011; Oncel and Sabankay 2012; Giannelli and Torzillo 2012). For a cost-effective production strategy, outdoor productions are also considered using sunlight for illumination.

Hydrogen is considered to be a promising energy source for future civilization. However renewable production technologies are lack of expertise to compete with existing ones. With this regard, the economics of a sustainable process seems to be vital to benefit the sustainable and promising characteristics. In order to generate a compatible technology; the challenges in hydrogen production technology should be re-considered. The main challenge to commercialize biohydrogen is the oxygen sensitivity of hydrogenase enzymes (Oncel 2015). Because of this criterion, large-scale open atmosphere production cannot be done for biohydrogen production phase. Thus closed well-sealed culture chambers are required.

With the known biohydrogen production, cell-free hydrogen production using hydrogenase enzymes, screening of potential strains, and gene editing to obtain a super strain capable of mass amounts of hydrogen production are trending. The recent tools on metabolomic studies seem to be promising to understand complete mechanism of hydrogen production as well.

9.9 Conclusion

Hydrogen production via biological sources has been a tremendous subject in order to reveal the potential algal species, considering commercialization aiming a green and clean fuel production. In this manner photobiological hydrogen production from microalgae has been an emerging area; first microalgae use direct biophotolysis and the activity of hydrogenases is higher than the others. As an enzymatic perspective, the structure behind the microalgal hydrogen production is dependent on the activation and specific activity of [FeFe] hydrogenases. In this context any attempt to increase the activity of hydrogenases may also increase the yield of hydrogen production. Keeping this theory in mind, today not only [FeFe] hydrogenases in microalgae but also [NiFe] hydrogenases in cyanobacteria and nitrogenases in heterocyst forming nitrogen-fixing cyanobacteria are also of importance considering structural, functional, and regulation arrangements.

References

Adams MWW (1990) The structure and mechanism of iron hydrogenases. Biochim Biophys Acta 1020:115–145

Antal TK, Krendeleva TE, Laurinavichene TV, Makarova VV, Ghirardi ML, Rubin AB, Tsygangov AA, Seibert M (2003) The dependence of algal H_2 production on photosystem II and O_2 consumption in sulfur-deprived *Chlamydomonas reinhardtii* cells. BBA 1607:153–160

Antal TK, Krendeleva TE, Rubin AB (2011) Acclimation of green algae to sulfur deficiency: underlying mechanisms and application for hydrogen production. Appl Microbiol Biotechnol 89:3–15

Aparico PJ, Azuara MP, Antonio B, Fernandez VM (1985) Effect of light intensity and oxidized nitrogen sources on hydrogen production by *Chlamydomonas reinhardtii*. Plant Physiol 78:803–806

Apte SK, Prabhavathi N (1994) Reaaragements of nitrogen fixation (*nif*) genes in heterocystous cyanobacteria. Bioscience 19(5):579–602

Baebprasert W, Jantaro S, Khetkorn W, Lindblad P, Incharoensakdi A (2011) Increased H_2 production in the cyanobacterium *Synechocystis* sp. strain PCC6803 by redirecting the electron supply via genetic engineering of the nitrate assimilation pathway. Metab Eng 13(5):610–616

Barz M, Beimgraben C, Staller T, Germer F, Opitz F, Marquardt C, Schwarz C, Gutekunst K, Vanselow KH, Schmitz R, LaRoche J, Schulz R, Appel J (2010) Distribution analysis of hydrogenases in surface waters of marine and freshwater environments. PLoS One 5:13849

Beer LL, Boyd ES, Peters JW, Posewitz MC (2009) Engineering algae for biohydrogen and biofuel production. Curr Opin Biotechnol 20:264–271

Belay A (2013) Biology and industrial production of *Arthrospira* (*Spirulina*). Handbook of microalgal culture: applied phycology and biotechnology, 2nd edn. Wiley, West Sussex, pp 339–358

Benemann JR (1997) Feasibility analysis of photobiological hydrogen production. Int J Hydrog Energy 22(10/11):979–987

Berggren G, Adamska A, Lambertz C, Simmons TR, Esselborn J, Attal M, Gambarelli S, Mouesca JM, Reijerse E, Lubitz W, Happe T, Artero V, Fontecave M (2013) Biomimetic assembly and activation of [FeFe]-hydrogenases. Nature 499:66–70

Bergman B, Carpenter EJ (1991) Nitrogenase confined to randomly distributed trichomes in the marine cyanobacterium *Trichodesmium thiebautii*. J Phycol 27:158–165

Bishop NI, Gaffron H (1963) photoreduction at λ705 mμ in adapted algae. Biochem Biophys Res Commun 8:471–476

Blaby IK, Blaby Haas CE, Tourasse N, Hom EFY, Lopez D, Aksoy M, Grossman A et al (2014) The *Chlamydomonas* genome project: a decade on. Trends Biotechnol 19(10):672–680

Bochenek M, Etherington GJ, Koprivova A, Mugford ST, Bell TG, Malin G, Kopriva S (2013) Transcriptome analysis of the sulfate deficiency response in the marine microalgae *Emiliania huxleyi*. New Phytol 199:650–662

Borowitzka AM (2013) Energy from microalgae: a short history. In: Borowitzka MA, Moheimani NR (eds) Algae for biofuels and energy. Springer, Dordrecht, pp 1–15

Bui ETN, Johnson PJ (1996) Identification and characterization of [Fe]-hydrogenases in the hydrogenosome of *Trichomonas vaginalis*. Mol Biochem Parasitol 76(1–2):305–310

Cao X, Wu X, Ji C, Yao C, Chen Z, Li G, Xue S (2008) Comparative transcriptional study on the hydrogen evolution of marine microalga *Tetraselmis subcordiformis*. Biochim Biophys Acta 1777:410–416

Carrieri D, Wawrousek K, Eckert C, Yu J, Maness PJ (2011) The role of bidirectional hydrogenases in cyanobacteria. Bioresour Technol 102:8368–8377

Casalot L, Rousset M (2001) Maturation of [NiFe] hydrogenases. Trends Microbiol 9(5):228–237

Cavazza C, Martin L, Mondy S, Gaillard J, Rater P, Fontecilla-Camps JC (2008) The possible role of an [FeFe]-hydrogenase-like protein in the plant responses to changing atmospheric oxygen levels. J Inorg Biochem 102:1359–1365

Chen HC, Melis A (2004) Localization and function of SulP, a nuclear encoded chloroplast sulfate permease in *Chlamydomonas reinhardtii*. Planta 220(2):198–210

Chen M, Zhao L, Sun YL, Cui SX, Zhang LF, Yang B, Wang J, Kuang TY, Huang F (2010) Proteomic analysis of hydrogen photoproduction in sulfur-deprived *Chlamydomonas* cells. J Proteome Res 9:3854–3866

Chenevier P, Mugherli L, Darbe S, Darchy L, Dimanno S, Tran PD, Valentino F, Iannello M, Volbeda A, Cavazza C, Artero V (2013) Hydrogenase enzymes: application in biofuel cells and inspiration for the design of noble-metal free catalysts for H_2 oxidation. R Chim 16:491–505

Chochois V, Dauvillée D, Beyly A, Tolleter D, Cuiné S, Timpano H, Ball S, Cournac L, Peltier G (2009) Hydrogen production in *Chlamydomonas*, photosystem II-dependent and-independent pathways differ in their requirement for starch metabolism. Plant Physiol 151:631–640

Corr MJ, Murphy JA (2011) Evolution in the understanding of [Fe]-hydrogenase. Chem Soc Rev 40:2279–2292

Dickson DJ, Page CJ, Ely RL (2009) Photobiological hydrogen production from *Synechocystis sp.* PCC 6803 encapsulated in silica sol gel. Int J Hydrog Energy 34:204–215

Doebbe A, Rupprecht J, Beckmann J, Mussgnug JH, Hallmann A, Hankamer B, Kruse O (2007) Functional integration of the HUP1 hexose symporter gene into the genome of *C. reinhardtii*, impacts on biological H_2 production. J Biotechnol 131(1):27–33

Dos Santos PC, Igarashi RY, Lee HI, Hoffman BM, Seefeldt LC, Dean DR (2005) Substrate interactions with the nitrogenase active site. Acc Chem Res 38:208–214

Faraloni C, Torzillo G (2010) Phenotypic characterization and hydrogen production in *Chlamydomonas reinhardtii* QB binding D1 protein mutants under sulfur starvation, changes in chlorophyll fluorescence and pigment composition. J Phycol 46:788–799

Fedorov AS, Kosourov S, Ghirardi MI, Seibert M (2005) Continuous hydrogen photoproduction by *Chlamydomonas reinhardtii* using a novel two-stage, sulfate-limited chemostat system. Appl Biochem Biotechnol 121:403–412

Finazzi G, Furia A, Barbagallo RM, Forti G (1999) State transitions, cyclic and linear transport and photophorylation in *Chlamydomonas reinhardtii*. BBA 1413:117–129

Florence M, Dubini A, Seibert M, Posewitz MC, Grossman AR (2007) Anaerobic acclimation in Chlamydomonas reinhardtii: anoxic gene expression, hydrogenase induction, and metabolic pathways. J Biol Chem 282:25475–25486

Florin L, Tsokoglou A, Happe T (2001) A novel type of iron hydrogenase in the green alga *Scenedesmus obliquus* is linked to the photosynthetic electron transport chain. J Biol Chem 276:6125–6132

Forestier M, King P, Zhang L, Posewitz M, Schwarzer S, Happe T (2003) Expression of two [Fe]-hydrogenases in *Chlamydomonas reinhardtii* under anaerobic conditions. Eur J Biochem 270:2750–2758

Friedrich B, Buhrke T, Burgdorf T, Lenz O (2005) A hydrogen-sensing multiprotein complex controls aerobic hydrogen metabolism in *Ralstonia eutropha*. Biochem Soc Trans 33:97–101

Gaffron H (1944) Photosynthesis photoreduction and dark reduction of carbon dioxide in certain algae. Biol Rev Camb Phil Soc 19:1–20

Gaffron H, Rubin J (1942) Fermentative and photochemical production of hydrogen in algae. J Gen Physiol 26:219–240

Ghirardi M, Mohanty P (2010) Oxygenic hydrogen production- current status of the technology. Curr Sci India 98:499–507

Ghirardi ML, Togasaki R, Seibert M (1997) Oxygen sensitivity of algal H_2-production. Appl Biochem Biotechnol 63–65:141–151

Ghirardi ML, Zhang L, Lee JW, Flynn T, Seibert M, Greenbaum E, Melis A (2000) Microalgae, a green source of renewable H_2. TIBTECH 18:506–511

Ghirardi ML, Cohen J, King P, Schulten K, Kim K, Seibert M (2006) [FeFe]-hydrogenases and photobiological hydrogen production. SPIE 6340:U257–U262

Giannelli L, Torzillo G (2012) Hydrogen production with the microalga *Chlamydomonas reinhardtii* grown in a compact tubular photobioreactor immersed in a scattering light nanoparticle suspension. Int J Hydrog Energy 37:16951–16961

Gibbs M, Gfeller RP, Chen C (1986) Fermentative metabolism of *Chlamydomonas reinhardtii*: III. Photoassimilation of acetate. Plant Physiol 82:160–166

Gimpel JA, Specht EA, Georgianna DR, Mayfield SP (2013) Advances in microalgae engineering and synthetic biology applications for biofuel production. Curr Opin Biotechnol 17:489–495

Ginovska-Pangovska B, Ho MH, Linehan JC, Cheng Y, Dupuis M, Raugei S, Shaw W (2014) Molecular dynamics study of the proposed proton transfer pathways in [FeFe] hydrogenase. Biochim Biophys Acta 1873:131–138

Girbal L, von Abendroth G, Winkler M, Benton PMC, Meynial-Salles I, Croux C et al (2005) Homologous and heterologous overexpression in *Clostridium acetobutylicum* and characterization of purified clostridial and algal Fe-only hydrogenases with high specific activities. Appl Environ Microbiol 71:2777–2781

Godaux D, Emonds-Alta B, Berne N, Ghysels B, Alric J, Remacle C, Cardol P (2013) A novel screening method for hydrogenase-deficient mutants in *Chlamydomonas reinhardtii* based on *in vivo* chlorophyll fluorescence and photosystem II quantum yield. Int J Hydrog Energy 38:1826–1836

Godman JE, Molnar A, Baulcombe DC, Bakl J (2010) RNA silencing of hydrogenase-like genes and investigation of their physiological roles in the green alga *Chlamydomonas reinhardtii*. BioChem J 431:345–351

Gomes de Oliviera Dal'Molin C, Quek LE, Palfreyman RW, Nielsen LK (2011) AlgaGEM – a genome-scale metabolic reconstruction of algae based on the *Chlamydomonas reinhardtii* genome. BMC Genomics 12(4):5

Greenbaum E (1982) Photosynthetic hydrogen and oxygen production: kinetic studies. Science 196:879–880

Greenbaum E, Guillard RRL, Sunda WG (1983) Hydrogen and oxygen photoproduction by marine algae. Photochem Photobiol 37:649–655

Greening C, Cook GM (2014) Integration of hydrogenase expression and hydrogen sensing in bacterial cell physiology. Curr Opin Microbiol 18:30–38

Grossman AR (2000) Acclimation of *Chlamydomonas reinhardtii* to its nutrient environment. Protist 151:201–224

Grossman AR (2005) Paths toward algal genomics. Plant Physiol 137:410–427

Guaernieri MT, Pienkos PT (2013) Algal omics: unlocking bioproduct diversity in algae cell factories. Photosynth Res. doi:10.1007/s11120-014-9989-4

Guan Y, Deng M, Yu X, Zhang W (2004) Two-stage photo-biological production of hydrogen by marine green alga *Platymonas subcordiformis*. Biochem Eng J 19:69–73

Gugger MF, Hoffmann L (2004) Polyphyly of the true branching cyanobacterial (stigonematales). Int J Syst Evol Microsc 54:349–357

Gust D, Moore TA, Moore AL (2009) Solar fuels via artificial photosynthesis. Acc Chem Res 42(12):1890–1898

Hallenbeck PC (2011) Hydrogen production by cyanobacteria. In: Hallenbeck PC (ed) Microbial technologies at advanced biofuel production. Springer US, Boston, pp 15–28

Hallenbeck PC, Abo-Hashesh M, Ghosh D (2012) Strategies for improving biological hydrogen production. Bioresour Technol 110:1–9

Hambourger M, Gervaldo M, Svedruzic D, King PW, Gust D, Ghirardi M, Moore AL, Moore TA (2008) [FeFe]-hydrogenase-catalyzed H_2 production in a photoelectrochemical. Biofuels Cell J Am Chem Soc 130:2015–2022

Happe T, Hemsheimer A (2011) Metalloprotein mimics – old tools in a new light. Trends Biotechnol 32(4):170–176

Happe T, Kaminski A (2002) Differential regulation of the Fe hydrogenase during anaerobic adaptation in the green alga *Chlamydomonas reinhardtii*. Eur J Biochem 269:1022–1032

Happe T, Naber JD (2003) Isolation, characterization and N-terminal amino acid sequence of hydrogenase from the green alga *Chlamydomonas reinhardtii*. Eur J Biochem 214:475–481

Happe T, Mosler B, Naber JD (1994) Induction, localization and metal content of hydrogenase in the green alga *Chlamydomonas reinhardtii*. Eur J Biochem 222:769–774

Healey FP (1970) Hydrogen evolution by microalgae. Planta (Berl) 91:220–226

Heinekey MD (2009) Hydrogenase enzymes: recent structural studies and active site models. J Organomet Chem 694:2671–2680

Hemaiswarya S, Raja R, Ravikumar R, Carvalho IS (2013) Mechanism of Action of Probiotics. Braz Arch Biol Technol 56:113–119

Hemschemeier A, Happe T (2011) Alternative photosynthetic electron transport pathways during anaerobiosis in the green alga *Chlamydomonas reinhardtii*. Biochim Biophys Acta 1807:919–926

Hemschemeier A, Fouchard S, Cournac L, Peltier G, Happe T (2008) Hydrogen production by *Chlamydomonas reinhardtii*: an elaborate interplay of electron sources and sinks. Planta 227:397–407

Hoffman BM, Lukoyanov D, Dean DR, Seefeldt LC (2013) Nitrogenase: a draft mechanism Acc. Chem Res 46(2):587–595

Horch M, Lauterbach L, Lenz O, Hildebrandt P, Zebger I (2012) NAD(H)-coupled hydrogen cycling – structure–function relationships of bidirectional [NiFe] hydrogenases. FEBS Lett 586:545–556

Horner DS, Heil B, Happe T, Embley TM (2002) Iron hydrogenases-ancient enzymes in modern eukaryotes. Trends Biotechnol 27(3):148–153

Howard JB, Rees DC (1996) Structural basis of biological nitrogen fixation. Chem Rev 96:2965–2982

Hu Y, Ribbe MW (2013) Nitrogenase assembly. Biochim Biophys Acta 1827:1112–1122

Jamers A, Blust R, De Coen W (2009) Omics in algae: paving the way for a systems biological understanding of algal stress phenomena? Aquat Toxicol 92:114

Jones AK, Lamle SE, Pershad HR, Vincent KA, Albracht SPJ, Armstrong FA (2003) Enzyme electrokinetics: electrochemical studies of the anaerobic interconversions between active an inactive states of *Allochromatium vinosum* [NiFe] hydrogenase. J Am Chem Soc 125:8505–8514

Kallas T, Rebitre MC, Rippka R, Tandaeu de Marsa N (1983) The structural *nif* genes of the cyanobacterial *Gloeothece* sp. and *Calothrix* sp. share homology with those of *Anabaena* sp., but the *Gloeothece* genes have a different arrangement. J Bacteriol 155:427–431

Kalyanasundaram K, Graetzel M (2010) Artificial photosynthesis: biomimetic approaches to solar energy conversion and storage. Curr Opin Biotechnol 21:298–310

Kamp C, Silakov A, Winkler M, Reijerse EJ, Lubitz W, Happe T (2008) Isolation and first EPR characterization of the [FeFe]-hydrogenases from green algae. Biochim Biophys Acta 1777:410–416

Kessler E (1962) Hydrogenase und H_2 Stoffwechsel bei algen vortr Gesamptgebiet. Bot NF 1:62–101

Kessler E (1974) Hydrogenase, photoreduction and anaerobic growth. In: Stewart WDP (ed) Algal physiology and biochemistry. Blackwell, Oxford, pp 456–473

Khetkorn W, Baebprasert W, Lindblad P, Incharoensakdi A (2012) Redirecting the electron flow towards the nitrogenase and bidirectional Hox-hydrogenase by using specific inhibitors results in enhanced H_2 production in the cyanobacterium *Anabaena siamensis* TISTR 8012. Bioresour Technol 118:265–271

Kim DH, Kim MS (2011) Hydrogenases for biohydrogen production. Bioresour Technol 102:8423–8431

Kima JP, Kang CD, Park TY, Kim MS, Sim SJ (2006) Enhanced hydrogen production by controlling light intensity in sulphur deprived *Chlamydomonas reinhardtii* culture. Int J Hydrog Energy 31:1585–1590

King PW, Posewitz MC, Ghirardi ML, Seibert M (2006) Functional studies of [FeFe] hydrogenase maturation in an *Escherichia coli* biosynthetic system. J Bacteriol 188:2163–2172

Kojima E, Lin B (2004) Effect of partial shading on photoproduction of hydrogen by *Chlorella*. J Biosci Bioeng 97:317–321

Kosourov SN, Seibert M (2009) Hydrogen photoproduction by nutrient-deprived *Chlamydomonas reinhardtii* cells immobilized within thin alginate films under aerobic and anaerobic conditions. Biotechnol Bioeng 102:50–58

Kosourov S, Patrusheva E, Ghirardi ML, Seibert M, Tsygankov A (2007) A comparison of hydrogen photoproduction by sulfur-deprived *Chlamydomonas reinhardtii* under different growth conditions. J Biotechnol 128:776–787

Kosourov SN, Batyrova KA, Petushkova EP, Tsygankov AA, Maria LG, Seibert M (2012) Maximizing the hydrogen photoproduction yields in *Chlamydomonas reinhardtii* cultures: the effect of the H2 partial pressure. Int J Hydrog Energy 37:8850–8858

Kruse O, Rupprecht J, Bader KP, Thomas-Hall S, Schenk PM, Finazzi G, Hankamer B (2005) Improved photobiological H_2 production in engineered green algal cells. J Biol Chem 280:34170–34177

Kufryk G (2013) Advances in utilizing cyanobacteria for hydrogen production. Adv Microbiol 3:60–68

Kützing FT (1849) Species algarum. FA Brockhaus, Leipzig

Lambertz C, Chernev P, Klingan K, Leidel N, Sigfridsson KGV, Happe T, Haumann M (2014) Electronic and molecular structures of the active-site H-cluster in [FeFe]-hydrogenase determined by site-selective X-ray spectroscopy and quantum chemical calculations. Chem Sci 5:1187–1203

Lamle SE, Halliwell LM, Armstrong FA, Albracht SP (2003) The electrochemical interconversions between the active and inactive states of a [NiFe]-hydrogenase; implications for the development of a bio-fuel cell. J Inorg Biochem 96:174

Laurinavichene T, Tolstygina I, Tsygankov A (2004) The effect of light intensity on hydrogen production by sulfur-deprived *Chlamydomonas reinhardtii*. J Biotechnol 114:143–151

Lee HS, Vermaas WFJ, Rittman BE (2010) Biological hydrogen production: prospects and challenges. Trends Biotechnol 28(5):262–271

Leite GB, Hallenbeck PC (2014) Engineered cyanobacteria: research and application in bioenergy. In: Gupta VK, Kubicek CP, Saddler J, Xu F, Tuohy MG (eds)

Bioenergy research: advances and applications. Elsevier, Oxford. doi:10.1016/B978-0-444-59561-4.00022-X

Leon-Banares R, Gonza D, Galvan A, Fernandez E (2004) Transgenic microalgae as green cell-factories. Trends Biotechnol 22(1):45–52

Lindahl PA (2012) Metal–metal bonds in biology. J Inorg Chem 106:172–178

Lindblad P, Christensson K, Lindberg P, Fedorov A, Pinto F, Tsygankov A (2002) Photoproduction of H_2 by wild type *Anabaena* PCC 7120 and a hydrogen uptake deficient mutant: from laboratory experiments to outdoor culture. Int J Hydrog Energy 27:1271–1281

Lopez D, Casero D, Cokus SJ, Merchant SS, Pellegrini M (2011) Algal functional annotation tool: a web based analysis suite to functionally interpret large gene lists using integrated annotation and expression data. BMC Bioinforma 12:282

Lubitz W, Ogata H, Rüdiger O, Reijerse E (2014) Hydrogenases. Chem Rev 114:4081–4148

Ludwig M, Schulz-Friedrich R, Appel J (2006) Occurrence of hydrogenases in cyanobacterial and anoxygenic photosynthetic bacteria: implications for phylogenetic origin of cyanobacterial and algal hydrogenases. J Mol Evol 63:758–768

Marr AC, Spencer DJE, Schroder M (2001) Structural mimics for the active site of [NiFe] hydrogenase. Coord Chem Rev 219–221:1055–1074

Martens R, Liese A (2004) Biotechnological applications of hydrogenases. Curr Opin Biotechnol 15:343–348

Masojidek J, Torzillo G, Koblizek M (2013) Photosynthesis in microalgae. In: Richmond A, Hu Q (eds) Handbook of microalgal culture: applied phycology and biotechnology, 2nd edn. Wiley, Oxford, pp 21–36

Mathews J, Wang G (2009) Metabolic pathway engineering for enhanced biohydrogen production. Int J Hydrog Energy 34:7404–7416

Matthew T, Zhou W, Rupprecht J, Lim L, Thomas-Hall SR, Doebbe A, Kruse O, Hankamer B, Marx UC, Smith SM, Schenk PM (2009) The metabolome of *Chlamydomonas reinhardtii* following induction of anaerobic H_2 production by sulfur depletion. Curr Opin Biotechnol 284:23415–23425

May P, Christian JO, Kempa S, Walther D (2009) ChlamyCyc: an integrative systems biology database and web-portal for Chlamydomonas reinhardtii. BMC Genomics 10:209

Mayfield SP, Franklin SE (2005) Expression of human antibodies in eukaryotic micro-algae. Vaccine 23:1828–1832

McGlynn SE, Shepard EM, Winslow MA, Naumov AV, Duschene KS, Posewitz MC, Broderick WE (2008) FEMS Lett 582:2183–2187

Mckinlay JB, Harwood CS (2010) Photobiological production of hydrogen gas as a biofuel. Curr Opin Biotechnol 21:244–251

Melis A (2002) Green alga hydrogen production: progress, challenges and prospects. Int J Hydrog Energy 27:1217–1228

Melis A (2007) Photosynthetic H_2 metabolism in Chlamydomonas reinhardtii (unicellular green algae). Planta 226:1075–1086

Melis A (2009) Solar energy conversion efficiencies in photosynthesis: minimizing the chlorophyll antenna to maximize efficiency. Plant Sci 177:272–280

Melis A, Happe T (2001) Hydrogen production. Green algae as a source of energy. Plant Physiol 127:740–748

Melis A, Zhang L, Forestier M, Ghirardi M, Seibert M (2000) Sustained photobiological hydrogen gas production upon reversible inactivation of oxygen evolution in the green alga Chlamydomonas reinhardtii. Plant Physiol 122:127–135

Merchant SS et al (2007) The Chlamydomonas genome reveals the evolution of key animal and plant functions. Science 318:245–250

Meuser JE, Boyd ES, Ananyey G, Karns D, Radakovits R, Narayana Murthy UM, GHirardi ML, Dismukes C, Peters JW, Posewitz MC (2011) Evolutionary significance of an algal gene encoding an [FeFe]-hydrogenase with F-domain homology and hydrogenase activity in Chlorella variabilis NC64A. Planta 234:829–843

Mevarech M, Rice D, Haselkorn R (1980) Nucleotide sequence of a cyanobacterial nifH gene coding for nitrogenase reductase. Proc Natl Acad Sci 77(11):6476–6480

Meyer TJ (1989) Chemical approaches to artificial photosynthesis. Acc Chem Res 22(163):170

Meyer J (2007) [FeFe] hydrogenases and their evolution: a genomic perspective. Cell Mol Life Sci 64:1063–1084

Miyake J, Miyake M, Asada Y (1999) Biotechnological hydrogen production: research for efficient light energy conversion. J Biotechnol 70:89–101

Mulder DW, Shepard EM, Meuser JE, Joshi N, King PW, Posewitz MC, Broderick JB, Peters JB (2011) Insights into [FeFe]-hydrogenase structure, mechanism, and maturation. Structure 19:1038–1052

Mulligan ME, Haselkorn R (1989) Nitrogen fixation (nif) genes of the cyanobacterial Anabaena species strain PCC7120. J Biol Chem 264(32):19200–19207

Mus F, Cournac L, Cardettini V, Caruana A, Peltier G (2005) Inhibitor studies on non-photochemical PQ reduction and H_2 photoproduction in Chlamydomonas reinhardtii. Biochim Biophys Acta 1708:322–332

Nguyen AV, Thomas-Hall SR, Malnoe A, Timmins M, Mussgnug JH, Rupprecht J, Kruse O, Hankamer O, Schenk PM (2008) Transcriptome for photobiological hydrogen production induced by sulfur deprivation in the green alga Chlamydomonas reinhardtii. Eukaryot Cell 7(11):1965–1979

Nguyen HM, Baudet M, Cuine S, Adriano JM, Barthe D, Billon E, Bruley C, Beisson F, Peltier G, Ferro M, Li-Beisson Y (2011) Proteomic profiling of oil bodies isolated from the unicellular green microalga Chlamydomonas reinhardtii: with focus on proteins involved in lipid metabolism. Proteomics 11:4266

Nicolet Y, Piras C, Legrand P, Hatchikian CE, Fontecilla-Camps JC (1999) Desulfovibrio desulfuricans iron hydrogenase: the structure shows unusual coordination to an active site Fe binuclear center. Structure 7:13–23

Nicolet Y, Lemon BJ, Fontecilla-Camps JC, Peters JW (2000) A novel FeS cluster in Fe-only hydrogenases. TIBS 25:138–153

Nicolet Y, de Lacey AL, Vernede X, Fernandez VM, Hatchikian CE, Fontecilla-Camps JE (2001) Crystallographic and FTIR spectroscopic evidence of changes in Fe coordination upon reduction of the active site of the Fe-only-hydrogenase from Desulfovibrio desulfuricans. J Am Chem Soc 123:596–1601

Nicolet Y, Cavazza C, Fontecilla-Camps JC (2002) Fe only hydrogenases: structure, function and evolution. J Inorg Biochem 91:1–8

Nicolet Y, Fontecilla-Camps JC, Fontecava M (2010) Maturation of [FeFe]-hydrogenases: structures and mechanisms. Int J Hydrog Energy 35:10750–10760

Oh YK, Raj SM, Jung GY, Park S (2011) Current status of the metabolic engineering of microorganisms for biohydrogen production. Bioresour Technol 102:8357–8367

Okhi Y (2014) Synthetic analogous of the active sites of nitrogenase and [NiFe] hydrogenase. Bull Chem Soc Jpn 87(1):119

Oliveira Dal'Molin CG, Quek L-E, Palfreyman R, Nielsen L (2011) AlgaGEM–a genomescale metabolic reconstruction of algae based on the Chlamydomonas reinhardtii genome. BMC Genomics 12:S5

Oncel S (2013) Microalgae for a macro energy world. Renew Syst Energy Rev 26:241–264

Oncel SS (2015) Biohydrogen from microalgae, uniting energy, life, and green future. In: Kim SK (ed) Handbook of marine microalgae. Biotechnology Advances Academic Press, pp 159–196

Oncel S, Kose A (2014) Comparison of tubular and panel type photobioreactors for biohydrogen production utilizing Chlamydomonas reinhardtii considering mixing time and light intensity. Bioresour Technol 151:265–270

Oncel S, Sabankay M (2012) Microalgal biohydrogen production considering light energy and mixing time as the two key features for scale-up. Bioresour Technol 121:228–234

Oncel S, Kose A, Faraloni C, Imamoglu E, Elibol M, Torzillo G, Vardar Sukan F (2014) Biohydrogen production using mutant strains of Chlamydomonas reinhardtii: the effects of light intensity and illumination patterns. Biochem Eng J 92:47–52

Peters JW, Szilagyi RK (2006) Exploring new frontiers of nitrogenase structure and mechanism. Curr Opin Chem Biol 10:101–108

Peters JW, Lanzilotta WN, Lemon BJ, Seefeldt LC (1998) X-ray crystal structure of the Fe-only hydrogenase (CpI) from Clostridium pasteurianum to 1.8 angstrom resolution. Science 282:1853–1858

Peters JW, Szilagyi RK, Naumov A, Douglas T (2006) A radical solution for the biosynthesis of the H-cluster of hydrogenase. FEBS Lett 580:363–367

Pierik AJ, Hulstein M, Hagen WR, Albracht SPJ (1998) A low-spin iron with CN and CO as intrinsic ligands forms the core of the active site in [Fe]-hydrogenases. Eur J Biochem 258:572–578

Polle JEW, Kanakagiri S, Jin ES, Masuda T, Melis A (2002) Truncated chlorophyll antenna size of the photosystems – a practical method to improve microalgal productivity and hydrogen production in mass culture. Int J Hydrog Energy 27:1257–1264

Posewitz MC, King PW, Smolinski SL, Zhang L, Seibert M, Ghirardi ML (2004) Discovery of two novel radical S-adenosylmethionine proteins required for the assembly of an active [Fe] hydrogenase. J Biol Chem 279:25711–25720

Postgate JR, Eady RR (1988) The evolution of biological nitrogen fixation. In: Bothe H, de Bruijn FJ, Newton WE (eds) Nitrogen fixation: hundred years after. Gustav Fischer, Stuttgart, pp 31–40

Prince RC, Kheshgi HD (2005) The photobiological production of hydrogen: potential efficiency and effectiveness as a renewable fuel. Crit Rev Microbiol 31:19–31

Rees DC, Howard JB (2000) Nitrogenase: standing at the crossroads. Curr Opin Chem Biol 4:559–566

Rice D, Mazur BJ, Haselkorn R (1982) Isolation and physical mapping of nitrogen fixation genes from the cyanobacterium Anabaena 7120. J Biol Chem 257(21):13157–13163

Rismandi Yazdi H, Haznedaroglu BZ, Bibby K, Peccia J (2011) Transcriptome sequencing and annotation of the microalgae Dunaliella tertiolecta: pathway description and gene discovery for production of next-generation biofuels. BMC Genomics 12:148

Rolland N, Atteia A, Decottignies P, Garin J, Hippler M, Kreimer G, Lemaire SD, Mittag M, Wagner V (2009) Chlamydomonas proteomics. Curr Opin Microbiol 12(3):285–291

Roseboom W, de Lacey AL, Fernandez WM, Hatchikian CE, Albracht SPJ (2006) The active site of the [FeFe]-hydrogenase from Desulfovibrio desulfuricans. II. Redox properties, light sensitivity and CO-ligand exchange as observed by infrared spectroscopy. J Biol Inorg Chem 11:102–118

Rosenberg JN, Oyler GA, Wilkinson L, Betenbaugh MJ (2008) A green light for engineered algae: redirecting metabolism to fuel a biotechnology revolution. Curr Opin Biotechnol 19:430–436

Rubach JK, Brazzolotto X, Gaillard J, Fontecave M (2005) Biochemical characterization of the HydE and HydG iron-only hydrogenase maturation enzymes from Thermatoga maritima. FEBS Lett 579:5055–5060

Rubio LM, Ludden PW (2005) Maturation of nitrogenase: a biochemical puzzle. J Bacteriol 187(2):405–414

Rupprecht J (2009) From system biology to fuel-Chlamydomonas reinhardtii as a model for a systems biology approach to improve biohydrogen production. J Biotechnol 142:10–20

Sakurai H, Masukawa H, Kitashima M, Inoue K (2013) Photobiological hydrogen production: bioenergetics and challenges for its practical application. J Photochem Photobiol C 17:1–25

Saleem M, Chakrabarti M, Raman AAA, Hasan DB, Dauad WMAW, Mustafa A (2012) Hydrogen production by Chlamydomonas reinhardtii in a two-stage process with and without illumination at alkaline pH. Int J Hydrog Energy 37:4930–4934

Schawarz G, Mendel RR, Ribbe MW (2009) Molybdenum cofactors, enzymes and pathways. Nature 460(13). doi: 10.1038/nature08302

Schmitter JM, Jacquot JP, de Lamotte-Guery F, Beauvallet C, Dutka S, Gadal P et al (1998) Purification, properties and complete amino acid sequence of the ferredoxin from a green alga, Chlamydomonas reinhardtii. Eur J Biochem 172:405–412

Scoma A, Krawietz D, Faraloni C, Giannelli L, Happe T, Torzillo G (2012) Sustained H_2 production in a Chlamydomonas reinhardtii D1 protein mutant. J Biotechnol 157:613–619

Seefeldt LC, Hoffman BM, Dean DR (2012) Electron transfer in nitrogenase catalysis. Curr Opin Chem Biol 16:19–25

Shafaat HS, Rüdriger O, Ogata H, Lubitz W (2013) [NiFe] hydrogenases: a common active site for hydrogen metabolism under diverse conditions. Biochim Biophys Acta 1827:986–1002

Shepard EM, Boyd ES, Broderick JB, Peters JW (2011) Biosynthesis of complex iron-sulfur enzymes. Curr Opin Chem Biol 15:319–327

Skjanes K, Pinto FL, Lindblad P (2010) Evidence for transcription of three genes with characteristics of hydrogenases in the green alga Chlamydomonas noctigama. Int J Hydrog Energy 35:1074–1088

Spolaore P, Joannis-Cassan C, Duran E, Isambert A (2006) Commercial applications of microalgae. J Biosci Bioeng 101:87–96

Srirangan K, Pyne ME, Chou CP (2011) Biochemical and genetic engineering strategies to enhance hydrogen production in photosynthetic algae and cyanobacterial. Bioresour Technol 102:8589–8604

Stephenson M, Stickland SH (1931) Hydrogenase: a bacterial enzyme activating molecular hydrogen. I. The properties of hydrogenases. Biochem J 25:205–214

Sun L, Akermark B, Ott S (2005) Iron hydrogenase active site mimics in supramolecular systems aiming for light-driven hydrogen production. Coord Chem Rev 249:1653–1663

Surzycki R, Cournac L, Peltier G, Rochaix JD (2007) Potential for hydrogen production with inducible chloroplast gene expression in Chlamydomonas. Proc Natl Acad Sci U S A 104:17548–17553

Switzer L (1981) Spirulina: the whole food revolution. Proteus Corporation Banta Books, Toronto

Sybirna K, Ezanno P, Baffert C, Leger C, Bottin H (2013) Arginine171 of Chlamydomonas reinhardtii [FeeFe] hydrogenase HydA1 plays a crucial role in electron transfer to its catalytic center. Int J Hydrog Energy 38:2998–3002

Tamagnini P, Axelsson R, Lindberg P, Oxelfelt F, Wünschiers R, Lindblad P (2002) Hydrogenases and hydrogen metabolism of cyanobacteria. Microbiol Mol Biol Rev 66(1):1–20

Tamagnini P, Leitao E, Oliveira P, Ferreira D, Pinto F, Harris DJ, Heidorn T, Lindblad P (2007) Cyanobacterial hydrogenase: diversity, regulation and applications. FEMS Microbiol Rev 31(6):692–720

Taylor B, Lee C, Bunt J (1973) Nitrogen-fixation associated with the marine blue-green alga, Trichodesmium, as measured by the acetylene-reduction technique. Arch Mikrobiol 88:205–212

Thompson AW, Zehr JP (2013) Cellular interactions: lessons from the nitrogen fixing cyanobacteria. J Phycol 49:1024–1035

Torzillo G, Seibert M (2013) Hydrogen production by microalgae. In: Richmond A, Hu Q (eds) Handbook of microalgal culture: applied phycology and biotechnology, 2nd edn. Wiley, Oxford, pp 417–444

Torzillo G, Scoma A, Faraloni C, Gianelli L (2014) Advances in the biotechnology of hydrogen production with the microalga Chlamydomonas reinhardtii. Crit Rev Biotechnol:1–12. doi: 10.3109/07388551.2014.900734

Tóth SZ, Schansker G, Garab G, Strasser RJ (2007) Photosynthetic electron transport activity in heat-treated barley leaves, the role of internal alternative electron donors to photosystem II. BBA Bioenerg 1767:295–305

Tsygankov AA (2007) Nitrogen fixing cyanobacteria: a review. Appl Biochem Microbiol 43(3):250–259

Tsygankov A, Kosourov S, Seibert M, Ghirardi ML (2002) Hydrogen photoproduction under continuous illumination by sulfur deprived, synchronous Chlamydomonas reinhardtii cultures. Int J Hydrog Energy 27:1239–1244

Vignais PN, Billoud B, Meyer J (2001) Classification and phylogeny of hydrogenases. FEMS Microbiol Rev 25:455–501

Volbeda SA, Fontecilla-Camps JC (2005) Structure–function relationships of nickel–iron sites in hydrogenase and a comparison with the active sites of other nickel–iron enzymes. Coord Chem Rev 249:1609–1619

Volbeda A, Charon MH, Piras C, Hatchikian EC, Frey M, Fontecilla-Camps JC (1995) Cystal structure of [NiFe] hydrogenase. Nature 373:580–587

Walker TL, Collet C, Purton S (2005) Algal transgenics in the genomic era. J Phycol 41:1077–1093

Weckwerth W (2011) Green systems biology – from single genomes, proteomes and metabolomes to ecosystems research and biotechnology. J Proteomics 875:284–305

Wilmotte A, Herdman M (2001) In: Boone DR, Castenholz RW (eds) Phylogenetic relationships among the cyanobacteria based on 16S rRNA sequences, Bergey's manual of systematic bacteriology, 2nd edn. Springer-Verlag, New York, pp 487–493

Winkler M, Heil B, Happe T (2002a) Isolation and molecular characterization of the [Fe]-hydrogenase from the unicellular green alga Chlorella fusca. Biochim Biophys Acta 1576:330–334

Winkler M, Hemscheimer A, Gotor C, Melis A, Happe T (2002b) [Fe]-hydrogenases in green algae: photo-fermentation and hydrogen evolution under sulfur deprivation. Int J Hydrog Energy 27(11–12): 1431–1439

Winkler M, Kuhlgert S, Hippler M, Happe T (2009) Characterization of the key step for light-driven hydrogen evolution in green algae. J Biol Chem 284(52):36620–36627

Winkler M, Esselborn J, Happe T (2013) Molecular basis of [FeFe]-hydrogenase function: an insight into the complex interplay between protein and catalytic cofactor. Bioch Biophys Acta 1827:974–985

Wykoff DD, Davies JP, Melis A, Grossman AR (1998) The regulation of photosynthetic electron transport during nutrient deprivation in Chlamydomonas reinhardtii. Plant Physiol 117:129–139

Yacoby I, Tegler LT, Pochekailov S, Zhang S, King PW (2012) Optimized expression and purification for high-activity preparations of algal [FeFe]-hydrogenase. PLoS One 7:35886

Young JPW (1992) Phylogenetic classification of nitrogenfixing organisms. In: Stacey G, Evans HJ, Burris RH (eds) Biological nitrogen fixation. Chapman & Hall, New York, pp 43–86

Zehr JP, Jenkins BD, Short SM, Steward GF (2003) Nitrogenase gene diversity and microbial community structure: a cross-system comparison. Environ Microbiol 7:539–554

Zhang L, Happe T, Melis A (2002) Biochemical and morphological characterization of sulfur-deprived and H_2-producing Chlamydomonas reinhardtii (green alga). Planta 214:552–561

Zhang L, He M, Liu J (2014) The enhancement mechanism of hydrogen photoproduction in Chlorella protothecoides under nitrogen limitation and sulfur deprivation. Int J Hydrog Energy 39:8969–8976

Zhao Y, Bian SM, Zhou HN, Huang JF (2006) Diversity of nitrogenase systems in diazotrophs. Plant Biol 48(7):745–755

Biohydrogen Production Scenario for Asian Countries

10

Rupam Kataki, Rahul S. Chutia, Neon J. Bordoloi, Ruprekha Saikia, Debashis Sut, Rumi Narzari, Lina Gogoi, G.N. Nikhil, Omprakash Sarkar, and S. Venkata Mohan

Abstract

Despite continuous advancement in energy technologies, the greenhouse gas and pollutant emission due to combustion of fossil fuel is increasing day by day due to its growing demand. With the growing worldwide concern regarding increasing global climate change and depleting energy source, it has become the necessity of the hour to generate fuel with safer, efficient, economic, and reasonably environmental-friendly technology. To address this issue, a variety of efficient end-use technologies and alternative fuels have been proposed; this includes compressed natural gas; reformulated gasoline or diesel; methanol; ethanol; synthetic liquids from natural gas, biomass, or coal; and hydrogen. In this regard hydrogen has emerged as a promising option since it offers to solve various important societal impacts of fuel use at the same time. Hydrogen (H_2) produced through wastewater treatment using biological routes (dark and photo-fermentation) can be considered as a renewable and sustainable resource. Negative-valued wastewater contains high levels of biodegradable organic material with net positive energy and minimizes the economics of H_2 production and treatment cost. This chapter mainly focuses on the global biohydrogen research trend specifically in Asian countries. Bibliometric and scientometric analysis performed with ISI Web of Knowledge [Thomson Reuters] documented significant increments in publications wherein India stands top in biohydrogen production using wastewater. Current status and road map showed that China followed by other Asian countries have significantly contributed towards H_2 production. Future perspective suggests for integrative H_2 production strategies such as microbial electrolysis, polyhydroxyalkanoate (PHA) production, bioaugmentation, and metabolic engineering to overcome some of the limitations for process scale-up.

R. Kataki (✉) • R.S. Chutia • N.J. Bordoloi
R. Saikia • D. Sut • R. Narzari • L. Gogoi
Biofuel Laboratory, Department of Energy, Tezpur University, Tezpur 784028, India
e-mail: rupamkataki@gmail.com

G.N. Nikhil • O. Sarkar • S. Venkata Mohan
Bioengineering and Environmental Sciences (BEES), CSIR-Indian Institute of Chemical Technology (CSIR-IICT), Hyderabad 500 007, India

A. Singh, D. Rathore (eds.), *Biohydrogen Production: Sustainability of Current Technology and Future Perspective*, DOI 10.1007/978-81-322-3577-4_10

10.1 Present Scenario

Increasing gaps between the energy requirement of the industrialized world and an inability to replenish needs from limited energy sources have resulted in a steep increase in fossil fuel utilization. This has not only put a severe strain on the depleting fossil fuels but also resulted in an alarming increase in pollution levels across the globe. An ever-increasing level of greenhouse gases (GHGs) from the combustion of fossil fuels in turn aggravated the perils of global warming. Combustion of fossil fuels adds about 6 gigatons of carbon per year in the form of carbon dioxide to the atmosphere (IPCC 2006). At present, the concentration of CO_2 is found to be exceeding 350 ppm where it can potentially intensify the greenhouse effect by raising the global temperature. The limited availability of global oil reserves and concerns about climate change from greenhouse gas emissions instigated marked interest in the development of clean and renewable energy alternatives to satisfy the growing energy demands. Therefore, diversification of energy resources is an essential requirement in the present-day energy scenario (Nouni 2012). Moreover, rapid development of alternative, renewable, carbon-neutral, and eco-friendly fuels is of paramount importance to fulfill the burgeoning energy demands.

Although H_2 is the most common element on earth, it does not occur in elemental form. The industrial production of molecular H_2 is mainly from fossil sources through steam reforming of natural gas, water-splitting electrolysis process, and as a by-product from some industrial processes. Currently, global H_2 production exceeds 1 billion m^3/day of which 48 % is produced from natural gas, 30 % from oil (often on-site in refineries), 18 % from coal, and the remaining (4 %) by water electrolysis (de Jong 2008). The production of H_2 from fossil fuels is accompanied by the production of greenhouse gases, namely, CO_2, CH_4, etc. At present, H_2 synthesis from biomass/waste through biological routes has emerging interest due to its renewability and sustainable nature. Microorganisms have flexible and diverse metabolic machinery to convert/synthesize a variety of organics to various forms of bioenergy. Establishing a practicable link between terminal electron acceptor-limited microorganism and an electron sink is the main basis for many of the bioenergy generation processes (Madsen 2008). At present, extensive as well as intense research is being focused toward bioenergy generation from renewable resources throughout the world.

10.2 Biological Hydrogen Production

Hydrogen production has diverse routes, viz., physical, chemical, biological, and thermochemical. Biological H_2 production is mainly contributed by anaerobic fermentation, which is broadly classified into two main categories: light independent and light dependent. Dark fermentation is a light-independent process that employs both obligate anaerobes and facultative bacteria for H_2 production from a variety of potentially biodegradable substrates, including wastewater. The generation of H_2 accompanies formation of soluble metabolic products as organic acids (e.g., acetate or butyrate) and alcohols (e.g., acetone, butanol) (Singh et al. 2010). Amassing of organic acids results in a sharp drop in pH and subsequent inhibition of bacterial H_2 production and finally, resulting in low yields. Biophotolysis is a light-dependent process wherein green algae and cyanobacteria undertake direct/indirect biophotolysis to produce H_2 by utilizing inorganic CO_2 in the presence of sunlight and water. Photosynthetic bacteria manifest H_2 production in the presence of light by photo-fermentation of organic substrates. It is essential to note that the biochemistry and microbial metabolism involved vary significantly based on the function of biocatalyst, operating conditions, microenvironment, and substrate/feedstock (Srikanth et al. 2009b).

Generation of H_2 via biological routes is relatively pollutant-free, requires low energy inputs, and is therefore considered as a prospective replacement to the conventional physicochemical methods (Mohan 2010). In recent years, there are increasing research activities in the domain of "biohydrogen," as shown by the increasing number of peer-reviewed articles with "biohydrogen" in the title. Bibliometric and scientometric analysis performed in April 2015 with ISI Web of Knowledge [Thomson Reuters] documented significant increments in publications on biohydrogen production in Asian countries since 2000. About 160 publications were documented in the year 2014 with citations of about 4000. Out of the total records on biohydrogen, most of them relate to the energy and fuels (>800 records), and among the top ten Asian countries, People's Republic of China is the leading country with more than 500 records (Fig. 10.1).

10.2.1 Photobiological Process

There are three physiologically distinct types of photosynthetic microorganisms, viz., unicellular green algae, cyanobacteria, and photosynthetic bacteria (Das and Veziroğlu 2001). H_2 production in photosynthetic microorganism depends on the type of photosynthetic machinery (anoxygenic and oxygenic) wherein the mechanism is marginally different (Richmond and Hu 2013). The light energy generates proton gradient and supplies electrons either by water-splitting reaction (direct photolysis) or from photosystem II (PSII)-independent process originating from the starch metabolism (indirect photolysis) (Allakhverdiev et al. 2010). In green algae and cyanobacteria, direct photolysis via oxygenic photosynthesis involves light as a driving force for PSII resulting in the production of reducing equivalents that oxidizes water into electrons, protons, and O_2 (Doebbe et al. 2010; Krassen et al. 2009). In both cases, ferredoxin in reduced form serves as an electron donor for [FeFe] hydrogenases followed by the reducing powers that are transferred to the chlorophyll α dimer (P700) residing in photosystem I (PSI) that gets excited by light absorption yielding electrons at a potential of −1.32 V (vs SHE) (Krabben et al. 2000). Finally, these electrons flow at a potential of −0.45 V through the internal electron transport chain to the iron-sulfur clusters located at the acceptor site of PSI (Jordan et al. 1998). In both the photolysis processes, PSII-driven water splitting converts reducing equivalents to H_2 by the [FeFe] hydrogenase (*Hyd*A). Hydrogenase acts as a proton/electron release valve by recombining proton and electron to produce H_2. Hydrogenase activity depends on the reducing equivalents derived directly from the photosynthetic water splitting (driven by PSII) or indirectly by the metabolism of organic matter.

Microalgae have extremely active [FeFe]-hydrogenase enzyme with a high conversion ability (12–14 %) of solar energy to molecular H_2 for the oxidation of water in the chloroplast (Melis 2009; Melis and Happe 2001). However, during direct biophotolysis, oxygen is a dominant suppressor of the hydrogenase enzyme. As a result, direct biophotolysis operated for short periods upon the start of illumination inactivates the H_2 production process. In the indirect photosynthesis, protons and electrons generated from water are stored in the form of starch (synthesized from photophosphorylation). Then, the electrons flow into the plastoquinone pool and followed by *Hyd*A via PSI under certain stress conditions (Melis 2007). Cyanobacteria and green algae accumulate reserve compounds in the Calvin cycle and at night these compounds provide energy for the cell metabolism (Krassen et al. 2009). Cessation of a photosynthetic light reaction generates excess reductant that converts into H_2 by hydrogenase observed during the shift from aerobic to anaerobic environment. Two kinds of O_2-sensitive [FeFe] hydrogenases (*Hyd*A1 and *Hyd*A2) induced proton reduction to H_2 under anaerobic conditions (Doebbe et al. 2010). The combination of processes of photosynthesis and H_2 evolution is an indirect process and time delayed (Krassen et al. 2009). Cyanobacteria have one additional mechanism for H_2 production, that is, via heterocyst during N_2 fixation. Alternatively, H_2 production catalyzed by nitrogenase during nitrogen fixation is extremely O_2 sensitive. Cyanobacteria have a specific mechanism for protecting nitrogenases from O_2 through the localization of nitrogenase in the heterocyst (Vyas and Kumar 1995).

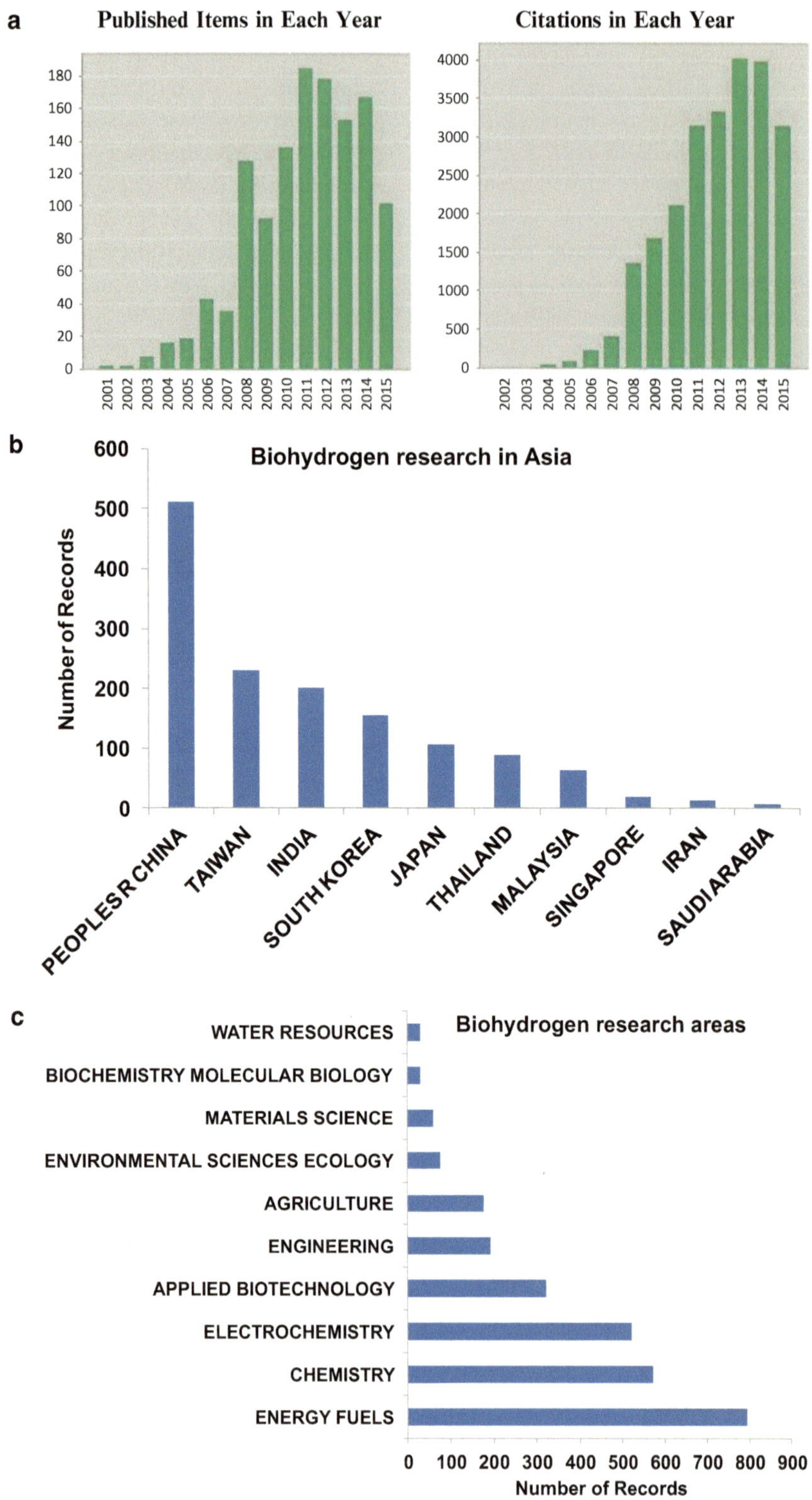

Fig. 10.1 Bibliometric and scientometric analysis for the topic "biohydrogen" taken from ISI Web of Knowledge [Thomson Reuters] – (**a**) publications and citation index; (**b**) number of records from Asian countries; (**c**) biohydrogen research areas

Under anaerobic conditions, photosynthetic bacteria use sunlight as energy source and produce H_2 by degrading the organic molecules. As photosynthetic bacteria do not require water as a source of electrons, it can easily circumvent the O_2 sensitivity issue that adversely affects the enzyme activity. Photosynthetic bacteria are capable in utilizing both the visible (400–700 nm) and the near-infrared (700–950 nm) spectrum. Anoxygenic photosynthesis in photosynthetic bacteria is advantageous for H_2 production, as it uses neither water as an electron source nor produces O_2 (Blankenship et al. 1995). Light absorption by a dimer of bacteriochlorophyll (*BChl*) simulates the reaction forming bacteriopheophytin (*BPh*) (Berg et al. 2002). The electron flow from *BPh* to quinine pool (QA) and then to the cytochrome subunit of the reaction center generates proton gradient driving the ATP generation and finally reduces to H_2 (Blankenship et al. 1995). The efficiency of light energy conversion to H_2 by photosynthetic bacteria is much higher than cyanobacteria because of the less quantum of light energy requirement compared to water photolysis (Batyrova et al. 2012; Melis 2012; Vyas and Kumar 1995). The photo-fermentation process is more feasible because of the ability of photosynthetic bacteria to trap energy from a wide range of light spectrum without producing oxygen and its versatility in utilizing various substrates.

The reduced ferredoxin serves as an electron donor for the [FeFe] hydrogenases in both oxygenic and anoxygenic photosynthesis, along with hydrogenases and nitrogenases that play a major role in H_2 production. The main drawback is that the O_2 liberated from the photolysis of water inhibits these two enzymes. Holding back the O_2 feedback mechanism is essentially required to enhance H_2 production. A reduction in sulfate concentration during algal growth resulted in decrement in photosynthesis, which reduced 90 % of O_2 production. This is sufficient for the hydrogenase enzyme to divert protons and electrons to yield H_2 for a longer period of time (Melis et al. 2000). The molecular architecture of the photosynthetic membrane makes it promising to transmit photosynthetically generated reducing equivalents from PSI to H_2 production (Millsaps et al. 2001). The PSI is a robust molecular photovoltaic device located on the non-appraised region of the thylakoid membrane (Lee et al. 2000). Studies have reported precipitating platinum on the stromal side of the photosynthetic thylakoid membrane, which is at the site of electron emergence from the PSI reaction center (Greenbaum 1985). The platinized chloroplast thylakoids are capable of trapping these electrons facilitating simultaneous photoevolution of H_2 and O_2 (Greenbaum 1988a, b). Photosynthetic membranes have the capability to transform with metals other than platinum, such as osmium and ruthenium. Doping of PSI with platinum showed no inhibitory effect over excitation transfer dynamics and/or the reaction center pigment (Lee et al. 1995). Functional nanoscale surface metallization at the reducing ends of isolated PSI was reported by substituting the negatively charged hexachloroplatinate ($[PtCl6]^{-2}$) with negatively charged ferredoxin, the naturally occurring water-soluble electron carrier in photosynthesis (Millsaps et al. 2001). Visible light-induced enzymatic H_2 production with the platinum colloid using the photosensitization of Mg chlorophyll *a* (Mg Chl *a*) was also reported (Saiki and Amao 2004). Mg Chl-*a* acts as an effective photosensitizer with an absorption maximum at 670 nm. However, most of the photobiological processes for H_2 production have major fundamental limitations, and practical engineering issues need to be resolved prior to the implementation in the real field (Hallenbeck and Benemann 2002). Bibliometric and scientometric analysis performed in April 2015 with ISI Web of Knowledge [Thomson Reuters] documented gradual improvement in publications on biohydrogen and photo-fermentation in Asian countries since 2007. The citation index also increased gradually and more than 170 citations were recorded in the year 2014 (Fig. 10.2).

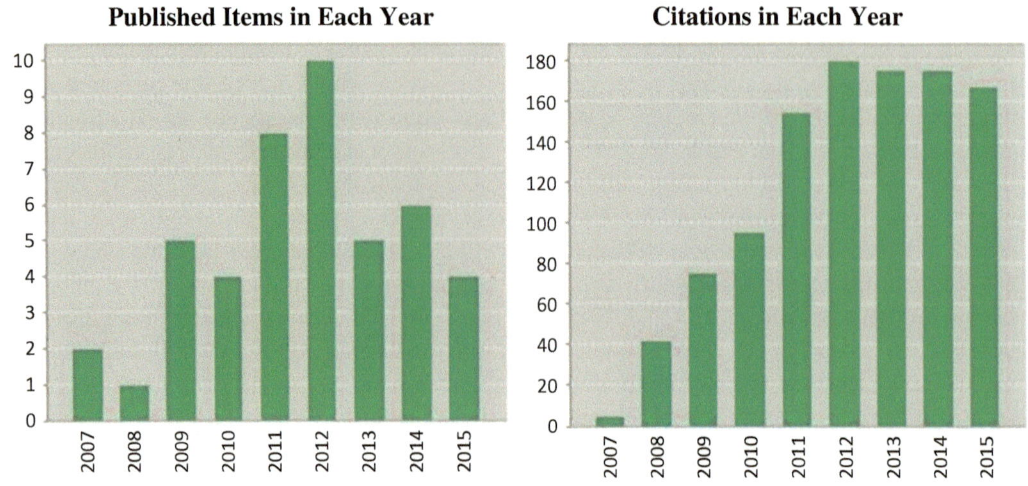

Fig. 10.2 Bibliometric and scientometric analysis for the topic "biohydrogen and photo-fermentation" taken from ISI Web of Knowledge [Thomson Reuters] representing publications and citation index

10.2.2 Dark Fermentation

Glycolysis is the primary biochemical pathway wherein the substrate is metabolized to pyruvate, which is the central molecule for microbial fermentation. The fate of pyruvate is diverse under anaerobic fermentation conditions. Pyruvate enters the acidogenic pathway and generates VFA, namely, acetic acid, propionic acid, butyric acid, malic acid, and so on, along with H_2 as by-product (Eqs. 10.1, 10.2, 10.3, 10.4, and 10.5). Both obligate and facultative acidogenic bacteria (AB) can catalyze H_2 production from organic substrates (Sarkar et al. 2013):

$$C_6H_{12}O_6 + 2H_2O \rightarrow 2CH_3.COOH + 2CO_2 + 4H_2\ (\text{acetic acid}) \tag{10.1}$$

$$C_6H_{12}O_6 \rightarrow CH_3.CH_2.CH_2.COOH + 2CO_2 + 2H_2\ (\text{butyric acid}) \tag{10.2}$$

$$C_6H_{12}O_6 + 2H_2 \rightarrow 2CH_3.CH_2.COOH + 2H_2O\ (\text{propionic acid}) \tag{10.3}$$

$$C_6H_{12}O_6 + 2H_2 \rightarrow COOH.CH_2.CH_2O.COOH + CO_2\ (\text{malic acid}) \tag{10.4}$$

$$C_6H_{12}O_6 \rightarrow CH_3.CH_2OH + CO_2\ (\text{ethanol}) \tag{10.5}$$

Facultative anaerobes convert pyruvate into acetyl-CoA and formate by the action of pyruvate formate lyase and then H_2 is produced by formate hydrogen lyase (Vardar-Schara et al. 2008). Obligate anaerobes convert pyruvate into acetyl-CoA and CO_2 through pyruvate ferredoxin oxidoreductase; this oxidation process requires the reduction of ferredoxin (Fd) (Kraemer and Bagley 2007). The proton-reducing reactions facilitate the generation of H_2, a common fermentation by-product during electron acceptor-limited microbial processes (Madsen 2015). During anaerobic fermentation, interconversion of metabolites takes place during substrate degradation that increases the availability of reducing equivalents inside the cell. The protons from redox mediators (NADH/FADH) gets detached in the presence of the NADH-dehydrogenase enzyme and gets reduced to H_2 in the presence of the hydrogenase enzyme with the help of the electrons donated by the oxidized ferredoxin (cofactor). On the other hand, membrane-bound protein complexes (NADH dehydrogenase and cytochrome b–c_1) and mobile carrier proteins (quinine and cytochrome C) facilitate the electron transport through the quinone (Q) pool. The continuous interconversions of Q and protons (from the cytosol) to QH_2 and QH_2 to Q and protons facilitate the electron transfer to the cytochrome b–c_1 complex (Cyt bc_1) and further to the cytochrome aa_3. Finally, from the cytochrome aa_3, the electron is transferred to the iron-containing protein Fd. This reduced Fd donates electrons to the active site component of the hydrogenase enzyme, which reduces the protons with this electrons producing H_2 (Vardar-Schara et al. 2008).

Fermentative conversion of an organic substrate to its products involves a cascade of interconnected biochemical reactions, viz., hydrolysis, acidogenesis, acetogenesis, and methanogenesis. The complex organic compounds are degraded to monomers during hydrolysis by hydrolytic microorganisms. Then, acidogenic bacteria ferment the monomers into mixture of low molecular weight volatile organic acids along with H_2 (Eqs. 10.1, 10.2, 10.3, 10.4, and 10.5) (Nikhil et al. 2014a). The reversible conversion of acetate production from H_2 and CO_2 by acetogens and homoacetogens can be also considered for H_2 production. Finally, the acetoclastic methanogens convert these organic acids to CH_4 and CO_2 during methanogenesis (Venkata Mohan 2010). Acetogens grow in syntrophic association with the hydrogenotrophic methanogens and keep H_2 partial pressure low enough to allow acidogenesis to become thermodynamically favorable for interspecies H_2 transfer. Henceforth, the methanogenic activity needs to be suppressed to make H_2 a major metabolic by-product (Goud et al. 2014b; Sarkar et al. 2016).

Dark fermentation is rapidly gaining importance as a practically viable strategy among the other biological routes of H_2 production, especially with the application of wastewater as a substrate associated with the usage of mixed consortia as a biocatalyst (Angenent et al. 2004). This robust process is relatively less energy intensive, with fewer eco-footprints, capable of utilizing wide range of biodegradable substrates and operation at ambient conditions. With these inherent striking features make it practically more feasible for the mass production of H_2 (Venkata Mohan 2010). When the metabolic pathway is such that it favors the production of acetic acid, the stoichiometric yield of H_2 is 4 mol for each mole of glucose (i.e., 544 mL H_2/g hexose at 25 °C), whereas the yield of H_2 is 2 mol for a mole of glucose (i.e., 272 mL H_2/g hexose at 25 °C) when the final product is butyric acid. The actual H_2 yield is lower than the theoretical yield, that is, 2 mol per mol of glucose (i.e., 272 mL H_2/g hexose at 25 °C), because H_2 is a by-product, and only a part of the substrate is metabolized and the rest is for biomass growth (i.e., 272 mL H_2/g hexose at 25 °C) (Guwy et al. 2011; Mohan 2009). There are strategies to achieve higher H_2 yields and production rates, viz., optimization of design and operation of bioreactors, inoculum enrichment, pretreatment of substrate, etc. (Mohan et al. 2007). Recently, there has been growing interest on coupled processes to obtain a higher H_2 yield by integrating dark fermentation with processes like photo-fermentation or bioelectrochemical systems (Mohanakrishna et al. 2010a; Mohanakrishna and Venkata Mohan

2013). Besides production of biohythane, a coupled dark fermentative-methanogenic stage has also been a popular choice that increases the sustainability of the coupled process by improving the energy recovery from the acidogenic residues. Bibliometric and scientometric analysis performed in April 2015 with ISI Web of Knowledge [Thomson Reuters] documented gradual increase in publications on dark fermentative-biohydrogen production in Asian countries. The citation index also increased gradually and more than 1000 citations were recorded in the year 2014 (Fig. 10.3).

10.2.3 Microbial Electrolysis

Microbial electrolysis is a process that involves combination of microbial metabolism and bioelectrochemical reactions facilitating the conversion of electron equivalents in organic compounds into H_2 (Liu et al. 2005; Rozendal et al. 2006). A microbial electrolysis cell (MEC) resembles a microbial fuel cell wherein the disparity exists with the necessity of a poising external potential to facilitate the conversion of biodegradable organic substrates into H_2. Protons transferred to the cathode are reduced to form H_2 in the presence of electrons coming from the anode under an applied voltage, which is essentially required to cross the endothermic barrier to form H_2 (Babu et al. 2013a, b). The standard redox potential for the reduction of protons to H_2 is −0.414 V. A potential greater than 0.11 V in addition to that generated by bacteria (−0.3 V) facilitates good H_2 production at the cathode (Cheng and Logan 2007; Logan and Grot 2005). This approach provides a route for extending H_2 production to pass through the endothermic barrier imposed by the microbial formation of fermentation products (Logan et al. 2008), and the potential required is relatively low compared to the theoretically applied voltage of 1.23 V for water electrolysis (Rozendal et al. 2007). Acetate (−0.279 V) can be converted to H_2 (−0.414 V) in a cathodic reaction against the thermodynamic gradient with the application of a relatively small voltage (−0.135 V) (Hu et al. 2008). In practice, a relatively higher voltage than this is required due to overpotentials created by physicochemical and microbial factors (Hallenbeck 2011).

Application of an external voltage results in the selective growth of electrochemically active bacteria, which can effectively sink electrons (Nikhil et al. 2015). Research on MEC documented more than 90 % of H_2 recovery as against 33 % with the dark fermentation process (Cheng and Logan 2007). MEC showed a capability of

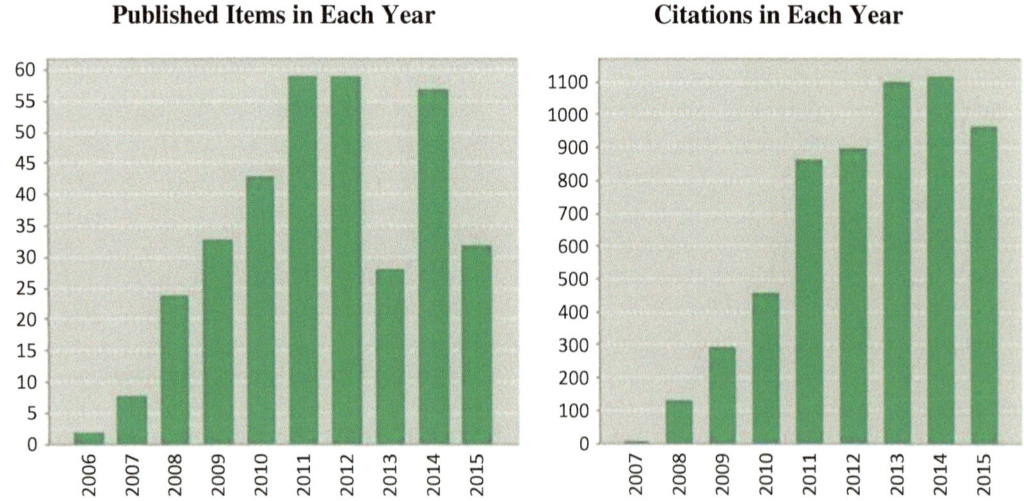

Fig. 10.3 Bibliometric and scientometric analysis for the topic "biohydrogen using dark fermentation" taken from ISI Web of Knowledge [Thomson Reuters] – indicating publications and citation index

converting a wide variety of soluble organic matter to H_2 or methane production with simultaneous wastewater treatment (Clauwaert and Verstraete 2009). Acidogenic effluents rich in fatty acids can be the primary substrate for H_2 production. Low-energy consumption compared to conventional water electrolysis, high product (H_2) recovery, and substrate degradation than the dark fermentation process are some of the potential benefits that make MEC an alternate process (Mohan and Babu 2011).

Yet, different parameters affect the performance of MEC, viz., biocatalyst, electrode materials, membrane, applied potential, and substrate composition and its loading rate and reactor configuration. MEC operated in a dual chamber allows separate capture of H_2 (cathode) and CO_2 (anode) and prevents fouling of the cathode by anodic bacteria. However, separation leads to pH changes due to acidification of the anode chamber by the production of protons and alkalization of the cathode chamber due to proton consumption (Hallenbeck 2011). Lowering the applied potential and eliminating the membrane create a single-chamber MEC (Hu et al. 2008). Operation of single-chamber MEC reduces some of the inherent (internal resistances (Hallenbeck 2011; Mohan et al. 2013). Much research is being focused on working with wastewater as a substrate for operating MEC (Dictor et al. 2010; Escapa et al. 2009; Wagner et al. 2009). Integrating MEC with other processes is also gaining much attention (Cusick et al. 2010; Lalaurette et al. 2009; Mohanakrishna and Venkata Mohan 2013; Wang et al. 2011). A number of challenges need to be addressed before MEC can be applied on a practical level. A low applied voltage with an elevated current density is an essential challenge for moving bench-scale MEC to commercial applications (Lee and Rittmann 2009). Integrating MEC with the effluent treatment plants to produce biohydrogen generation is a recent interest among the research fraternity (Mohan 2010; Pandey et al. 2013).

10.2.4 Integration for Hybrid Process

Dark fermentation process generates soluble metabolic products (SMPs) such as volatile fatty acids and alcohols during the metabolism of organic substrates. The process becomes unfavorable for H_2 production and leads to inhibition due to accumulation of VFA in the system. Employing process integration is one of the promising options to make the process economically feasible (Mohanakrishna and Venkata Mohan 2013). Combination of dark and photo-fermentation could achieve a theoretical maximum yield of 12 mol H_2/mol hexose. Therefore, a two-stage process, i.e., dark fermentation followed by photo-fermentation, has been considered as an effective and efficient system to increase H_2 yield to enhance energy recovery from organic wastewater and lower COD in the process effluents. Various integration strategies are reported for the utilization of acid-rich effluents as the substrate, viz., anaerobic dark fermentation for H_2 integrated with the methanogenesis, photo-fermentation, bioelectricity using microbial fuel cell, and production of value-added products like bioplastics (Reddy et al. 2012; Venkateswar Reddy et al. 2014). Photosynthetic bacteria are capable of utilizing organic acids as carbon and light as energy sources for H_2 production. Integrating heterotrophic dark fermentation with photo-heterotrophic/photo-fermentation processes in two stages results in additional H_2 production (Chandra et al. 2015; Chandra and Venkata Mohan 2014). Sometimes, H_2 production is thermodynamically feasible only if there is an additional energy input that can be in the form of electricity in microbial electrolysis cell (Babu et al. 2013a).

10.2.4.1 Biohythane

The mixture of H_2 and methane is called Hythane™, HCNG, or methagen (Ljunggren and Zacchi 2010). The suggested H_2 content with CH_4 ranges between 10 and 25 % by volume (Fulton et al. 2010). In comparison, the H_2-

specific calorific value of 119,930 kJ/kg is nearly two and half folds higher than CH_4 (50,020 kJ/kg) (Bauer and Forest 2001). The recent advantages in bioenergy research allow the use of biological process, i.e., dark fermentation for the production of both these gases using waste, as a feedstock (Jia et al. 2014; Mohan et al. 2008). Dual-stage dark fermentation process was reported for the production of H_2 and CH_4 (mixture also called as biohythane) from biomass/biowaste (Monlau et al. 2015). This biological method guarantees the regulation of $H_2/(CH_4 + H_2)$ ratio by regulating the ambiance of microbial fermentation and also accounting for waste remediation simultaneously. Recently, Pasupuleti and Mohan (2015) reported the production of biohythane in a single-stage biosystem using spent wash as feedstock, integrating wastewater remediation as integral part. The distillery spent wash had the potential to produce approximately 1100 million cubic meters of biogas making it a potential feedstock for biohythane production (Pasupuleti et al. 2014).

10.2.4.2 Hybrid Dark-Photo Fermentation System

Photosynthetic bacteria, certain purple non-sulfur (PNS) bacteria, and green algae such as *Chlorella* can readily utilize the VFA generated from the acidogenesis process to produce additional H_2 (Srikanth et al. 2009a, b). Henceforth, integration of anoxygenic photo-fermentation with dark fermentation will have dual advantages of increased H_2 production along with treatment efficiency (Chandra and Venkata Mohan 2014; Chandra et al. 2015). Photo-fermentation has constraints like light penetration problems; complex nutritional, obligate environmental conditions; and susceptibility for contamination (Liu et al. 2009; Özkan et al. 2012; Ozmihci and Kargi 2011). Substrate inhibition is the major obstacle when effluent containing high concentration of VFA is used as the sole substrate for the photo-fermentation process (Chen et al. 2010). Ammonium ions in acidogenic feed can also inhibit photo-fermentation process that suppress the nitrogenase enzyme, which is responsible for catalyzing H_2 production in the photosynthetic bacteria (Akköse et al. 2009; Pekgöz et al. 2011; Wang et al. 2009). Biohydrogen production by coculture of anaerobic and photosynthetic bacteria in single stage has been studied which reported higher H_2 production yield (4.5 mol/mol glucose) by coculture of *C. butryricum* and *Rhodobacter* sp. as compared to single-stage dark fermentation (1.9 mol/mol glucose) and sequential two-step fermentation (3.7 mol/mol glucose) of starch. Similarly, higher H_2 yields from different substrates were reported by cocultures of *R. marinum* and *V. fluvialis* as compared to *R. marinum* alone. Better H_2 yield (60 %) was observed in combined fermentation by *Lactobacillus amylovorus* and *R. marinum* from algae biomass in comparison to sequential two-stage fermentation (45 %) (Rai et al. 2012, 2014).

10.2.4.3 Bioaugmentation

The conventional anaerobic bioreactor was shifted metabolically from methanogenesis to acidogenesis to produce H_2 as a main product by applying bioaugmentation strategy (Goud et al. 2014a; Venkata Mohan et al. 2007). Studies have reported that the H_2 production rate improved significantly after augmenting with acidogenic bacteria. Cocultures of *Clostridium acetobutylicum* X9 and *Ethanoigenens harbinense* B49 augmentation revealed considerable improvement in both hydrolysis and consequent H_2 production (Ren et al. 2008). Augmentation of a constructed microbial consortium (*Enterobacter cloacae* IIT-BT 08/*Citrobacter freundii* IIT-BT L139/*Bacillus coagulans* IIT-BT S1) showed an improvement in H_2 production (Kotay and Das 2010). This approach can be applied to the full-scale anaerobic reactors producing CH_4 to shift toward H_2 production. Application of this strategy can advance process efficiency within a short period. Bioaugmentation improves the H_2-producing capacity of the system and system stability during operation. However, the success of this strategy depends on a number of factors, such as survivability and persistence of augmented biocatalysts in the system, operating conditions, and substrate composition and nature, as well as assortment of native microflora of the system (Chandra and Mohan 2014; Lu et al. 2009).

10.3 Waste Remediation vs. Biohydrogen Production

Rapid growth in industrialization and urbanization is generating enormous magnitudes of wastewater. The regulatory obligation for their treatment prior to disposal perceptibly makes wastewater an ideal commodity to produce renewable energy in the form of H_2 by anaerobic treatment. The intrinsic advantage of wastewater is its biodegradable organic fraction associated with its inherent net positive energy (Venkata Mohan et al. 2013). Wastewater as a resource can meet a major portion of the world's energy demand – if it could be transformed to economically useful energy forms (Lin et al. 2012). Utilization of wastewater as a potential substrate for H_2 generation through biological routes documented considerable interest due to its sustainable nature. Harnessing of H_2 from wastewater will significantly reduce the cost of overall wastewater treatment process (Jia et al. 2014; Laurinavichene et al. 2012; Mohanakrishna et al. 2010b; Mohanakrishna and Venkata Mohan 2013). Wastewater treatment is an energy-intensive process that increases the monetary burden on the effluent treatment plant (ETP) operators, especially pertaining to the industry (Venkata Mohan and Pandey 2013). Finding ways to produce/recover useful products or value addition through wastewater remediation is gaining implication in the modern times. In the perspective of environmental sustainability, negative-valued wastewater can be considered as a prospective substrate/feedstock for biological H_2 production by simultaneously achieving pollution control (Lin et al. 2012). Reducing the wastewater treatment cost by generating bioenergy, such as H_2 gas, from the organic matter present in wastewater is a sustainable opportunity (Sivaramakrishna et al. 2014).

Renewable biohydrogen-producing technologies have the potential to become cost competitive as they can use low-value waste as feedstock, e.g., municipal, agricultural, and industrial organic waste and wastewater (Venkata Mohan et al. 2013). The last decade witnessed considerable efforts on the application of various wastewaters from domestic and industrial as potential substrates for the production of H_2 through biological machinery, mainly through light-driven and light-independent fermentation processes. Simple sugars to complex effluents and agricultural and food industry wastes rich in carbohydrates were also evaluated for H_2 production. When wastewater is being used as a substrate, substrate degradation efficiency is also important, along with H_2 production, when process efficiency is considered. A trade-off exists between technical efficiency based on H_2 production and substrate removal in alliance with operating conditions. A neutral pH is ideal for substrate degradation, while an acidic pH helps H_2 production. Balancing the conditions for combined performance is especially important in sustaining the economicX feasibility and ecological adequacy of the process (Nikhil et al. 2014b). A schematic layout of strategies could be applied to recover bioenergy and value-added products along with wastewater treatment. Bibliometric and scientometric analysis performed in April 2015 with ISI Web of Knowledge [Thomson Reuters] documented significant publications in biohydrogen production using wastewater since 2003 among the Asian countries. More than 30 publications with citations of above 700 were documented in the year 2014. About 150 records attribute to energy and fuel area with India and China are on lead positions as noticed from the figures (Fig. 10.4).

Theoretically, 1 kg of glucose ($C_6H_{12}O_6$) contains 1.066 kg of chemical oxygen demand (COD) (937.5 g or 5.2 mol of glucose equal to 1 kg of COD). By dark fermentation, 1 mol (180 g) of glucose can produce 4 and 2 mol of molecular H_2 based on acetate and butyrate pathways, respectively. By the photo-fermentation pathway, 1 mol of glucose can produce 12 mol of H_2. Theoretically, 1 kg COD can produce 20.83 mol of H_2. According to the ideal gas law (PV=RT) at STP [standard temperature (300 K/27 °C) and pressure (1 atm)], 1 mol of H_2 occupies 22.4 l volume. Accordingly, 1 mol glucose (192 g COD) can produce 89.6 l of H_2. Therefore, 1 kg COD (5.2 mol glucose) can produce 20.83 mol of H_2 (466.6 l of H_2/41.6 g of H_2).

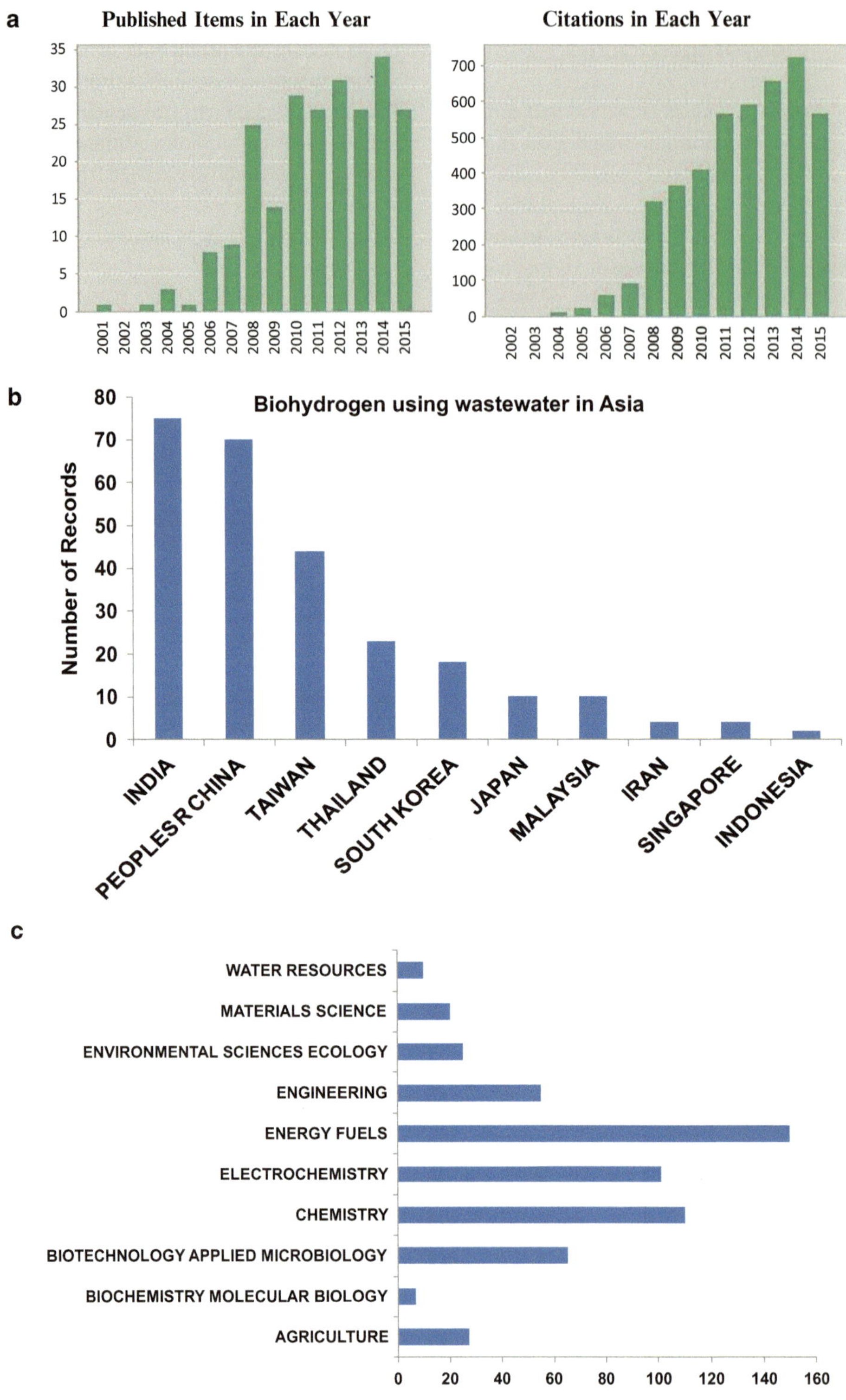

Fig. 10.4 Bibliometric and scientometric analysis for the topic "biohydrogen using wastewater" taken from ISI Web of Knowledge [Thomson Reuters] – (**a**) publications and citation index; (**b**) number of records from Asian countries; (**c**) research areas involved in biohydrogen research

When 40 % of COD removal efficiency was considered for H_2 production, the dark fermentation process can produce 125 g of H_2 and photo-fermentation conversion can yield 16.6 g of H_2. The food processing industry in India is producing ~3,000,000 × 10^5 l of wastewater per year. It contains an average COD of 20 g/l, accounting for a total of 6000 × 10^6 g of COD per year. Photo-fermentative process can produce about 300 × 10^6 kg of H_2 per year (with 40 % removal rate) which accounts for $ 1200 million per year (at a rate of $ 4 per kg H_2). Similarly, the same wastewater can generate a revenue of $80 million (5 × 10^6 kg of H_2) per year by dark fermentative route (on 40 % removal basis) (Dahiya et al. 2015; Venkata Mohan et al. 2013).

10.4 Thermochemical Route

Thermochemical conversion routes for biohydrogen production mainly include combustion, liquefaction, pyrolysis, and gasification processes. During combustion the biomass is burnt in air at temperatures around 800–1000 °C by using stoves, furnaces, boilers or steam turbines, etc., to release hot gases Agarwal et al. (2013, 2015). Biomass with moisture content lower than 50 % (Orecchini and Bocci 2007) is feasible for combustion process. Energy efficiency during combustion process is very low (10–30 %) and the pollutants are emitted as by-products. Therefore, combustion process cannot be considered as a suitable option for H_2 production (Ni et al. 2006). In liquefaction process biomass is heated in an inert atmosphere at temperature range of about 525–600 K in water under a pressure of 5–20 MPa. In the liquefaction process, the solvent or catalyst can be added whenever it is required. The disadvantages of the liquefaction process can be listed as (a) difficult to reach the operation conditions and (b) H_2 production is lower. Therefore, this process is not considered as a favorable one for the production of hydrogen. Pyrolysis is the most promising thermochemical conversion process for the H_2 production, where biomass is heated at a temperature of 650–800 K under a pressure of 0.1–0.5 MPa in an inert atmosphere. During the process biomass is converted into liquid product (usually called bio-oil), solid products (called as biochar), and gaseous compounds (H_2, CH_4, CO, CO_2, and some other gases) (Jalan and Srivastava 1999). Methane and other hydrocarbon vapors produced during pyrolysis can be steam reformed to increase the production of H_2. Application of water-gas shift reaction also increases the production of H_2. Oily products obtained during pyrolysis can also be used for the production of H_2 (Evans et al. 2003). Pyrolytic liquid can be separated as water-soluble and water-insoluble fractions based on their solubility in water. The soluble fraction can be used in the production of H_2, while the insoluble fraction can be used for adhesive formulation. Incorporation of catalyst into a pyrolytic reactor increases the production of gaseous compounds. Some of the most conventional and active tar-cracking catalysts are Ni-based catalyst (Asadullah et al. 2001; Saxena et al. 2008); dolomite (Narvaez et al. 1997); Y-type zeolite (Williams and Brindle 2002); K_2CO_3, Na_2CO_3, and $CaCO_3$ (Chen et al. 2003); and various metal oxides (Al_2O_3, SiO_2, ZrO_2, CeO_2 (Saxena et al. 2008), TiO_2 (Sutton et al. 2002), and Cr_2O_3 (Chen et al. 2003)). Chlorides, carbonates, and chromates are some inorganic salts which have beneficial effect on H_2 production (Ni et al. 2006). Gasification of biomass (moisture content less than 35 %) is also used to produce gaseous product which can be steam reformed to produce H_2 gas, and this can be further improved by using water-gas shift reactions (Demirbas 2002). Unwanted tar and ash formed during gasification may cause a major problem which can be reduced by using proper design and control of the reactor, using a catalyst, fractionation, and leaching technique (Wornat et al. 1995; Arvelakis and Koukios 2002; Ni et al. 2006). A novel gasification method, viz., hydrogen production by reaction integrated novel gasification (HyPr-RING), was proposed by Lin et al. (2001), which is an integration of the water-hydrocarbon reaction, water-gas shift reaction, and absorption of CO_2 and other pollutants in a single reactor under both subcritical and supercritical water conditions. The reaction that occurs during HyPr-

RING is an exothermic process, and it yields high amount of H_2 at relatively lower temperature of 923–973 K and also avoids the formation of tar and char simultaneously. In general, H_2 gas is produced along with other gas constituents, and therefore its separation and purification are necessary. At the present time, various methods are successfully developed for H_2 gas purification such as CO_2 absorption, drying/chilling, and membrane separation (Lin et al. 2001; Reij et al. 1998). It is expected that H_2 from biomass by using thermochemical conversion processes, mainly the new developed gasification processes, may be helpful for large-scale production of H_2.

10.5 Feedstocks for Biohydrogen Production

Hydrogen can be produced from various feedstocks. The measures that will define a feedstock for biohydrogen production include availability, carbohydrate content, cost of production, and biodegradability. Due to their simple structures and easy biodegradability, simple sugars like glucose, sucrose, and lactose are favored as a substrate for the production of H_2. The prospective feedstocks for production of H_2 could be broadly classified as biomass which includes energy crops; food crops and algae; waste materials such as waste sludge, industrial wastewater, agricultural and food industry waste, etc.; and carbohydrates.

Production of H_2 can be done from various feedstocks like alage (Sarkar et al. 2015), corn stover (Cao et al. 2009; Datar et al. 2007), beer lees (Cui et al. 2010), rice (Lo et al. 2010; Lalitha Devi et al. 2015) and wheat (Kongjan and Angelidaki 2010) straw, cassava stillage (Luo et al. 2010), rice and wheat bran (Noike et al. 2002), sugarcane (Pattra et al. 2008) and sweet sorghum (Panagiotopoulos et al. 2010), and bagasse and potato steam peels (Mars et al. 2010). Moreover, waste materials such as agricultural and food wastes (de Vrije et al. 2002), waste fibers, industrial wastes, and organic waste water (Ueno et al. 2007) can also be used for H_2 production. But, there are some difficulties in using cellulose or hemicellulose directly as a carbon source for biofuel production due to which pretreatment and hydrolysis are required for fermentative biofuel production from the lignocellulosic feedstock (Cheng et al. 2011). Other feedstocks used for producing biofuel include lignocellulosic materials (e.g., grasses, straw, wood), starch based (e.g., barley, corn, sweet potato, sweet sorghum, and wheat), and sucrose based (e.g., molasses and sugarcane). Kanai et al. (2005) reported the suitability of application of thermophilic microorganisms like *Thermococcus kodakarensis* KOD1, *Clostridium thermolacticum*, and *Clostridium thermocellum* JN4 for cellulosic biohydrogen production (Magnusson et al. 2008; Cavinato et al. 2011).

Nowadays, microalgal biomass is gaining prominence as one of the most exciting feedstocks for biofuel production due to its rapid growth rate, cultivability (no soil is needed), high capturing ability of greenhouse gases, and short harvesting cycle of 1–10 days (Harun et al. 2010; Cheng et al. 2011; Wijffels and Barbosa 2010) and considered as third-generation feedstock. Various researchers have reported that microalgae are very helpful in H_2 production, such as *Chlorococcum littorale* (Schnackenberg et al. 1996; Ueno et al. 2001), *Chlamydomonas reinhardtii* (Ghirardi et al. 2000), *Chlorella fusca* (Winkler et al. 2002), *Scenedesmus obliquus* (Florin et al. 2001; Gouveia and Oliveira 2009), and *Platymonas subcordiformis* (Guan et al. 2004). Though microalgae are gaining prominence as a more competent feedstock for biohydrogen production with properties like high growth and CO_2 fixation rates and predominance of hexose in the carbohydrate content, there still remain some difficulties in efficient collection and mass production of microalgae which prevent it to be used as a third-generation feedstock for biohydrogen production. In order to resolve these problems, more advanced research efforts are still needed.

10.6 An Overview of Latest Trends in Biohydrogen Production: World's Perspective

Till the beginning of the twenty-first century, the entire world has seen an increased trend toward the use of petroleum energy for transportation

(Loppacher and Kerr 2005). In order to meet the increasing demand of energy, fossil fuel has been extracted indiscriminately which has raised the concern over global warming due to increasing concentrations of carbon dioxide (CO_2) in the atmosphere (Balat 2009). With environmental, economical, and political concerns coupled with rapid depletion of petroleum, there is a growing interest for alternate source of energy such as biofuels, bioethanol, biodiesel, and hydrogen. H_2 production from renewable sources such as agricultural or other waste streams contributes to energy production possibility which not only increases the flexibility and improves the economics of distributed, centralized, and semi-centralized reforming but also lowers net greenhouse gas emissions without employing any carbon sequestration technologies (Levin and Chahine 2010).

H_2 is produced through various biochemical and thermochemical routes that emerged as few of the most promising alternative energy technologies. It is a secondary form of energy (Veziroğlu and Şahi 2008; Balat 2008). It is considered to be the cleanest fuel and 2.75 times greater than hydrocarbon fuels with energy yield of 122 kJ/g (Kapdan and Kargi 2006). The use of H_2 as a fuel for transportation and stationary applications has grabbed a worldwide attention; it is being explored for use in combustion engines and fuel cell electric vehicles (Cherry 2004). It can be stored chemically or physiochemically in various solid and liquid compounds such as metal carbon nanostructures, hydrides, alanates, borohydrides, light hydrocarbons, methane, and methanol (Shakya et al. 2005).

At present, most of H_2 is produced through thermo-catalytic and gasification processes using natural gas; heavy oils and naphtha are the next largest sources , followed by coal (Mohan et al. 2007); and only 4 % and 1 % are generated from water using electricity and biomass, respectively (Das et al. 2008). Thus, it can be said that 95 % of H_2 is produce from fossil fuel-based processes (Balat 2010). Therefore, to reduce the net CO_2 emission in the atmosphere, the use of biomass instead of fossil fuels can provide a sustainable option to produce H_2 (Larsen et al. 2004).

The current merchant (purchased from H_2 producers) and captive (consumed by H_2 producer) market of H_2 is used mostly in oil refining, food production, metal treatment, and fertilizer manufacture (Lipman 2011). Its applications in chemical processing, petroleum recovery and refining, metal production and fabrication, aerospace, and fuel cells are well established. H_2 is also used as a reducing and hydrogenating agent, shielding gas, food additive, rotor coolant, etc. At present, the largest demand for H_2 is observed in petroleum refinery and ammonia production. The market size of global H_2 production was estimated to be 53 million metric tons in 2010, in which 12 % is shared by merchant H_2 and the rest with captive production. The H_2 production market in terms of value was estimated to be $82.6 billion in 2010. The global H_2 production volume is forecasted to grow by a compound annual growth rate of 5.6 % during 2011–2016 due to decreasing sulfur levels in petroleum products, lowering crude oil quality, and rising demand of H_2-operated fuel cell applications. The Asia and Oceania region is the largest market with 39 % of global production share in 2010, accounting for a production of 21 million metric tons of H_2 (Markets and Markets 2011).

High effective octane number, fast burning speed, high energy density, and zero ozone forming potential make H_2 a very suitable and special transportation fuel (Balat 2008). Due to the penetration of H_2 in the transportation sector and its potential use for refining high-sulfur crude oils, the demand for H_2 is expected to grow exponentially in the near future Elliott (2000). It is estimated that to fuel 100 million fuel cell-powered cars, about 40 million tonnes of H_2 per year would be required (Dalcor and Camford 2005).

Dalcor and Camford (2005) reported that during the production of H_2 (about 59 %) from steam methane reforming (SMR) of natural gas, about 30 million tonnes of CO_2 per year is emitted in the atmosphere. Though some amount of CO_2 can be recovered and used as an industrial gas or for oil and gas recovery, most of it is vented. Due to the proximity of Western H_2 plants to favorable geological storage sites, the CO_2 generated during H_2 production from hydrocarbon sources is a good candidate for sequestration (Lindsay et al. 2009). The fastest growing demand for H_2 comes from the oil and gas industry (for the upgrading of heavy oil from the oil sand developments in Alberta). The current surplus of H_2 is either used to supplement furnace fuel requirements in the vicinity of production or is vented to the atmosphere. A 200,000-tonne surplus could power about 500,000–900,000 vehicles. Roughly 1 tonne of H_2 can fuel two to four fuel cell vehicles (which require pure H_2 as fuel) for 1 year or one urban transit bus for about 45 days (Dalcor and Camford 2005).

10.7 Status and Road Map Toward Biohydrogen Production in Asian Countries

10.7.1 Korea

In terms of energy consumption, the rank of Korea is tenth in the world. Korea is strongly dependent on energy import and imports more than 97 % of its total energy consumption. Developing hydrogen energy technology in Korea has great potential to cope with the nation's energy security and to establish future economic growth (Lee et al. 2008).

Presently, Korea's main focus is on the development of hydrogen energy technology R&D for the sustainable development and low carbon green society. Probably, after 2020, hydrogen will be used in portable power generation systems, micro-power systems, and applications in the fields of transportation, residential and industrial and distributed generation systems (Laurikko 2006; Lee et al. 2010).

In Korea, the Ministry of Science and Technology (MOST, www.most.go.kr) and the Ministry of Commerce, Industry and Energy (MOCIE. www.mocie.go.kr) are the two main government agencies to develop future energy technologies involved in the field of hydrogen and fuel cell development. Between them they have formed National RD&D Organization for Hydrogen and Fuel Cells (www.h2fc.or.kr). MOST launched the Hydrogen Energy R&D Centre [www.h2.re.kr5] in 2003 in their "21st Century Frontier Programme."

The Korean government has mainly concentrated on the strategic investment for the fields of hydrogen and fuel cells. In the field of developments in hydrogen energy fuel cells and different storage systems, researches are going on in various universities and research institutes in Korea (Laurikko 2006).

10.7.2 China

To sustain its increasing energy needs, China requires an alternative energy system.

It is the world's second largest energy-consuming and the third largest energy-producing country. China represents one of the largest potential markets for fuel cells in the world. In 1996, China first received an international grant, from the Global Environment Facility (GEF), for fuel cell vehicle deployment. In all the countries in Asia, especially in China, government incentives and public policies play an important role in the development of the H_2 economy. The main source for H_2 production in China is generally based on the residential sector. One of the research programs in China, named the National Basic Research Program (NBCP), primarily concentrates on H_2 production, storage, and transportation on the industrial scale. Some of the NBCP projects are in the applications of fuel cells and on the lab-scale production of H_2 from water with solar energy. China utilizes fuel cells for light-duty buses, minivans, and cars in collaboration with some other countries. Another program named the National High-Technology Development Program (NHTDP) addresses fossil fuel H_2 and fuel cell technology and advanced H_2 generation for motor applications (ODEC). In 2002, the Chinese government declared that they will finance approximately 18 million dollars for the development of fuel cells, especially PEMFC by funding the Dalian Institute of Chemical Physics (DICP).

In 2003, the DICP provided a new 75 kW polymer electrolyte membrane stack to Tsinghua

University, and this stack is utilized in a bus for transportation (Haslam et al. 2012). Additionally, China made approximately 120 million dollars of investments in fuel cell-powered automobiles and has many institutes that specialize in H_2-based fuel cells. The Shanghai municipal government in China has some projects for the R&D of fuel cells that spends approximately 12 million dollars per year (Geiger 2003). The National Development and Reform Commission (NDRC) is a sector of the State Council of China that focuses on sustainable progress in China for a cleaner and pollution-free path to H_2. Long-term plans focus on a H_2 economy that will most likely be realized after 2050. Two main cities in China, Beijing and Shanghai, have been nominated for demonstrations of fuel cell buses by the Global Environment Facility (GEF). One of the policies, named the Green Power System, in Shanghai is mainly focused on research into renewable H_2 production (Beser and Padilla 2003). One of the top companies in China, named Shanghai Shen-Li High Tech. Co. Ltd., produces H_2 power and utilizes it in the development of H_2 fuel cell cars in collaboration with the Shanghai Automobile Industry (Green Car Congress). In China, the Chinese Ministry of Science and Technology spends approximately 9.4 million dollars for H_2-based fuel cell automobiles. Shanghai is working on its own H_2 infrastructure project and has started to produce H_2 for fuel cell buses in the city. The supply of H_2 fuel is very easily available compared with other cities because of the infinite and elastic fuel sources. Some chemical companies in Shanghai produce H_2 as an industrial by-product, and this production significantly satisfies the needs of short-term users in the city (Haslam et al. 2012).

China has a strong research base, with all of its top-tier universities conducting fuel cell research. Tsinghua University is the home headquarters of mainland China's premier H_2 association. Wuhan University of Technology maintains a robust fuel cell program as part of its curriculum. Tsinghua University in China has some projects and basic research intended for production, storage, and transportation of H_2, for fuel cell engines and for the development of PEM fuel cells. The China Association for Hydrogen Energy also promotes the path to a renewable H_2 economy by considering H_2 to be the ultimate fuel for fuel cells for various applications (Shi 2006). One important program in China, known as the MOST973 program, spends 5.6 million dollars in the development of H_2 storage materials, membranes, etc.; GEF committed to a 5-year fuel cell bus demonstration project in Shanghai and Beijing with an estimated cost of $32 million (US). In addition to operating in rural areas, H_2 and fuel cells also operate in Guangzhou, Beijing, Shanghai, Suzhou City, and the Fengxian District (Zhang and Maruyama 2001).

In China, 20 Passat Lingyu FCVs were used as demonstration passenger cars in the Summer Olympic 2008. In 2010, the expo Shanghai showcased H_2 and fuel cell technology in the marketplace. In June 2011, 40 top scientists from around China met in Beijing for a 3-day roundtable discussion about H_2 and fuel cells, and they decided to ask the government to support H_2 and fuel cell research before the 2015 commercialization for FCEV.

10.7.3 India

In India the status of H_2 energy is in research, development, and demonstration level. The National Hydrogen Energy Board (NHEB) funded by the Ministry of New and Renewable Energy has approved the National Hydrogen Energy Road Map in 2006 to abridge technological gaps regarding various aspects of H_2 such as H_2 production, its storage, and its utilization for power generation using fuel cell technologies and internal combustion engines (http://mnre.gov.in/file-manager/UserFiles/faq_hydrogenenergy.htm). The objectives of this program are:

- To study and evaluate the feasibility of production of hydrogen by various processes/technologies, especially based on renewable energy methods
- To develop materials, processes, systems, and subsystems for the storage of hydrogen

- To support projects on utilization of hydrogen as a fuel for stationary, automobile, and portable applications
- To support projects for the development of hydrogen infrastructure in public-private participation made for production, storage, and applications of hydrogen, including safety, standards and codes, capacity building, and public awareness
- To support demonstration projects relating to production, storage, and applications of hydrogen (http://www.mnre.gov.in/schemes/new-technologies/hydrogen -energy/)

Laboratory-scale prototypes of H_2-fueled motorcycles, engine-generator sets, three wheelers, and water/methanol electrolyser for the production of H_2 have been developed. Banaras Hindu University, Varanasi, has developed and demonstrated about 15 H_2-fueled motorcycles in its campus. In Faridabad and Delhi, H_2-blended compressed natural gas fuel dispensing facilities have been set up. Along with the Ministry of Science and Technology, the Ministry of Petroleum and Natural Gas, CSIR Laboratories, Indian Space Research Organisation, oil and gas companies, Defense Research and Development Organization, Department of Atomic Energy, and private sector automobile companies are also involved in the research and development and demonstration program related to H_2 (http://mnre.gov.in/file-manager/UserFiles/faq_hydrogenenergy.htm).

Currently in India much of the activities are focused toward research and development of materials, processes, and pilot plant for production, storage, and use of H_2 as a fuel. Various projects have been taken up to develop and demonstrate the applications of H_2 for power generation and transportation. Also steps toward the development of man power and their training have been taken up in this direction.

Since 2006–2007, a total of 54 RD&D projects are being supported by the Ministry of New and Renewable Energy (MNRE), 38 projects out of which in the area of H_2 production, its storage, and applications and 16 projects related to different fuel cell technologies. During the 11th Plan Period (from 2007–2008 to 2011–2012), 44 projects (26 %) out of 169 new RD&D projects supported by MNRE were related to H_2 and fuel cell which signifies the extent of the support provided to H_2 energy and fuel cell activities. About Rs. 118 crore has been allocated to support H_2 energy and fuel cell projects by the MNRE (Nouni 2012).

The status of H_2-based technologies in India can be categorized in three parts:

Production: Petroleum refining and fertilizer and chemical industries in India currently produce H_2 commercially. In chlor-alkali industries, H_2 is produced as a by-product. In addition to that, a limited amount of H_2 is also produced through electrolysis for commercial use. A study conducted by the University of Petroleum and Energy Studies that estimated a theoretical H_2 production and consumption in the country during 2007–2008 is shown in Table 10.1.

As per information compiled by the Alkali Manufacturers Association of India, about 0.0081 MMT of surplus H_2 was available from chlor-alkali units during 2010–2011(Nouni 2012).

The institute/organizations involved in biohydrogen production are listed in Table 10.2.

Storage: Storage of H_2 for the development and viability of H_2-fueled vehicles is still a challenging aspect to achieve a H_2-based economy. According to NHEB for driving range of about 500 km, the approximate range should

Table 10.1 Theoretically estimated amount of hydrogen produced and consumed in India during 2007–2008

Sector	Estimated production during 2007–2008 (million tonne/year)	Utilization during 2007–2008 (million tonne/year)
Fertilizer industry	1.99	1.99 (captive use)
Petroleum refineries	1.69	1.462 (captive use)
Chlor alkali industry	0.073	0.064
Total	3.753	3.516

Source: http://mnre.gov.in/file-manager/UserFiles/faq_hydrogenenergy.htm

Table 10.2 The institute/organizations involved in biohydrogen production in India

Institute/organization	Work done
Indian Oil Corporation Limited (IOCL), Dwarka, New Delhi	A demonstration project for on-site hydrogen production using alkaline electrolyser of 5 N cu m/h capacity, blending it with compressed natural gas and dispensing of H-CNG was commissioned
Electrical Research and Development Association (ERDA), Vadodara	Developed a prototype demonstration project for wind
	Hydrogen-based stand-alone electrical generation
Indian Institute of Science (IISc), Bangalore	Developing oxy-steam gasification unit using an open top downdraft gasification system for hydrogen production rate of about 0.1 kg/kg biomass at various steam-to-biomass ratios
National Institute of Technology (NIT), Rourkela	Developing a bench-scale fluidized bed gasifier of 5 kW capacity for hydrogen production rate of about 0.09 kg/kg of feedstock
Solar Energy Centre (SEC), Gwalpahari	Working on generation of hydrogen from solar energy by using PV-generated electricity for operating an electrolyser
Indian Institute of Technology (IIT), Kharagpur	A pilot plant of 800 l capacity for biohydrogen production had been installed
Indian Institute of Chemical Technology (IICT), Hyderabad	Undertaken work relating to catalyst development and bench-scale reactor development for hydrogen production studies from biomass-derived glycerol during 2008–2011
Central Institute of Mining and Fuel Research (CIMFR), Dhanbad	Developing a novel process for the production of hydrogen from renewable and fossil fuel-based liquid and gaseous hydrocarbons, by the nonthermal plasma reformation technique

Source: Nouni (2012)

be 5–13 kg. H_2 can be stored in three forms: gaseous, liquid, or as a solid in combination with a metal hydride. The suitability of storage method depends on its economic criteria, end use, environmental issues, and safety aspects. Most of the R&D work has been focused on intermetallic hydrides, liquid organic hydrides, metal hydrides, complex hydrides, etc. Meanwhile during this 11th Plan Period, three R&D projects have been completed, while seven new R&D projects were sanctioned related to H_2 storage (Nouni 2012; Gupta 2012).

The organization involved with H_2 storage are Indian Institutes of Technology, Chennai/Guwahati; Banaras Hindu University (BHU), Varanasi; National Environmental Engineering Research Institute (NEERI), Nagpur; National Institute of Technology (NIT), Tiruchirappalli; International Advanced Research Centre for Powder Metallurgy and New Materials (ARCI), Hyderabad; Indian Institute of Technology Bombay, Mumbai; Thiagarajar College of Engineering, Madurai; and Non-Ferrous Materials Technology Development Centre (NFTDC), Hyderabad (http://www.eai.in/ref/ae/hyn/hyn.html; http://www.mnre.gov.in/schemes/new-technologies/hydrogen-energy/).

Hydrogen-fueled vehicles: The final stage of H_2 energy is its utilization as a fuel, and a tremendous amount of efforts have been targeted on the development of internal combustion engines by modifying petrol, diesel, and gaseous engines so that it can perform well with H_2. The prototypes of such engines (small single-cylinder engines) were modified and integrated with a 2.5 kVA alternator by IIT, Delhi, to generate power. In association with Mahindra and Mahindra (M&M), IIT, Delhi, also developed a H_2-fueled engine for three wheelers. Meanwhile BHU , Varanasi, has modified petrol-driven motorcycles and three wheelers to run with H_2 fuel (Nouni 2012; www.energyaccess.in).

After the acceptance of National Hydrogen Road Map, the first and foremost task was to introduce H_2 as a fuel; hence blending of H_2 was considered to be one of the most suitable options available to gain experience regarding its production, storage, dispensing, and use of H_2. Therefore, to implement the use of H_2, up to 30 % of fuel was being implemented by the Society of Indian Automobile Manufacturers (SIAM) and R&D Centre of IOCL. In this direction two cars, two three wheelers, one cargo vehicle, and two minibuses have been developed by five automobile companies, i.e., Bajaj Auto Limited, Ashok Leyland Limited, Tata Motors, Volvo Eicher, and M&M. IOCL conducted a performance and emission test on this vehicles, and according to the reports, the optimum percentage of blending of H_2 by volume with CNG is 18 %. Currently, field endurance test (30,000 km for three wheelers and 50,000 km for other vehicles) are being performed on these vehicles (Nouni 2012).

The institutes involved with designing and development of H_2-fueled vehicles are Bharat Heavy Electricals Ltd. (BHEL), Hyderabad; Indian Institutes of Technology, Delhi/Chennai/Kanpur; Society of Indian Automobile Manufacturers (SIAM), New Delhi; Indian Oil Corporation Ltd. (IOCL), Faridabad; Mahindra & Mahindra, Chengalpattu; Banaras Hindu University (BHU), Varanasi; Indian Institute of Technology, Kanpur; Annamalai University, Annamalai Nagar; Electrical Research and Development Association , Vadodara; and University of Petroleum and Energy Studies (UPES), Dehradun (www.mnre.gov.in; Nouni 2012).

10.7.4 Japan

Japan is one of the most motivated countries in Asia in the development of a renewable H_2 economy in the implementation of short-term and long-term plans (Pudukudy et al. 2014; Okano 2002). The production of H_2 by reforming of natural gas and water electrolysis was employed as a short-term plan, and water photolysis through the thermochemical route is the long-term plan. This country is also considering biomass in the H_2 production plan (Romeri 2004). The Japanese government also directs to fund Japanese automakers and spends 380 million dollars per year on research, progress, and commercialization of fuel cells (Haslam et al. 2012). In 1991, the Policy Study Group for Fuel Cell Commercialisation was introduced by the Ministry of the Economy, Trade and Industry (METI) of Japan (Maruta 2005; TakaharaI 2005) with a main aim to execute fuel cell technologies based on H_2 (Pudukudy et al. 2014). METI aimed to produce hydrogen-based fuel cell vehicles on the road and hoped for 15 million vehicles by 2030. The Japan Ministry of the Environment in Japan plans to produce H_2 from seawater, which uses electricity generated from wind (Laurikko 2006).

10.7.5 Malaysia

Among the Asian countries, Malaysia has wide range of renewable and nonrenewable sources of energy. Presently, Malaysia is searching for an enhanced renewable H_2 economy . Malaysia is blessed with oil and gas resources, and some oil resources are exploited in the production of biofuels, which are another form of renewable energy such as H_2 (Zhou and Thomson 2009). Fundamentals of a renewable H_2 economy are based on H_2 and fuel cells and the country has already spent a large amount of money in the development of this field.

From 1996 to 2007, the Ministry of Science, Technology and Innovation spent 9.7 million dollars for fuel cell research and from 2002 to 2007, 2 million dollars for H_2 production and storage research. In Malaysia, different H_2 production technologies are broadly divided into two categories: one from renewable resources and the other from nonrenewable resources (Iyuke et al. 2003). The nonrenewable sources mainly include the steam methane reforming (SMR) method. The country focuses its main attention on the biomass resources as an origin of H_2 production from renewable sources in Malaysia. Various technologies employed for this purpose are gasification, pyrolysis, fermentation, biological water-gas shift reaction, etc. Some researchers carried out research work on palm oil for its potential use as a source in the gasification reaction for the production of H_2 (Yong et al. 2007). Other methods include water electrolysis with the electricity produced by solar and wind resources.

Presently, the steam methane reforming (SMR) process plays a major role in the key development in the production of industrial grade H_2 in Malaysia. Lots of studies on the progress of H_2 production in Malaysia are still in progress. A number of research works suggested that biomass can replace the fossil fuel in accordance with price and eco-friendly issues, in the present situation in Malaysia (Iyuke et al. 2003; Shafie et al. 2012; Mohammed et al. 2011). Various researchers observed that H_2 production from palm oil waste by using biomass gasification and dark fermentation techniques needs additional development. Some universities, mainly Universiti Kebangsaan Malaysia (UKM) and the Universiti Teknologi Malaysia (UTM), primarily work in the field of development of H_2 energy fuel cells and different storage systems. Malaysia's Eco-House, situated at the UKM, mainly focuses on solar-H_2 technology. The Eco-House is usually based on the photovoltaic electricity production and storage with a H_2 generator and a fuel cell that stores and regenerates electricity for residential applications. The Fuel Cell Institute at UKM and Universiti Malaysia Terengganu (UMT) are involved in H_2 production through the autothermal catalytic reforming of methane and methanol and H_2 storage in nanostructured carbon. Extensive studies on fuel cell development have also been performed in these universities. H_2 production, purification, storage, applications, demonstrations, and other topics related to the development of H_2 economy are especially focused on UTM (Kamarudin et al. 2003; Sari et al. 2013). Synergy program in Singapore focused on the development of clean energy projects for stationary and transportation applications and is advancing in using H_2 in transportation applications. Based on the economic nature and available resources, most of the Asian countries have its own R&D provisions for H_2 fuel-based mobile and stationary applications (Dunn 2002; Zhou and Thomson 2009).

10.8 Future Perspective

Focused and multidisciplinary research on biohydrogen production integrated with wastewater remediation is underway. This process possesses certain inherent limitations, namely, low substrate conversion efficiency, accumulation of carbon-rich acid intermediates, dynamic buffering, and redox change, which need considerable attention (Venkata Mohan et al. 2010a, b). Fundamental understanding of the potential limiting factors is essential to overcome these limitations in the direction of enhancing process efficiency (Arimi et al. 2015). Process engineering and optimization of operational factors govern the performance of any biological system and have considerable influence on fermentative H_2 production. Understanding the biochemistry and microbiological aspects based on the functional role of membrane components and mechanism of proton reduction, community analysis, culture development aspects, and design and development of efficient bioreactors for both dark and photo-fermentation operations are some of the key areas where considerable focus is required. Light utilization is the key factor in the photo-fermentation process, and developing effective photo-bioreactors with efficient internal light distribution characteristics is especially important for the industrial scale application (Venkata Mohan et al. 2013). Optimization of process parameters is essential for up-scaling of the technology. Unutilized residual organic fraction remaining as a soluble fermentation product after the acidogenic process is one of the key limitations that needs significant attention. Integration approaches toward the utilization of acid-rich wastewater with simultaneous bioenergy recovery can be effectively and completely established for economic viability of the process toward commercialization (Ghimire et al. 2015). Waste/wastewater can be utilized in a biorefinery approach in different modes that is valourized into a gamut of high value biobased products (Venkata Mohan et al., 2016). Metabolic engineering is one of the promising areas that can be used advantageously to enhance the H_2 production rate. Both basic and applied researches are on the way to gain more insight into the process for understanding and establishing optimized conditions. Several novel approaches have been proposed in recent years to surpass some of the persistent drawbacks. Biohydrogen technology requires multidisciplinary research to make the process environmentally sustainable and economically viable (Venkata Mohan et al. 2013; Venkata Mohan and Pandey 2013).

10.9 Conclusion

Global biohydrogen research trend is considerably improving from the last few years, but further research is necessary to improve the biohydrogen production and to understand the impact of the composition of the substrate on biohydrogen performances. The biological processes involved are restricted by the composition of the organic waste, and they are highly dependent on the operating conditions such as low pH, low partial pressure, high temperature, and acclimated microbial communities. These parameters affect not only the yields of biohydrogen in mixed culture but also redirect by-product spectrum and impact the structure of the microbial communities. Within the context of the countries studied in this chapter, it is clear that Japan remains at the forefront of fuel cell technology both in terms of patent activity and academic research. The data shows that China and Korea have been successful in catching up to the better-established programs in Japan. In Asian countries, the biohydrogen research is appreciated distinctively with increasing publications. It is important to note that hydrogen can be produced from a wide variety of feedstocks available almost everywhere. It is clear that as the technologies develop and mature, hydrogen may become the most abundant fuel available. There are many processes under development which will have a minimal environmental impact. The development of these technologies may decrease the world's dependence on fuels that come primarily from unstable regions. The two common challenges for the biological hydrogen-producing systems are the relatively low hydrogen yield and production rate. Enhancement in hydrogen yield may be possible by using suitable microbial strain, process modification, efficient bioreactor design, and also genetic and molecular engineering technique, to redirect metabolic pathway. Extensive researches in the past two decades have reviewed promising prospect of biohydrogen production. There have been substantial improvement and development in both the yield and volumetric production rates of hydrogen fermentations. Scaling up of the process to pilot or large scale to generate baseline engineering data will sustain the technology with respect to commercialization. Interaction between the research community and industry from time to time will help understand the requirements and design the technology which holds the key to successful commercialization of biohydrogen production process .

Acknowledgements The authors (SVM, GNN, OS) wish to thank financial support from Ministry of New and Renewable Energy (MNRE), Government of India and Council for Scientific and Industrial Research (CSIR) in the form of research grants as MNRE Project No. 103/131/2008-NT, XII five year network project (SETCA (CSC-0113)), respectively.

References

Agarwal M, Tardio J, Mohan SV (2013) Biohydrogen production from kitchen based vegetable waste: effect of pyrolysis temperature and time on catalysed and non-catalysed operation. Bioresour Technol 130:502–509

Agarwal M, Tardio J, Mohan SV (2015) Effect of pyrolysis parameters on yield and composition of gaseous products from activated sludge: towards sustainable biorefinery. Biomass Conv Bioref 5:227–235

Akköse S, Gündüz U, Yücel M, Eroglu I (2009) Effects of ammonium ion, acetate and aerobic conditions on hydrogen production and expression levels of nitrogenase genes in Rhodobacter sphaeroides OU 001. Int J Hydrog Energy 34:8818–8827

Allakhverdiev SI et al (2010) Photosynthetic hydrogen production. J Photochem Photobiol C: Photochem Rev 11:101–113

Angenent LT, Karim K, Al-Dahhan MH, Wrenn BA, Domíguez-Espinosa R (2004) Production of bioenergy and biochemicals from industrial and agricultural wastewater. Trends Biotechnol 22:477–485

Arimi MM, Knodel J, Kiprop A, Namango SS, Zhang Y, Geißen S-U (2015) Strategies for improvement of biohydrogen production from organic-rich wastewater: a review. Biomass Bioenergy 75:101–118

Arvelakis S, Koukios E (2002) Physicochemical upgrading of agroresidues as feedstocks for energy production via thermochemical conversion methods. Biomass Bioenergy 22:331–348

Asadullah M, Fujimoto K, Tomishige K (2001) Catalytic performance of Rh/CeO2 in the gasification of cellulose to synthesis gas at low temperature. Ind Eng Chem Res 40:5894–5900

Babu ML, Sarma P, Mohan SV (2013a) Microbial electrolysis of synthetic acids for biohydrogen production: influence of biocatalyst pretreatment and pH with the function of applied potential. J Microb Biochem Technol. S6:003, http://dx.doi.org/10.4172/1948-5948

Babu ML, Subhash GV, Sarma P, Mohan SV (2013b) Bioelectrolytic conversion of acidogenic effluents to biohydrogen: an integration strategy for higher substrate conversion and product recovery Bioresour Technol 133:322 331

Balat M (2008) Potential importance of hydrogen as a future solution to environmental and transportation problems. Int J Hydrog Energy 33:4013–4029

Balat M (2009) Bioethanol as a vehicular fuel: a critical review. Energy Sources Part A 31:1242–1255

Balat M (2010) Thermochemical routes for biomass-based hydrogen production. Energy Sources Part A Recovery Utilization Environ Effects 32:1388–1398

Batyrova KA, Tsygankov AA, Kosourov SN (2012) Sustained hydrogen photoproduction by phosphorus-deprived Chlamydomonas reinhardtii cultures. Int J Hydrog Energy 37:8834–8839

Bauer C, Forest T (2001) Effect of hydrogen addition on the performance of methane-fueled vehicles. Part I: effect on SI engine performance International. J Hydrog Energy 26:55–70

Berg JM, Tymoczko JL, Stryer L (2002) Glycolysis and gluconeogenesis. W. H. Freeman and Company, New York

Beser J, Padilla B (2003) A New Mexico hydrogen cluster opportunity assessment. Final report prepared for the New Mexico Economic Development. LA-UR-04-2146

Blankenship RE, Olson JM, Miller M (1995) Antenna complexes from green photosynthetic bacteria. In: Anoxygenic photosynthetic bacteria. Springer, Dordrecht, pp 399–435

Cao XX, Huang X, Liang P, Xiao K, Zhou Y, Zhang X, Logan BE (2009) A new method for water desalination using microbial desalination cells. Environ Sci Technol 43:7148–7152

Cavinato C, Bolzonella D, Fatone F, Cecchi F, Pavan P (2011) Optimization of two-phase thermophilic anaerobic digestion of biowaste for hydrogen and methane production through reject water recirculation. Bioresour Technol 102:8605–8611

Chandra R, Venkata Mohan SV (2014) Enhanced biohydrogenesis by co-culturing photosynthetic bacteria with acidogenic process: augmented dark-photo fermentative hybrid system to regulate volatile fatty acid inhibition. Int J Hydrog Energy 39:7604–7615

Chandra R, Nikhil G, Mohan SV (2015) Single-stage operation of hybrid dark-photo fermentation to enhance biohydrogen production through regulation of system redox condition: evaluation with real-field wastewater. Int J Mol Sci 16:9540–9556

Change IPoC (2006) 2006 IPCC guidelines for national greenhouse gas inventories. Intergovernmental Panel on Climate Change. http://www.ipcc-nggip.iges.or.jp/public/2006gl/

Chen G, Andries J, Spliethoff H (2003) Catalytic pyrolysis of biomass for hydrogen rich fuel gas production. Energy Convers Manag 44:2289–2296

Chen C-Y, Yeh K-L, Lo Y-C, Wang H-M, Chang J-S (2010) Engineering strategies for the enhanced photo-H 2 production using effluents of dark fermentation processes as substrate. Int J Hydrog Energy 35:13356–13364

Cheng S, Logan BE (2007) Sustainable and efficient biohydrogen production via electrohydrogenesis. Proc Natl Acad Sci 104:18871–18873

Cheng CH, Hsu SC, Wu CH, Chang PW, Lin CY, Hung CH (2011) Quantitative analysis of microorganism composition in a pilot- scale fermentative biohydrogen production system. Int J Hydrog Energy 36:14153–14161

Cherry RS (2004) A hydrogen utopia? Int J Hydrog Energy 29:125–129

Clauwaert P, Verstraete W (2009) Methanogenesis in membraneless microbial electrolysis cells. Appl Microbiol Biotechnol 82:829–836

Cui MJZL, Yuan XH, Zhi LL, Wei JQ (2010) Shen biohydrogen production from poplar leaves pretreated by different methods using anaerobic mixed bacteria. Int J Hydrogen Energy 35:4041–4047

Cusick RD, Kiely PD, Logan BE (2010) A monetary comparison of energy recovered from microbial fuel cells and microbial electrolysis cells fed winery or domestic wastewaters. Int J Hydrog Energy 35:8855–8861

Dalcor and Camford (2005) A study conducted for Natural Resources Canada by Dalcor Consultants Ltd. and Camford Information Services, Canadian hydrogen survey - 2004/2005 ⟨http://www.iphe.net/IPHErestrictedarea/Steeringkyoto/9-14-day1/2-1-6%20Japan.pdf⟩

Dahiya S, Sarkar O, Swamy Y, Mohan SV (2015) Acidogenic fermentation of food waste for volatile fatty acid production with co-generation of biohydrogen. Bioresour Technol 182:103–113

Datar R, Huang J, Maness PC, Mohagheghi A, Czernik S, Chornet E (2007) Hydrogen production from the fermentation of corn stover biomass pretreated with a steam-explosion process. Int J Hydrogen Energy 32:932–939

Das D, Veziroğlu TN (2001) Hydrogen production by biological processes: a survey of literature. Int J Hydrog Energy 26:13–28

Das D, Khanna N, Veziroğlu NT (2008) Recent developments in biological hydrogen production processes. Chem Ind Chem Eng Q 14:57–67

de Jong W (2008) Sustainable hydrogen production by thermochemical biomass processing. Chapter 6:185–225, CRC Press, Taylor and Francis, Florida

de Vrije T, de Haas GG, Tan GB, Keijsers ERP, Claassen PAM (2002) Pretreatment of Miscanthus for hydrogen production by Thermotoga elfii. Int J Hydrog Energy 27:1381–1390

Demirbaş A (2002) Hydrogen production from biomass by the gasification process. Energy Sources 24:59–68

Dictor M-C, Joulian C, Touzé S, Ignatiadis I, Guyonnet D (2010) Electro-stimulated biological production of hydrogen from municipal solid waste. Int J Hydrog Energy 35:10682–10692

Doebbe A et al (2010) The interplay of proton, electron, and metabolite supply for photosynthetic H2 production in Chlamydomonas reinhardtii. J Biol Chem 285:30247–30260

Dunn S (2002) Hydrogen futures: toward a sustainable energy system. Int J Hydrog Energy 27:235–264

Elliott D (2000) Renewable energy and sustainable futures. Futures 32:261–274

Escapa A, Manuel M-F, Morán A, Gómez X, Guiot S, Tartakovsky B (2009) Hydrogen production from glycerol in a membraneless microbial electrolysis cell. Energy Fuel 23:4612–4618

Evans R et al (2003) Hydrogen from biomass-catalytic reforming of pyrolysis vapors. In: US DOE hydrogen, fuel cells & infrastructure technologies program—2003 annual merit review meeting. http://energy.gov/sites/prod/files/2014/03/f12/hfcit_annual_progress_report_2003.pdf

Florin L, Tsokoglou A, Happe T (2001) A novel type of iron hydrogenase in the green alga Scenedesmus obliquus is linked to the photosynthetic electron transport chain. J Biol Chem 276:6125–6132

Fulton J, Marmaro RW, Egan GJ (2010) System for producing a hydrogen enriched fuel. Google Patents. http://www.google.co.in/patents/US7721682

Geiger S (2003) Fuel cells in China-a survey of current developments. Fuel cells today. Elsevier, Amsterdam, The Netherlands. http://www.hydrogenambassadors.com/china2004/images/Article_680_chinasurvey1003.pdf

Ghimire A, Frunzo L, Pirozzi F, Trably E, Escudie R, Lens PNL, Esposito G (2015) A review on dark fermentative biohydrogen production from organic biomass: process parameters and use of by-products. Appl Energy 144:73–95

Ghirardi ML, Zhang L, Lee JW, Flynn T, Seibert M, Greenbaum E, Melis A (2000) Microalgae: A green source of renewable H2. Trends Biotechnol 18:506–511

Goud RK, Sarkar O, Chiranjeevi P, Venkata Mohan S (2014a) Bioaugmentation of potent acidogenic isolates: a strategy for enhancing biohydrogen production at elevated organic load. Bioresour Technol 165:223–232

Goud RK, Sarkar O, Mohan SV (2014b) Regulation of biohydrogen production by heat-shock pretreatment facilitates selective enrichment of Clostridium sp. Int J Hydrog Energy 39:7572–7586

Gouveia L, Oliveira AC (2009) Microalgae as a raw material for biofuels production. J Ind Microbiol Biotechnol 36:269–274

Greenbaum E (1985) Platinized chloroplasts: a novel photocatalytic material. Science (New York, NY) 230:1373–1375

Greenbaum E (1988a) Energetic efficiency of hydrogen photoevolution by algal water splitting. Biophys J 54:365–368

Greenbaum E (1988b) Interfacial photoreactions at the photosynthetic membrane interface: an upper limit for the number of platinum atoms required to form a hydrogen-evolving platinum metal catalyst. J Phys Chem 92:4571–4574

Guan Y, Deng M, Yu X, Zhang W (2004) Two-stage photo-biological production of hydrogen by marine green alga Platymonas subcordiformis. Biochem Eng J 19:69–73

Gupta BR (2012) Indian association for hydrogen energy and advanced materials HEAM NEWS. Future prospects of hydrogen, energy as alternative fuel in India

Guwy A, Dinsdale R, Kim J, Massanet-Nicolau J, Premier G (2011) Fermentative biohydrogen production systems integration. Bioresour Technol 102:8534–8542

Hallenbeck PC (2011) Microbial paths to renewable hydrogen production. Biofuels 2:285–302

Hallenbeck PC, Benemann JR (2002) Biological hydrogen production; fundamentals and limiting processes. Int J Hydrog Energy 27:1185–1193

Harun R, Danquah MK, Forde GM (2010) Microalgal biomass as a fermentation feedstock for bioethanol production. J Chem Technol Biotechnol 85:199–203, http://www.marketsandmarkets.com/Market-Reports/hydrogen-generation-market-494.html

Haslam GE, Jupesta J, Parayil G (2012) Assessing fuel cell vehicle innovation and the role of policy in Japan, Korea, and China. Int J Hydrog Energy 37:14612–14623

Hu H, Fan Y, Liu H (2008) Hydrogen production using single-chamber membrane-free microbial electrolysis cells. Water Res 42:4172–4178

Iyuke SE, Mohamad AB, Kadhum AAH, Daud WR, Rachid C (2003) Improved membrane and electrode assemblies for proton exchange membrane fuel cells. J Power Sources 114:195–202

Jalan R, Srivastava V (1999) Studies on pyrolysis of a single biomass cylindrical pellet—kinetic and heat transfer effects. Energy Convers Manag 40:467–494

Jia X et al (2014) Integration of fermentative biohydrogen with methanogenesis from fruit–vegetable waste using different pre-treatments. Energy Convers Manag 88:1219–1227

Jordan R, Nessau U, Schlodder E (1998) Charge recombination between the reduced iron-sulphur clusters and P700+. In: Photosynthesis: mechanisms and effects. Springer, Dordrecht, pp 663–666

Kamarudin SK, Daud WRW, Mohammad AW, Som AM, Takriff MS (2003) Design of a tubular ceramic membrane for gas separation in a PEMFC system. Fuel Cells 3:189–198

Kanai T, Imanaka H, Nakajima A, Uwamori K, Omori Y, Fukui T, Atomi H, Imanaka T (2005) Continuous hydrogen production by the hyperthermophilic archaeon, Thermococcus kodakaraensis KOD1. J Biotechnol 116:271–282

Kapdan IK, Kargi F (2006) Bio-hydrogen production from waste materials. Enzyme Microb Technol 38:569–582

Kongjan P, Angelidaki I (2010) Extreme thermophilic biohydrogen production from wheat straw hydrolysate using mixed culture fermentation: Effect of reactor configuration. Bioresour Technol 101:7789–7796

Kotay SM, Das D (2010) Microbial hydrogen production from sewage sludge bioaugmented with a constructed microbial consortium. Int J Hydrog Energy 35:10653–10659

Krabben L et al (2000) Influence of the axial ligands on the spectral properties of P700 of photosystem I: a study of site-directed mutants. Biochemistry 39:13012–13025

Kraemer JT, Bagley DM (2007) Improving the yield from fermentative hydrogen production. Biotechnol Lett 29:685–695

Krassen H, Schwarze A, Friedrich B, Ataka K, Lenz O, Heberle J (2009) Photosynthetic hydrogen production by a hybrid complex of photosystem I and [NiFe] hydrogenase. ACS Nano 3:4055–4061

Lalaurette E, Thammannagowda S, Mohagheghi A, Maness P-C, Logan BE (2009) Hydrogen production from cellulose in a two-stage process combining fermentation and electrohydrogenesis. Int J Hydrog Energy 34:6201–6210

Lalitha Devi G, Sukumaran RK, Venkata Mohan S, Sajna K, Sarkar O, Pandey A (2015) Rice straw hydrolysate to fuel and volatile fatty acid conversion by Clostridium sporogenes BE01: bio-electrochemical analysis of the electron transport mediators involved. Green Chem 17(5):3047–3058

Larsen HH, Feidenhans'l RK, Sønderberg Petersen L (2004) Risø energy report 3. Hydrogen and its competitors. Risoe-R; No. 1469(EN). Forskningscenter Risoe, Roskilde

Laurikko J (2006) Transport-related hydrogen activities in Asia. Report within PREMIA WP2. International activities on alternative motor fuels

Laurinavichene TV, Belokopytov BF, Laurinavichius KS, Khusnutdinova AN, Seibert M, Tsygankov AA (2012) Towards the integration of dark- and photo-fermentative waste treatment. 4. Repeated batch sequential dark- and photofermentation using starch as substrate International. J Hydrog Energy 37:8800–8810

Lee H-S, Rittmann BE (2009) Significance of biological hydrogen oxidation in a continuous single-chamber microbial electrolysis cell. Environ Sci Technol 44:948–954

Lee JW, Lee I, Laible PD, Owens TG, Greenbaum E (1995) Chemical platinization and its effect on excitation transfer dynamics and P700 photooxidation kinetics in isolated photosystem I. Biophys J 69:652–659

Lee I, Lee JW, Stubna A, Greenbaum E (2000) Measurement of electrostatic potentials above oriented single photosynthetic reaction centers. J Phys Chem B 104:2439–2443

Lee SK, Mogi G, Kim JW (2008) The competitiveness of Korea as a developer of hydrogen energy technology: the AHP approach. Energy Policy 36:1284–1291

Lee SK, Mogi G, Lee SK, Hui K, Kim JW (2010) Econometric analysis of the R&D performance in the national hydrogen energy technology development for measuring relative efficiency: the fuzzy AHP/DEA integrated model approach. Int J Hydrog Energy 35:2236–2246

Levin DB, Chahine R (2010) Challenges for renewable hydrogen production from biomass. Int J Hydrog Energy 35:4962–4969

Lin S-Y, Suzuki Y, Hatano H, Harada M (2001) Hydrogen production from hydrocarbon by integration of water-carbon reaction and carbon dioxide removal (HyPr-RING method). Energy Fuel 15:339–343

Lin C-Y, Lay C-H, Sen B, Chu C-Y, Kumar G, Chen C-C, Chang J-S (2012) Fermentative hydrogen production from wastewaters: a review and prognosis. Int J Hydrog Energy 37:15632–15642

Lindsay I, Lowe C, Reddy S, Bhakta M, Balkenende S (2009) Designing a climate friendly hydrogen plant. Energy Procedia 1:4095–4102

Lipman T (2011) An overview of hydrogen production and storage systems with renewable hydrogen case studies. Prepared for the clean energy states alliance, Montpelier, Vermont citing US DOE data 11

Liu H, Grot S, Logan BE (2005) Electrochemically assisted microbial production of hydrogen from acetate. Environ Sci Technol 39:4317–4320

Liu B-F, Ren N-Q, Ding J, Xie G-J, Cao G-L (2009) Enhanced photo-H 2 production of R. faecalis RLD-53 by separation of CO 2 from reaction system. Bioresour Technol 100:1501–1504

Ljunggren M, Zacchi G (2010) Techno-economic analysis of a two-step biological process producing hydrogen and methane. Bioresour Technol 101:7780–7788

Lo YC, Chen CY, Lee CM, Chang JS (2010) Combining dark-photo fermentation and microalgae photosynthetic processes for high-yield and CO2- free biohydrogen production. Int J Hydrogen Energy 35:10944–10953

Logan B, Grot S (2005) A bioelectrochemically assisted microbial reactor (BEAMR) that generates hydrogen gas. Patent Appl 60:022

Logan BE, Call D, Cheng S, Hamelers HV, Sleutels TH, Jeremiasse AW, Rozendal RA (2008) Microbial electrolysis cells for high yield hydrogen gas production from organic matter. Environ Sci Technol 42:8630–8640

Loppacher LJ, Kerr WA (2005) Can biofuels become a global industry?: government policies and trade constraints. Energy Polit 5:7–27

Lu Y et al (2009) Characteristics of hydrogen and methane production from cornstalks by an augmented two- or three-stage anaerobic fermentation process. Bioresour Technol 100:2889–2895

Luo G, Xie L, Zou Z, Wang W, Zhou Q (2010) Evaluation of pretreatment methods on mixed inoculum for both batch and continuous thermophilic biohydrogen production from cassava stillage. Bioresour Technol 101:959–964

Madsen EL (2008) Environmental Microbiology: From Genomes to Biogeochemistry. Blackwell Publishing, Malden

Madsen EL (2015) Environmental microbiology: from genomes to biogeochemistry. Wiley & Sons, Inc., Hoboken, New Jersey

Magnusson L, Islam R, Sparling R, Levin D, Cicek N (2008) Direct hydrogen production from cellulosic waste materials with a single-step dark fermentation process. Int J Hydrogen Energy 33:5398–5403

Markets and Markets (2011) Hydrogen generation market-by merchant & captive type, distributed & centralized generation, application & technology-trends & global forecasts (2011–2016). Report code: EP1708

Mars AE, Veuskens T, Budde MAW, van Doeveren PFNM, Lips SJ, Bakker RR, De Vrije T, Claassen PA (2010) Biohydrogen production from untreated and hydrolyzed potato steam peels by the extreme thermophiles Caldicellulosiruptor saccharolyticus and Thermotoga neapolitana. Int J Hydrogen Energy 35:7730–7737

Maruta A (2005) Japan's hydrogen and fuel cells projects. Hann over fair. International conference hydrogen & fuel cells on their way to commercialisation. http://www.fair-pr.com/hm05/conference/maruta.pdf

Melis A (2007) Photosynthetic H2 metabolism in Chlamydomonas reinhardtii (unicellular green algae). Planta 226:1075–1086

Melis A (2009) Solar energy conversion efficiencies in photosynthesis: minimizing the chlorophyll antennae to maximize efficiency. Plant Sci 177:272–280

Melis A (2012) Photosynthesis-to-fuels: from sunlight to hydrogen, isoprene, and botryococcene production. Energy Environ Sci 5:5531–5539

Melis A, Happe T (2001) Hydrogen production. Green algae as a source of energy. Plant physiology 127:740–748

Melis A, Zhang L, Forestier M, Ghirardi ML, Seibert M (2000) Sustained photobiological hydrogen gas production upon reversible inactivation of oxygen evolution in the green alga chlamydomonas reinhardtii. Plant Physiol 122:127–136

Millsaps JF, Bruce BD, Lee JW, Greenbaum E (2001) Nanoscale photosynthesis: photocatalytic production of hydrogen by platinized photosystem I reaction centers. Photochem Photobiol 73:630–635

Mohammed M, Salmiaton A, Azlina WW, Amran MM, Fakhru'l-Razi A, Taufiq-Yap Y (2011) Hydrogen rich gas from oil palm biomass as a potential source of renewable energy in Malaysia. Renew Sustain Energy Rev 15:1258–1270

Mohan SV (2009) Harnessing of biohydrogen from wastewater treatment using mixed fermentative consortia: process evaluation towards optimization. Int J Hydrog Energy 34:7460–7474

Mohan SV (2010) Waste to renewable energy: a sustainable and green approach towards production of biohydrogen by acidogenic fermentation. In: Sustainable biotechnology. Springer, Dordrecht, pp 129–164

Mohan SV, Babu ML (2011) Dehydrogenase activity in association with poised potential during biohydrogen production in single chamber microbial electrolysis cell. Bioresour Technol 102:8457–8465

Mohan SV, Babu VL, Sarma P (2007) Anaerobic biohydrogen production from dairy wastewater treatment in sequencing batch reactor (AnSBR): effect of organic loading rate. Enzyme Microb Technol 41:506–515

Mohan SV, Mohanakrishna G, Sarma P (2008) Integration of acidogenic and methanogenic processes for simultaneous production of biohydrogen and methane from wastewater treatment. Int J Hydrog Energy 33:2156–2166

Mohan SV, Srikanth S, Velvizhi G, Babu ML (2013) Microbial fuel cells for sustainable bioenergy generation: principles and perspective applications. In: Biofuel technologies. Springer, Dordrecht, pp 335–368

Mohanakrishna G, Venkata Mohan S (2013) Multiple process integrations for broad perspective analysis of fermentative H 2 production from wastewater treatment: technical and environmental considerations. Appl Energy 107:244–254

Mohanakrishna G, Goud RK, Mohan SV, Sarma P (2010a) Enhancing biohydrogen production through sewage supplementation of composite vegetable based market waste. Int J Hydrog Energy 35:533–541

Mohanakrishna G, Mohan SV, Sarma P (2010b) Utilizing acid-rich effluents of fermentative hydrogen production process as substrate for harnessing bioelectricity: an integrative approach. Int J Hydrog Energy 35:3440–3449

Monlau F, Kaparaju P, Trably E, Steyer J-P, Carrere H (2015) Alkaline pretreatment to enhance one-stage CH 4 and two-stage H 2/CH 4 production from sunflower stalks: mass, energy and economical balances. Chem Eng J 260:377–385

Narváez I, Corella J, Orío A (1997) Fresh tar (from a biomass gasifier) elimination over a commercial steam-reforming catalyst. Kinetics and effect of different variables of operation. Ind Eng Chem Res 36:317–327

Ni M, Leung DY, Leung MK, Sumathy K (2006) An overview of hydrogen production from biomass. Fuel Process Technol 87:461–472

Nikhil G, Mohan SV, Swamy Y (2014a) Behavior of acidogenesis during biohydrogen production with formate and glucose as carbon source: substrate associated dehydrogenase expression. Int J Hydrog Energy 39:7486–7495

Nikhil GN, Venkata Mohan S, Swamy YV (2014b) Systematic approach to assess biohydrogen potential of anaerobic sludge and soil rhizobia as biocatalysts: influence of crucial factors affecting acidogenic fermentation. Bioresour Technol 165:323–331

Nikhil G, Mohan SV, Swamy Y (2015) Applied potentials regulate recovery of residual hydrogen from acid-rich effluents: influence of biocathodic buffer capacity over process performance. Bioresour Technol 188:65–72

Noike T, Takabatake H, Mizuno O, Ohba M (2002) Inhibition of hydrogen fermentation of organic wastes by lactic acid bacteria. Int J Hydrogen Energy 27: 1367–1371

Nouni M (2012) Hydrogen energy and fuel cell technology: recent developments and future prospects in India. Renew Energy Akshay Urja 5:10–15

Okano K (2002) Introduction to the Hydrogen Energy Systems Society of Japan. HESS. www.hpath.org/resources/path-newsletter-02-11-01.pdf

Orecchini F, Bocci E (2007) Biomass to hydrogen for the realization of closed cycles of energy resources. Energy 32:1006–1011

Özkan E, Uyar B, Özgür E, Yücel M, Eroglu I, Gündüz U (2012) Photofermentative hydrogen production using dark fermentation effluent of sugar beet thick juice in outdoor conditions. Int J Hydrog Energy 37:2044–2049

Ozmihci S, Kargi F (2011) Dark fermentative biohydrogen production from waste wheat starch using co-culture with periodic feeding: effects of substrate loading rate. Int J Hydrog Energy 36:7089–7093

Panagiotopoulos IA, Bakker RR, de Vrije T, Koukios EG, Claassen PAM (2010) Pretreatment of sweet sorghum bagasse for hydrogen production by Caldicellulosiruptor saccharolyticus. Int J Hydrogen Energy 35:7738–7747

Pandey A, Chang J-S, Hallenbeck PC, Larroche C (2013) Biohydrogen. Elsevier, Amsterdam

Pasupuleti SB, Mohan SV (2015) Single-stage fermentation process for high-value biohythane production with the treatment of distillery spent-wash. Bioresour Technol 189:177–185

Pasupuleti SB, Sarkar O, Mohan SV (2014) Upscaling of biohydrogen production process in semi-pilot scale biofilm reactor: evaluation with food waste at variable organic loads. Int J Hydrog Energy 39:7587–7596

Pattra S, Sangyoka S, Boonmee M, Reungsang A (2008) Bio-hydrogen production from the fermentation of sugarcane bagasse hydrolysate by Clostridium butyricum. Int J Hydrogen Energy 33:5256–5265

Pekgöz G, Gündüz U, Eroğlu I, Yücel M, Kovács K, Rákhely G (2011) Effect of inactivation of genes involved in ammonium regulation on the biohydrogen production of Rhodobacter capsulatus. Int J Hydrog Energy 36:13536–13546

Pudukudy M, Yaakob Z, Mohammad M, Narayanan B, Sopian K (2014) Renewable hydrogen economy in Asia–Opportunities and challenges: an overview. Renew Sustain Energy Rev 30:743–757

Rai PK, Singh S, Asthana R (2012) Biohydrogen production from cheese whey wastewater in a two-step anaerobic process. Appl Biochem Biotechnol 167:1540–1549

Rai PK, Singh S, Asthana R, Singh S (2014) Biohydrogen production from sugarcane bagasse by integrating dark-and photo-fermentation. Bioresour Technol 152:140–146

Reddy MV, Nikhil G, Mohan SV, Swamy Y, Sarma P (2012) Pseudomonas otitidis as a potential biocatalyst for polyhydroxyalkanoates (PHA) synthesis using synthetic wastewater and acidogenic effluents. Bioresour Technol 123:471–479

Reij MW, Keurentjes JT, Hartmans S (1998) Membrane bioreactors for waste gas treatment. J Biotechnol 59:155–167

Ren N, Wang A, Gao L, Xin L, Lee D-J, Su A (2008) Bioaugmented hydrogen production from carboxymethyl cellulose and partially delignified corn stalks using isolated cultures. Int J Hydrog Energy 33:5250–5255

Richmond A, Hu Q (2013) Handbook of microalgal culture: applied phycology and biotechnology. Wiley & Sons, Inc., Hoboken, New Jersey

Romeri M (2004) Hydrogen:a new possible bridge between mobility and distributed generation (CHP). World Energy Conference (WEC). http://www.worldenergy. org/weceis/congress/papers/romeriv0904.pdf

Romeri V, Sim R (2004) Hydrogen: a new possible bridge between mobility and distributed generation (CHP). In: 19th world energy congress, pp 5–9

Rozendal RA, Hamelers HV, Euverink GJ, Metz SJ, Buisman CJ (2006) Principle and perspectives of hydrogen production through biocatalyzed electrolysis. Int J Hydrog Energy 31:1632–1640

Rozendal RA, Hamelers HV, Molenkamp RJ, Buisman CJ (2007) Performance of single chamber biocatalyzed electrolysis with different types of ion exchange membranes. Water Res 41:1984–1994

Saiki Y, Amao Y (2004) Visible light-induced enzymatic hydrogen production from oligosaccharides using Mg chlorophyll-a and platinum colloid conjugate system. Int J Hydrog Energy 29:695–699

Sari R, Yaakob Z, Ismail M, Daud WRW, Hakim L (2013) Palladium–alumina composite membrane for hydrogen separator fabricated by combined sol–gel, and electroless plating technique. Ceram Int 39:3211–3219

Sarkar O, Goud RK, Subhash GV, Mohan SV (2013) Relative effect of different inorganic acids on selective enrichment of acidogenic biocatalyst for fermentative biohydrogen production from wastewater. Bioresour Technol 147:321–331

Sarkar O, Agarwal M, Kumar AN, Mohan SV (2015) Retrofitting hetrotrophically cultivated algae biomass as pyrolytic feedstock for biogas, bio-char and bio-oil production encompassing biorefinery. Bioresour Technol 178:132–138

Sarkar O, Kumar AN, Dahiya S, Krishna KV, Yeruva DK, Mohan SV (2016) Regulation of acidogenic metabolism towards enhanced short chain fatty acid biosynthesis from waste: metagenomic profiling. RSC Adv 6:18641–18653

Saxena R, Seal D, Kumar S, Goyal H (2008) Thermochemical routes for hydrogen rich gas from biomass: a review. Renew Sustain Energy Rev 12:1909–1927

Schnackenberg J, Ikemoto H, Miyachi S (1996) Photosynthesis and hydrogen evolution under stress conditions in a CO 2-tolerant marine green alga, Chlorococcum littorale. J Photochem Photobiol B Biol 34:59–62

Shafie S, Mahlia T, Masjuki H, Ahmad-Yazid A (2012) A review on electricity generation based on biomass residue in Malaysia. Renew Sustain Energy Rev 16:5879–5889

Shakya B, Aye L, Musgrave P (2005) Technical feasibility and financial analysis of hybrid wind–photovoltaic system with hydrogen storage for Cooma. Int J Hydrog Energy 30:9–20

Shi D (2006) Chinese hydrogen update. Ministry of Science and Technology of China. In: 6th IPHE steering committee meeting Reykjavik, Iceland

Singh S, Sudhakaran AK, Sarma PM, Subudhi S, Mandal AK, Gandham G, Lal B (2010) Dark fermentative biohydrogen production by mesophilic bacterial consortia isolated from riverbed sediments. Int J Hydrog Energy 35:10645–10652

Sivaramakrishna D, Sreekanth D, Sivaramakrishnan M, Kumar BS, Himabindu V, Narasu ML (2014) Effect of system optimizing conditions on biohydrogen production from herbal wastewater by slaughterhouse sludge. Int J Hydrog Energy 39:7526–7533

Srikanth S, Mohan SV, Devi MP, Babu ML, Sarma P (2009a) Effluents with soluble metabolites generated from acidogenic and methanogenic processes as substrate for additional hydrogen production through photo-biological process. Int J Hydrog Energy 34:1771–1779

Srikanth S, Mohan SV, Devi MP, Peri D, Sarma P (2009b) Acetate and butyrate as substrates for hydrogen production through photo-fermentation: process optimization and combined performance evaluation. Int J Hydrog Energy 34:7513–7522

Sutton D, Kelleher B, Ross JR (2002) Catalytic conditioning of organic volatile products produced by peat pyrolysis. Biomass Bioenergy 23:209–216

Takahara I (2005) Japan's approach to commercialisation of fuel cell or hydrogen technology. IPHE. Steering committee meeting

Ueno Y, Fukui H, Goto M (2007) Operation of a two-stage fermentation process producing hydrogen and methane from organic waste. Environ Sci Technol 41:1413–1419

Ueno Y, Haruta S, Ishii M, Igarashi Y (2001) Microbial community in anaerobic hydrogen-producing microflora enriched from sludge compost. Appl Microbiol Biotechnol 57:555–562

Vadiee A, Yaghoubi M, Sardella M, Farjam P (2015) Energy analysis of fuel cell system for commercial greenhouse application–a feasibility study. Energy Convers Manag 89:925–932

Vardar-Schara G, Maeda T, Wood TK (2008) Metabolically engineered bacteria for producing hydrogen via fermentation. J Microbial Biotechnol 1:107–125

Venkata Mohan S (2010) Waste to renewable energy: a sustainable and green approach towards production of biohydrogen by acidogenic fermentation. In: Sustainable biotechnology: renewable resources and new perspectives. Springer, 129–164

Venkata Mohan S, Srikanth S, Babu ML, Sarma PN (2010a) Insight into the dehydrogenase catalyzed redox reactions and electron discharge pattern during fermentative hydrogen production. Bioresour Technol 101:1826–1833

Venkata Mohan S, Raghavulu SV, Goud RK, Srikanth S, Babu VL, Sarma PN (2010b) Microbial diversity analysis of long term operated biofilm configured anaerobic reactor producing biohydrogen from wastewater under diverse conditions. Int J Hydrog Energy 35:12208–12215

Venkata Mohan S, Pandey A (2013) Chapter 1 – biohydrogen production: an introduction. In: Larroche AP-SCCH (ed) Biohydrogen. Elsevier, Amsterdam, pp 1–24

Venkata Mohan S, Mohanakrishna G, Veer Raghavulu S, Sarma PN (2007) Enhancing biohydrogen production from chemical wastewater treatment in anaerobic sequencing batch biofilm reactor (AnSBBR) by bioaugmenting with selectively enriched kanamycin resistant anaerobic mixed consortia. Int J Hydrog Energy 32:3284–3292

Venkata Mohan S, Chandrasekhar K, Chiranjeevi P, Suresh Babu P (2013) Chapter 10 – biohydrogen production from wastewater. In: Larroche AP-SCCH (ed) Biohydrogen. Elsevier, Amsterdam, pp 223–257

Venkata Mohan S, Nikhil GN, Chiranjeevi P, Reddy CN, Rohit MV, Kumar AN, Sarkar O (2016) Waste biorefinery models towards sustainable circular bioeconomy: critical review and future perspectives. Bioresour Technol.doi:10.1016/j.biortech.2016.03.130

Venkateswar Reddy M, Chitanya DNSK, Nikhil GN, Venkata Mohan S, Sarma PN (2014) Influence of co-factor on enhancement of bioplastic production through wastewater treatment. Clean–Soil Air Water 42:809–814

Veziroğlu TN, Şahi S (2008) 21st century's energy: hydrogen energy system. Energy Convers Manag 49:1820–1831

Vyas D, Kumar H (1995) Nitrogen fixation and hydrogen uptake in four cyanobacteria. Int J Hydrog Energy 20:163–168

Wagner RC, Regan JM, Oh S-E, Zuo Y, Logan BE (2009) Hydrogen and methane production from swine wastewater using microbial electrolysis cells. Water Res 43:1480–1488

Wang B, Wan W, Wang J (2009) Effect of ammonia concentration on fermentative hydrogen production by mixed cultures. Bioresour Technol 100:1211–1213

Wang A, Sun D, Cao G, Wang H, Ren N, Wu W-M, Logan BE (2011) Integrated hydrogen production process from cellulose by combining dark fermentation, microbial fuel cells, and a microbial electrolysis cell. Bioresour Technol 102:4137–4143

Williams PT, Brindle AJ (2002) Catalytic pyrolysis of tyres: influence of catalyst temperature. Fuel 81:2425–2434

Wijffels RH, Barbosa MJ (2010) An outlook on microalgal biofuels. Science 329:796–799

Winkler M, Heil B, Heil B, Happe T (2002) Isolation and molecular characterization of the [Fe]-hydrogenase from the unicellular green alga Chlorella fusca. Biochim Biophys Acta 1576:330–334

Wornat MJ, Hurt RH, Yang NY, Headley TJ (1995) Structural and compositional transformations of biomass chars during combustion. Combust Flame 100:131–143

Yong TLK, Lee KT, Mohamed AR, Bhatia S (2007) Potential of hydrogen from oil palm biomass as a source of renewable energy worldwide. Energy Policy 35:5692–5701

Zhang Z, Maruyama A (2001) Towards a private–public synergy in financing climate change mitigation projects. Energy Policy 29:1363–1378

Zhou A, Thomson E (2009) The development of biofuels in Asia. Appl Energy 86:S11–S20

http://www.greencarcongress.com/2013/03/ballard-20130313.html

http://investing.businessweek.com/research/stocks/private/snapshot.asp?privcapId=23713224

www.energyaccess.in

Waste-to-Hydrogen Energy in Saudi Arabia: Challenges and Perspectives

11

R. Miandad, M. Rehan, O.K.M. Ouda, M.Z. Khan, K. Shahzad, I.M.I. Ismail, and A.S. Nizami

Abstract

Hydrogen (H_2) has emerged as a promising alternative fuel that can be produced from renewable resources including organic waste through biological processes. In the Kingdom of Saudi Arabia (KSA), the annual generation rate of municipal solid waste (MSW) is around 15 million tons that average around 1.4 kg per capita per day. Similalry, a significant amount of industrial and agricultural waste is generated every year in KSA. Most of these wastes are disposed in landfills or dumpsites after partial segregation and recycling and without material or energy recovery. This causes environmental pollution and release of greenhouse gas (GHG) emissions along with public health problems. Therefore, the scope of producing renewable H_2 energy from domestic and industrial waste sources is promising in KSA, as no waste-to-energy (WTE) facility exists. This chapter reviews the biological and chemical ways of H_2 production from waste sources and availability of waste resources in KSA.

11.1 Introduction

The current world population of 7.2 billion is projected to increase by 1 billion till 2025 with an annual growth rate of 1 % (WHO 2014). The Global South i.e. developing Asia, Middle East, Africa, and Latin American countries is the place where most of this growth will occur due to rapid growth in urbanization and population. As a result, the energy demand is increasing significantly in developing countries, especially in Asia (Ouda et al. 2013), which is expected to increase by 46–58 % with an annual rate of 3.7 % till 2025 (FAO 2010; US-EIA 2007). Fossil fuels are the

R. Miandad • M. Rehan • K. Shahzad
I.M.I. Ismail • A.S. Nizami (✉)
Center of Excellence in Environmental Studies (CEES), King Abdul Aziz University, Jeddah, Saudi Arabia
e-mail: Nizami_pk@yahoo.com; anizami@kau.edu.sa

O.K.M. Ouda
Department of Civil Engineering, Prince Mohamed Bin Fahd University, Al-Khobar, Saudi Arabia

M.Z. Khan
Environmental Research Laboratory, Department of Chemistry, Aligarh Muslim University, Aligarh 202 002, Uttar Pradesh, India

A. Singh, D. Rathore (eds.), *Biohydrogen Production: Sustainability of Current Technology and Future Perspective*, DOI 10.1007/978-81-322-3577-4_11

most relied choice at the moment to fulfill the energy demands (Demirbas et al. 2016). As a consequence, existing reserves of fossil fuels sources are depleting along with global climate change. Therefore, the renewable energy sources are getting more attention to fill the ever increasing energy demand-supply gap. The advances in technologies with lower operational cost and governmental incentives are increasing the growth in renewable energy sector (Nizami et al. 2015a; Nizami et al. 2016; Ouda et al. 2016; Sadaf et al. 2015). Moreover, the national and international protocols such as Kyoto Protocol and Agenda 21 are adding momentum to move from fossil fuels-based economies toward renewable fuels-based economies (Tawabini et al. 2014).

The high growth rate of population and urbanization along with raised living standards is also resulting in excessive generation of municipal solid waste (MSW) worldwide. In the next 15 years, the world's average generation rate of MSW will increase from 1.2 to 1.4 kg per capita per day. Currently, around 2.4 billion tons of MSW is generated every year worldwide that will reach up to 2.6 billion tons by 2025 (UD 2012). The MSW management is not only important for sanitation purposes but also for generation of energy and recyclable materials for revenue and environmental protection (Rathi 2006). Therefore, the sustainable management of MSW is one of the national policy agenda of most developed nations to protect the public health, aesthetic places, and land-use resources (Ouda et al. 2013).

11.1.1 Energy Perspectives of the Kingdom of Saudi Arabia (KSA)

The Kingdom of Saudi Arabia (KSA) is located in the Middle East and lies between 16° 22′ and 32° 14′ north latitudes and 34° 29′ and 55° 40′ east longitudes. KSA is one of the world's largest crude oil producer country. A large socio-economic development has occurred since the last four decades due to oil-export revenue. KSA population is increasing at an annual rate of 3.4 % coupled with high-living standards and urbanization growth (Nizami et al. 2015b). The increase in urbanization was observed from 50 to 80 % from 1975 to 2000, while it has reached up to 85 % in 2010 (Aga et al. 2014). The most urbanized cities of KSA are Riyadh (the capital of KSA), Jeddah, Dammam, Makkah, Medina, Al-Hasa, and Al Taif with the population of 5.2, 3.4, 2, 1.7, 1.2, 1.1, and 1 million respectively (CDSI, 2010). Energy demand of KSA has increased significantly with a rate of 5.6 % from 2006 to 2010 (MEP 2010). The current electricity demand in the country is about 55 GW that is expected to surpass 120 GW by 2032 (Ouda et al. 2015). At present, fossil fuels are the only source to meet all energy requirements of the country. KSA's government has initiated a program called King Abdullah City of Atomic and Renewable Energy (KACARE) to utilize the indigenous renewable energy resources through science, research, and industry. The ambition of KACARE program is to generate half of the electricity from renewable energy sources, including solar, wind, nuclear, geothermal, and waste-to-energy (WTE) by 2032 (KACARE 2012).

11.1.2 Waste Generation in KSA

In KSA, the generation rate of MSW is about 15 million tons per year with an average rate of 1.4 kg per capita per day (Ouda et al. 2015). The Ministry of Municipalities and Local Affair regulates the management of MSW in the country that includes waste collection and disposal to landfill sites (Nizami et al. 2015c). The landfill requirement is extremely high with 2.8 million m^2 per year (Ouda et al. 2013). Metals and cardboard are the recycled materials (10–15 % of total MSW) regulated by informal sector (Khan and Kaneesamkandi 2013). Most of the in-use landfills are approaching to their full capacities and resulting in waste leachate, sludge, odor, and greenhouse gas (GHG) emissions (Ouda and Cekirge 2014). WTE technologies are widely used to recover energy and value-added products (VAP) from different fractions of MSW. The examples of WTE technologies include pyrolysis, gasification, anaerobic digestion (AD), incineration, plasma arc gasification, refuse derived fuel (RDF), and transesterification (Gardy et al.

2014; Ouda et al. 2016; Tahir et al. 2015). In KSA, there is no such WTE or material recovery facility (MRF) exists (Nizami et al. 2015b).

11.1.3 Aim of the Chapter

This chapter in its first part reviews the biological and chemical H_2 production processes and their advantages and disadvantages, potential waste substrates for H_2 production, and technological advances and challenges. In second part, a review of the available waste resources in KSA for H_2 production is carried out with an ambition to explore indigenous sources of renewable energy and solve the waste management problems.

11.2 Waste-to-Hydrogen Energy

H_2 is one of the most abundantly available elements on the earth with highest energy content per unit weight (142 KJ/g) and efficiency of producing electricity (Bhutto et al. 2011). It can be stored in liquid and gas forms and can be converted into different forms of energy. This makes H_2 a promising alternative fuel and future energy carrier (Table 11.1). Annually, around 500 billion m^3 of H_2 is produced globally with 10 % growth rate (Winter 2005); of which 40 % is produced from natural gas, 30 % from heavy oils and naphtha, 18 % from coal, 4 % from electrolysis, and 1 % from biomass (Nath and Das 2003; Kapdan and Kargi 2006). Most of the H_2 applications are currently limited to industrial sector, where H_2 is used in petrochemical manufacturing, glass purification, hydrogenation of unsaturated fats and vegetable oil and steel processing, and desulfurization and reformulation of gasoline in refineries (Kotay and Das 2008; Kapdan and Kargi 2006). Moreover, H_2 is used in metallurgical processes in heat-treating applications for the removal of oxygen (O_2) as O_2 scavenger. Collectively, 49 % of the produced H_2 is used in ammonia production, 37 % is utilized in petroleum refining, 8 % is used in methanol production, and 6 % is utilized in various small applications (Konieczny et al. 2008).

H_2 can be produced from renewable and nonrenewable sources (Fig. 11.1). A source of energy is required for each process of H_2 production in the form of heat or electrolyte (Bhutto et al. 2011). The most common technique for H_2 production is the reforming of natural gas (Holladay et al. 2009), while the typical methods of H_2 production are fossil fuel non-catalytic partial oxidation and auto-thermal reforming. To further improve these methods, membrane processes, methane selective oxidation, and oxidative dehydrogenation procedures are adopted (Armor 1999). However in recent years, the H_2 production from biological methods using renewable resources has gained significant attention (Kapdan and Kargi 2006).

11.2.1 Biological Methods of H_2 Production from Waste

The biological processes that produce H_2 from renewable sources include direct photolysis, indirect photolysis, photo-fermentation, dark fermentation, and microbial electrolysis (bioelectrohydrogenesis). Each process has advantages and disadvantages based on substrate type, process mechanisms, end products, and energy requirements (Table 11.2).

The mechanism of biological H_2 production was first discovered by Hans Gaffron in the early 1940s, when he found that green algae can either consume H_2 as an electron donor in carbon dioxide (CO_2) fixation process or produce H_2 under anaer-

Table 11.1 Characteristics of H_2

Characteristics	Unit	Values
Boiling point	K	20.3
Liquid density	kg/m^3	71
Gas density	kg/m^3	0.08
Heat of vaporization	kJ/kg	444
Lower heating value (mass)	MJ/kg	120
Lower heating value (liquid, volume)	MJ/m^3	8960
Diffusivity in air	cm^2/s	0.63
Lower flammability limit	vol. % (in air)	4
Upper flammability limit	vol. % (in air)	75
Ignition temperature in air	°C	585
Ignition energy	MJ	0.02
Flame velocity	cm/s	270

Sequeira and Santos 2010

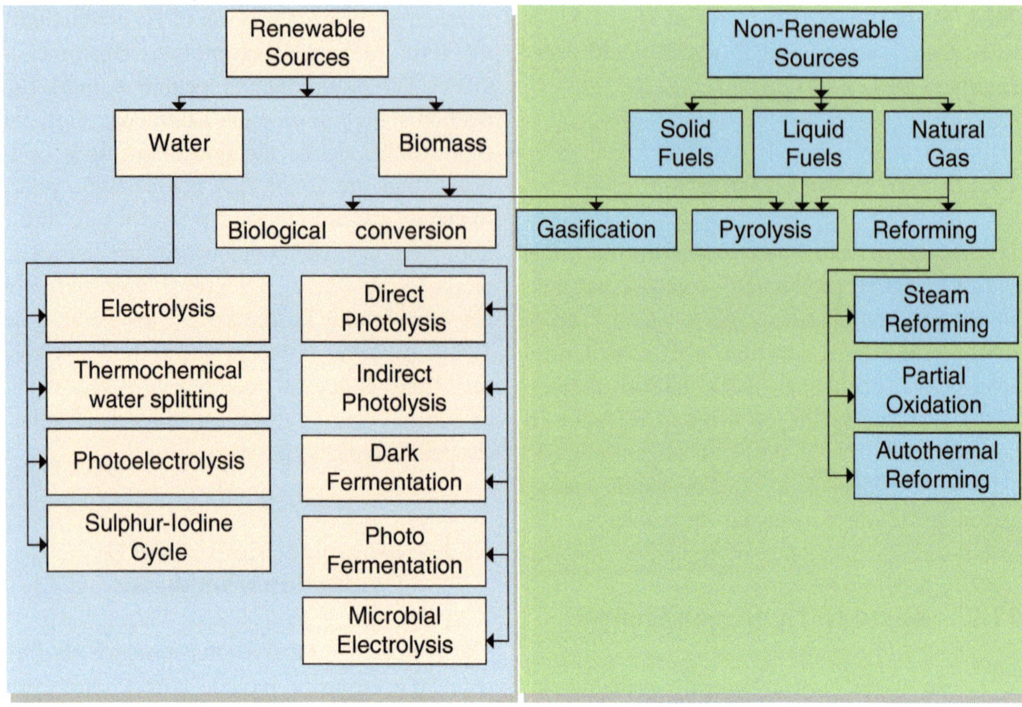

Fig. 11.1 H_2 production processes from renewable and non-renewable sources (Armaroli and Balzani 2011)

obic conditions in both dark and light (Kumar and Das 2000a, b; Benemann 1996, 1997). H_2 can be produced by a number of microorganisms through enzymatic activities (Table 11.3). These microorganisms produce H_2 in a variety of different ways due to their diversity in microbial physiology and metabolism. All of these processes offer advantages over the conventional H_2 production processes in terms of lower catalyst cost and less energy consumption by using microbial cells and mesophilic operation, respectively (Hallenbeck et al. 2009). The enzymes that control H_2 production are called hydrogenase and nitrogenase (Lindberg et al. 2004).

11.2.1.1 Direct Biophotolysis

Direct biophotolysis process involves solar energy to split water molecules into hydrogen ions and oxygen by photosynthesis. These hydrogen ions are then converted into H_2 by the hydrogenase enzymes (Table 11.4). Different types of algae and cyanobacteria species have been used in producing H_2, such as *Chlamydomonas reinhardtii* (Momirlan and Veziroglu 2005; Holladay et al. 2009), *Scenedesmus obliquus* (Das and Veziroglu 2008), *Chlorococcum littorale* (Hallenbeck and Benemann 2002; Hallenbeck et al. 2009), *Platymonas subcordiformis* (Kumar and Das 2000a, b), *Chlorella fusca* (Gaffron and Rubin 1942), *Anabaena* sp., *Oscillatoria* sp., *Calothrix* sp., *Synechococcus* sp., *Gloeobacter* sp. (Melis and Happe 2001), *Anabaena cylindrica* (Schulz 1996; Melis and Happe 2001), *Anabaena variabilis* (Benemann 1997; Miura 1995; Ni et al. 2006; Greenbaum et al. 1983; Ghirardi et al. 2000), etc.

Photoautotrophic microorganism such as cyanobacteria or green algae is equipped with chlorophyll A and other pigments to absorb sunlight and use photosynthetic systems (PSI and PSII) to carry out oxygenated photosynthesis. The photons having wavelength shorter than 680 nm are absorbed by the pigments in PSII and generate strong oxidant, which is capable of splitting water into protons H+, electrons e-, and O_2. These electrons are transferred to PSI through a series of electron carrier. Similarly, photons having wavelength under 700 nm are absorbed by

Table 11.2 Advantages and disadvantages of biological H_2 production methods

	Advantages	Disadvantages
Direct biophotolysis	From water and sunlight, H_2 is directly produced	High intensity of light is required
		O_2 and H_2 are produced simultaneously and O_2 can be dangerous for the system
	An increase of tenfolds in conversion energy in comparison to biomass (trees, crops, grasses, etc.)	Even low concentrations of O_2 can disturb hydrogenase (e.g. green algae)
		Photochemical efficiency is lower
Indirect biophotolysis	H_2 is produced by cyanobacteria using water	To stop degradation of H_2, hydrogenase enzymes are eliminated
	N_2 is fixed from atmosphere	In gas mixture, around 30 % of O_2 is present
Photo-fermentation	Associated bacteria can use a wide spectral light energy	H_2 is produced at a slow rate
	Different organic wastes can be used	On nitrogenase, O_2 affects negatively
	Substrate conversion efficiencies are high	Only 1–5 % conversion efficiency is achieved
	A wide range of substrates can be degraded	Due to toxic nature of the substrate (effluent), pretreatment is required
		High investment costs due to expensive setup installations, and large reactor surface areas
Dark fermentation	It is a simpler and less expensive process	For hydrogenase, O_2 is inhibitor
		Lower yield of H_2 is achieved
	A high rate of H_2 production is achieved	The process becomes thermodynamically unfavorable with the increase of H_2 pressure
	H_2 can be produced 24 by 7 without light	
	As substrates, a wide range of carbon sources can be used	CO_2 is required to separate from the gas mixture
Microbial electrolysis	Electricity or H_2 can be made directly from waste sources	The involved metabolic pathways are not well defined
	For H_2 economy, it is a promising approach	Most of the studies focused on only mixed cultures that are used in already enriched and active microbial fuel cells
	Wastewater, especially effluents with low organic content can be used	
	Sustainable and effective process	A low volumetric H_2 production occurs when the power densities at the electrode surface becomes low
		A high voltage negatively affects energy efficiency

Bhutto et al. 2011; Kotay and Das 2008; Levin et al. 2004; Das and Veziro 2001

pigments of PSI to further enhance the energy level of electrons. These electrons then reduce oxidized ferredoxin (Fd) and/or nicotinamide adenine dinucleotide phosphate (NADP+) into their reduced forms. It leads to proton gradient across the cellular membrane that drives adenosine triphosphate (ATP) production through ATP synthase enzyme. Under special conditions, hydrogenase or nitrogenase enzymes utilize the reducing equivalents to reduce protons for molecular hydrogen evolution (Yu and Takahashi 2007).

Table 11.3 Comparison of various biological H_2 production methods

Process	Organism involved	Maximum reported rate (mmol H_2/L h^{-1})
Direct biophotolysis	*Chlamydomonas reinhardtii*	0.07
Indirect biophotolysis	*Anabaena variabilis*	0.36
Photo-fermentation	*Rhodobacter sphaeroides*	0.16
Dark fermentation	*Enterobacter cloacae* DM 11, *Clostridium* sp. strain no. 2	64.5–75.6
Two-stage fermentation (dark-fermentation+ photolysis)	*Enterobacter cloacae*	47.9–51.2
	DM 11+ *Rhodobacter*	
	sphaeroides OU 001	
	51.20	
	Mixed microbial flora +	
	Rhodobacter sphaeroides	
	OU 001	

Kotay and Das 2008; Bhutto et al. 2011; Hallenbeck and Benemann 2002

Table 11.4 Chemical reactions involved in the biological H_2 production methods

Process
Direct biophotolysis
$2\ H_2O + light \rightarrow 2\ H_2 + O_2$
Indirect biophotolysis
(a) $6H_2O + 6CO_2 + light \rightarrow C_6H_{12}O_6 + 6O_2$
(b) $C_6H_{12}O_6 + 2H_2O \rightarrow 4H_2 + 2CH_3COOH + 2CO_2$
(c) $2CH_3COOH + 4H_2O + light \rightarrow 8H_2 + 4CO_2$
Overall reaction
$12H_2O + light \rightarrow 12\ H_2 + 6O_2$
Photo-fermentation
$CH_3COOH + 2H_2O + light \rightarrow 4H_2 + 2CO_2$
Dark fermentation
$C_6H_{12}O_6 + 2H_2O \rightarrow 2CH_3COOH + 4H_2 + 2CO_2$
Microbial electrolysis
$C_6H_{12}O_6 + 2H_2O \rightarrow 4H_2 + 2CO_2 + 2CH_3COOH$
Anode: $CH_3COOH + 2H_2O \rightarrow 2CO_2 + 8e- + 8H+$
Cathode: $8H+ + 8e - \rightarrow 4H_2$

Bhutto et al. 2011; Levin et al. 2004; Das and Veziroglu 2001; Franks et al. 2009

One of the main challenges of this process is to remove the produced O_2, as it inhibits hydrogenase enzyme activity and therefore limits H_2 production (Hallenbeck and Benemann 2002).

Theoretically, direct biophotolysis process is an economical and sustainable method for H_2 production by utilizing renewable resources like algae, light, and water. However, the process becomes unavailable at commercial scale due to challenges such as strong inhibition effect of evolved O_2 on enzymatic activity (Bhutto et al. 2011), generation and impact of highly explosive H_2-O_2 mixtures, low conversion efficiencies in different light intensities, low H_2 production rates, no waste utilization, and limitations in designing large-scale reactors (Wang and Wan 2009; Das and Veziroglu 2008; Yetis et al. 2000). Therefore, photo-fermentation and dark-fermentation processes have advantages for treating waste with H_2 production (Tables 11.2, 11.3, and 11.4).

11.2.1.2 Indirect Biophotolysis

Indirect biophotolysis process involves two stages coupled with CO_2 fixation; O_2 is released in the first stage with CO_2 fixation and H_2 is produced in the second stage (Momirlan and Veziroglu 2005). The reduced substrates such as glycogen and starch accumulate during the photosynthetic O_2 evolution and CO_2 fixation stage, which are then used in anaerobic conditions to produce H_2 and CO_2 in the second stage. The conversion of these substrates into H_2 in the second step is carried out by the algae or other organisms like photosynthetic or fermentative bacteria. This separation of O_2 and H_2 production reactions not only helps to overcome the two major problems of direct biophotolysis such as enzyme deactivation and production of explosive gas mixture, but also makes H_2 purification relatively easier by using conventional separation methods (Yu and Takahashi 2007). The processes of direct biophotolysis can either be carried out in a single reactor producing O_2 and H_2 in an alternating cycle or in separate reactors like open ponds and photo bioreactors.

11.2.1.3 Photo-Fermentation

Photo-fermentation utilizes photosynthetic bacteria to quantitatively produce H_2 and CO_2 from various organic substrates, especially organic acids such as acetate and butyrate mediated by nitrogenase, using light energy. This is a simple process and has advantage of high H_2 production rate due to bacterial ability in degrading a wide range of substrates. This process has been demonstrated to produce H_2 from various organic acids, food, agricultural waste (Hallenbeck and Benemann 2002), and industrial wastewater (Yildiz et al. 1994; Davila-Vazquez et al. 2008). However, pretreatment of substrates originating from waste sources may require the removal of any possible toxicity and dirty color.

11.2.1.4 Dark-Fermentation

The dark-fermentation process is a stable process of H_2 production due to anaerobic conditions. The bacteria are grown in the dark on carbohydrate-rich substrates to generate electrons. These electrons are then taken up by O_2 in aerobic conditions, while they are used by protons to be reduced to H_2 molecules in anaerobic conditions. One of the main drawbacks of this process is the production of gas mixture of H_2 and CO_2 with possible CH_4, CO, and H_2S that is technically a challenge when used in fuel cells.

A variety of substrates can be used in dark-fermentation process including simple sugar, starch containing waste, cellulose containing wastes, food industry wastes, wastewater, and waste sludge (Kapdan and Kargi 2006). The process parameters are critical to be controlled for efficient H_2 production, such as process temperature, pH, substrate type and composition, type of organic acid produced, reactor configuration, pressure, and residence time (Geelhoed et al. 2010).

11.2.1.5 Microbial Electrolysis

The microbial electrolysis is a bio-electrochemical system that generates electrical current to reduce protons to H_2 in a process called bioelectrohydrogenesis. The microbial electrolysis cell (MEC) consists of four parts that are anodic chambers, cathodic chambers, electronic separator, and external electrical power source (Liu et al. 2005; Hamelers et al. 2010). The reaction typically utilizes acetate as the electron donor to be oxidized. The electrons and protons are then combined to produce H_2 at the cathode (Table 11.4).

This process can utilize wastewater and agro-industrial residues containing biopolymers like cellulose and starch to produce H_2. The reactions carried out in MEC can be catalyzed both by microorganisms and chemicals like platinum and nickel. The MEC can achieve much higher H_2 yields (80–100 %) in comparison to fermentative H_2 production (<33 %) (Table 11.5), since it uses electric current to overcome the energy barriers for oxidation of the substrate (Gralnick and Newman 2007). The bio-electrochemical systems recover almost 90 % of the energy using acetate as a substrate that is three times greater than the fermentation process (Catal et al. 2015; Hu et al. 2008). However, the overall performance of the MEC depends upon the physiology of the microorganisms as well as the physicochemical transport processes. Various types of losses such as ohmic resistance, concentration, and conductivity associated with these bio-electrochemical systems adversely affect the overall H_2 production rates (Logan et al. 2008). Thus, there still remains a great challenge to keep the electrical potential in balance at both the bioanode and biocathode chambers (Liu et al. 2005).

Table 11.5 Comparison of the energy parameters of different biological techniques used for H_2 production

Systems	Energy efficiency (%)	References
Direct biophotolysis	0.07 mmol H_2/(l h)	Kotay and Das (2008)
Indirect biophotolysis	0.36 mmol H_2/(l h)	Kotay and Das (2008)
Photo-fermentation	0.16 mmol H_2/(l h)	Bhutto et al. (2011)
Dark fermentation	64.5–75.6 mmol H_2/(l h)	Bhutto et al. (2011)
Two-stage fermentation (dark-fermentation+ photolysis)	47.9–51.2 mmol H_2/(l h)	Hallenbeck and Benemann (2002)
Microbial Electrosynthesis System (MES)	90% at the rate of 0.5 kWh/m^3-H_2	Catal et al. (2015)

11.2.2 Chemical Ways of H_2 Production from Waste

11.2.2.1 Pyrolysis

Pyrolysis is a thermal degradation process that converts the carbonaceous substrates into liquid oil, solid residue (char), and gases in the absence of oxygen at temperatures of 300–650 °C (Manara and Zabaniotou 2012). This is a highly flexible process that can be optimized in accordance to the desired products such as liquid fuel, gases, and char from the specific substrate. This optimization is carried out by regulating the composition of the substrate, temperature and retention time variations in the reaction chamber, catalyst, and substrate particle size. The liquid oil and char are the most investigated products of pyrolysis that are mainly dependent on prevailing temperature conditions (Rehan et al. 2016). The liquid oil as a fuel source is the main product when process is carried out at temperatures of 400–550 °C, while gases are higher when process temperature is greater than 700 °C. Similarly, substrate resident time (from few seconds to 2 h) in the reaction chamber is an important parameter affecting the yields of final products (Chen et al. 2014).

H_2 can be produced from pyrolysis of different substrates such as sludge, legume straw (Li et al. 2004; Zhang et al. 2011), wheat straw (Hornung et al. 2009), crude beech-wood oil (Davidian et al. 2007), nutshell, olive husk, grape residue, beech wood, straw pellet, and waste plastics (Di Blasi et al. 1999). Kasakura and Hiraoka (1982) reported that 5.5 vol.% H_2 is produced from the sludge pyrolysis with 3.65 vol.% CO. According to Chen et al. (2014), the increase of temperature from 500 to 700 °C also increased the gases production from 30–35 to 45–50 vol.%, while Demirbas and Arin (2004) reported the increase of gases from 27–41 to 41–55 vol.% with increase of temperature from 377 to 752°C.

In pyrolysis process, the use of catalyst can also increase the gases production. It is reported that by using Ni/Al_2O_3 and $Ni\text{-}K/La_2O_3\text{-}Al_2\text{-}O_3$ catalysts, H_2 production is increased by 45–50%. Moreover, the substrate particle size increased the gases yield at 28.2, 38.5, 15, 18, 5, and 8 mol% of H_2, CO, CO_2, CH_4, C_2H_6, and C_2H_4 respectively, with particle size of 0.45–0.90 mm. Fixed-bed reactor, free-fall reactor, and fluidized bed reactor are commonly used reactor configuration for pyrolysis to produce H_2 (Hornung et al. 2009; Li et al. 2004).

11.2.2.2 Gasification

Gasification is a thermophilic process used to convert the carbonaceous substrates into H_2 at temperature of 800–900 °C in a controlled-oxygen environment (Uddin et al. 2013). Different substrate sources such as palm oil waste (Inayat et al. 2012), meat and bone waste (Soni et al. 2009), wood sawdust (Wu et al. 2011), plastic residue (Czernik and French 2006), rich husk (Karmakar and Datta 2011), pellets (Ruoppolo et al. 2012), and pig compost (Wang et al. 2013) are utilized for H_2 production.

Different catalysts are used to increase the process yield. The use of Ni/MCM-41 catalyst with sawdust in a two-stage fixed bed reactor increased H_2 production from 30.1 to 50.6 vol.% (Wu et al. 2011). In another study by Wu and Williams (2009), H_2 production from polypropylene (PP) plastic was significantly increased with the use of Ni/Al_2O_3 catalyst. Asadullah et al. (2001) reported that the use of Rh/CeO_2 catalyst produced 1290 H_2 per μmol that can be further increased with the increase in temperature. The use of cedar wood with $Rh/CeO_2/SiO_2$ catalyst also increased the H_2 production at the temperature of 550–700 °C. Lv et al. (2003) reported that with the increase in temperature from 700 to 900 °C, H_2 production was increased from 22 to 71 g per kg.

11.3 Waste-to-Hydrogen Energy Potential in KSA

Cost effectiveness, availability, high carbon contents, and biodegradability are the key parameters in the selection of H_2-producing substrate. Glucose, sucrose, and lactose are the preferable sources of substrate for H_2 production (Kapdan and Kargi 2006). Currently there is no WTE or waste-to-hydrogen facility exists in KSA, since all of the collected wastes are disposed in landfills or dumpsite untreated. Following is the review of the potential available H_2-producing waste sources in KSA.

11.3.1 Organic Fraction of MSW in KSA

In 2014, the total MSW generation in KSA was around 15 million tons with an average rate of 1.4 kg per capita per day (Maria 2013). This waste rate is estimated to become double (around 30 million tons) by 2033. The overall generated waste in KSA consists of up to 75 % organic waste, including food waste as the largest waste stream (50.6 % of total MSW) with amount of 7.7 million tons per year and generation rate of 0.71 kg per capita per day (Fig. 11.2). The food waste contains rice, meat, bakery products, and fat with percentages of 38.7, 25, 18.7, and 13 %, respectively. Moreover, bones, fruits, and vegetables are present in food waste with a percentage value of 2.2 %. The chemical composition of food waste shows high percentage of moisture content (38.4 %), carbohydrates (25.6 %), proteins (17.3 %), fats (15.3 %), ash (3.2 %), and fibers (0.3 %) (Abu-Qudais and Abu-Qdais 2000; Khan and Kaneesamkandi 2013; Alruqaie and Alharbi 2012). The other waste fractions of MSW include paper (12 %), plastic (17.4 %), glass (3 %), cardboard (6.6 %), wood waste (2 %), metals (1.9 %), textile (1.9 %), aluminum (0.8 %), leather (0.1 %), and others (3.7 %) (Fig. 11.2).

The food waste in three large cities of KSA (i.e., Riyadh, Jeddah, and Dammam) is exceeding 6 million tons per year (Maria 2013). During the month of 2014 Ramadan, 5 thousand tons of food was wasted in first 3 days only in Makkah municipality (Irfan 2014). The alarming news is the wastage of 35–40 % cooked rice annually in KSA with a total loss of 1.6 billion SR (Saudi Gazette 2014). Such waste composition with high fraction of organic contents (up to 75 %) makes it very suitable substrate for biological processes such as dark-fermentation and two-stage anaerobic dark and photo-fermentation to produce H_2. For instance, the starch present in food waste can be hydrolyzed to glucose and maltose using enzymes followed by conversion to organic acids and then to H_2 (Kapdan and Kargi 2006). Han and Shin (2004) used food waste in dark-fermentation process using a leaching bed reactor for H_2 production. The results showed that the process was suitable as initial step for H_2 production followed by the methanogenesis process, like a two-stage anaerobic process.

11.3.2 Plastic Waste and Used Oil in KSA

Plastic waste is the second-largest waste stream of MSW in KSA with an annual production of 2.7 million tons with an average rate of 0.3 kg per capita per day (Nizami et al. 2015a, b, c; Ouda et al. 2013; Ouda and Cekirge 2014). Currently, only informal sector is involved in plastic waste recycling. However, most of the collected plastic wastes are disposed in landfills or dumpsites untreated. In addition to environmental problems, plastic waste causes operational overburden from its collection to final disposal due to its clogging and non-biodegradable nature (Nizami et al. 2016; Ouda et al. 2016). Recycling through conventional-mechanical techniques such as sorting, grinding, washing, and extraction can recycle only 15–20 % of plastics. Moreover, the plastic waste is polluted with dirt, aluminum foils, food waste, and soil.

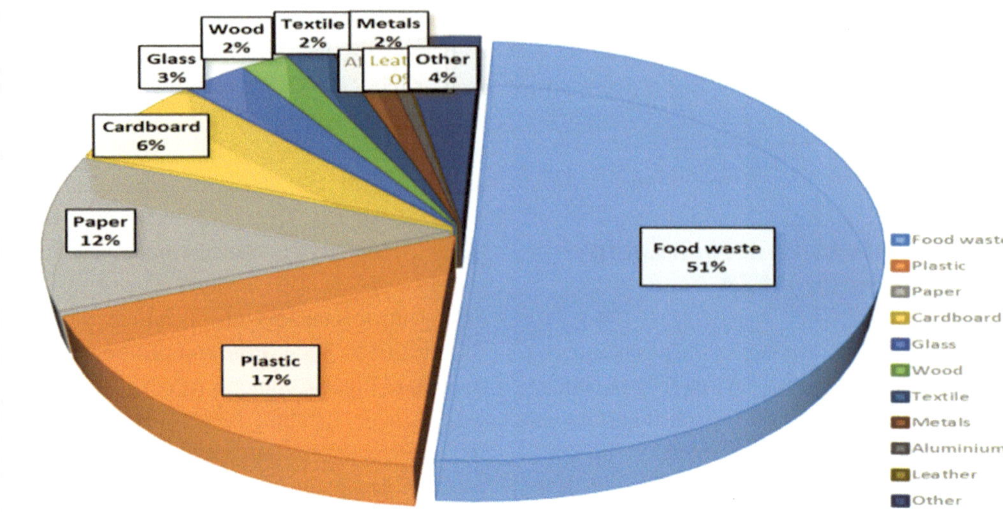

Fig. 11.2 MSW composition in KSA (Nizami et al. 2015a, b, c; Khan and Kaneesamkandi 2013; Ouda et al. 2016; Nizami et al. 2016)

Thermal and catalytic pyrolysis and gasification are one of the WTE technologies used to convert plastic waste into liquid fuel and syngas, respectively (Yuan 2006). H_2 is produced from syngas, as it is a combination of H_2 and CO. A large fraction of the country's MSW is also consisted of used oil from household, restaurants, and automobile industry. This waste oil can also be used to produce H_2 using pyrolysis and gasification technologies. Kim (2003) used shredded waste tires and waste oil in gasification process as substrates to produce H_2.

11.3.3 Slaughterhouse Waste in KSA

Millions of animals are slaughtered in KSA every year during the pilgrimage (Hajj) and Ramadan (month of fasting) periods. For example, in 2014 Hajj season, more than 2.5 million animals were slaughtered in KSA to perform Hajj rituals (Amtul 2014). Typically, 12 % waste per body weight is produced from sheep and goat slaughtering, whereas cattle slaughtering produces 38 % waste per body weight (Singh 2013). This waste includes rumen, blood, offal materials, bones, and tallow. Although there is little information available for these waste quantities, but it is evident that animal related-waste quantities are huge in KSA. The slaughterhouse waste has been used in many studies for producing H_2 in a two-stage fermentation process (Gomez et al. 2006, 2009). Gomez et al. (2009) produced H_2 from sludge and slaughterhouse waste and found that the H_2 production was more stable from the process when slaughterhouse waste was used than the sludge waste only.

11.3.4 Agricultural Waste in KSA

It is estimated that more than 440 million tons of agriculture residue is produced in KSA every year and most of them is incinerated or disposed of inefficiently (Sadik et al. 2010). Among the agricultural waste, most share comes from date palm trees. There are around 23 million date palm trees in KSA, which produce 780 thousand tons of agriculture residues per year (Al-Abdoulhadi et al. 2011). The agricultural waste can be used as a substrate in the chemical processes such as pyrolysis and gasification and biological processes such as anaerobic dark and photo-fermentation for H_2 production. In biological processes, the substrate is pretreated using waste grinding, delignification, and hydrolysis. The hydrolyzed waste is then converted into organic acids and finally into H_2 (Kapdan and Kargi 2006).

11.3.5 Industrial Waste in KSA

In KSA, there are more than 12 thousand industries working in different sectors that produce large quantities of waste and waste sludge on daily basis (Ouda and Cekirge 2014). In a study on Jeddah industrial sludge, it was estimated that 120 tons of sludge is produced every day on dry solid basis. In 2013, around 200 dry tons of sludge was produced every day in Riyadh. Collectively, the projected amounts of sludge in KSA will be around 1.6 thousand and 1.8 thousand dry tons per day by 2020 and 2025, respectively (SAWEA 2013). The waste sludge contains large quantities of carbohydrates and proteins; thus can be used in biological processes such as photo-fermentation and two-stage anaerobic dark and photo-fermentation for H_2 production (Kapdan and Kargi 2006).

11.4 Challenges and Perspectives of Waste-to-Hydrogen Energy Production

11.4.1 Challenges

Projections regarding the shortage of fossil fuel reserves in the 21st century enforce researchers to think for alternative renewable energy sources. This has increased the significance of H_2 production processes. However to shift from fossil fuel-based economy to H_2 energy-based economy, efforts are required to overcome the challenges of H_2 production pathways toward optimizing the production processes (Table 11.6). According to Momirlan and Veziroglu (2002), H_2 energy production challenges lie in investment cost, storage and delivery, conversion, and end-use applications (Table 11.6). Moreover, according to Hallenbeck and Benemann (2002), utilizing solar energy as a renewable source in biological conversion methods is limited due to its low density and diffusive nature.

The advantage of biological processes is the high efficiency of H_2 production by utilizing waste materials that are often considered refuse or garbage. However, the challenges in these biological processes are lacking information about microbial activities and their sensitivities to O_2 and H_2, as they affect the process yield (Table 11.6). Moreover, the microbial inherent properties, limitation of photosynthetic efficiency, and hydrogenase catalytic functions are the microbial challenges of H_2 production (Laurinavichene et al. 2006). According to Levin et al. (2004), the rate of H_2 production in biological processes is low and requires further process optimization. Kim (2003) examined different carbonaceous wastes for H_2 production through steam reforming. These renewable sources are quite cheap but require high-process temperature (1200 °C). On

Table 11.6 Challenges in H_2 production

Barriers in basic science
Organism
Bacteria do not produce more than 4 mol H_2/mol glucose naturally
Enzyme
Hydrogenase over expression not stable, O_2 sensitivity and H_2 feedback inhibition
Barriers in fermentation process
Substrate
High cost of suitable substrate (glucose) and low yield using renewable biomass
Strain
Lack of suitable-industrial strain
Process
Commercially feasible product yield, incomplete substrate utilization, and sustainable process sterilization
Barriers in engineering aspects
Reactor design
Lack of kinetics/appropriate reactor design for H_2 production and light intensity in case of photo-bioreactor
Thermodynamics
Thermodynamic barrier NAD(P)H→H_2 (+4.62 kJ/mol)
H_2 storage
H_2 purification/separation and its storage

Bhutto et al. 2011; Kapdan and Kargi 2006; Das and Veziroglu 2001; Kotay and Das 2008; Hong et al. 2013; Rathore et al. 2016 the contrary, H_2 production through electrolysis can be a cleaner way; however this can be only applied to areas where electricity is cheap (Kapdan and Kargi 2006). Moreover, demineralization of water is required to avoid corrosion and deposition on electrodes (Armor 1999).

11.4.2 Perspectives

The eco-friendly and sustainable methods of generating H_2 energy will only be possible, if produced from renewable sources (Fig. 11.3). The H_2 economy is a sustainable future energy system that will produce electricity and energy carriers utilizing renewable sources. High demand of H_2 energy brings the attention of researchers for the development of cost-effective and efficient methods for producing H_2 as a renewable and sustainable fuel (Kapdan and Kargi 2006).

The comparison of H_2 production by different processes using conventional fuels and biological systems was carried out by Tanisho (1996), Benemann (1997), Bockris (1981), Kumar and Das (2000a, b), and Benemann (1997). The use of fermentative process can be better than photosynthetic process for H_2 production, as it produces various kinds of value-added fatty acids during the process such as lactic acid, acetic acid, and butyric acid. However, if these acids are not separated from photosynthetic process, it may cause water pollution. According to Das and Veziroglu (2001), biological processes are not energy intensive, as they can be carried out at ambient temperature.

Gases produced via biological processes are H_2 (60–90 v/v) with impurities of CO_2 and O_2.

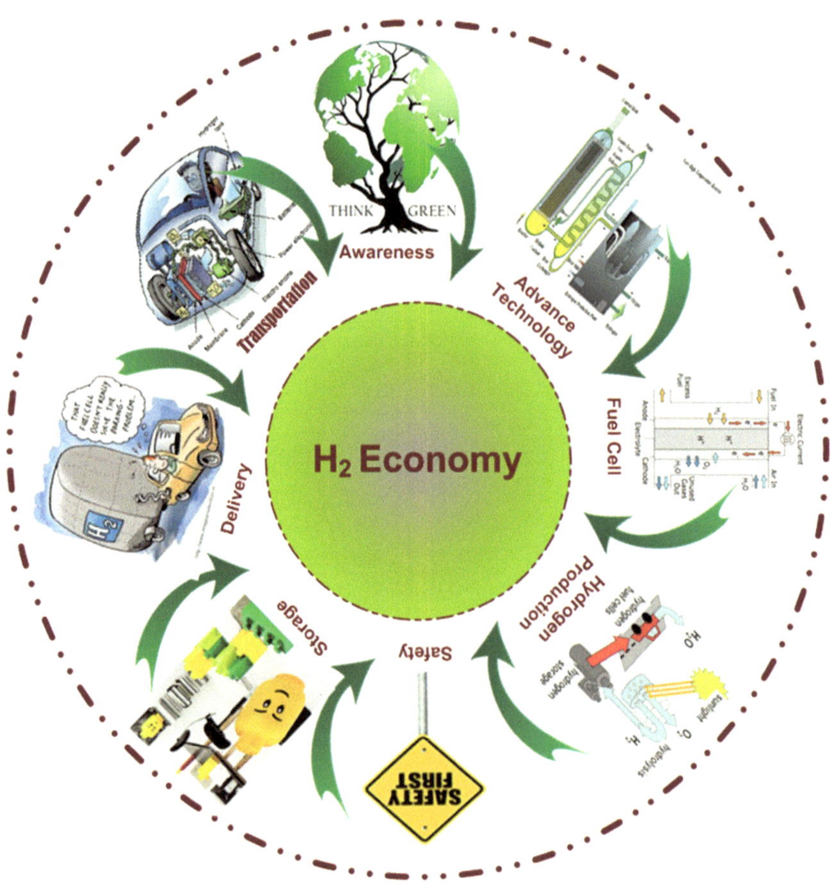

Fig. 11.3 An overview of various drivers for H_2 economy

CO_2 is water soluble and can be reused as fire extinguisher. A 50 % w/v solution of potassium hydroxide (KOH) is a good CO_2 adsorbent, thus can be used for CO_2 removal from H_2. The presence of O_2 may cause fire hazards, thus can be removed by using an alkaline pyrogallol solution. The moisture contents reduce the heating values of H_2, thus can be removed by drying and chilling process (Das and Veziro 2001).

H_2 production is efficiently carried out at small scale using biological systems where substrate is easily accessible to reduce energy consumption and transportation cost. However, its production at industrial scale is still facing process and economic constraints. Therefore, there is a strong need of metabolically engineered microorganism to enhance H_2 production ($\geq$4 mol H_2/mol of glucose) for large-scale H_2 production (Maness et al. 2009). Engineering work is also required to further upgrade the bioreactors for H_2 production at commercial scale. Moreover, detailed life cycle assessment (LCA) studies are required to evaluate the sustainability of biological and chemical H_2 production methods from waste sources in terms of economic, environmental and energy balances (Rathore et al. 2016; Shahzad et al. 2015; Nizami and Ismail 2013; Nizami et al. 2016; Miandad et al. 2016).

References

Abu-Qudais M, Abu-Qdais HA (2000) Energy content of municipal solid waste in Jordan and its potential utilization. Energy Convers Manag 4:983–991

Aga O, Ouda OKM, Raza SA (2014) Investigating waste to energy potential in the Eastern Region, Saudi Arabia, Renewable Energy for Developing Countries (REDEC 2014), November 26–27, 2014, Beirut, Lebanon

Al-Abdoulhadi IA, Dinar HA, Ebert G, Buttner C (2011) Effect of salinity on leaf growth, leaf injury and biomass production in date palm cultivators. Indian J Sci Technol 4(11):1542–1546

Alruqaie IM, Alharbi BH (2012) Environmental advantage assessment of recycling food waste in Riyadh, Saudi Arabia. Res J Environ Sci 6:230–237

Amtul S (2014) Kingdom: 2.5m animals sold in Saudi Arabia during hajj. Saudi Gazette. Available from: http://www.saudigazette.com.sa

Armaroli N, Balzani V (2011) The hydrogen issue. ChemSusChem 4:21–38

Armor JN (1999) The multiple roles for catalysis in the production of H2. Appl Catal A: Gen 176:159–176

Asadullah M, Fujimoto K, Tomishige K (2001) Catalytic performance of Rh/CeO2 in the gasification of cellulose to synthesis gas at low temperature. Ind Eng Chem Res 40:5894–5900

Benemann J (1996) Hydrogen biotechnology: progress and prospects. Nat Biotechnol 14:1101–1103

Benemann JR (1997) Feasibility analysis of photobiological hydrogen production. Int J Hydrog Energy 22:979–987

Bhutto AW, Bazmi AA, Kardar MN, Yaseen M, Zahedi G, Karim S (2011) Developments in Hydrogen Production through Microbial Processes; Pakistan's Prospective. Int J Chem Environ Eng 2(3):189–205

Bockris JO'M (1981) The economics of hydrogen as a fuel. Int J Hydrog Energy 6:223–241

Catal T, Lesnik KL, Liu H (2015) Suppression of methanogenesis for hydrogen production in single chamber microbial electrolysis cells using various antibiotics. Bioresour Technol 187:77–83

CDSI, Central Department of Statistics & Information (2010) Population & Housing Census for 1431 A.H (2010 A.D) Findings 2010. Available online at: http://www.cdsi.gov.sa/

Chen D, Yin L, Wang H, He P (2014) Pyrolysis technologies for municipal solid waste: a review. Waste Manag 34:2466–2486

Czernik S, French RJ (2006) Production of hydrogen from plastics by pyrolysis and catalytic steam reform. Energy Fuels 20:754–758

Das D, Veziro TV (2001) Hydrogen production by biological processes: a survey of literature. Int J Hydrog Energy 26:13–28

Das D, Veziroglu TN (2008) Advances in biological hydrogen production processes. Int J Hydrog Energy 33:6046–6057

Davidian T, Guilhaume N, Iojoiu E, Provendier H, Mirodatos C (2007) Hydrogen production from crude pyrolysis oil by a sequential catalytic process. Appl Catal Environ 73:116–127

Davila-Vazquez G, Arriaga S, Alatriste-Mondragon F, De-Leo'n-Rodrı'guez R, Rosales-Colunga LM, Razo-Flores E (2008) Fermentative biohydrogen production: trends and perspectives. Rev Environ Sci Biotechnol 7:27–45

Demirbas A, Arin G (2004) Hydrogen from biomass via pyrolysis: relationships between yield of hydrogen and temperature. Energy Sources 26:1061–1069

Demirbas A, Rehan M, Al-Sasi BO, Nizami AS (2016) Evaluation of natural gas hydrates as a future methane source. Pet Sci Technol. doi:10.1080/10916466.2016.1185442

FAO: Food and Agriculture Organization (2010) Energy supply and demand: trends and prospects. Report available online: ftp://ftp.fao.org/docrep/fao/010/i0139e/i0139e03.pdf

Franks AE, Nevin KP, Jia H, Izallalen M, Woodard TL, Lovley DR (2009) Novel strategy for three-dimensional real-time imaging of microbial fuel cell communities: monitoring the inhibitory effects of proton accumulation within the anode biofilm. Energy Environ Sci 2:113–119

Gaffron H, Rubin J (1942) Fermentative and photochemical production of hydrogen in algae. J Gen Physiol 26:219–240

Gardy J, Hassanpour A, Lai X, Rehan M (2014) The influence of blending process on the quality of rapeseed oil-used cooking oil biodiesels. Int Sci J (J Environ Sci) 3:233–240

Ghirardi ML, Zhang L, Lee JW, Flynn T, Seibert M, Greenbaum E (2000) Microalgae: a green source of renewable h2. Trends Biotechnol 18:506–511

Gomez X, Moran A, Cuetos MJ, Sanchez ME (2006) The production of hydrogen by dark fermentation of municipal solid wastes and slaughterhouse waste: a two-phase process. J Power Sources 157:727–732

Gomez X, Cuetos MJ, Prieto JI, Moran A (2009) Biohydrogen production from waste fermentation: mixing and static conditions. Renew Energy 34:970–975

Gralnick JA, Newman DK (2007) Extracellular respiration. Mol Microbiol 65:1–11

Greenbaum E, Guillard RRL, Sunda WG (1983) Hydrogen and oxygen photoproduction by marine algae. Photochem Photobiol 37:649–655

Hallenbeck PC, Benemann JR (2002) Biological hydrogen production; fundamentals and limiting processes. Int J Hydrog Energy 27:1185–1193

Hallenbeck PC, Ghosh D, Skonieczny MT, Yargeau V (2009) Microbiological and engineering aspects of biohydrogen production. Indian J Microbiol 49:48–59

Hamelers HVM, Ter-Heijne A, Sleutels THJA, Jeremiasse AW, Strik DPBTB, Buisman CJN (2010) New applications and performance of bio-electrochemical systems. Appl Microbiol Biotechnol 85:1673–1685

Han SK, Shin HS (2004) Biohydrogen production by anaerobic fermentation of food waste. Int J Hydrog Energy 29:569–577

Holladay JD, Hu J, King DL, Wang Y (2009) An overview of hydrogen production technologies. Catal Today 139:244–260

Hong Y, Nizami AS, Pourbafrani M, Saville BA, MacLean HL (2013) Impact of cellulase production on environmental and financial metrics for lignocellulosic ethanol. Biofuels Bioprod Biorefin 7:303–313

Hu H, Fan Y, Liu H (2008) Hydrogen production using single-chamber membrane-free microbial electrolysis cells. Water Res 42:4172–4178

Inayat A, Ahmad MM, Mutalib M, Yusup S (2012) Process modeling for parametric study on oil palm empty fruit bunch steam gasification for hydrogen production. Fuel Process Technol 93:26–34

Irfan M (2014) Massive wastage 'Unacceptable'. Arab News. Friday 4, July 2014. Available from: www.arabnews.com

KACARE presentation. Renewable energy-waste to energy. A pillar of the sustainable energy kingdom. In: First international environment conference, Yanbu Al Sinaiyah, KSA: King Fahd Civic Centre 20–21 November 2012

Kapdan IK, Kargi F (2006) Bio-hydrogen production from waste materials. Enzyme Microb Technol 38:569–582

Karmakar M, Datta A (2011) Generation of hydrogen rich gas through fluidized bed gasification of biomass. Bioresour Technol 102:1907–1913

Kasakura T, Hiraoka M (1982) Pilot plant study on sewage sludge pyrolysis—I. Water Res 16:1335–1348

Khan MSM, Kaneesamkandi Z (2013) Biodegradable waste to biogas: renewable energy option for the Kingdom of Saudi Arabia. Int J Innov Appl Stud 4(1):101–113

Kim HY (2003) A low cost production of hydrogen from carbonaceous wastes. Int J Hydrog Energy 28:1179–1186

Konieczny A, Mondal K, Wiltowski T, Dydo P (2008) Catalyst development for thermocatalytic decomposition of methane to hydrogen. Int J Hydrog Energy 33:264–272

Kotay SM, Das D (2008) Biohydrogen as a renewable energy resource—prospects and potentials. Int J Hydrog Energy 33:258–263

Kumar N, Das D (2000a) Biological hydrogen production in a packed bed reactor using agroresidues as solid matrices. In: Proceedings of 13th WHEC, Beijing, 2000

Kumar N, Das D (2000b) Enhancement of hydrogen production by enterobacter cloacae 08. Process Biochem 35:589–593

Laurinavichene TV, Fedorov AS, Ghirardi ML, Seibert M, Tsygankov AA (2006) Demonstration of sustained hydrogen photo-production by immobilized, sulfur-deprived chlamydomonas reinhardtii cells. Int J Hydrog Energy 31(5):659–667

Levin DB, Pitt L, Love M (2004) Biohydrogen production: prospects and limitations to practical application. Int J Hydrog Energy 29:173–185

Li S, Xu S, Liu S, Yang C, Lu Q (2004) Fast pyrolysis of biomass in free-fall reactor for hydrogen-rich gas. Fuel Process Technol 85:1201–1211

Lindberg P, Lindblad P, Cournac L (2004) Gas exchange in the filamentous cyanobacterium Nostoc punctiforme strain ATCC 29133 and its hydrogenase-deficient mutant strain NHM5. Appl Environ Microbiol 70(4):2137–2145

Liu H, Grot S, Logan B (2005) Electrochemically assisted microbial production of hydrogen from acetate. Environ Sci Technol 39:4317–4320

Logan B, Call D, Cheng S, Hamelers HM, Sleutels Tom HJA, Jeremiasse A, Rozendal A (2008) Microbial electrolysis cells for high yield hydrogen gas production from organic matter. Environ Sci Technol 42:8631–8640

Lv P, Chang J, Xiong Z, Huang H, Wu C, Chen Y (2003) Biomass air-steam gasification in a fluidized bed to produce hydrogen-rich gas. Energy Fuels 17:677–682

Manara P, Zabaniotou A (2012) Towards sewage sludge based biofuels via thermochemical conversion—a review. Renew Sustain Energy Rev 16:2566–2582

Maness PC, Yu J, Eckert C, Ghirardi ML (2009) Photobiological hydrogen production—prospects and challenges. Microbe 4(6):659–667

Maria RK (2013) Processing of food wastes. Chapter-3, pp 110–117. Adv Food Nutr Res. doi:10.1016/S1043-4526(09)58003-5

Melis A, Happe T (2001) Hydrogen production. Green algae as a source of energy. Plant Physiol (Rockv) 127:740–748

MEP: Ministry of Economy and Planning (2010) The Nine Development Plan 2010–2014. Riyadh: Ministry of Economy and Planning Documents

Miandad R, Barakat MA, Aburiazaiza AS, Rehan M, Nizami AS (2016) Catalytic pyrolysis of plastic waste: a review. Accepted and in Press in Process Safety and Environmental Protection. DOI: 10.1016/j.psep.2016.06.022

Miura Y (1995) Hydrogen production by biophotolysis based on microalgal photosynthesis. Process Biochem 30:1–7

Momirlan M, Veziroglu TN (2002) Current status of hydrogen energy. Renew Sustain Energy Rev 6:141–179

Momirlan M, Veziroglu TN (2005) The properties of hydrogen as fuel tomorrow in sustainable energy system for a cleaner planet. Int J Hydrog Energy 30:795–802

Nath K, Das D (2003) Hydrogen from biomass. Curr Sci (Bangalore) 85:265–271

Ni M, Leung DYC, Leung MKH, Sumathy K (2006) An overview of hydrogen production from biomass. Fuel Process Technol 87:461–472

Nizami AS, Ismail IMI (2013) Life cycle assessment of biomethane from lignocellulosic biomass. Chapter in Singh A et al (eds) Life cycle assessment of renewable energy sources, green energy and technology. Springer, London, doi:10.1007/978-1-4471-5364-1_4

Nizami AS, Rehan M, Ouda OKM, Shahzad K, Sadef Y, Iqbal T et al (2015a) An argument for developing waste-to-energy technologies in Saudi Arabia. Chem Eng Trans 45:337–342. http://dx.doi.org/10.3303/CET1545057

Nizami AS, Ouda OKM, Rehan M, El-Maghraby AMO, Gardy J, Hassanpour A et al (2015b) The potential of Saudi Arabian natural zeolites in energy recovery technologies. Energy 2015:1–10. http://dx.doi.org/10.1016/j.energy.2015.07.030

Nizami AS, Rehan M, Ismail IMI, Almeelbi T, Ouda OKM (2015c) Waste biorefinery in Makkah: a solution to convert waste produced during Hajj and Umrah Seasons into wealth. In: Conference: 15th scientific symposium for Hajj, Umrah and Madinah visit, Held in May 2015 in Madinah, Saudi Arabia. doi:10.13140/RG.2.1.4303.6560

Nizami AS, Shahzad K, Rehan M, Ouda OKM, Khan MZ, Ismail IMI, Almeelbi T, Basahi JM, Demirbas A (2016) Developing waste biorefinery in Makkah: A way forward to convert urban waste into renewable energy. Appl Energy. http://dx.doi.org/10.1016/j.apenergy.2016.04.116

Ouda OKM, Cekirge HM, Raza SA (2013) An assessment of the potential contribution from waste-to-energy facilities to electricity demand in Saudi Arabia. Energy Convers Manag 75:402–406

Ouda OKM, Raza SA, Al-Waked R, Al-Asad JF, Nizami A-S (2015) Waste-to-energy potential in the Western Province of Saudi Arabia. J King Saud Univ Eng Sci. doi:http://dx.doi.org/10.1016/j.jksues.2015.02.002

Ouda OKM, Raza SA, Nizami AS, Rehan M, Al-Waked R, Korres NE (2016) Waste to energy potential: a case study of Saudi Arabia. Renew Sustain Energy Rev 61:328–340. http://dx.doi.org/10.1016/j.rser.2016.04.005

Rathi S (2006) Alternative approaches for better municipal solid waste management in Mumbai, India. J Waste Manag 26(10):1192–1200

Rathore D, Nizami AS, Pant D, Singh A (2016) Key issues in estimating energy and greenhouse gas savings of biofuels: challenges and perspectives. Biofuel Res J 10:380–393

Rehan M, Nizami AS, Shahzad K, Ouda OKM, Ismail IMI, Almeelbi T, Iqbal T, Demirbas A (2016) Pyrolytic liquid fuel: a source of renewable energy in Makkah. Energy Sources Part A. doi:10.1080/15567036.2016.1153753

Ruoppolo G, Ammendola P, Chirone R, Miccio F (2012) H2 -rich syngas production by fluidized bed gasification of biomass and plastic fuel. Waste Manag 32:724–732

Sadaf Y, Nizami AS, Batool SA, Chaudhary MN, Ouda OKM, Asam ZZ, Habib K, Rehan M, Demibras A (2015) Waste-to-energy and recycling value for developing integrated solid waste management plan in Lahore. Energy Sources Part B. doi:10.1080/1556249.2015.105295

SAWEA: Saudi Arabian Water Environmental Association (2013) Wastewater sludge treatment and disposal in KSA. A presentation given by Miahona. Available from: http://www.sawea.org/pdf/waterarabia2013/Sesssion_C/WASTEWATER_SLUDGE_TREATMENT_AND_DISPOSAL%20IN%20KSA.pdf

Schulz R (1996) Hydrogenases and hydrogen production in eukaryotic organisms and cyanobacteria. J Mar Biotechnol 4:16–22

Sequeira CA, Santos DM (2010) Hydrogen production. Ciência & Tecnologia dos Materiais 22(3–4):76–86

Shahzad K, Rehan M, Ismail IMI, Sagir M, Tahir MS, Bertok B, Nizami AS (2015) Comparative life cycle analysis of different lighting devices. Chem Eng Trans 45:631–636. doi:10.3303/CET1545106

Soni C, Wang Z, Dalai A, Pugsley T, Fonstad T (2009) Hydrogen production via gasification of meat and bone meal in two-stage fixed bed reactor system. Fuel 88:920–925

Tahir MS, Shahzad K, Shahid Z, Sagir M, Rehan M, Nizami AS (2015) Producing methane enriched biogas using solvent absorption method. Chem Eng Trans 45:1309–1314. doi:10.3303/CET1545219

Tawabini BS, Ouda OKM, Raza SA (2014) Investigating of waste to energy potential as a renewable energy resource in Al-Hasa Region, Saudi Arabia, 5th international symposium on energy from biomass and waste (VENICE 2014), 17–20 November 2014, Venice, Italy

The study on wastewater treatment and water reuse in Saudi-Aramco, Saudi Arabia, commissioned by the ministry of economy, trade and industry, January 2012

UD: Urban Development Series-Knowledge Papers (2012) A global review of solid waste management. Available Online: http://siteresources.worldbank.org/INTURBANDEVELOPMENT/Resources/336387-1334852610766/Chap3.pdf

Uddin MN, Daud WWMA, Abbas HF (2013) Potential hydrogen and non-condensable gases production from biomass pyrolysis: insights into the process variables. Renew Sustain Energy Rev 27:204–224

US-EIA: US Energy Information Administration (2007) http://www.eia.gov/forecasts/aeo/archive.cfm. Accessed on 25 Jan 2015

Wang J, Wan W (2009) Factors influencing fermentative hydrogen production: a review. Int J Hydrog Energy 34(2):799–811

Wang J, Xu S, Xiao B, Xu M, Yang L, Liu S, Hu Z, Guo D, Hu M, Ma C, Luo S (2013) Influence of catalyst and temperature on gasification performance of pig compost for hydrogen-rich gas production. Int J Hydrog Energy 38:14200–14207

WHO: World Health Organization (2014) Available Online, http://www.who.int/gho/urban_health/situation_trends/urban_population_growth_text/en/

Winter CJ (2005) Into the hydrogen energy economy-milestones. Int J Hydrog Energy 30:681

Wu C, Williams PT (2009) Hydrogen production by steam gasification of polypropylene with various nickel catalysts. Appl Catal Environ 87:152–161

Wu C, Wang L, Williams PT, Shi J, Huang J (2011) Hydrogen production from biomass gasification with Ni/MCM-41 catalysts: influence of Ni content. Appl Catal Environ 108:6–13

Yetis M, Gunduz U, Eroglu I, Yucel M, Turker L (2000) Photoproduction of hydrogen from sugar refinery wastewater by rhodobacter sphaeroides O.U. 001. Int J Hydrog Energy 25(11):1035–1041

Yildiz FH, Davies JP, Grossman AR (1994) Characterization of sulfate transport in chlamydomonas reinhardtii during sulfur limited and sulfur-sufficient growth. Plant Physiol (Rockv) 104:981–987

Yu J, Takahashi P (2007) Biophotolysis-based hydrogen production by cyanobacteria and green microalgae. In: Méndez-Vilas A (ed) Communicating current research and educational topics and trends in applied microbiology. Formatex, Badajoz, pp 79–89

Yuan X (2006) Converting waste plastics into liquid fuel by pyrolysis: developments in China. In: Scheirs J, Kaminsky W (eds) Feedstock recycling and pyrolysis of waste plastics. Wiley, Changsha, pp 729–755

Zhang B, Xiong S, Xiao B, Yu D, Jia X (2011) Mechanism of wet sewage sludge pyrolysis in a tubular furnace. Int J Hydrog Energy 36:355–363

Biohydrogen Economy: Challenges and Prospects for Commercialization

12

Mona Sharma and Anubha Kaushik

Abstract

Biohydrogen has several properties that project it as a potential alternative to our largely used fossil fuels. However, its production technology must overcome a number of limitations before it could successfully compete in the fuel market and be deployed on a large scale. Although the potential of microbes to produce biohydrogen has raised a lot of excitement, yet a lot needs to be done on technical, economical, policy making, and legislative fronts to make it replace the existing fossil fuels. Review of published literature suggests that the biohydrogen production system holds great promise for industrial application. In this chapter, we have attempted to elucidate the major challenges being faced for using biohydrogen on a commercial scale by making an assessment of its economics taking into account the different processes like production, storage, transportation, and delivery to the user and also assess its future commercialization perspectives.

12.1 Current Energy Scenario and Biohydrogen as Potential Sustainable Energy Alternative

Considering the fast pace of economic growth of modern societies and large-scale dependence on fossil fuels for meeting the energy demands, some viable alternative fuel has become the need of the day so that the adverse impacts of fossil fuel consumption on environment could be averted or reduced. Rising global population and rapid economic growth have led to increase in demand of global energy. As per the current policies, the world's need for energy will be almost

M. Sharma (✉)
Department of Environmental Sciences, School of Earth, Environment and Space Studies, Central University of Haryana, Mahendergarh, India
e-mail: monasharma140@gmail.com

A. Kaushik
University School of Environmental Management, GGS Inderprastha University, New Delhi 110078, India
e-mail: aks.es.10@gmail.com

A. Singh, D. Rathore (eds.), *Biohydrogen Production: Sustainability of Current Technology and Future Perspective*, DOI 10.1007/978-81-322-3577-4_12

60 % more in 2035 than today and out of which, 35 % will be that required for China and India together (World Energy Outlook 2010). Transportation is using 27 % of the primary energy and is also one of the fastest growing sectors (Hunt et al. 2014).

As fossil fuels are limited resources, their continuous use is not sustainable (Srivastava and Prasad 2000), and emissions of greenhouse gases (GHGs) like carbon dioxide (CO_2) are increased by their combustion. About 45 % of the identified world oil reserves has been combusted and has caused an increase of 20 % of atmospheric CO_2 since 1900 (Keeling and Whorf 2005). Conservative estimates suggest that demand for oil will outstrip its supply by 2050 (Holmes and Jones 2003). A sustainable energy economy is therefore needed, which makes it very important to switch over to an alternative fuels, one that is not limited in supply and one that is environmentally benign. Reduction in ecological footprint of future energy generation should be based upon a multifaceted approach like wind, nuclear, solar energy, and biohydrogen along with fossil fuels through which carbon is sequestered (Hoffart et al. 2002; Pacala and Socolow 2004; Patil 2007).

Hydrogen has several desirable characteristics including its high-energy content that increases its possibility of becoming the fuel of the future. Energy content of 1 kg hydrogen is 141.9 MJ (39.4 kWh), which is much higher in comparison to that of methane or natural gas (55.5 MJ), diesel/fuel oil (48 MJ), LPG (46.4 MJ), kerosene (46 MJ), gasoline (44.4 MJ), coal (24 MJ), etc. (https://en.wikipedia.org/wiki/Energy_density). Many researchers have therefore proposed H_2 as an energy carrier for the upcoming economy. However, a renewable and continuous supply of hydrogen is required to power this economy. Thus, production of hydrogen in a sustained manner using biomass of different types is a technological advancement as well as a challenge.

Various biological methods of biohydrogen production may be preferred to chemical methods because of the possibility to use sunlight and carbon dioxide by photosynthetic bacteria, cyanobacteria, and microalgae for conversion to hydrogen under moderate and environmentally benign conditions. Widely available primary energy sources including natural gas, coal, biomass, organic wastes, solar, geothermal, wind, hydro, or nuclear power can produce H_2, enabling a more diverse primary supply for fuels.

H_2 can be utilized in fuel cells as well as internal combustion engines with high conversion efficiency without any emissions of greenhouse gases and air pollutants. It would be possible to reduce emission of air pollutants and GHGs if hydrogen is produced from renewable sources, nuclear energy, or fossil fuels with appropriate capture and sequestration of carbon. Several studies have identified H_2 as the safest and pollution-free fuel during the past decade, and its demand has been increasing day-by-day (Sathyaprakasan and Kannan 2015).

Biohydrogen as an alternative fuel, however, faces both technical and economic issues that must be addressed and overcome before its commercialization. Technologies for hydrogen production, storage, and distribution have evolved over the years, but there is a need to adapt the same for use in an energy system. New hydrogen energy infrastructure has to be developed that would involve big expenses and logistic problems in corresponding supply and demand in the transition. Though there are rapid progressions in H_2 technologies, technical and economic issues have to be solved for economically competitiveness.

This chapter looks mainly into commercialization and the economic aspects of different biological methods of hydrogen production, storage, and transportation for assessing its future as a fuel.

12.2 Scope of Biohydrogen Production

Various countries have started taking initiatives to promote biohydrogen production as it is an alternative source of energy. Despite considerable research and significant progress in improving the rate of biological H_2 production, there is still a very little literature available on its economics for commercialization. Various microorganisms are capable to produce H_2 from the available renewable substrate under moderate conditions which make such biological strategies potentially competitive with chemical process like gasification and reforming (Chong et al. 2009).

Biohydrogen processes are known to be "CO_2 neutral," as fueled by carbohydrates which originate from CO_2 fixation by photosynthesis. Furthermore, biohydrogen is free of CO_2 and "hydrogen sulfide emissions, and no pretreatment is required for use in fuel cells for generation of electricity" (Macaskie et al. 2005).

Unlike other hydrogen production processes, biohydrogen production technology has the possibility of using renewable energy sources like biomass (Chong et al. 2009). Biohydrogen can be generated from different feeds like food wastes (Ferchichi et al. 2005; Karlsson et al. 2008), agricultural residues (Nath and Das 2003; Hawkes et al. 2008), and effluents from various industrial processes such as refining sugar (Yetis et al. 2000; Ren et al. 2006), distilling alcohol (Sasikala et al. 1992), olive processing (Erog˘lu et al. 2004), cheese production (Davila-Vazquez et al. 2008a), and tofu production (Zhu et al. 1995, 2002). Cyanobacterial biomass has been used for producing H_2, and the spent biomass from the photobioreactor has been used to bioremediate textile wastewater with dual benefits of energy production and savings in the cost of waste disposal (Kaushik et al. 2011; Mona et al. 2011a, b, c, d, 2013).

Biohydrogen production rates of various biological systems were compared by Levin et al. (2004) based on calculations of the size of the bioreactors needed to power proton exchange membranes (PEMs) of fuel cells, which coupled with biohydrogen production may prove useful in generating electricity on a commercial scale. There is a great need to develop synergy between biohydrogen scientists and fuel cell engineers, because practical applications would be possible only when the bioreactors are able to produce sufficient H_2 to the fuel cells for 24 h on a continuous basis.

12.3 Economics of Biohydrogen Production

Designs of the bioreactor and the production systems selected for the generation of biohydrogen play a decisive role in determining the cost of the production method. In order to assess the most optimal production method, various aspects including hydrogen yield, energy requirements, and ease of production are important considerations. Some of the cost estimates projected by different biohydrogen researchers are discussed here for a comparative cost scenario that would help assess the prospects of biohydrogen for application in the future.

12.3.1 Cost of Algal Hydrogen Production

An estimate of the costs involved in various components of hydrogen production process using algae, Amos (2004) gave such estimates, which are shown in Table 12.1. The table shows the capital cost for a 300 kg^{-1} stand-alone system with the pond area of 110,000 m^2, 10 cm pond depth, containing a truncated antennae mutant with 0.2 g/L cell concentration cost for a site in Phoenix, Arizona (Amos 2004).

Table 12.1 Algal hydrogen production capital costs

	Cost ($)
Algal ponds	1,100,000
Pressure swing adsorption (PSA) unit	121,000
Pressure swing adsorption (PSA) compressor	359,000
Storage compressor	578,000
High-pressure storage	913,000
Equipment cost	3,071,000
Engineering and construction	1,423,000
Contractor fees and contingency	674,000
Total investment	5,168,000

Adopted from Amos (2004)

12.3.2 Cost of Biohydrogen Production in Comparison to Conventional Methods

The cost of H_2 generated from various types of biological processes as compared to costs involved in the production of gasoline, ethanol, biodiesel, and natural gas has recently been compared by Karthic and Joseph (2012), which has been reproduced in Table 12.2.

From the table, it follows that hydrogen production by most of the available methods except pyrolysis is rather high in comparison to conventional methods of fuel production. It is therefore strongly recommended that intensive R&D should be focused to make biological hydrogen production technologies cost-effective.

Also, when the hydrogen production is based on renewable technologies, it avoids fuel price uncertainties.

12.3.3 Cost of Biohydrogen Production Based on Microbial Culture, Reactor Design, Capital Investments, and Operation

Very recently, an attempt has been made to work out the economics of different hydrogen-producing processes in bioreactors based on parameters like type and amount of the culture, reactor layout, capital, and operating costs for power, labor, water, and general supplies as important determinants of overall biohydrogen production cost (Sathyaprakasan and Kannan 2015).

The assumptions and estimations made by these researchers to calculate production cost of biological hydrogen through various processes are described here. Calculations are based on cost estimates for different parameters, and comparative estimates are represented in Table 12.3.

Table 12.2 Unit cost comparison of hydrogen production processes with conventional processes

Name of the process	Raw materials	Energy content of the fuel (MJ kg^{-1})	Unit cost of energy content of the fuel US \$ GJ^{-1}
Photobiological H_2 hydrogen	H_2O, organic acids	142	0.0106
Fermentative hydrogen	Molasses	–	0.0106
Pyrolysis for hydrogen production	Coal, biomass	–	0.00424
Hydrogen from water electrolysis	H_2O	–	0.01166
Hydrogen from nuclear energy	Electrolysis and water splitting	–	0.01272–0.02014
H_2 by biomass gasification	Biomass	–	0.04664–0.08692
H_2 from wind energy	Wind mill	–	0.03604
H_2 from photovoltaic power station	Solar energy	–	0.04452
H_2 from thermal decomposition of steam	H_2O	–	0.01378
H_2 from photochemical	Organic acids	–	0.02226
Gasoline	Crude petroleum	43.1	0.00636
Fermentative ethanol	Molasses	26.9	0.03339
Biodiesel	Jatropha seed	37.0	0.00424
Natural gas	Raw natural gas	33–50	0.0106

Adopted from Karthic and Joseph (2012)

Table 12.3 Comparison of biohydrogen production using different methods

Process	Direct biophotolysis	Indirect biophotolysis	Photofermentation	Dark fermentation
Costs in million USD				
Power cost	2.40	2.50	2.50	2.50
Water cost	0.03	0.03	0.0	0.03
Labor cost	50,469.30	65.10	23.03	17.83
General supply cost	7760.66	17.32	3.51	2.70
Culture production cost	5807.97	8.22	2.70	2.01
Glucose substrate	0.0	0.0	144.19	867.18
Gas separation and handling cost	23.03	0.08	0.14	0.05
Subtotal operating cost	64,063.38	93.25	176.10	892.31
Contingency 10 %	6406.35	9.31	17.62	89.22
Total operating cost	70,469.73	102.56	193.69	981.53
Cost per gigajoule of hydrogen	11,185.67	16.28	30.74	155.79
Cost per kilogram of hydrogen	1342.27	1.96	3.70	18.70

Adopted from Sathyaprakasan and Kannan (2015)

12.3.3.1 Estimating Amount of Culture Requirement

Rate of hydrogen production by the process of direct biophotolysis is estimated to be at 0.07 mmoles L^{-1} h^{-1} by Resnick (2004).

Hydrogen production rate would be 1.68 mmol H_2 L^{-1} d^{-1} if the reactor works on a daily basis without any break. Around 0.86 standard cubic foot (SCF) is equivalent to one mole of hydrogen. For 50 million SCF H_2 d^{-1}, the requirement of culture has been calculated to be 35.7 × 109 L. A cubic foot at a temperature of 21 °C and a pressure of 101.35 kPa are defined as a standard cubic foot of gas. The cost of culture production is assumed to be 10 % of the cost of labor and general supplies.

The rate of production of indirect biophotolysis is reported on an average as 0.36 mmol $H_2L^{-1}h^{-1}$. Assuming that the microalgae can produce H_2 at this rate continuously for 24 h during a year, the daily production rates would be 8.64 mmoles H_2 $L^{-1}d^{-1}$. Requirement of culture for 50 million SCF of H_2 d^{-1} has been calculated to be 6.73 × 106 m^3 by the same authors.

Hydrogen production rates by photofermentation process reported on an average are 153 mmoles H_2 L^{-1} h^{-1}. Daily production rates of 3.672 mol H_2 L^{-1} d^{-1} can be achieved (Miyake et al. 1999; Nath and Das 2003). For the production of 50 million SCF of H_2 d^{-1}, the requirement of culture has been calculated to be 15,760 m^3 (Sathyaprakasan and Kannan 2015).

Hydrogen production at a rate of 121 mmol L^{-1} h^{-1} through dark fermentation in *Clostridium, Enterobacter, Bacillus, Thermoanaero* bacterium (Das and Veziroglu 2001). With the same assumption that this production rate can be maintained 24 h a day, the daily production rate is calculated as 2.904 mmoles H_2 L^{-1} d^{-1}. The requirement of culture to produce 50 million SCF of H_2 d^{-1} has been calculated to be 19,938 m^3 (Sathyaprakasan and Kannan 2015).

12.3.3.2 Estimating Cost of Reactor Layout

Sathyaprakasan and Kannan (2015) have given the estimated costs for hydrogen production by considering a tubular bioreactor design employed for the production of hydrogen. There is a series of units in the reactor; each unit contains eight tubes (20 m long, 4 cm dia). The tube volume is 0.025 m^3, and eight such tubes would hold 0.201 m^3 of the culture. To accommodate the required amount of the culture, 117.5 million tubes are used with an area of 109 m^2 per tube, and the total area comes out around 19.35 billion m^2.

Benemann (1996) and Hallenbeck and Benemann (2002) suggested open photobioreactors with algal culture and a dark anaerobic reactor for indirect biophotolysis. The photobioreactors produce glucose from carbon dioxide, water, and several other nutrients. These algae are then passed into a dark anaerobic reactor in order to yield hydrogen and some amount of acetic acid derivative. These are further sent to a photobioreactor which produces hydrogen by fermenting the acetic acid derivative. In order to recycle the process, the algal biomass is reintroduced into the open pond. Sathyaprakasan and Kannan (2015) estimated the number of open ponds (length 100 m, width 10 m, depth 25 m) as 27,000 to hold 6.73 million liters of the culture that is required for the process. Considering a 2 m space to be left between two ponds, an area of 1224 m^2 per pond and 8490 acres of land for the whole setup was calculated.

A dark fermentation reactor is also needed in this process in which the alga has to be moved in stages. An assumption is made that 20% of the culture is moving through the dark fermenter at a time, representing 1.35 million m^3 of culture. For an array of 200 cylindrical bioreactors (height 10 m, radius 15 m), an area of 57 acres of land was calculated by the same authors.

A photoreactor system is required for production of additional hydrogen by the breakdown of acetate. Such a layout is estimated to require an additional 2120 acres of land. The total surface area as footprint of the open photobioreactor system would therefore be 10,700 acres of land. Glucose source is required in the process of photofermentation which has a high hydrogen production rate. For holding 20,600 m^3 of culture, a total of 858,800 reactor units are required with a land area of 1700 acres (Sathyaprakasan and Kannan 2015).

Costs of various materials used for the manufacture of the reactors are glass ($51.61), plexiglass ($15.47), clear PVC ($5.56), and blue tarp ($0.46).

12.3.3.3 Estimating Capital Cost

In the case of a direct photolytic reactor, the requirement of a capital investment is to the tune of 1.9 trillion dollars for an estimated area of 19.3×109 m^2. The total cost is $153 billion/$m^2$ by assuming cost of per acre land as $19,999 and by subjecting the cost to a scaling factor of 0.9.

The primary components involved in the capital investment required for the indirect biophotolysis plant are open ponds, dark fermenters, and closed photo-fermenter.

The estimated cost for the open pond is $7 m^{-2} (Resnick 2004) and the total capital investment is $240 million with a cost variation at scaling factor 0.9. Dark fermenter has higher cost than that of a pond ($7 m^{-2}) and lower than that of a photo-fermenter ($110 m^{-2}). The estimated cost for the construction of 200 dark fermenters is $10.5 million (Resnick 2004; Benemann 1997). The cost of the closed photo-fermenter system is $172, and for this plant, the total capital investment is $300 million.

By assuming a 20-year linear depreciation of the investment, the estimated total annual capital cost of this approach is $2.40 per gigajoule (Sathyaprakasan and Kannan 2015). The cost is significantly less in comparison to that of direct biophotolysis, where the annual capital cost per gigajoule is $1220.

The calculated cost of the bioreactor for the process of fermentation is $100 m^{-2}, the expenses go up to $62.60 per reactor because activated carbon is needed in this process (Sathyaprakasan and Kannan 2015).

12.3.3.4 Estimation of Electricity Cost

Out of its total output, 3% of the reactor is used as power, and the assumed industrial electricity price is estimated as $0.0478 kWh^{-1}. Electricity is also consumed in various other processes like mixing/pumping, pressure swing adsorption (PSA) compressor, and storage compressor, and their annual cost is $12,000, $6000, and $18,000, respectively (Amos 2004).

12.3.3.5 Cost Estimates for Water

About 65 million moles of water will be required per day for the conversion process by considering the efficiency as 90%. This is equivalent to 432,000 cubic meters of water annually (Resnick 2004; Akkerman et al. 2002). For direct biophotolysis, the annual cost of water used will come up to $22,000 by assuming a cost of $550 for 100 cubic meters. This cost calculation method is also applied for the remaining three processes (Sathyaprakasan and Kannan 2015).

12.3.3.6 Estimates of Labor Cost and General Supplies

It is accepted that full-time on-site supervision is not required in hydrogen production system of algae, and just 25 % time is adequate at the site of production (Amos 2004).

According to certain studies, estimated total labor cost for a closed plant is \$16,000 per hectare. The labor cost will be \$26,000 per hectare by assuming 10 % maintenance per hectare and fully loaded cost of \$100,000. The bioreactors occupying 1.94 million hectares would have a corresponding labor cost of \$50.4 billion (Sathyaprakasan and Kannan 2015).

Thus, the direct biophotolysis method appears to be a less economical choice for hydrogen production. Requirement of large land leads to significant increase in labor costs. The annual cost of general supplies for a reactor is taken as \$4000. Applying this value for 1.94 million ha, the cost has been estimated as \$7.75 billion annually (Sathyaprakasan and Kannan 2015).

12.3.3.7 Cost for Gas Separation and Handling

The estimated cost of pressure swing adsorption (PSA) process for separation of hydrogen from other gases is about 3 % of the total capital investment. For the construction of a biohydrogen production plant that would cost around \$50 million is expected to have PSA component costing \$1.48 million.

12.3.3.8 Expenditure on Glucose as Substrate

For the process of fermentation, glucose is required as substrate, and it is estimated that a little less than 175 million kg of glucose is needed each year for these processes. Considering the cost of laboratory grade glucose as \$75 ton^{-1}, an annual cost of \$144 million is estimated for the total glucose required as substrate for hydrogen fermenter plant.

The photofermentation plants have a good production rate, but there are difficulties in its scale-up. Because of high hydrogen production rates and smaller carbon footprint, the dark fermentation process is considered as an efficient method. However, to drive this process, biomass is required, and its cost is quite significant. In order to reduce the cost, it is critical to identify some cheaper sources of biomass.

Glucose is needed only for the fermentation processes, and glucose requirement for dark fermentation is higher than that for photofermentation. Although all other costs involved in the process of fermentation are comparable, it is the cost of glucose alone which makes photofermentation process more economical over dark fermentation.

Indirect biophotolysis has high cost in which requirements of supplies are more, being a continuous process. The costs of the indirect biophotolysis plant become more reasonable due to higher rate of production and lesser land requirement. By dividing the cost per gigajoule value by the calorific value of hydrogen, the cost of per kilogram of hydrogen is found to be 120 MJ kg^{-1}.

Although the photofermentation plant shows impressive hydrogen production rates in the lab, still there is difficulty in scaling-up the reactor. The cost of fermenters can be cut down by the identification of low cost carbon sources. The setting-up of a hydrogen production plant depends upon the costs as well as the availability of fresh water.

12.3.3.9 Estimating Operating Costs

Operating costs for the system are \$4000 for PSA operating, \$55,000 for maintenance and washing, and \$1,360,000 for capital-related charges annually (Amos 2004). Electricity for mixing, periodic washing of the pond covers, maintenance, electricity for the compressors, and PSA operating costs are included in these costs. The total algal hydrogen production capital cost and their comparison are listed in Tables 12.1 (Amos 2004) and 13.2 (Sathyaprakasan and Kannan 2015).

12.4 Hydrogen Storage Technologies and Economic Aspects

Energy storage issues are becoming more relevant as the share of green energy is becoming significant in some countries. Storage is an important issue for green energy because as we know that we cannot always have the required sunshine and the blowing wind. The major stumbling block for making renewable energy practical and dependable has been a yearlong production, and storage issues, particularly for days when the sun is not shining and the wind is not blowing, but now these goals seem achievable with some new technologies (Luoma 2009).

H_2 though considered by many as the alternative energy carrier since 15 years is yet to go a long way before commercialization. Major development efforts are required for storage of hydrogen for transport applications. It includes advanced conventional storage technologies like compressed gaseous storage and liquid storage in vacuum super-insulated tanks.

Although several promising concepts are there, these are in every preliminary stage like compressed gas, liquid hydrogen, metal hydride, and underground storage.

There are certain merits and demerits in each alternative. For example, liquid hydrogen has high density for storage, but it requires an insulated storage container and an energy-intensive liquefaction process. Underground storage was found to be the cheapest method on consideration of the major capital and operating costs over a range of production rates and storage times; underground storage was found to be the least expensive method (Amos 1998), and liquid hydrogen was found more advantageous over compressed gas as it has longer storage times.

For large quantities of hydrogen, pipeline delivery is found to be the affordable option, whereas for lesser quantities of H_2, theliquid form is more suitable for longer delivery distances.

12.4.1 Capital Cost of Storage Equipment

12.4.1.1 Estimates of Compressed Gas

Cost of a compressor is based on the amount of work done by it depending on flow rate and the pressure of inlet and outlet. Reciprocating compressors are widely used for H_2 applications, but centrifugal compressors are also an option. Timmerhaus and Flynn (1989) showed that reciprocating compressors are having higher efficiencies but have a cost about 50 % more than that of a comparable centrifugal compressor. The cost of a compressor is high because of high-operating pressures (Garret 1989). While comparing the cost of different sizes (10–28,300 kW) of compressors Amos (1998) found the cost to vary from \$650–\$6600 kW^{-1} to (\$440–\$4900 hp^{-1}) showing the larger compressors to be several times cheaper than smaller compressors on a unit basis.

12.4.1.2 Cost of Liquid Hydrogen

Liquid hydrogen stations are similar to tube trailer stations (Miller et al. 2006), but their storage component, which is a very low-temperature cryogenic tank, is filled on-site from a liquid hydrogen delivery truck. Liquid hydrogen stations can provide the fuel to a large number of vehicles, and generally they are more cost-effective when located in the vicinity of a liquid hydrogen plant.

On the basis of hydrogen production rate, the capital cost of a liquid hydrogen plant can be estimated, and range of sizing exponents for liquid hydrogen plants is 0.6–0.7 (Garret 1989; Cuoco et al. 1995). The total capital cost breakup as 10 % for planning, 60 % for equipment, and 30 % for construction was given by Zittel and Wurster (1996).

Capital costs for liquid hydrogen facilities of varying capacities were compared by Amos (1998) who reported that the cost ranged from \$25,600 to \$118,000 $kg^{-1}\ h^{-1}$ as the size of the liquefier varied from 1500 kg/h to 170 kg/h.

12.4.1.3 Cost of Metal Hydride

In case of the metal hydrides, capital cost includes the storage material, the pressure container, and the heat exchanger (integrated) that are used for the purpose of cooling and heating, during absorption and desorption processes, respectively (Schwarz and Amonkwah 1993). There are some cases in which the gas has to be compressed, due to availability of some particular properties of the hydride used. Economy of scale-up for metal hydride storage does not matter significantly as much of the capital cost is for the hydride material (Carpetis 1994).

Table 12.4 provides some estimated cost for metal hydride storage. Values of small hydride units for H_2 vary from \$820 kg^{-1} to \$60,000 kg^{-1}. There are units of metal hydrides for storage of more than 25 kg hydrogen (Hydrogen Components, Inc. 1997).

12.4.1.4 Estimates for Underground Storage

Underground storage method is the most efficient and low-cost method that is useful for storing large amount of hydrogen gas. Costs for storage vary upon whether there is a suitable natural cavern or rock formation, or whether a cavern is to be mined. Another low-cost option is used for natural gas wells which are abundantly available, followed by solution salt mining method and hard rock mining method.

Taylor et al. (1986) estimated the solution mining and hard rock mining costs as \$23 m^{-3} and \$34–84 m^{-3}, respectively, depending upon the depth. The cost of an underground natural gas storage system (with 89 km of high-pressure pipeline, 1930 kW compressor, a solution-mined cavern of 800 million SCF, and a working volume roughly equivalent to 2 million kg of H_2) installed by New York State Electric and Gas (NYSEG) was \$57.2 million (New York State Electric and Gas; NYSEG 1996a, b).

One additional expense for underground storage is incurred due to the value of the cushion gas that remains when the storage system is at the end of its discharge cycle. Brine can also be used to displace this gas at an additional expense for pumping and storing the brine solution (Taylor et al. 1986).

Table 12.5 describes estimates of some underground biohydrogen storage devices showing price range \$2.50–18.90 kg^{-1}. This is an order of magnitude less than liquid hydrogen storage and two orders of magnitude less than compressed gas aboveground storage.

12.4.1.5 Major Storage Issues

There is a need for storage of biological hydrogen to maintain reliable daily delivery of hydrogen during the night or during periods of poor sunlight when there is no hydrogen production.

Another problem is generally faced when the hydrogen storage tanks get full; all additional hydrogen produced that day is lost. Most often, tanks having 70 % of maximum capacity or more are used for hydrogen. During several points, it is observed that the storage levels get down to 30 %, so it would not be practical to reduce the on-site storage below a level, or else one would run out of hydrogen on those days.

Table 12.4 Capital costs for metal hydride

Size (kg)	Cost (\$)	Cost kg^{-1} H_2 (\$ kg^{-1})	References
–	–	\$1765	Carpetis (1994)
–	–	\$2100–\$2600	Carpetis (1994)
0.036	\$2150	\$60,000	Hydrogen Components Inc. (1997)
0.089–8.9	–	\$820–\$1300	Oy (1992)
8.9–890	–	\$1400–\$1800	Oy (1992)
2.7	\$8500–\$33,000	\$3150–\$12,200	Zittel and Wurster (1996)
0.089 0.2	\$6000–	\$3000–\$10,000	Zittel and Wurster (1996)

Adopted from Amos (1998)

Table 12.5 Cost of underground biohydrogen storage

Size (kg)	Cost kg^{-1}[a] (\$ kg^{-1})	Source
–	\$10.00	Carpetis (1994)
8.9–890	\$2.50–7.00	Oy (1992)
–	\$6.30–18.90	Taylor et al. (1986)
2,000,000	\$28.60[b]	NYSEG (1996c)

[a]All costs are adjusted to USD (1995)
[b]Includes 89 km of pipeline
Adopted from Amos (1998)

Economics can be improved in two ways by connecting a system to a pipeline. First, there is no need for high-pressure storage since the hydrogen is dumped directly into the pipeline on its production, and the pipeline itself provides some storage due to changes in operating pressure. Second, production is no longer limited by storage capacity; hence, the rate of hydrogen production per year is higher.

12.5 Biohydrogen Applications for Transportation

Various hydrogen supply pathway application for different types of transport systems depends to a great extent on the type of storage facility for hydrogen available on board. Compressed gaseous hydrogen storage is suitable for cars, vans for delivery, buses available for public, tramways, and regional trains as well as for boats and small ships. Storage of liquid hydrogen is generally found suitable for such applications and also for aircrafts and large ships. Hydrogen storage for trucks that have long driving distances exceeding 1000 km does not seem very feasible (Altmann et al. 2004).

Compressed gaseous hydrogen storage can be used both for decentralized and large centralized on-site productions. In the latter case, the market needs to be developed to such an extent as to allow for short transport distances by using pipeline or in liquid form using truck. Large liquefaction plants are required for storage of liquid hydrogen. One plant may supply several filling stations supporting up to hundreds of consumers for small-scale consumption in cars, buses, tramways, or small ships. A large dedicated plant requires to supply hydrogen for small number of large consumers like aircraft.

Distribution options include pipeline delivery and road trailer of hydrogen, which are a commercial reality since many decades. Decentralized generation of hydrogen can be done by making use of already available infrastructures for electricity, natural gas, or biomass transport. The latter is becoming more and more acceptable in hydrogen business.

Compression and liquefaction are included for conditioning of hydrogen, both of which are commercially feasible but need considerable development.

For road transport, proton exchange membrane (PEM) fuel cell is now being viewed as the most promising hydrogen application technology. It requires advance R&D for advancement in materials and components. Medium-temperature membranes need to be developed most importantly for better performance (Altmann et al. 2004).

12.5.1 Major Issues in Introducing Hydrogen as Fuel

The following are the major critical issues for the introduction of hydrogen in transport:

1. The cost of the vehicle based on fuel cell and the cost of hydrogen as a fuel are expected to come down with improvement of technologies and are optimistically put at achieving the goals of \$50–100 kWh^{-1} for vehicle and fuel productions (Altmann et al. 2004).
2. The performance of fuel cell and hydrogen-based vehicles can match that of conventional technologies. However, everything else being the same, hydrogen-based technologies do not still offer enough advantages to shift user choice. It is obvious that in order to be competitive, they have to provide comparable performance at comparable cost, with accessible and reliable infrastructures.
3. Major associated challenges are distribution and storage issues.. A major prerequisite is the development of a wide network of refueling stations, but the acceptability and demand are crucial before it takes off. For the justification of the investment costs, the cost of hydrogen distribution has to be kept low, and introduction has to be massive.
4. Environmental profits of using hydrogen have to be highlighted. Solutions based on electrolysis would only be beneficial for the environment as long as the electricity used for the electrolysis is produced from the fuels

which are carbon-free. The introduction of hydrogen in transport would be feasible only if there is use of low-cost renewables for electricity generation or by using high-performance fuel cells.
5. The hydrogen industry would depend upon the national plan and policy. The key industrial stakeholders like car manufacturers, refineries and fuel providers, infrastructure providers, etc. will invest in a new and novel technology only if the future market prospects are clear.
6. In certain areas, rules and legislation could also influence user choices, by promoting the use of hydrogen, penalizing emissions of carbon dioxide, or limiting the use of conventional technologies.

12.5.2 Capital Costs of Transportation Equipment

12.5.2.1 Compressed Gas Transport Costs

Capital costs of tube trailer depend upon the following: (i) operating pressure of the truck, (ii) the storage capacity of each trailer, and (iii) the distance to the user site. If we increase the capacity of a tube trailer, that would improve operation but also increase the cost of each truck. For local delivery, the same truck can make several trips back and forth between the production site and the customer site, but for long distances, each truck might be able to make only one or two deliveries per day.

A 460 kg hydrogen vessel for a tube trailer containing 16 tubes was calculated as $340,000. The cost of a truck to go with it was additionally $110,000 (Taylor et al. 1986).

12.5.2.2 Compressed Gas Pipeline Costs

Hydrogen pipelines (0.25–0.30 m dia) are constructed using commercial steel and operate at 1–3 MPa. Natural gas pipes for comparison are constructed as large as 2.5 m in dia and have work at 7.5 MPa (Hart 1997). Cost of the pipeline installation depends upon the natural gas construction prices. Installation costs of some recent Trans Canada Pipeline Projects show that over a varying length of 40.2–731 km pipeline, the cost varies from $132,000 km^{-1} to $1,250,000 km^{-1} (Amos 1998).

Compressor power and maintenance costs are the major operating cost for hydrogen pipelines. Some hydrogen losses may occur in the piping network, but for natural gas piping systems, these losses are reported to be less than 1 % (Hart 1997).

An estimate of the cost of piping hydrogen from North Africa to Central Europe (3300 km) including compression costs was around $0.9–$1.20 kg^{-1}, while another estimate from the United States put the cost much less at $0.39/kg (Report to Congress 1995). Two more studies by Oy (1992) and Johannsen (1993) also reported that the cheapest means of transporting hydrogen for large quantities are pipelines except for transport across the ocean, when liquid hydrogen transport is the cheapest.

12.5.2.3 Liquid Hydrogen Transport Costs

Total capital costs of liquid hydrogen transport involve mainly the insulated tank trailer plus the cost of the cab transport. Although hydrogen is not transported overseas, a hydrogen barge is expected to cost 3.5–4 times as much as a liquefied natural gas barge (Carpetis 1994).

The liquid hydrogen truck transport costs include the same charges as for gas transport (fuel, driver wages, and maintenance charges) and additionally include boil-off losses during transport. Expected boil-off losses during transfer between tanks are 10–20 %, but can be as high as 50 % (Huston 1984; Johannsen 1993; Taylor et al. 1986).

According to one source, the estimated long-distance transportation cost of liquid hydrogen from Africa to Europe would be $1.80–$2.10 kg^{-1} (Johannsen 1993). Another source mentions that shipping liquid hydrogen across the Atlantic would triple its price (Oy 1992).

12.5.2.4 Metal Hydride Transport Costs

The major cost for transportation of hydrogen using metal hydrides is the capital cost of metal hydride and containers. Once filled, the hydride containers can be shipped with charges depending on the distance and weight.

There is a continued research in area of long-range transport of hydrogen using barges or ships. Canada has large hydroelectric resources that could be used to produce hydrogen. Hence, it has designed several vessels for carrying hydrogen across the Atlantic. One design uses five barges contained on a single ship, designed to go for 50 days without venting hydrogen.

Two other designs include a ship with four spherical dewars, each holding 3.5 million kg of hydrogen and a single-hull design capable of carrying 7 million kg of hydrogen with a boil-off rate of 0.2–0.4 % d^{-1} (Hart 1997). Another option proposed was the use of airplanes to deliver hydrogen over great distances to reduce transport times and consequently reduce boil-off losses. Efforts are underway to reduce transfer losses in hydrogen transportation, with a goal of reducing the losses to 8 % (Johannsen 1993).

It may be concluded that liquid hydrogen transport by truck is the cheapest alternative for transportation, except when large quantities of hydrogen need to be transported, where pipeline delivery is more competitive. Because the major expense is on installing the pipeline and not the pipeline cost itself, a larger pipeline can be installed to handle multiple users at about the same cost. Very little energy is required to pump the hydrogen through the pipeline.

In all cases, except pipeline delivery, a minimum transport cost is associated with each delivery method for a given distance. Costs of rail car vary little with production rate and distance, because these get fully utilized due to long-transit times associated with rail transport.

However, with liquid hydrogen, the effect is small compared to compressed gas because the driver is carrying more hydrogen per trip. One hydrogen tanker can carry more than 20 times the amount of hydrogen as a tube trailer.

At a medium production rate of 450 kg h^{-1} and a 160 km delivery distance, liquid hydrogen trucking is the cheapest means of transport, but metal hydride would also compete because of its high-storage density.

The price of one liquid hydrogen tanker with cab is estimated at $500,000, the price of 15 tube trailers with cabs is about $3.75 million, and the price of six metal hydride transports is $6.9 million to transport the same amount of hydrogen over the same distance.

12.6 Combined Storage and Transportation Costs

There are three major factors which play an important role in reducing the cost for hydrogen delivery, i.e., rate of hydrogen production, delivery distance, and time of the storage which may be interdependent on each other. As time of the hydrogen storage may depend on delivery distance. If a small hydrogen unit is producing one truck load of hydrogen gas for every 3 days, there would be a need for 2 days of storage if the site of consumer is far away. On the other hand, if the delivery distance is less, the truck may make a small distance trip everyday without much elaborate storage facilities.

Options for transport and delivery can also be combined. For instance, metal hydride delivery could be compatible with compressed gas storage, underground storage, or even liquid hydrogen storage. Another option, i.e., pipeline transport without any storage may also be an option.

Based on production rate and delivery frequency, the storage time is planned, and total cost including depreciation is estimated. When the costs of storage and delivery are added, the benefit of liquid hydrogen becomes apparent. For production rates of 450 kg h^{-1}, 1 day of storage and a 160 km of delivery distance, liquid hydrogen is only slightly cheaper than metal hydride transport, but at a longer distance of 1600 km liquid hydrogen is four times economic than metal hydride and seven times cheaper than compressed gas.

12.7 Conclusion and Future Prospects

It may be concluded from the foregoing review and discussion that hydrogen production is still in preliminary stage, and a number of issues related to its production still have to be addressed before it is accepted as an alternative fuel of the future. Hydrogen molar yield and the feedstock cost are two major issues that have to be addressed. The main challenge in fermentative hydrogen production is that only 15 % of the organic source energy could be obtained in the form of biohydrogen (Logan 2004). The goal for fermentation technology set by the US Department of Energy is to obtain 6 mols of H_2 per mol of glucose by 2018 (Davila-Vazquez et al. 2008b), for which intensive research is in progress. Improvement in hydrogen production is also being tested by gene manipulation focusing on endogenous gene disruption (Datsenko and Wanner 2000). Hydrogen molar yields may be increased by recent developments in the field of metabolic engineering by discovering new pathways to make full use of the 12 mols of hydrogen in one mole of glucose (Davila-Vazquez et al. 2008b).

There are two major factors that would have an important bearing on the cost of biohydrogen production for commercial use. Cost of the photobioreactor has to be brought down, which will depend on appropriate and less expensive materials to be used in the fabrication of the photobioreactors. Another important cost factor revolves around the storage system required for a standalone system. Conventional high-pressure compression and storage are full-grown technologies and cost reductions in this area would come only through better or new technology. The underground storage technology is useful for large quantities of gas or long-term storage of hydrogen. Compressed gas is useful for small quantities of gas, high cycle times or short storage times and metal hydrides are used for small quantities of gas.

For economic viability of biohydrogen technology, there is an urgent need that its production rate should match the rate of electron transfer for bacteria or other microorganisms. To achieve this goal, more R&D is required in the field of genetic engineering to increase electron transfer rate in the biological system used and produce more hydrogen. Cost of the installed reactor has to be near \$1 m^{-2} for commercial use of this new fuel.

References

Akkerman I, Janssen M, Rocha J, Wijffels RH (2002) Photobiological hydrogen production: photochemical efficiency and bioreactor design. Int J Hydrog Energy 27:1195–1208

Altmann M, Schmidt P, Wurster R, Zerta M, Zittel W (2004) Potential for hydrogen as a fuel for transport in the long term (2020 – 2030), Report EUR 21090, European Commission, Joint Research Centre, Institute for Prospective Technological Studies, Seville, Spain.

Amos WA (1998) Costs of storing and transporting hydrogen. Report by National Renewable Energy Laboratory 1617 Cole Boulevard Golden, Colorado, U.S. Prepared under Task No. HY914041. NREL/TP-570-25106.

Amos WA (2004) Updated cost analysis of photobiological hydrogen production from *chlamydomonas reinhardtii* green algae. Milestone completion report by National Renewable Energy Laboratory, Cole Boulevard Golden, Colorado, U.S. NREL/MP-560 35593.

Benemann JR (1996) Hydrogen biotechnology: progress and prospects. Nat Biotechnol 14:1101–1103

Benemann JR (1997) Feasibility analysis of photobiological hydrogen production. Int J Hydrog Energy 22:979–987

Carpetis C (1994) A technology and costs of hydrogen storage. TERI Inf Dig Energy 4(1):1–13

Chong M-L, Sabaratnam V, Shirai Y, Hassan MA (2009) Biohydrogen production from biomass and industrial wastes by dark fermentation. Int J Hydrog Energy 34(8):3277–3287

Cuoco A, Sgalambro G, Paolucci MD, Alessio L (1995) AIs photovoltaic hydrogen in Italy competitive with traditional fossil fuels? Energy 20(12):1303–1309

Das D, Veziroglu TN (2001) Hydrogen production by biological processes: a survey of literature. Int J Hydrog Energy 26:13–28

Datsenko KA, Wanner BL (2000) One-step inactivation of chromosomal genes in *Escherichia coli* K-12 using PCR products. Proc Natl Acad Sci U S A 97(12):6640–6645

Davila- Vazquez G, Arriaga S, Alatrriste-Mondragon F, de Leon-Rodriguez A, Rosales- Colunga LM, Razo-Flores E (2008) Fermentative biohydrogen production: trends and perspectives. Rev Environ Sci Biotechnol 7:27–45

Davila-Vazquez G, Alatriste-Mondrago'n F, de Leo'n Rodrı'guez A, Razo-Flores E (2008) Fermentative hydrogen production in batch experiments using lactose, cheese whey and glucose: influence of initial substrate concentration and pH. Int J Hydrog Energy 33(19):4989–4997

Erog˘lu E, Gu¨ndu¨z U, Yu¨cel M, Tu¨rker L, Erog˘lu I (2004) Photobiological hydrogen production by using olive mill wastewater as a sole substrate source. Int J Hydrog Energy 29:163–171

Ferchichi M, Crabbe E, Hintz W, Gil GH, Almadidy A (2005) Influence of culture parameters on biological hydrogen production by Clostridium saccharoperbutylacetonicum ATCC 27021. World J Microbiol Biotechnol 21(6–7):855–862

Garret DE (1989) Chemical engineering economics. Von Nostrand Reinhold, New York

Hallenbeck PC, Benemann JR (2002) Biological hydrogen production: fundamentals and limiting processes. Int J Hydrog Energy 27:1185–1193

Hart D (1997) Hydrogen power: the commercial future of the ultimate fuel. Financial Times Energy Publishing, London

Hawkes FR, Forsey H, Premier GC, Dinsdale RM, Hawkes DL, Guwy AJ, Maddy J, Cherryman S, Shine J, Auty D (2008) Fermentative production of hydrogen from a wheat flour industry co-product. Bioresour Technol 99:5020–5029

Hoffart MI, Caldeira K, Benford G, Criswell DR, Green C, Herzog H, Jain AK, Kheshgi HS, Lackner KS, Lewis JS, Lightfoot HD, Manheimer W, Mankins JC, Mauel ME, Perkins LJ, Schlesinger ME, Volk T, Wigley TML (2002) Advanced technology paths to global climate stability: energy for a greenhouse planet. Science 298:981–987

Holmes B, Jones N (2003) Brace yourself for the end of cheap oil. New Sci 179(2406):9. https://en.wikipedia.org/wiki/Energy_density

Hunt RW, Zavalin A, Bhatnagar A, Chinnasamy S (2014) In: Gikonyo B (ed) Advances in biofuel production: algae and aquatic plants. Apple Academic Press, Toronto, pp 1–363

Huston EL (1984) A liquid and solid storage of hydrogen. In: Proceedings of the 5th World hydrogen energy conference. Vol. 3; 15–20 Jul 1984, Toronto, Canada

Hydrogen Components, Inc. (1997). Fuel Cell Store, Web page. Littleton, Colorado

Johannsen TB (1993) Renewable energy: sources for fuels and electricity

Karlsson A, Vallin L, Ejlertsson J (2008) Effects of temperature, hydraulic retention time and hydrogen extraction rate on hydrogen production from the fermentation of food industry residues and manure. Int J Hydrog Energy 33(3):953–962

Karthic P, Joseph S (2012) Comparison and limitations of biohydrogen production processes. Res J Biotechnol 7(2):59–71

Kaushik A, Mona S, Kaushik CP (2011) Integrating photobiological hydrogen production with dye-metal bioremoval from simulated textile wastewater. Bioresour Technol 102:9957–9964

Keeling CD, Whorf TP (2005) Atmospheric CO_2 records from sites in the SIO air sampling network. Trends: a compendium of data on global change. Carbon dioxide information analysis center. O ak Ridge National Laboratory, US Department of Energy, Oak Ridge

Levin DB, Pitt L, Love M (2004) Biohydrogen production: prospects and limitations to practical application. Int J Hydrog Energy 29:173–185

Logan BE (2004) Extracting hydrogen and electricity from renewable resources. Environ Sci Technol 38(9):160A–167A

Luoma JR (2009) Report on the challenge for green energy: how to store excess electricity

Macaskie LE, Baxter-Plant VS, Creamer NJ, Humphries AC, Mikheenko IP, Mikheeno PM, Penfold DW, Yong P (2005) Applications of bacterial hydrogenases in waste decontamination, manufacture of novel bionanocatalysts and in sustainable energy. Biochem Soc Trans 33(Pt1):76–79

Miller M, Weinert J, Nicholas M (2006) Clean hydrogen for transportation applications: report California PATH Working Paper UCB-ITS-PWP-2006-5 NREL/TP-570-25106

Miyake J, Miyake M, Asada Y (1999) Biotechnological hydrogen production: research for efficient light energy conversion. J Biotechnol 70:89–101

Mona S, Kaushik A, Kaushik CP (2011a) Waste biomass of *Nostoc linckia* as adsorbent of crystal violet dye: optimization based on statistical model. Int Biodeterior Biodegrad 65:513–521

Mona S, Kaushik A, Kaushik CP (2011b) Biosorption of reactive dye by waste biomass of *Nostoc linckia*. Ecol Eng 37:1589–1594

Mona S, Kaushik A, Kaushik CP (2011c) Biosorption of chromium(VI) by spent cyanobacterial biomass from a hydrogen fermentor using box-behnken model. Int Biodeterior Biodegrad 65:656–663

Mona S, Kaushik A, Kaushik CP (2011d) Sequestration of Co(II) from aqueous solution using immobilized biomass of *Nostoc linckia* waste from a hydrogen bioreactor. Desalination 276:408–415

Mona S, Kaushik A, Kaushik CP (2013) Prolonged hydrogen production by *Nostoc* in photobioreactor and multi-stage use of the biological waste for column biosorption of some dyes and metals. Biomass Bioenergy 50:27–35

Nath K, Das D (2003) Hydrogen from biomass. Curr Sci 85(3):265–271

New York State Electric & Gas (1996a) ANYSEG completes seneca lake storage project

New York State Electric & Gas (1996b) ASeneca lake storage project completed

New York State Electric & Gas (1996c) ANYSEG brings natural gas to New York's North Country

Oy N (1992) New hydrogen technologies project report: evaluation of techniques

Pacala S, Socolow R (2004) Stabilization wedges: solving the climate problem for the next 50 years with current technologies. Science 305:968–972

Patil V (2007) The relevance of biofuels. Curr Sci 92:707

Ren N, Li J, Wang Y, Liu S (2006) Biohydrogen production from molasses by anaerobic fermentation with a pilot-scale bioreactor system. Int J Hydrog Energy 31:2147–2157

Report to Congress (1995) Hydrogen: technology and policy. 95–540 SR

Resnick RJ (2004) Economics of biological methods ofhydrogen production. Master's Thesis submitted in the Massachusetts Institute of Technology, Massachusetts Ave, Cambridge

Sasikala K, Ramana CV, Rao PR (1992) Photoproduction of hydrogen from the waste water of a distillery by Rhodobacter sphaeroides O.U. 001. Int J Hydrog Energy 17(1):23–27

Sathyaprakasan P, Kannan G (2015) Economics of biohydrogen production. Int J Environ Sci Dev 6(5):352–356

Schwarz JA, Amonkwah KAG (1993) Hydrogen storage systems. U.S. Geological Survey, Washington, DC

Srivastava A, Prasad R (2000) Triglycerides-based diesel fuels. Renew Sust Energ Rev 4:111–133

Taylor JB, Alderson JEA, Kalyanam KM, Lyle AB, Phillips LA (1986) A technical and economic assessment of methods for the storage of large quantities of hydrogen. Int J Hydrog Energy 11(1):5–22

Timmerhaus C, Flynn TM (1989) Cryogenic Engineering. Plenu Press, New York

World Energy Outlook (2010) International Energy Agency, Paris, France

Yetis M, Gu¨ndu¨z U, Erog˘lu I, Yu¨cel M, Tu¨rker L (2000) Photoproduction of hydrogen from sugar refinery wastewater by Rhodobacter sphaeroides O.U. 001. Int J Hydrog Energy 25:1035–1041

Zhu H, Miyake J, Tsygankov AA, Asada Y (1995) Hydrogen production from highly concentrated organic wastewater by photosynthetic bacteria & anaerobic bacteria. Water Treat 10:61–68

Zhu HG, Ueda S, Asada Y, Miyake J (2002) Hydrogen production as a novel process of wastewater treatment—studies on tofu wastewater with entrapped R. sphaeroides and mutagenesis. Int J Hydrog Energy 27(11–12):1349–1357

Zittel W, Wurster R (1996) Hydrogen in the energy sector. Ludwig Bolkow-ST Report. http://www.hyweb.de/knowledge/w-i-energie W-eng

13 Comparative Environmental Life Cycle Assessment of Biohydrogen Production from Biomass Resources

Christina Wulf, Lisa Thormann,
and Martin Kaltschmitt

Abstract

The goal of this chapter is to assess environmental impacts of biohydrogen production regarding anthropogenic climate change, emissions with an acidification impact and further impact categories by means of a life cycle approach. In conjunction with reducing the use of fossil resources, there is a need to prioritize those technologies that will provide the least impact on the environment. Thus, a variety of processes of hydrogen production derived from biomass feedstock are investigated related to environmental effects. This case study considers biohydrogen production derived from biomass sources from forestry and short rotation coppice (SRC), herbaceous biomass (i.e. wheat straw), energy crops (mainly maize and grain) and biowaste in Germany. The technology with the most promising results regarding the environmental impact is steam methane reforming (SMR) of a substrate mix from nonfood substrates compared to steam methane reforming of natural gas.

13.1 Introduction

Today, technological efficiency and economic profitability solitarily determine the implementation of a transportation fuel on the market. However, environmental aspects gain substantial public interest. Nowadays, sustainable development of the transportation system is a primary goal of the European society (The European Parliament and the Council of the European Union 2009b) as well as an increased share of biofuels within the European transportation system.

C. Wulf (✉)
Institute of Environmental Technology and Energy Economics, Hamburg University of Technology, Eissendorfer Str. 40, 21073 Hamburg, Germany

The Institute of Energy and Climate Research – Systems Analysis and Technology Evaluation, Forschungszentrum Jülich GmbH, 52425 Jülich, Germany
e-mail: christina.wulf@tuhh.de

L. Thormann • M. Kaltschmitt
Institute of Environmental Technology and Energy Economics, Hamburg University of Technology, Eissendorfer Str. 40, 21073 Hamburg, Germany

A. Singh, D. Rathore (eds.), *Biohydrogen Production: Sustainability of Current Technology and Future Perspective*, DOI 10.1007/978-81-322-3577-4_13

To achieve these goals, the EU has implemented the renewable energy directive (RED) aiming a share of 5 % renewable energy within the transportation sector until 2020 to decrease the undesirable effects of the European mobility sector on the worldwide climate caused by, for example, greenhouse gas (GHG) emissions (European Commission 2012). With the increasing diversification of transportation fuels, biohydrogen has the potential to become a common fuel in the mobility sector in the future; even though, the source (i.e. energy carrier) of renewable energy is not specified by the EU legislation. Additionally, most European governments promote an increase in biofuels as an instrument to reduce GHG emissions within the transportation sector to contribute to the goal of the European Union to decrease GHG emissions by 6 % in the year 2020 compared to the 2010 level. In contrast to traditional transportation fuels derived from fossil energy sources, the emissions of biofuels need to have a potential to save at least 35 % of the GHG emissions (The European Parliament and the Council of the European Union 2009a, b).

Hence, this chapter follows a case study design with an in-depth analysis of various biohydrogen production pathways. The biomass resources assessed include the major types of biomass resources available in Germany. The present research focusses on answering the following question: Can hydrogen be used as an energy carrier, i.e. transportation fuel, of the future under environmental paradigms?

Biohydrogen production pathways represent in themselves complex systems that are challenging to analyse. The production pathways consider (1) biomass cultivation, (2) its transport to the hydrogen production site and (3) preprocessing and conversion technologies. The life cycle assessment (LCA)'s comprehensive approach aims to point out determining factors assessing potential environmental aspects associated with the final product biohydrogen. However, not all categories of environmental impacts practicable in LCA are (1) relevant in biohydrogen production, (2) provide sufficient data for a reliable LCA or (3) are non-site-specific (Joint Research Centre – Institute for Environment and Sustainability 2011). Therefore, this chapter preferably applied for cases where general data is adequate and appropriate.

Generally, biohydrogen can be produced based on chemical, thermochemical, biological, biochemical and biophotolytical conversion processes from all types of biomass resources (Demirbas 2009). Thermochemical biohydrogen production includes the reforming of biogas (Alves et al. 2013; Zech et al. 2015) as well as the direct gasification of woody biomass (Kalinci et al. 2012; Koroneos et al. 2008; Zech et al. 2015), whereas the biological, biochemical and biophotolytical conversion processes comprise a big variety of technologies (Manish and Banerjee 2008). Mostly discussed are the dark fermentation and the photofermentation (Djomo and Blumberga 2011; Ochs et al. 2010). Possible feedstocks for these processes are not only organic wastes and high-carbohydrate biomass but also cultivated algea (Ferreira et al. 2012, 2013).

Important applications of hydrogen are primarily within the chemical industry (i.e. ammonia and methanol syntheses, oil refining, other petrochemical processes) with considerable market impact. Nevertheless, a variety of applications can be found illustrating the political interest for the use of hydrogen within the transport sector in the future (Gandía et al. 2013). At present, the worldwide production of hydrogen is between $50 \cdot 10^6$ and $80 \cdot 10^6$ t/a (United States Department of Energy 2013; Eni S.p.A. 2013); being a secondary energy carrier, hydrogen provided from organic matter or fossil energy sources has a low heating value of 120 MJ/kg and a density of 0.0899 kg/m^3 (Lauermann et al. 2000).

Several approaches exist to assess environmental impacts of energy production and use (e.g. biofuels). The methodological approach taken in this chapter is a LCA. An LCA is a systematic, standardized analytical method that aids to identify and evaluate environmental impacts of a process or multiple competing processes (DIN EN ISO 14040; DIN EN ISO 14044). The benefit of using an LCA approach in assessing the environmental impacts of biohydrogen production is first and foremost that it provides a comprehensive overview of all processes (i.e. cradle to grave approach) as well as impacts involved.

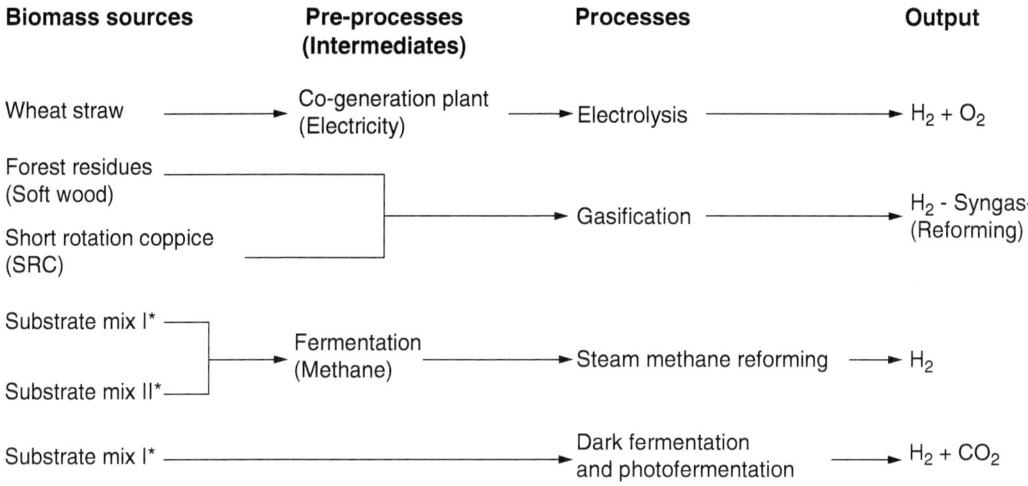

Fig. 13.1 Process flow sheet of a variety of biohydrogen production pathways (*Table 13.1)

Whereas previous research displays merely the environmental impact/performance (e.g. climate change) from biohydrogen production (Djomo and Blumberga 2011; Kabir and Kumar 2011; Zech et al. 2013), this chapter studies six different biohydrogen production pathways in regard to a variety of technologies (Fig. 13.1):

1. Alkaline water electrolysis with electricity from biomass cogeneration plants fired with straw
2. Steam reforming of biomethane (biogas production by fermentation of a typical substrate mix I (Table 13.1))
3. Steam reforming of biomethane (biogas production by fermentation of a typical substrate mix II (Table 13.1) with nonfood substances)
4. Gasification of softwood using the fast internal circulating fluidized bed (FICFB) gasifier
5. Gasification (FICFB) of wood from short rotation coppice (SRC)
6. Dark fermentation combined with photofermentation from substrate mix I (Table 13.1)

The focus on this study lies on the variety of biohydrogen production technologies and biomass feedstock rather than finding the best feedstock for one technology. Already the most appropriate technology for one biomass feedstock has been chosen.

13.2 Methodology

In undertaking this chapter, a standardized LCA methodology was employed to assess the initial research questions. LCA is a technique for analysing the potential environmental impacts associated with a product or service over its entire life (i.e. from 'cradle to grave') (Finnveden et al. 2009). By identifying energy and materials consumed as well as waste streams and emissions released to the environment, an LCA allows an identification of opportunities for decreasing the environmental impact of biohydrogen production routes (Joint Research Centre – Institute for Environment and Sustainability 2011).

The life cycle assessment of the biohydrogen production pathways is accomplished in line with the international standards ISO 14040 (DIN EN ISO 14040) and 14044 (DIN EN ISO 14044).

This section outlines the methodology implemented in this chapter in the following four sections: (1) goal and scope, (2) inventory analysis, (3) impact analysis and (4) interpretation (Fig. 13.2).

Table 13.1 Overview of main data for steam reforming of biomethane

Resource	Substrate mix[a]		Unit
Pre-chain	Fermentation		
Capacity		7	MW
Duration of life		20	a
Full load hours		7800	h
Biomass demand		88,000	t_{FM}/a
Electricity demand		4200	MWh/a
Demand of operating materials	Water	220	l/a
	$FeCl_3$	13,000	kg/a
	Triethylene glycol	2500	kg/a
	Propane	170,000	kg/a
Direct emissions	CH_4	8200	kg/a
	SO_2	6500	kg/a
Intermediate product	Biomethane		
Hydrogen production	Steam reforming		
Capacity		720	kg H_2/h
Duration of life		20	a
Full load hours		7500	h
Demand of energy carrier (biomethane)		46	kWh/kg H_2
Demand of energy carrier (German mix substrates)		39	kg/kg H_2
Demand of energy carrier (nonfood substrates)		78	kg/kg H_2
Demand of operating materials	Water	14	l/kg H_2
Direct emissions	NOx	2.4	g/kg H_2
	SO_2	0.12	g/kg H_2
	CH_4	48	g/kg H_2
	CO_2	460	g/kg H_2
Export credit	Heat	15	MJ/kg H_2

Kaltschmitt et al. (2012), Spath and Mann (2001), Cetinkaya et al. (2012), Pehnt (2002), Boyano et al. (2011), Scheftelowitz et al. (2013)

[a]Substrate mix I (energy crops): liquid manure 43.1 %, maize silage 38.5 %, grass silage 8.0 %, wheat silage 5.1 %, biowaste 4.5 %, wheat grain 0.7 %. Substrate mix II (nonfood): biowaste from municipal biowaste collection (kitchen and garden residues) 87.6 %, grass silage 7.4 %, liquid manure 4.9 %

13.2.1 Goal and Scope

In general, a clear definition of the goal of the LCA is the basic requirement for the performance of a consistent assessment. This goal description predefines the choice of system parameters, the complexity, respectively, the level of detail of the entire assessment as well as the system boundaries. A precise and clear definition of the goal is essential for achieving comparable and reliable results. Hence, important aspects are amongst others: (1) the selection of the alternatives to be assessed leading to an identical product or service, (2) the definition of a functional unit, (3) the determination of the complexity of the analysis, (4) the content of the inventory, (5) the impact analysis and (6) spatial and temporal conditions. The goal and scope definition provides the foundation the three subsequent LCA steps to be investigated (see below).

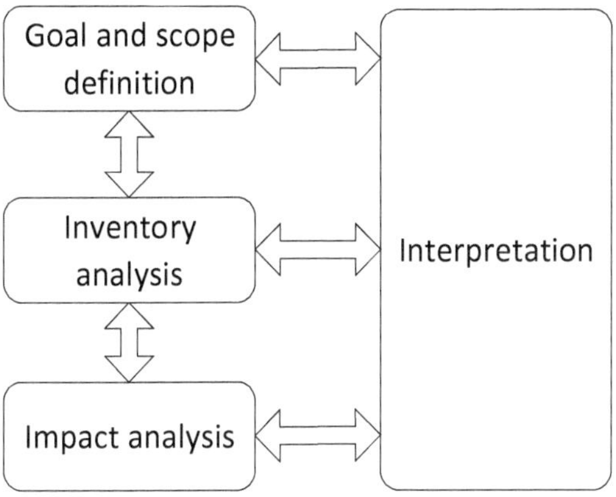

Fig. 13.2 Process flow sheet of an LCA approach ISO 14040 (DIN EN ISO 14040) and 14044 (DIN EN ISO 14044)

The goal of this analysis is to determine the environmental impacts of biohydrogen derived from agricultural (by-)products, forestry residues and energy crops in regard to different processing routes.

The system boundary includes overall energy and material demand for biohydrogen production covering all specific pre-chains for energy, biomass resources and other material provision as well as the overall material demand for building infrastructure.

The scope includes (1) biomass feedstock supply, (2) preprocessing and conversion technologies and (3) storage. Geographically this analysis is situated in Germany and the conditions of the German energy market are applied. Therefore, electricity needed within the different process steps was modelled according to the German electricity mix 2012 (Destatis 2012). The time horizon for the analysis is the year 2012.

The functional unit used is the energy content of the produced biohydrogen at the production site, in kilogramme (kg), based on the lower heating value (LHV). This corresponds to an energy content of 120 MJ. Furthermore, the following assumptions apply: ambient conditions (temperature ~15 °C), 1.5 MPa and a purity of 99.99 % at the factory gate, considering that the purity of hydrogen significantly affects the use in fuel cell vehicles (Boyd 2010).

In several process steps during biohydrogen production by-products may be produced. Therefore, credits were given for avoided production of the by-products of the biohydrogen production process wherever suitable. In case that no credit was applicable (e.g. wood residues and wood logs for non-energetic purposes were produced), allocation by either mass or volume was applied. In the case of heat as a by-product from cogeneration units, conventional allocation was not feasible. Hence, allocation by exergy was pertained. Therefore, the exergy of heat E was calculated including the energy content of heat Q and the temperature level according to Eq. (13.1). The heat flow has 363 K (T) and T_u represents the ambient air temperature. The exergy of electricity is described by its energy content:

$$E = Q(1 - T_u / T), \text{ with } T \text{ in } K \quad (13.1)$$

13.2.2 Inventory Analysis

In general, in a life cycle inventory analysis (LCIA), all input and output parameters (e.g. energy inputs, raw material inputs, products, coproducts, waste, emission to air) are collected and based on this all stages of the product life cycle are described. Therefore, a model process is developed to transfer the life cycle of a product or service into series of clearly defined processes that can be assessed. To reduce the complexity of the assessment and to allow the inventory analysis to be handled within given time constraints, all process chains are traced back until the point is reached where additional processes will no longer influence the overall results of the impact categories significantly.

The inventory analysis has been carried out based on commercially available software tools, i.e. Umberto. This software is connected with a database providing basic data (Swiss Centre for Life Cycle Inventories 2013).

13.2.3 Impact Analysis

In general, the LCA step, in the present chapter, selects suitable impact categories, category indicators and characterization models. In addition, this step classifies the life cycle inventory resulting in regard to one or multiple impact categories.

The impact categories analysed within this LCA are (1) climate change with the unit of CO_2-eq (carbon dioxide equivalent) according to IPCC (2007), (2) acidification of soil and aquatic ecosystems in SO_2-eq (sulphur dioxide equivalent) according to CML 2001 (Guinée et al. 2001), (3) eutrophication of terrestrial and aquatic ecosystems in PO_4^{3-}-eq (phosphate equivalent) also according to CML 2001 (Guinée et al. 2001), (4) human toxicity (non-cancer) in CTU_h (comparative toxic unit for humans) and (5) human toxicity (cancer) in CTU_h according to USEtox (Rosenbaum et al. 2008). For the analysis of USEtox, the metals have been excluded as their results are not so reliable as the results of other substances (Hauschild et al. 2012). With the impact categories (climate change, acidification, eutrophication), the main environmental damages caused by biomass cultivation in Germany are covered. USEtox has not been used very often yet. This study applies it for the assessment of biohydrogen production.

13.2.4 Interpretation

In general, a results interpretation from the inventory and impact analysis can be done by discussing all impact categories separately. Finally, some recommendations for improvements can be given if conceivable.

Significant issues based on the LCA were discussed alongside the data and results in Sect. 13.4.

13.3 Biohydrogen Production Pathways

The biohydrogen production pathways that are assessed in this chapter are described in the following section. All processes and assumptions described here are included in the LCA. The LCA methodology presented was applied to the case studies in the subsequent sections; especially the energy system varies for all process routes with regard to (1) a variety of biomass feedstock, (2) different energy technologies and (3) by-products.

The biomass resource has significant influence on the technologies applied due to its broad diversity. Most biomass feedstock are considered suitable for biohydrogen production (Kaltschmitt 2009). Currently, there is a wide variety of biomass resources in Germany and in Europe including easily accessible biomass resources on one hand and biomass waste streams derived from agriculture, forestry, industry and municipalities on the other hand (Kaltschmitt 2009). The biohydrogen production technologies applied (i.e. gasification, steam reforming, electrolysis) are comparable to fossil fuel processing pathways (i.e. biomethane to natural gas, wood chips to coal).

13.3.1 Steam Methane Reforming

In regard to monetary terms, the most efficient route to produce hydrogen is from hydrocarbon feedstock (Rostrup-Nielsen 2005). Natural gas is the most common source for the steam reforming process. Considering that it consists mainly of methane, it can be substituted with biomethane obtained from biogas. Generally, there is no difference between steam methane reforming originating from natural gas and biomethane because fossil methane and biomethane are chemically identical (Gellert 2013). Figure 13.3 displays the process flow sheet of steam methane reforming including energy flows and emissions to produce 1 kg of biohydrogen.

Biomass Resource The first substrate mix represents the actual German substrate mix utilized in existing biogas plants in 2012 (Scheftelowitz et al. 2013). This biomass mix contains mainly maize and other energy crops (Table 13.1). The second biomass mix contains only nonfood substrates (i.e. biowaste, grass, manure). The amount of substrate mix utilized varies between 39 kg

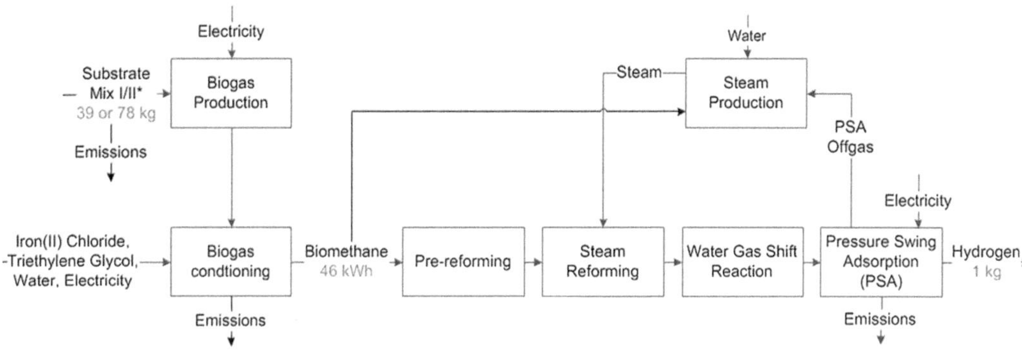

Fig. 13.3 Process flow sheet of steam reforming of biomethane derived from biogas (*Table 13.1)

(substrate mix I) and 78 kg (substrate mix II) with regard to the biomass resource for the production of 4.6 kg of biomethane for 1 kg of biohydrogen production.

Process Steps The process steps of producing biomethane include (1) fermentation of the organic substrate (with regard to the substrate mix in operation) in an anaerobic digestion plant via microorganisms converting biomass into biogas, (2) removal of hydrogen sulphide from biogas by inserting iron(II) chloride that binds sulphur and precipitates as iron sulphide, (3) separation of carbon dioxide by water scrubbing from the biogas, (4) drying of the methane with triethylene glycol adsorbing the water fraction to gain biomethane with the purity characteristics of natural gas (Roddy 2012) and (5) adding of propane from fossil sources to reach a similar Wobbe Index as natural gas (Scheftelowitz et al. 2013) to be allowed to blend the gas into the existing natural gas grid. If biomethane is available in the natural gas grid, it can be removed at any location and converted into biohydrogen via steam methane reforming. The LCI of the biomethane production route refers to the process parameters applied in Kaltschmitt et al. (2012), Boyano et al. (2011) and Spath and Mann (2001). The digestate of biogas production can be used as fertilizers in agriculture. In this study, it was assumed that during cultivation of maize, grass and wheat, the digestate was used as fertilizer. All emissions occurring with the use of digestate are assigned to the cultivation of biomass.

The following process steps of steam biomethane reforming derived from biogas include (1) desulphurization of biomethane below the ppm level with a zinc oxide catalytic converter to avoid adverse reactions of the catalyst of the reforming process with sulphur, (2) reforming of biomethane by addition of steam to hydrogen and carbon monoxide-rich gas (i.e. temperature around 800 °C and pressure at 3 MP), (3) transformation of synthesis gas into hydrogen via water-gas shift reaction and finally (4) removal of carbon dioxide via pressure swing adsorption (PSA). Apart from biohydrogen, heat was produced as a by-product within this process (Gupta 2009). In Table 13.1 all relevant parameters for the biohydrogen production process are displayed.

Allocation Within this plant heat is produced as a side product. Since this thermal energy can be used in other applications to increase the overall efficiency of the process, a credit for the avoided production of heat from biogas in an industrial furnace is given in this assessment.

13.3.2 Gasification

There are a number of assumptions or in some cases constraints that shape the gasification process developed within this chapter. In line with the requirements for technical implementation of the process, the process parameters are as displayed in Fig. 13.4 and Table 13.2.

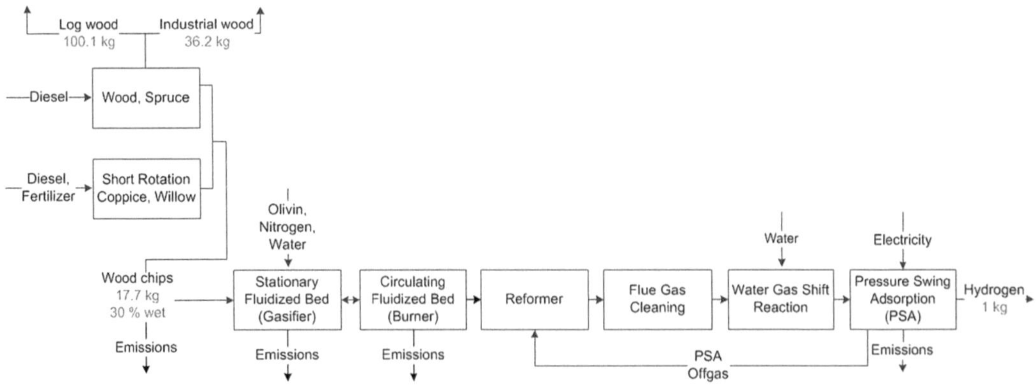

Fig. 13.4 Process flow sheet of gasification of wood chips from forest residues (FR) and short rotation coppice (SRC)

Table 13.2 Overview of the main data for the fast internal circulating fluidized bed (FICFB) process

Resource	Willow (SRC)/ spruce (forest residue)		Unit
Capacity		8.6	MW_{th}
Hydrogen capacity		280	kg H_2/h
Efficiency of gasification[a]		75	%
Duration of life		20	a
Full load hours		7500	h
Demand of wood (50 % water)		31	kg/kg H_2
Electricity demand		4.1	kWh/kg H_2
Demand of operating materials	Olivine	290	g/kg H_2
	N_2	390	g/kg H_2
	Water	5.3	l/kg H_2
Direct emissions	See text		
Ash		4.6	g/kg H_2

Hofbauer et al. (2002); Müller-Langer (2011), Gellert (2013), Swiss Centre for Life Cycle Inventories (2013)

[a]From dried biomass to synthesis gas, cold gas efficiency

Biomass Resources The biomass resources used for gasification are firstly forest residues (FR) from spruce analysed according to Swiss Centre for Life Cycle Inventories (2013) and secondly willow cultivated in short rotation coppice (SRC) analysed according to Belau (2012). In the first pathway, the fraction of forest wood is divided into wood for furniture production (i.e. primarily log), wood for other industries such as pulp and paper production (i.e. secondary log) and the remaining wood residues considered for energy production.

In both pathways feedstock was immediately converted into wood chips after collection below the size of <10 mm. Subsequently, the water content of fresh wood, i.e. ~wt50 % (Kaltschmitt 2009), was reduced by the use of natural air drying to ~30 wt%. Losses during storage and transportation were additionally taken into account (Seifert 2010).

The biomass feedstock was chosen according to an already existing electricity production plant in Güssing, Austria. In this plant a FICFB process was applied.

Process Steps The gasification process occurs in two coupled fluidized bed reactors. A fraction of the biomass resources were burned in the first reactor, i.e. circulating fluidized bed reactor at 900 °C to provide heat for the endothermic gasification, while in the second reactor, i.e. stationary fluidized bed, gasification proceeded at 850 °C. The un-vaporized fraction of the biomass resource (i.e. ~20 %) was reprocessed into the first reactor and burned accordingly. The silicate olivine was employed as fraction of the bed material, and nitrogen was added to the process with the purpose of reacting as a flushing and sealing gas. Then, the raw synthesis gas is cleaned and hydrogen is produced via a water-gas shift reaction within the steam reformer. Afterwards, the produced biohydrogen is purified within a PSA (Hofbauer et al. 2002) (Fig. 13.4).

All process parameters involved in performing the LCI are based on modelling results performed at TUHH (Gellert 2013); further significant data is summarized in Table 13.2.

For the direct emissions, it was considered that flue gas from the burning of wood chips (14.3 MJ wood chips burned/kg hydrogen) occurred.

Allocation Allocation of key biomass resources is based on volume in terms of the allocation factors (1) residue wood 1.00, (2) industry wood 2.04 and (3) log wood 5.65.

13.3.3 Electrolysis

Another biomass that can be converted into hydrogen is straw. For instance it can be used in a gasification or fermentation process to generate electricity from the biosynthesis gas. In an electrolyser the electricity splits water into hydrogen and oxygen. Likewise, it is possible to burn straw directly to produce both heat and electricity. At present, burning of straw is commonly used in Denmark (Skøtt 2011).

In this chapter, alkaline water electrolysis was taken into account as it can be considered one of the most customary processes to produce hydrogen (Millet and Grigoriev 2013). The most significant process steps are shown in Fig. 13.5.

Biomass Resources The biomass feedstock considered for electrolysis is wheat straw according to Belau (2012).

Process Steps In a cogeneration plant, chips from wheat straw were burned. The process medium (i.e. water) from the combustion was steamed and superheated to 540 °C and 6 MP. The electricity was generated via a turbine, and the remaining heat was utilized in a district heat network as it was planned for the straw-fired CHP plant in Midleton, Ireland (Erm21c Limited 2010). The emissions of the cogeneration and the used straw are divided between the heat and the electricity (see 'Allocation').

Subsequently, the electricity was used in an alkaline water electrolyser to convert deionized water via an electrochemical process into hydrogen. To ensure a defined conductibility, potassium hydroxide was added into the water. In the electrolyser module, the hydrogen was further processed and, thus, dried and purified by a deoxo dryer. The oxygen produced within this process step was released into the environment (Gupta 2009). All relevant process parameters are displayed in Table 13.3.

Allocation The output of grain per hectare in Germany was assumed to be 8.4 t_{FM}/(ha a) of grain (i.e. wheat) and 8.0 t_{FM}/(ha a) of straw. The straw-grain ratio considered is 1:0.95 (Belau 2012). The allocation was performed by mass. Additionally, an exergy allocation of the cogeneration plant that burned chips from wheat straw was considered. The ratio between electricity and heat was calculated with equation (14.1) to be 1:0.83.

13.3.4 Dark Fermentation Followed by Photofermentation

Fermentation is a biological process in which energy is extracted from a carbon source (Lee et al. 2010). Generally, a distinction between dark fermentation (i.e. non-phototrophic) and photofermentation (i.e. phototrophic) is made depending on the utilization of light as an energy source or not (Martinez-Merino et al. 2013). In this chapter, both types of fermentation were considered (Fig. 13.6).

Biomass Resource The biomass resource utilized is substrate mix I described in Table 13.1.

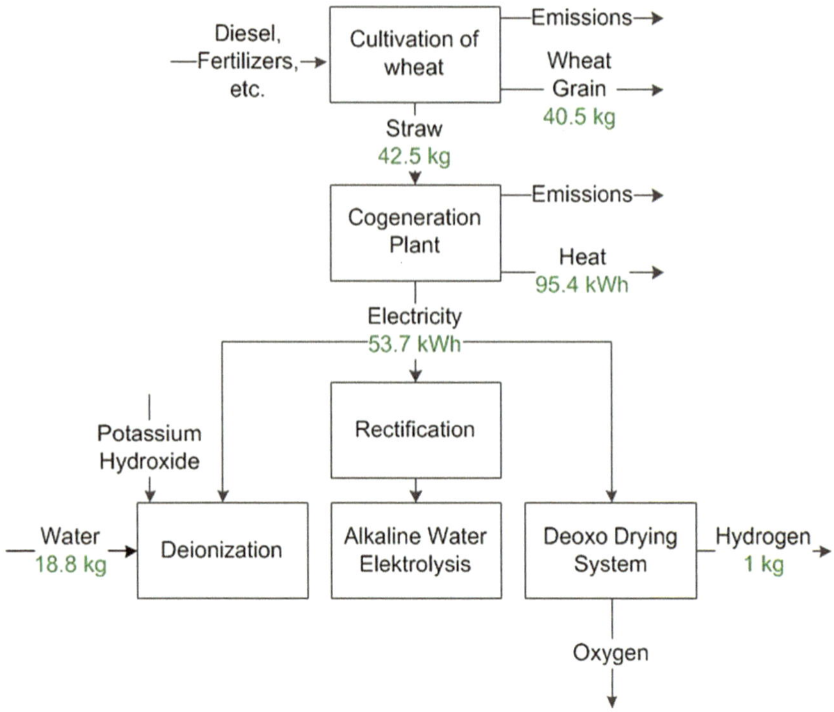

Fig. 13.5 Electrolysis of water with electricity derived from burning of straw

Process Steps Firstly, a pretreatment (i.e. enzymatic treatment, liquefaction and saccharification) of the substrate mix I is necessary to open up the structure of the biomass resource and to gain a homogenous biomass feedstock (Markowski et al. 2010). After pretreatment of the biomass, dark fermentation takes place. This process step is based on bacteria that grow in the dark on substrates rich in carbohydrates producing hydrogen via fermentation under anaerobic conditions (Das and Veziroglu 2008) and moderate temperatures (thermophile e.g. 70 °C). It is comparable to the fermentation in Sect. 14.3.1 without the methanation step. Subsequently, photofermentation (i.e. a light-dependent fermentation) takes place. This implies the splitting of organic acids derived from dark fermentation under anaerobic conditions into hydrogen in a photobioreactor. The accured sewage water has a higher BOD and a higher phospherus content than average sewage water, but little other contaminations. It can be fed to a municipal wastewater treatment plant. According to Ochs et al. (2010), the enzyme production can be neglected. All other process parameters are stated in Table 13.4.

Allocation For the combined process of dark fermentation and photofermentation, no allocation was applied, as all intermediates such as ethanol and lactic and butyric acids that occur are fed to the photofermentation.

Table 13.3 Overview of main data for electrolysis

Resource	Straw		Unit
Pre-chain	Cogeneration plant		
Capacity		18	MW
Duration of life		20	a
Full load hours		8000	h
Biomass demand		110,000	t_{FM}/a
Electricity production		140,000	MWh/a
Heat production		250,000	MWh/a
Electrical efficiency		33	%
Overall efficiency		88	%
Demand of operating materials	Neglected		
Direct emissions	NOx	400	mg/m³ Air
	SO_2	100	mg/m³ Air
	HCl	50	mg/m³ Air
	N_2O	11	mg/m³ Air
Residues	Ash	770,000	kg/a
Intermediate product	Electricity		
Hydrogen Production	Alkaline water electrolysis		
Capacity		55	kg H_2/h
Duration of life		20	a
Full load hours		8000	h
Demand of energy carrier (electricity)		54	kWh/kg H_2
Demand of straw (15 % water)		42	kg/kg H_2
Demand of operating materials	Water	19	l/kg H_2
	KOH	850	mg/kg H_2

Wulf and Kaltschmitt (2013), Belau (2012), Erm21e Limited (2010), Schmid (2011)

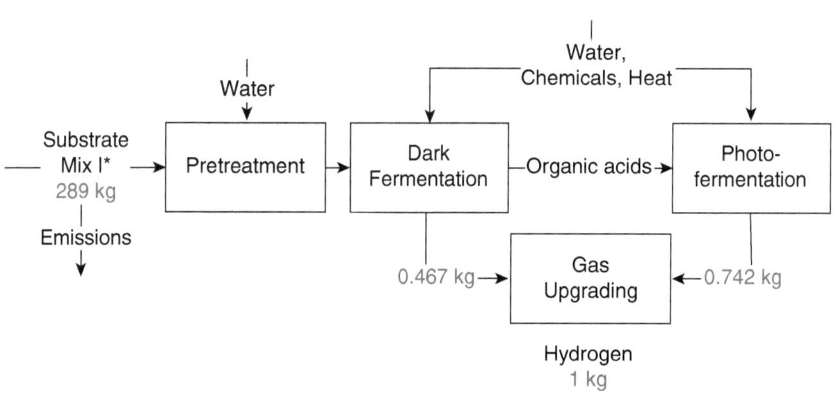

Fig. 13.6 Dark fermentation and photofermentation derived from substrate mix I (*Table 13.1)

Table 13.4 Overview of main data for dark fermentation and photofermentation

Resource	Substrate mix I[a]		Unit
Hydrogen capacity		60	kg H_2/h
Duration of life		20	a
Full load hours dark fermentation		7500	h
Full load hours photofermentation		3300	h
Demand of substrate		260	kg/kg H_2
Electricity demand		1.8	kWh/kg H_2
Heat demand		7.3	kWh/kg H_2
Demand of operating materials	KOH	1.5	kg/kg H_2
	K_2HPO_4	0.30	kg/kg H_2
	Water	240	l/kg H_2
Sewage water		24	l/kg H_2

Ochs et al. (2010), Markowski et al. (2010), Rechtenbach (2009)
[a]Substrate mix as per Table 13.1

13.4 Results and Discussion

According to the research question presented within the first subsection, the methodology was outlined within the second subsection and the data discussed within the third subsection. The impact categories considered are climate change, eutrophication of terrestrial and aquatic ecosystems, acidifying of soil and aquatic ecosystems and human toxicity divided into cancer and non-cancer effects. They were applied for the different pathways of biohydrogen production. The results are discussed and compared below.

13.4.1 Impact Categories

First, this section provides an overview of the results of different biohydrogen production pathways and a variety of feedstock with regard to multiple process steps applied. The results are divided into different categories:

1. Biomass includes all processes to cultivate and harvest the biomass resource (e.g. wood from short rotation coppice). Furthermore, emissions induced by intermediates (e.g. from electricity production from wheat straw) are included in this category.
2. Consumables cover all operating resources for the main process (e.g. water for steam methane reforming).
3. All substances emitted during the main process are called direct emissions (e.g. nitrogen oxides from steam methane reforming or wastewater treatment).
4. This category comprises all emissions caused by the materials and the energy related to the construction of the plant of the main process.
5. All emissions from materials and energy related to the demolition of the plant of the main process are summarized.

The results for the combined process of dark fermentation and photofermentation from substrate mix I were neglected in the figures below. This process is not yet applicable considering environmental impacts in regard to the impact categories applied. Already the high amount of biomass feedstock needed to produce one kilogramme of hydrogen implies a very high environmental impact. The main reason for the high demand is the low efficiency in the microbial conversion in the photobioreactor (i.e. photofermentation). Additionally the high amount of chemicals used for the processes, e.g. to maintain a good pH value for the dark fermentation and the dipotassium phosphate (K_2HPO_4) for the photofermentation, causes a high environmental impact. This effect has also been observed in literature before. Ochs et al. (2010) used biological waste (i.e. potato peels) as feedstock. Therefore, this production pathway has only an environmental impact from the process and its material and energy demand. The results vary roughly between 39 and 29 kg CO_2-eq/kg H_2. Also Djomo and Blumberga (2011) only used different kinds of biological waste and generated results between 5.0 and 5.8 kg CO_2-eq/kg H_2. However, their process consumed less chemicals than Ochs et al.'s. Manish and Banerjee (2008) used sugarcane as feedstock, but they did not take the cultivation of the biomass into account. Furthermore, in contrast to the other studies, no pretreatment of the biomass and no chemicals were considered. This results to GHG emissions of 3.4 kg CO_2-eq/kg H_2. In our study the results laid around 10 kg CO_2-eq/kg H_2 (without the impact of cultivating the biomass).

GHG Emissions Figure 13.7 shows the GHG emissions. This graphic is quite revealing in several ways. Compared to the other biohydrogen production pathways, the electrolysis of wheat straw shows significantly higher values of GHG emissions (i.e. 13.9 kg CO_2-eq/kg H_2) due to the limitations of combustion efficiencies (i.e. electrical power efficiency of 33.5 %, overall efficiency 87.7 % (Erm21c Limited 2010) and the electrolyser (efficiency of 60 %) (Schmid 2011)). In addition, the cultivation of cereals depends on the use of fertilizers that influence the amount of GHG emissions emitted as a result of nitrous oxide emissions on the field. As can be seen from Fig. 13.7, the origin of GHG emissions during electrolysis is almost exclusively caused by the biomass source. Another biomass source, e.g. wood chips, might lower the result significantly as it can be cultivated without fertilizer.

The lowest emissions could be achieved by gasification of forest residues (i.e. 3.7 kg CO_2-eq/kg H_2). Significantly higher GHG emissions were caused by gasification of biomass sources from short rotation coppice (i.e. 5.3 kg CO_2-eq/kg H_2) due to the use of fertilizers during SRC cultivation and of agricultural machinery. The results obtained from the gasification processes indicate that a significant amount of GHG emissions is caused by consumables during biohydrogen production. The GHG emissions of steam methane reforming in comparison with electrolysis of wheat straw were 56 % lower based on substrate mix I and 66 % lower based on substrate mix II.

The most surprising aspect of the data for steam methane reforming is the comparably small amount of GHG emissions, even though the composition of the biomass as the substrate mix I contains predominantly maize. Generally,

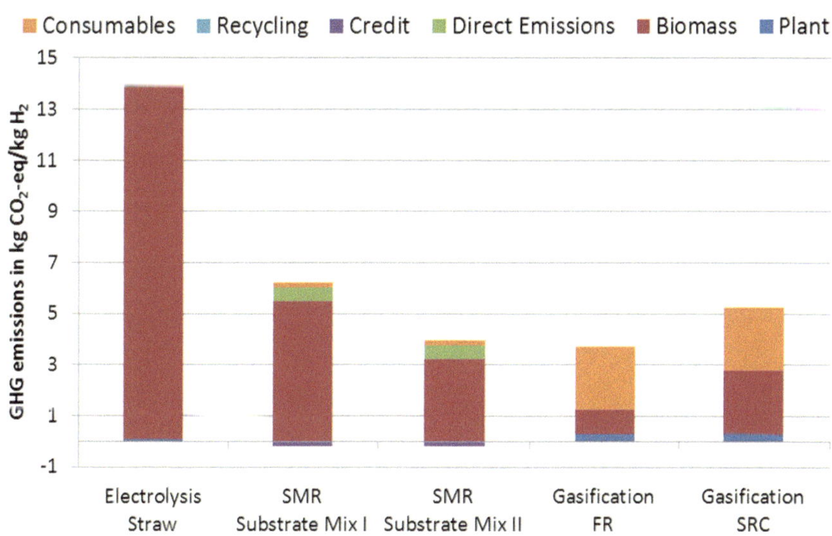

Fig. 13.7 Emissions of all greenhouse gases (GHG) for multiple biohydrogen production processes (*SMR* steam methane reforming, *FR* forest residues, *SRC* short rotation coppice)

maize causes high GHG emissions due to the use of fertilizers compared to biowaste per kilogramme substrate. However, the biogas yield from biowaste is significantly lower compared to maize (120 Nm^3/t_{FM} compared to 200 Nm^3/t_{FM}) (Kuratorium für Technik und Bauwesen in der Landwirtschaft e.V. 2014). The origin of GHG emissions during this biohydrogen production pathway is not limited to the biomass source but also to direct emissions because a small percentage of the methane emitted during the reforming process and because propane needed to be added (Sect. 13.3) to the biomethane before the feed-in into the natural gas grid. The fossil propane emits a significant amount of carbon dioxide during the reforming process.

GHG emissions are also analysed in several other studies. However, only few can be compared to these results directly. For example, regarding the gasification of woody biomass in Kalinci et al. (2012) as well as in Susmozas et al. (2013), the systems produce their own electricity from the syngas. This leads in most countries to much lower emissions than using electricity from the local grid. However, in practice this is not common because electricity from the grid is cheaper than producing the electricity internally. A comparison between the provisions of the biomass is still possible. Kalinci et al. (2012) used pinewood that emits between 0.32 and 0.37 kg CO_2-eq/kg H_2 which is around one third of the value given here for forest residues. Moreno and Dufour (2013) also investigated pinewood for gasification. They come to the conclusion that the provision of pinewood emits 0.72 kg CO_2-eq/kg H_2, which is much closer to the results obtained in this study. Koronoes et al. (2008) add a liquefaction of the hydrogen after the production which leads to much higher results (16.9 kg CO_2-eq/kg H_2) than in this study and the reader cannot deduct the value without liquefaction. They also analysed the gasification of biomass with an integrated process of electricity generation from the syngas and an electrolysis which is comparable to the electrolysis of electricity from burning straw discussed in this study. Also for this system, they used their internally produced electricity for all processes. As the provision of woody biomass results in much less GHG emissions compared to straw, it is comprehensible that this pathway emits very little (2.4 kg CO_2-eq/kg H_2) GHG emissions. The results from Zech et al. (2015) for the gasification of forest residues come very close to the results calculated here (4.1 kg CO_2-eq/kg H_2) as well as the results for the reforming of biogas from maize silage, manure and biowaste (5.2 kg CO_2-eq/kg H_2). This value lies between the results for the two discussed substrate mixes.

Emissions with an Acidifying Impact It is apparent from Fig. 13.8 that the amount of emis-

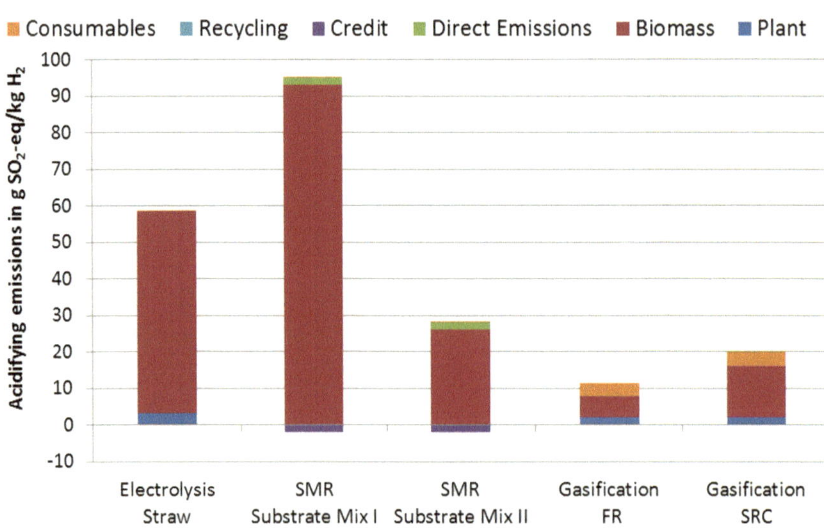

Fig. 13.8 Emissions with an acidifying impact for multiple biohydrogen production processes (*SMR* steam methane reforming, *FR* forest residues, *SRC* short rotation coppice)

sions with an acidifying effect of the different biohydrogen production pathways occurs in a broad range. The emissions vary between 11 and 93 g SO_2-eq/kg H_2. The mean score for acidifying emissions is the biomass source. The highest amount of acidifying emissions was produced by steam methane reforming due to the mix of biomass sources, and the lowest emissions occurred during gasification of wood chips derived from forest residues followed by the gasification of wood chips from short rotation coppice. The minor difference between the two gasification production pathways was caused by different machinery for harvesting of wood. The use of biomethane from substrate mix II caused considerably lower emissions (26 g SO_2-eq/kg H_2) compared to substrate mix I due to the smaller amount of direct emissions from mineral fertilizers and digestate releasing ammonia on the field. The correlation between biomass source and credit can be seen for both steam methane reforming routes. The credit was given to both process pathways but does not influence the overall results.

Moreno and Dufour (2013) are discussing acidifying emissions for the system described above. In their study, pinewood showed higher emissions (45 compared to 11 g SO_2-eq/kg H_2) than in this study. However, both studies agree that the provision of biomass causes the highest impact of the emissions. Susmozas et al. (2013) who used poplar from short rotation coppice came to the conclusion that the acidifying emissions are not as high as in this study (11 compared to 20 g SO_2-eq/kg H_2). This might be due to the usage of internally produced electricity. Koroneos et al. (2008) analysed even for their electrolysis system with internal electricity generation comparably high values (96 g SO_2-eq/kg H_2), which are difficult to comprehend.

Emissions with an Eutrophying Impact The emissions with an acidifying impact show slightly different trends as the emissions with an eutrophying impact (Fig. 13.9). The ratio of the respective values of the associated properties to each other is in the same range. The highest impact value in this category occurs during steam methane reforming of the substrate mix I (i.e. 27 g PO_4^{3-}-eq/kg H_2) due to the direct emissions of ammonia as a product of fertilizer decomposition mainly from the maize production and the impact of the electricity used for upgrading the biogas. The lowest impact was analysed for the gasification process route of forest residues (i.e. 12 g PO_4^{3-}-eq/kg H_2). The mean score for the two gasification process pathways are the consumables. This can be explained by the use of electricity, which is higher than for the other processes considered.

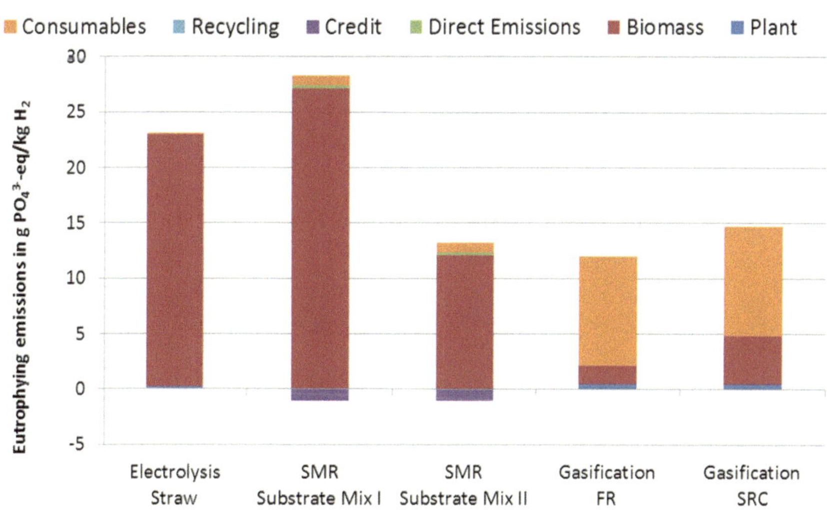

Fig. 13.9 Emissions with a eutrophying impact for multiple biohydrogen production processes (*SMR* steam methane reforming, *FR* forest residues, *SRC* short rotation coppice)

The results for eutrophying emissions from Moreno and Dufour (2013) as well as Susmozas et al. (2013) show the same trend as for the acidifying emissions. Moreno and Dufour (2013) calculated a higher impact (22 compared to 12 g PO_4^{3-}-eq/kg H_2), while Susmozas et al. (2013) calculated a lower impact (3 compared to 14 g PO_4^{3-}-eq/kg H_2). The values from Koroneos et al. (2008) for their electrolysis systems is with 17 g PO_4^{3-}-eq/kg H_2 between the others.

Emissions with a Human Toxicity Impact: Noncarcinogenic In Fig. 13.10 the reforming of biomethane from substrate mix I shows the highest results due to the emissions of carbon disulphide (a building block in organic chemistry) in the upstream chains of the biomass cultivation (e.g. fertilizer production, production of agricultural machinery). If credit for the export of heat from the reforming process (heat from biogas) would be taken into account, the reforming processes would reach much lower values of 1.9 10^{-10} and −1.1 10^{-10}, respectively. In the process described previously, chlorpyrifos (i.e. insecticide) (Edwards 2006) is the main cause for human toxicity causing 74 % of the damage within this category (credit). Thus, the gasification of SRC wood shows the highest human toxicity impact (non-cancer). Carbon disulphide accounts for 74 % of the overall emissions.

The variety of studies analysing USEtox is comparably small. For biohydrogen no other study was found. However, some aspects can still be discussed. Boulamanti et al. (2013) analysed different systems to produce electricity from maize and manure via fermentation and compared it with the average European electricity mix. They found out that the reference system has lower results than the biogas system. This is also true for the systems analysed here if you do not take the avoided heat production from biogas into account, see Sect. 13.4.2.

Emissions with Human Toxicity: Impact Carcinogenic The average scores of the human toxicity impact (carcinogenic) show that the comparison between the biohydrogen technologies applied significant differences. Electrolysis of wheat straw has still great impact on human toxicity (i.e. 2.1 10^{-10} CTU_h/kg H_2), whereas steam biomethane reforming (nonfood) gains values for human toxicity of 6.0 10^{-11} CTU_h/kg H_2 (Fig. 13.11). However, contrary to all expectations, the gasification of forest residues shows the largest impact on human toxicity due to the usage of forestry machinery such as power saws and thus emissions of dioxins, furans and formaldehydes – not as direct emissions in the forest but in the upstream chains of these processes. For all

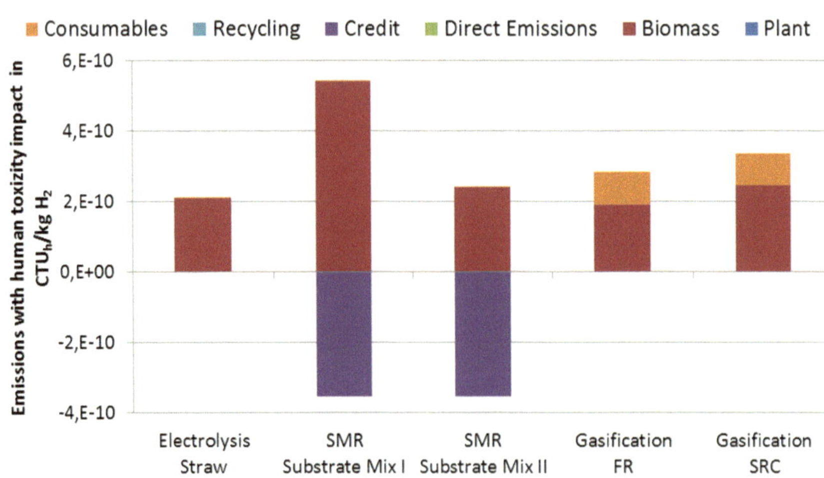

Fig. 13.10 Emissions with human toxicity impact (non-cancer) for multiple biohydrogen production processes (*SMR* steam methane reforming, *FR* forest residues, *SRC* short rotation coppice)

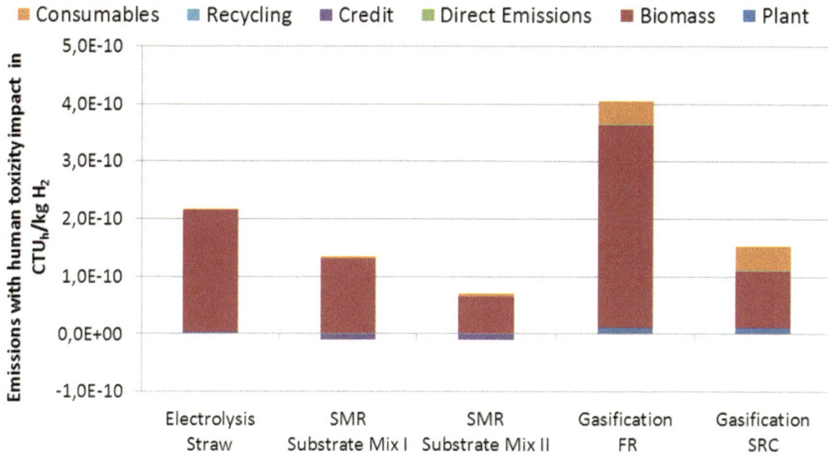

Fig. 13.11 Emissions with human toxicity impact (cancer) for multiple biohydrogen production processes (*SMR* steam methane reforming, *FR* forest residues, *SRC* short rotation coppice)

biohydrogen production pathways, the significant input factor was once again the biomass source.

For the two steam biomethane reforming process pathways, credit was given. For the gasification processes, the consumables were relevant for the quantity of toxic impacts assessed due to the high electricity demand.

Boulamanti et al. (2013) also identified formaldehydes as one of the main polluters for human toxicity.

13.4.2 Discussion

The output of the analysis considering the impact categories applied defines the environmental impacts of a variety of biohydrogen production pathways. For a comprehensive understanding, in Fig. 13.12 the biohydrogen production pathways are compared to a reference scenario. The reference scenario refers to SMR with natural gas as this process is the most common one for hydrogen mass production (Eni S.p.A. 2013).

The figure shows the GHG emissions arising from the conversion of straw during electrolysis are higher than the reference scenario SMR from natural gas. Nevertheless, all values from the other technologies displayed for GHG emissions are lower compared to SMR of natural gas.

The results for SMR of substrate mix II show extraordinarily small amounts in all impact categories (except for acidification), whereas the gasification requires further research to reduce the amount of consumables to produce less emissions.

13.5 Conclusion

The aim of this chapter is to highlight individual contribution of environmental impacts for providing biohydrogen with conventional and innovative production pathways from a variety of biomass sources. Thus, a life cycle assessment of different biohydrogen concepts was performed to examine environmental effects such as anthropogenic climate change, acidification of soil and aquatic ecosystems, eutrophication of terrestrial and aquatic ecosystems and human toxicity.

The investigation of biohydrogen production concepts has shown that the biomass source has significant influence on the environmental impact of biohydrogen production pathways. The most important findings are summarized below:

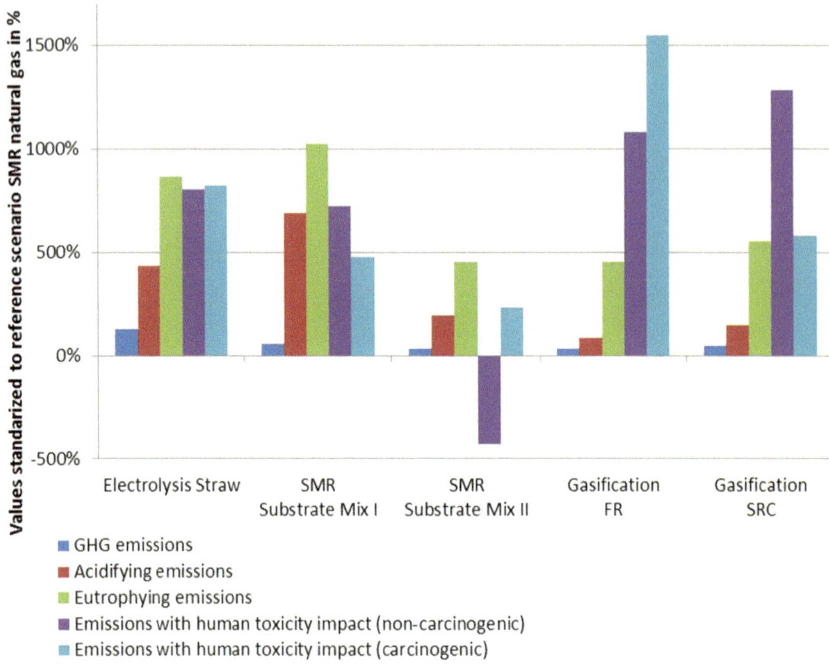

Fig. 13.12 Comparison of the biohydrogen production pathways to a reference scenario 'SMR natural gas' (*SMR* steam methane reforming, *FR* forest residues, *SRC* short rotation coppice)

1. Alkaline water electrolysis with electricity from biomass cogeneration plants fired with wheat straw shows the lowest potential for GHG reduction. The GHG emissions even increase compared to hydrogen production from natural gas as well as emissions with a toxic impact on humans. Probably a direct use of electricity and heat instead of electrolysis is more favourable because of the additional efficiency losses during the electrolysis.
2. For steam reforming of biomethane (biogas production by fermentation of substrate mix I), the main conclusion is the need for improvement of biomass cultivation (e.g. reduced use of fertilizer) as the biomass source accounts for most of the emissions related to the process.
3. The results of steam reforming of biomethane (biogas production by fermentation of substrate mix II) suggest that this process is currently most suitable for biohydrogen production in terms of the environmental impact as it can utilize waste products and gives credit to heat. However, the local availability of this nonfood biomass has to be taken into account.
4. The results for gasification of softwood using the fast internal circulating fluidized bed (FICFB) gasifier suggest that the amount of consumables need to be reduced significantly to improve the environmental performance.
5. Gasification (FICFB) of wood from short rotation coppice (SRC) seems to be a promising route if the emissions displayed in the category human toxicity can be decreased. These are mainly caused by the consumables and biomass cultivation.
6. While no reliable results for dark fermentation combined with photofermentation from substrate mix I are provided within the impact categories applied, more research is required to determine the efficiency of this technology to subsequently analyse the environmental impact.

Overall it can be concluded that the gasification and the reforming of biomass have the potential to be climate friendly. However, the cultivation of biomass has to be improved to reduce the effects of other environmental impacts.

References

Alves HJ, Bley Junior C, Niklevicz RR, Frigo EP, Frigo MS, Coimbra-Araújo CH (2013) Overview of hydrogen production technologies from biogas and the applications in fuel cells. Int J Hydrog Energy 38(13):5215–5225

Belau T (2012) Energiepflanzen: Daten für die Planung des Energiepflanzenanbaus, KTBL-Datensammlung mit Internetangebot, 2nd edn. Kuratorium für Technik und Bauwesen in der Landwirtschaft, Darmstadt

Boulamanti AK, Donida Maglio S, Giuntoli J, Agostini A (2013) Influence of different practices on biogas sustainability. Biomass Bioenergy 53:149–161

Boyano A, Blanco-Marigorta A, Morosuk T, Tsatsaronis G (2011) Exergoenvironmental analysis of a steam methane reforming process for hydrogen production. Energy 36(4):2202–2214

Boyd B (2010) Hydrogen Fuel Quality for FCVs: current status of efforts at CGA, Teleconference of the national hydrogen and fuel cells codes & standards coordinating committee

Cetinkaya E, Dincer I, Naterer GF (2012) Life cycle assessment of various hydrogen production methods. Int J Hydrog Energy 37(3):2071–2080

Das D, Veziroglu TN (2008) Advances in biological hydrogen production processes. Int J Hydrog Energy 2008(33):6046–6057

Demirbas A (2009) Biohydrogen: for future engine fuel demands. Springer, London

Destatis (2012) Gross electricity production in Germany from 2010 to 2012. available at: https://www.destatis.de/EN/FactsFigures/EconomicSectors/Energy/Production/Tables/GrossElectricityProduction.html. Accessed 12 June 2013

DIN EN ISO 14040, Environmental management – life cycle assessment – principles and framework, German and English version EN ISO 14040, vol. 2006, Beuth Verlag, Berlin

DIN EN ISO 14044, Environmental management – life cycle assessment – principles and framework. German and english version EN ISO 14044, vol. 2006, Beuth Verlag, Berlin

Djomo SN, Blumberga D (2011) Comparative life cycle assessment of three biohydrogen pathways. Bioresour Technol 102(3):2684–2694

Edwards D (2006) Memorandum: Finalization of Interim Reregistration Eligibility Decisions (IREDs) and Interim Tolerance Reassessment and Risk Management Decisions (TREDs) for the organophosphate pesticides, and completion of the tolerance reassessment and reregistration eligibility process for the organophosphate pesticides. US Environmental Protection Agency (EPA), Washington, DC

Eni S.p.A. (2013) Hydrogen production by SCT-CPO technology, available at: http://www.eni.com/en_IT/innovation-technology/technological-focus/produzione-idrogeno/produzione-idrogeno.shtml. Accessed 14 Oct 2014

Erm21c Limited (2010) Straw fired biomass CHP plant: preplanning information, Skibbereen.

European Commission (2012) Proposal for a Directive of the European Parliament and of the Council amending Directive 98/70/EC relating to the quality of petrol and diesel fuels and amending Directive 2009/28/EC on the promotion of the use of energy from renewable sources

Ferreira AF, Marques AC, Batista AP, Marques PASS, Gouveia L, Silva C (2012) Biological hydrogen production by *Anabaena* sp. – yield, energy and CO_2 analysis including fermentative biomass recovery. Int J Hydrog Energy 37(1):179–190

Ferreira AF, Ribeiro LA, Batista AP, Marques PASS, Nobre BP, Palavra AMF, da Silva PP, Gouveia L, Silva C (2013) A biorefinery from *Nannochloropsis* sp. microalga – energy and CO2 emission and economic analyses. Bioresour Technol 138:235–244

Finnveden G, Hauschild MZ, Ekvall T, Guinée J, Heijungs R, Hellweg S, Koehler A, Pennington D, Suh S (2009) Recent developments in life cycle assessment. J Environ Manage 91(1):1–21

Gandía LM, Arzamedi G, Dieguez PM (eds) (2013) Renewable hydrogen technologies: production, purification, storage, applications and safety. Elsevier Science, Burlington

Gellert S (2013) Thermochemische Herstellung von Wasserstoff aus Biomasse unter besonderer Berücksichtigung der Rohgasreformierung. Dissertation, Energietechnik, Technische Universität Hamburg-Harburg, Hamburg, 2013

Guinée J, Gorrée M, Heijungs R, Huppes G, Kleijn R, de Koning A, van Oers L, Wegener Sleeswijk A, Suh S, de Udo Haes H, de Bruijn H, van Duin R, Huijbregts M (2001) LCA – an operational guide to the ISO-standards: part 2b: operational annex. Universiteit Leiden, Leiden

Gupta RB (ed) (2009) Hydrogen fuel: production, transport, and storage. CRC Press/Taylor & Francis Group, Boca Raton

Hauschild MZ, Goedkoop M, Guinée JB, Heijungs R, Huijbregts MAJ, Jolliet O, Margni M, Schryver A, Humbert S, Laurent A, Sala S, Pant R (2012) Identifying best existing practice for characterization modeling in life cycle impact assessment. Int J Life Cycle Assess 18(3):683–697

Hofbauer H, Rauch R, Loeffler G, Kaiser S, Fercher E, Tremmel H (2002) Six years experience with the FICFB-gasification process. In: Palz W (ed) Biomass for energy, industry and climate protection: Twelfth European Biomass Conference; proceedings of the international conference held in Amsterdam, the Netherlands, 17–21 June 2002. ETA; WIP, Florence/Munich

IPCC (ed) (2007) Climate change 2007: the physical science basis. Contribution of working group I to the fourth assessment report of the Intergovernmental Panel on Climate Change. Cambridge University Press, New York

Joint Research Centre – Institute for Environment and Sustainability (2011) International reference life cycle data system (ILCD) handbook: general guide for life cycle assessment: provisions and action steps, Recommendations for Life Cycle Impact Assessment in the European context, EUR, Scientific and technical research series, vol 24571, First edition, Publications Office, Luxembourg

Kabir MR, Kumar A (2011) Development of net energy ratio and emission factor for biohydrogen production pathways. Bioresour Technol 102(19):8972–8985

Kalinci Y, Hepbasli A, Dincer I (2012) Life cycle assessment of hydrogen production from biomass gasification systems. Int J Hydrog Energy 37(19):14026–14039

Kaltschmitt M (ed) (2009) Energie aus Biomasse: Grundlagen, Techniken und Verfahren, 2nd edn. Springer, Heidelberg

Kaltschmitt M, Weinberg J, Stegelmeier M, Bohnenschäfer W, Gansler J, Ebert M (2012) Teilbericht B. Analyse der Erneuerbaren Energien am Wärmemarkt. In: Schultz R, Hochi J, Personn H (eds) Ökologische und ökonomische Optimierung des Wärmemarktes: Unter besonderer Berücksichtigung des Endenergiebedarfs und von Biogas/Bioerdgas, Schriftenreihe des Biogasrat e.V. Ibidem-Verlag, Stuttgart, pp 109–176

Koroneos C, Dompros A, Roumbas G (2008) Hydrogen production via biomass gasification—a life cycle assessment approach. Chem Eng Process 47(8):1261–1268

Kuratorium für Technik und Bauwesen in der Landwirtschaft e.V (KTBL) (2014) Wirtschaftlichkeitsrechner Biogas. Available at: http://daten.ktbl.de/biogas/showSubstrate.do?zustandReq=3#anwendung. Accessed 22 Oct 2014

Lauermann G, Häussinger P, Lohmüller R, Watson AM (2000) Hydrogen, 1. Properties and occurrence. In: Ullmann's encyclopedia of industrial chemistry. Wiley-VCH, Weinheim

Lee H-S, Vermaas WFJ, Rittmann BE (2010) Biological hydrogen production: prospects and challenges. Trends Biotechnol 28(5):262–271

Manish S, Banerjee R (2008) Comparison of biohydrogen production processes. Int J Hydrog Energy 33(1):279–286

Markowski M, Urbaniec K, Budek A, Trafczyński M, Wukovits W, Ljunggren M, Zacchi G (2010) Estimation of energy demand of fermentation-based hydrogen production. J Clean Prod 18:S81

Martinez-Merino V, Gil MJ, Cornejo A (2013) Biomass sources for hydrogen production. In: Gandía LM, Arzamedi G, Dieguez PM (eds) Renewable hydrogen technologies: production, purification, storage, applications and safety. Elsevier Science, Burlington, pp 87–110

Millet P, Grigoriev S (2013) Water electrolysis technologies. In: Gandía LM, Arzamedi G, Dieguez PM (eds) Renewable hydrogen technologies: production, purification, storage, applications and safety. Elsevier Science, Burlington, pp 19–41

Moreno J, Dufour J (2013) Life cycle assessment of hydrogen production from biomass gasification. Evaluation of different Spanish feedstocks. Int J Hydrog Energy 38(18):7616–7622

Müller-Langer F (2011) Analyse und Bewertung ausgewählter zukünftiger Biokraftstoffoptionen auf der Basis fester Biomasse. Dissertation, Umwelttechnik und Energiewirtschaft, Technische Universität Hamburg-Harburg, Hamburg, 2011

Ochs D, Wukovits W, Ahrer W (2010) Life cycle inventory analysis of biological hydrogen production by thermophilic and photo fermentation of potato steam peels (PSP). J Clean Prod 18:88–94

Pehnt M (2002) Ganzheitliche Bilanzierung von Brennstoffzellen in der Energie- und Verkehrstechnik, Forschungsbericht. VDI-Verlag, Düsseldorf

Rechtenbach D (2009) Fermentative Erzeugung von Biowasserstoff aus biogenen Roh- und Reststoffen, Dissertation, Umwelttechnik und Energiewirtschaft, Technische Universität Hamburg-Harburg, Stuttgart, 2009

Roddy J (2012) Biomass and biofuel production. In: Sayigh AAM (ed) Comprehensive renewable energy, vol 5. Elsevier Science, Amsterdam

Rosenbaum RK, Bachmann TM, Gold LS, Huijbregts M, Jolliet O, Juraske R, Koehler A, Larsen HF, MacLeod M, Margni M, McKone TE, Payet J, Schuhmacher M, Meent D, Hauschild MZ (2008) USEtox—the UNEP-SETAC toxicity model: recommended characterisation factors for human toxicity and freshwater ecotoxicity in life cycle impact assessment. Int J Life Cycle Assess 13(7):532–546

Rostrup-Nielsen T (2005) Manufacture of hydrogen. Catal Today 106(1–4):293–296

Scheftelowitz M, Schicketanz S, Reinicke F, Beil M (2013) Stromerzeugung aus Biomasse: Zwischenbericht

Schmid R (2011) Smart Grids meets green Transportation, 19.-22.6., Thessaloniki

Seifert M (2010) Methodische Vorgehensweise zur Standortidentifikation und Planung der Biomassebereitstellung für Konversionsanlagen am Beispiel von Bio-SNG-Produktionsanlagen, Dissertation, Umwelttechnik und Energiewirtschaft, Technische Universität Hamburg-Harburg, Hamburg, 2010

Skøtt T (2011) Straw to energy: status, technologies and innovation in Denmark 2011. Innovation Network for Biomass (INBIOM), Tjele

Spath PL, Mann MK (2001) Life cycle assessment of hydrogen production via natural gas steam reforming. National Renewable Energy Laboratory NREL, Golden

Susmozas A, Iribarren D, Dufour J (2013) Life-cycle performance of indirect biomass gasification as a green alternative to steam methane reforming for hydrogen production. Int J Hydrog Energy 38(24):9961–9972

Swiss Centre for Life Cycle Inventories (2013) ecoinvent Version 3

The European Parliament and the Council of the European Union (2009a) Directive 2009/28/EG on the promotion of the use of energy from renewable sources and amending and subsequently repealing Directives 2001/77/EC and 2003/30/EC, vol 2009

The European Parliament and the Council of the European Union (2009b) Directive 2009/30/EG amending Directive 98/70/EC as regards the specification of petrol, diesel and gas–oil and introducing a mechanism to monitor and reduce greenhouse gas emissions and amending Council Directive 1999/32/EC as regards the specification of fuel used by inland waterway vessels and repealing Directive 93/12/EEC

United States Department of Energy (2013) Report of the hydrogen production expert panel: a subcommittee of the hydrogen & fuel cell technical advisory committee. U.S. Department of Energy (DOE), Washington, DC

Wulf C, Kaltschmitt M (2013) Life cycle assessment of biohydrogen production as a transportation fuel in Germany. Bioresour Technol 150:466–475

Zech K, Grasman R, Oehmichen K, Kiendl I, Schmersahl R, Rönsch S, Weindorf W, Funke S, Michaelis J, Wietschel M, Seifert M, Müller-Langer F (2013) Hy-NOW: Evaluierung der Verfahren und Technologien für die Bereitstellung von Wasserstoff auf Basis von Biomasse

Zech K, Oehmichen K, Grasemann E, Michaelis J, Funke S, Seiffert M (2015) Technical, economic and environmental assessment of technologies for the production of biohydrogen and its distribution. Int J Hydrog Energy 40(15):5487–5495

Biohydrogen: Global Trend and Future Perspective

14

Ratan Singh, Anoop Singh, and Dheeraj Rathore

Abstract

Global utilisation of confined and traded hydrogen is projected to increase more than 300 billion cubic meters through 2018 with an annual growth rate of 3.5 %. The maximum share of growth through 2018 is likely to occur in China though the world's largest hydrogen consumption will continue with the USA. This view is supported by the published research in the recent past. Published articles on biohydrogen research are highest from China followed by the USA and India. Published data on biohydrogen also suggested that biohydrogen production from the Asian countries is mainly focusing on dark fermentation, whilst the European countries are focusing on dark and photofermentation. So far, the current biohydrogen production system is appropriate for decentralised small-scale systems, integrated with waste from agriculture and industries or from waste-processing facilities, using reactors operating with mixed microflora (aerobic, anaerobic, thermophilic, purple non-sulphur photosynthetic bacteria) or pure cultures enriched from natural sources. Seed inocula for biohydrogen production have been obtained from heat sludge, compost, waste water, food waste, etc. With many social, economic and environmental benefits, hydrogen energy is considered as a future of the sustainable energy source.

R. Singh • D. Rathore (✉)
School of Environment and Sustainable Development, Central University of Gujarat, Sector-30, Gandhinagar, Gujarat 382030, India
e-mail: dheeraj.rathore@cug.ac.in

A. Singh
Department of Scientific and Industrial Research (DSIR), Ministry of Science and Technology, Government of India, Technology Bhawan, New Delhi 110016, India
e-mail: apsinghenv@gmail.com

14.1 Introduction

Hydrogen gas is extensively used for various purposes in different industries, e.g. production of chemicals, desulphurisation and reformulation of gasoline in refineries, processing steel, production of electronic devices and food industry for hydrogenation agent of fats and oils. Besides this, hydrogen could be an ideal renewable energy

A. Singh, D. Rathore (eds.), *Biohydrogen Production: Sustainability of Current Technology and Future Perspective*, DOI 10.1007/978-81-322-3577-4_14

resource and an environmentally safe alternative of fossil fuel as combustion of H_2 does not contribute any harmful emissions like other fuels. There has been substantial progress towards the use of hydrogen as an energy source. With about 10 % annual growth rate, global hydrogen trade is around 50 million tonnes per year (Winter 2005). The National Hydrogen Program of the USA estimated around 10 % share of total energy will be from hydrogen by 2025 (Armor 1999). With reformed economic structure and strong support to progress in production technologies, hydrogen's share can reach 22.1 % in 2030 and become the dominating energy source from then onwards in Taiwan (Lee and Lee 2008). Executive summary of the National Hydrogen Energy Roadmap (2002) concluded that 'Widespread use of hydrogen will affect every aspect of the US energy system, from production through end-use. The individual segments of a hydrogen energy system production, delivery, storage, conversion, and end use applications are closely interrelated and interdependent'. Though most of the hydrogen supply in the world is by non-biological processes, biological hydrogen production is still in its infancy stage with a developing attitude. Biohydrogen has attracted global attention because of its potential to become an inexhaustible, low-cost and renewable source of clean energy (Show et al. 2012). The global hydrogen generation market value is expected to reach $138.2 billion by 2019, growing at a CAGR of 5.9 %, from 2014 to 2019 (http://www.marketsandmarkets.com).

14.2 Current Biohydrogen Research

The recent SCOPUS database survey on biohydrogen publication showed 7834 publications since 1984. The researchers get boosted to explore the biohydrogen production in 2003, and tremendous work has been reported after that (Fig. 14.1). Presently, 146 journals are publishing papers on biohydrogen-related research. The top five journals publishing on biohydrogen research include *International Journal of Hydrogen Energy*, *Bioresource Technology*, *Renewable and Sustainable Energy Reviews*, *Applied Microbiology and Biotechnology* and *Advanced Materials Research*. The researchers from China have published the maximum number of papers amongst the 100 countries working on biohydrogen. The USA is at the second place in publishing research on biohydrogen and India at the third place (Fig. 14.2).

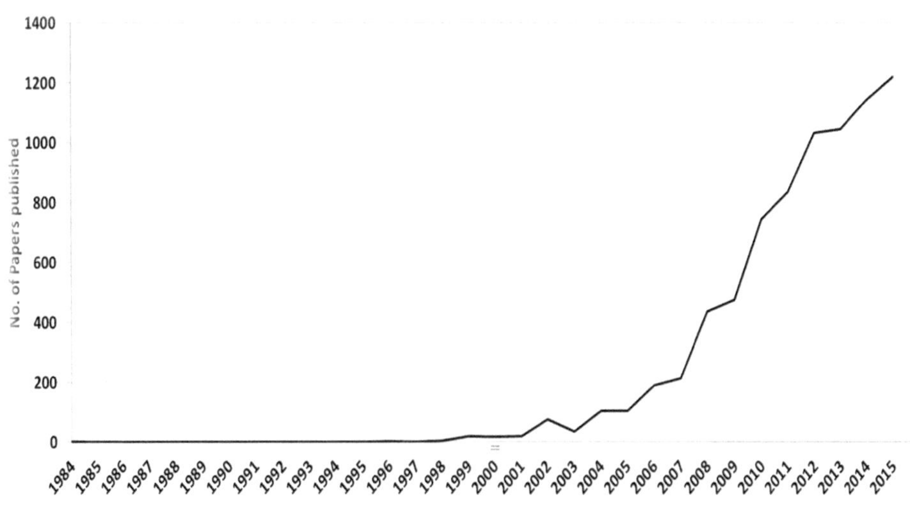

Fig. 14.1 Number of papers published on biohydrogen (Source: SCOPUS)

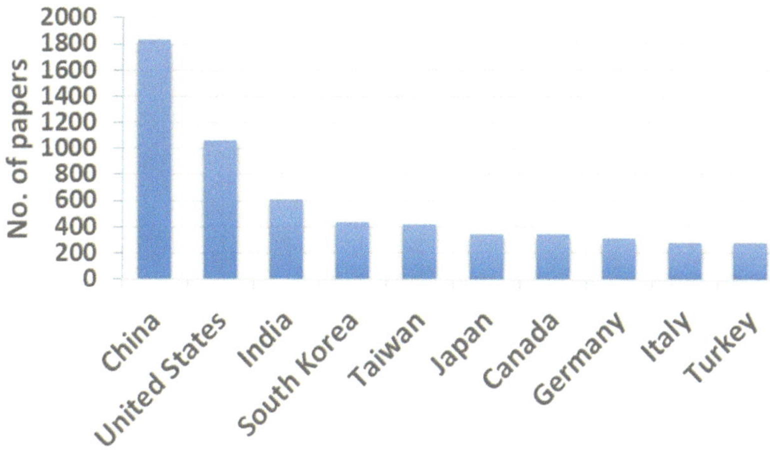

Fig. 14.2 Top ten countries publishing research on biohydrogen (Source: SCOPUS)

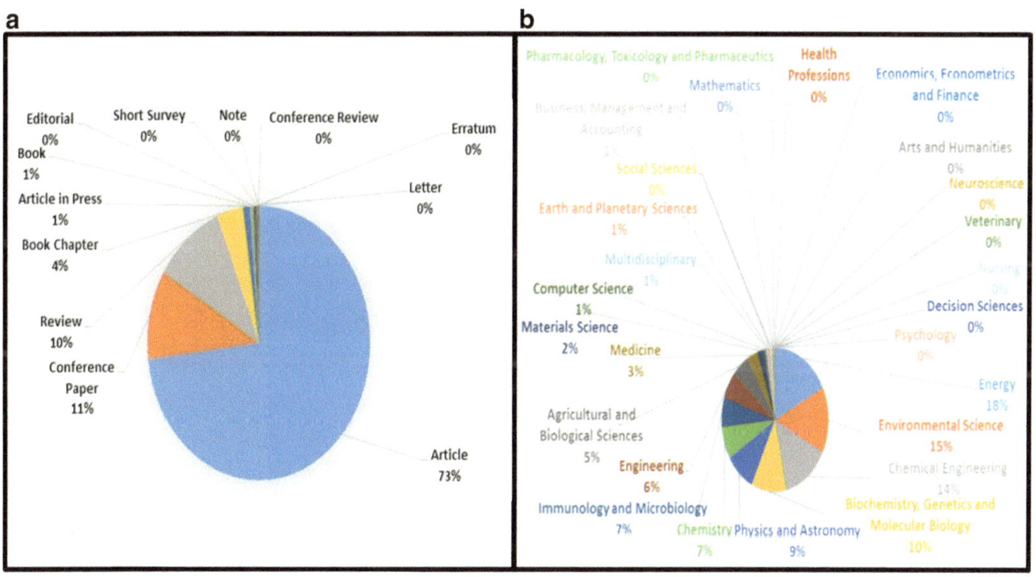

Fig. 14.3 Percentage-wise publication type (**a**) and publication in subject area (**b**) (Source: SCOPUS)

The survey also shows that maximum publications on biohydrogen are covered as articles (about 73 %) (Fig. 14.3a) and more than 50 % publications covered under energy, environmental sciences, chemical engineering and biochemistry, genetics and molecular biology subject area (Fig. 14.3b). This clearly shows that biohydrogen research is focused to explore biohydrogen as an alternative energy source.

14.3 Asia

The biohydrogen production studies in Asian countries mainly focused on utilisation of anaerobic fermentation technique (Table 14.1). Plant, microalgal and cyanobacterial biomass is traditionally being in use for biogas production. Microalgae and cyanobacteria have the advantage that they grow in a liquid medium including

Table 14.1 Biohydrogen production in Asian countries

Country	Substrate	Technology	Selected seed inoculum	Hydrogen production	References
India	Rice straw	Dark fermentation	Anaerobic mixed culture -	771 mL/L	Sen et al. (2016)
Thailand	Waste water from starch industry	Dark fermentation	Heat-treated mixed microorganism	250 ml (STP)/g COD degraded	Sinbuathong et al. (2015)
Thailand	Mixed carbon (glucose G, xylose X and acetic acid A)	Anaerobic fermentation	*Rhodobacter sphaeroides* S10	7.26 m H_2/g	Pattanamanee et al. (2015)
India	Waste water	Anaerobic fermentation	–	1.105 mmolH_2/m3/min	Parihar and Upadhyay (2015)
Malaysia	DL-7-Azatryptophan	Photofermentation	Cyanobacteria (*Anabaena variabilis* ATCC 29413)	97 mL H_2 mg chl a^{-1}	Salleh et al. (2015)
India	Food waste	Identical bench-scale anaerobic reactors	Consortium of anaerobic bacteria	30 % (6.3 g/L)	Dahiya et al. (2015)
Korea	Mixed liquor volatile suspended solid (MLVSS) and synthetic waste water	Anaerobic sequencing batch reactor	–	2.2 mol H_2/mol sucrose	Won and Lau (2015)
China	Sugarcane bagasse	Thermophilic, anaerobic fermentation	*Thermoanaerobacterium aotearoense* SCUT27/Δldh	1.86 mol H_2/mol total sugar	Lai et al. (2014)
India	Designed synthetic waste water	Single-chamber membrane-less MEC (microbial electrolysis cell)	Anaerobic mixed consortium	0.6 V	Babu et al. (2013)
Taiwan	Starch containing waste water from textile factory	Bioreactor	Soil, cow dung and sewage sludge (seed inoculation for bacterial culture)	0.97 mol H_2/mol hexose	Lay et al. (2012)
Malaysia	Effluent of tofu (bean curd), brewery industry, palm and olive oil mill	Dark fermentation	Photosynthetic purple non-sulphur bacteria	–	Wu et al. (2012)
Thailand	Waste water from cassava starch processing	Thermophilic fermentation	Thermophilic mixed culture (PK, SW, PR)	Respectively, 287, 264 and 232 mL H_2/g starch	Sompong et al. (2011)
Thailand	Sago starch in waste water	Thermophilic fermentation	Thermophilic mixed culture	422 mL-H_2/g-starch	Hasyim et al. (2011)

Thailand	Effluent from palm oil industry	Sequencing batch reactor (anaerobic)	*Thermoanaerobacterium thermosaccharolyticum* PSU-2 (concentration of 0, 10, 20 and 30 %)	44.6, 42.1, 39.5 and 41.3 %	Seengenyoung et al. (2011)
China	Starch-containing modified media	Proton exchange membrane fuel cell (PEMFC) (anaerobic fermentation)	Municipal waste water and sludge (seed inoculation for bacterial culture)	186 mL/g-starch	Wei et al. (2010)
Taiwan	Molasses	Anaerobic continuous flow hydrogen fermentor	*Clostridium sporosphaeroides* F52, *Clostridium tyrobutyricum* F4 and *Clostridium pasteurianum* F40	12–220 %	Hsiao et al. (2009)
India	Sugarcane distillery effluent	Anaerobic fermentation reaction	*Citrobacter freundii* 01, *Enterobacter aerogenes* E10 and *Rhodopseudomonas palustris* P2	2.76 mol mol[1] glucose	Vatsala et al. (2008)
China	Cattle waste water	Anaerobic fermentation	Natural mixed microfloral consortia	319 H_2/g	Tang et al. (2008)
China	Cornstalk waste	Anaerobic fermentation	Cow dung (seed inoculation for mixed anaerobic culture)	149.69 ml H_2g^{-1} TVS	Zhang et al. (2007)
Malaysia	Effluent from palm oil mill	Anaerobic fermentation reaction	Anaerobic microorganisms	4708 ml H_2/(l POME)	Atif et al. (2005)
Thailand	Cassava waste water	Dark anaerobic fermentation	Anaerobic seed sludge and *Rhodospirillum rubrum*	429 mL H_2 g^{-1} VSS	Reungsang et al. (2004)

a third dimension that is lacking in the plant, where the available space for cultivation is two-dimensional (Koutinas et al. 2016; Costa and Morais 2011). In addition, the microalgal biomass does not require pretreatment for removal of complex material (lignocelluloses) (Kumar and Singh 2016), whilst plant biomass requires pretreatment due to the presence of lignocellulose. Recovery of CO_2, produced from biogas, is another additional benefit obtained from anaerobic digestion of microalgae (Akil and Jayanthi 2014).

Akil and Jayanthi (2014) studied the potential of distillery waste water for biohydrogen production using dark fermentation in an anaerobic sequencing batch reactor. Experimental data showed the feasibility of biogas production along with substrate degradation with distillery waste water used as substrate.

Mullai et al. (2013a) conducted a batch experiment on the effect of initial pH and glucose concentration with nickel nanoparticle concentration on biohydrogen production at 30–35 °C temperature using anaerobic microflora. Result of the experiment showed higher biohydrogen yield to their corresponding control after using nickel nanoparticles with an average size of 13.64 nm. Biohydrogen yield was maximum (2.54 mol of hydrogen mol^{-1} of glucose) at optimum conditions. The optimisation of biohydrogen production was carried out by employing response surface methodology (RSM) with a central composite design (CCD). Nickel nanoparticle concentration was significant in optimisation of biohydrogen production, and it enhanced the biohydrogen production by 22.71 %.

Production of both, biodiesel and biohydrogen, from biomass could be considered as an economically feasible option for sustainable energy production. Nag Dasgupta et al. (2015) conducted an experiment with freshwater microalga *Scenedesmus* sp. NBRI012 and *Chlorella* sp. NBRI029 and reported that *Scenedesmus* sp. NBRI012 and *Chlorella* sp. NBRI029 exhibited high biomass yield (1.31 ± 0.11 and 2.62 ± 0.13 g L^{-1}, respectively) and lipid content (244.44 ± 12.3 and 587.38 ± 20.2 mg L^{-1}, respectively) in the presence of an organic carbon source (in this experiment acetate was used as carbon source). *Scenedesmus* sp. NBRI012 has shown the higher H_2 (maximum evolution of 17.72 % v/v) production. Deprivation of sulphur during the H_2 production was found to increase the lipid content (410.03 ± 18.5 mg L^{-1}) of the residual biomass. Interestingly, fatty acid profiling of the lipid extracted from *Scenedesmus* sp. NBRI012 has showed abundance of fatty acids with a carbon chain length of C16 and C18. Saponification value, iodine value and cetane number of biodiesel were found suitable to the range of Indian Standards (IS 15607), Brazilian National Petroleum Agency (ANP255) and the European biodiesel standard EN14214.

There has been a significant research on generating biohydrogen from waste water and residual biomass (Nath and Das 2004; Mohan 2015; Singh et al. 2015). Mullai et al. (2013b) demonstrated that sediments of mangrove hosted rich assemblages of microorganisms, predominantly of mixed bacterial cultures, which can be used efficiently for biohydrogen production in anaerobic dark fermentation. In a similar research finding, Mullai et al. (2013b) also reported that the addition of trace metal to the medium (100 mg L^{-1} $FeSO_4 \cdot 7H_2O$) enhanced the biohydrogen yield from 2.3 mol to 2.6 mol H_2 per mol of glucose. Hawkes et al. (2007) suggested that continuous dark fermentative hydrogen production technology may be suitable for commercial development of biohydrogen by using mixed microflora at mesophilic temperatures. Higher substrate concentrations could be more energy efficient, although there are product inhibition limitations (Dwidar et al. 2012). Inhibition by H_2 can be reduced by continuous extraction through membranes or by sparging and stirring. Investigated reactor types suggested that granules have the best performance with soluble substrate, for particulate feedstock biofilm reactors or continuous stirred-tank reactors (CSTRs) may be most successful. Hydrogen production is reported to proceed without methane production by maintaining reactor pH in the range between 4.5 and 6.7 (Hawkes et al. 2007).

Anaerobic digestion is a currently available technology, and the two-phase anaerobic treatment system is reported to give greater conversion efficiency than anaerobic digestion alone (Hawkes et al. 2007; Brown et al. 2008).

14.4 Europe

In European countries, researchers have used variable organic substrate available for the production of biohydrogen mainly using dark fermentation technique (Table 14.2). Continuous H_2 production was investigated over a period of nearly 200 days in a thermophilic semi-continuous dark fermentation process with no pH control. The highest H_2 yield of 121.45±44.55 N L H_2/kg VS was obtained at an organic loading rate (OLR) of 2.5 kg VS/m^3/day and a hydraulic retention time (HRT) of 4 days (Ghimire et al. 2015). The dark fermentation effluents mainly contained volatile fatty acids (VFAs) and alcohols as metabolites and unhydrolysed solid residues. The supernatant, after separation, was used to recover H_2 in a photofermentation using *Rhodobacter sphaeroides*. The solid residual fraction along with photofermentation effluent was converted into methane by anaerobic digestion. By combining dark fermentation and photofermentation, the H_2 yield from the food waste increased 1.75-fold. Moreover, by adding anaerobic digestion as a posttreatment of the dark fermentation residue, the total energy yield was substantially increased to reach 5.55 MJ/kg VS food waste added versus 3.55 MJ/kg VS food waste (Ghimire et al. 2015).

A sequential combination of biohydrogen and biomethanation production phases has the potential for even higher bioenergy recovery from organic waste water (Arimi et al. 2015). An economic evaluation of an integrated technology for industrial-scale new-generation biofuel production using whey, vinasse and lignocellulosic biomass as raw materials is reported by Koutinas and co-worker (2016). Anaerobic packed-bed bioreactors were used for organic acid production using initially synthetic media and then wastes. Butyric, lactic and acetic acids were predominately produced from vinasse, whey and cellulose, respectively. Mass balance was calculated for a 16,000 L daily production capacity. The investment needed for the installation of the factory was estimated to be about 1.7 million with depreciation expected at about 3 months. For cellulosics, the installation investment was estimated to be about sevenfold higher with depreciation at about 1.5 years (Koutinas et al. 2016).

Escherichia coli can perform at least two modes of anaerobic hydrogen metabolism, viz. respiratory hydrogen oxidation and fermentative hydrogen production, and expresses at least two types of hydrogenase activity. Respiratory hydrogen oxidation is catalysed by two 'uptake' hydrogenase isoenzymes, hydrogenase 1 and hydrogenase 2, and fermentative hydrogen production is catalysed by hydrogenase 3. Harnessing and enhancing the metabolic capability of *Escherichia coli* to perform anaerobic mixed acid fermentation is therefore an attractive approach for biohydrogen production from sugars (Redwood 2008). Orozco et al. (2010) evaluated two strains of *Escherichia coli* (strain MC4100 and a mutant strain IC007 derived from *E. coli* strain MC4100) for their hydrogen production potential using fermentation. In another experiment, Penfold et al. (2003) found that strain HD701 of *E. coli* can evolve more hydrogen than its parental wild type, strain MC4100. In this experiment, researchers changed the conditions systematically to elucidate the best conditions for optimum H_2 production by cell suspensions of this strain. They also reported that increasing the temperature from room temperature (18–20 °C) to 30 °C increased the rate of H_2 production. Increasing the cell concentration in resting cell suspensions gave in a higher rate of H_2 evolution. A 50% (v/v) inoculum was optimal, this producing 102.2±5.1 ml/L of H_2 as compared to 22.0±3.2 ml/L at a 10% inoculum. Stirring the medium doubled the rate of H_2 evolution. Hydrogen evolution was increased to 9 l using pH control at 5.5. This equates to a molar yield of 1 mol H_2 mol^{-1} glucose (Penfold et al. 2003).

Dried, sintered Pd biomaterials (Bio-Pd) were tested by Redwood and Macaskie (2006), as anodes in a proton exchange membrane (PEM) fuel cell for their ability to generate electricity

Table 14.2 Current status for biohydrogen production in European countries

Country	Substrate	Technology	Selected microbial utilisation	Hydrogen production	References
Ireland	Glucose and mannitol (volatile solid)	Co-fermentation	Microalgae (*Laminaria digitata*)	85.0 mL/g VS	Xia et al. (2016)
UK	Food waste or wheat feed	Anaerobic digestion	–	84.2 L H_2 kg^{-1}	Patterson et al. (2015)
Italy	Food waste	Dark fermentation (DF), photofermentation (PF) and anaerobic digestion (AD)	*Rhodobacter sphaeroides*	1.75-fold	Ghimire et al. (2015)
France	Buffalo slurry co-fermented with cheese whey and crude glycerol	Mixture design	Microbial community (F210)	117 mL H_2/g VS	Marone et al. (2015)
Sweden	$BG11_0$ medium	Photofermentation	Δ *hupW* strain of *Nostoc* PCC 7120 (filamentous heterocystous cyanobacterium)	4.85 mL H_2 L^{-1} h^{-1}	Nyberg et al. (2015)
Turkey	Acetate VSS (volatile suspended solid)	Photofermentation processes	*Rhodobacter capsulatus* DSM 1710	1.04 mmol/$L_{reactor}$·h	Akman et al. (2015)
Portugal	Sugars and small amounts of furfural and HMF (hydroxymethylfurfural)	Sequential batch fermentation	*Spirogyra* biomass and the subsequent fermentation by *Clostridium butyricum* DSM 10702	2.59 mol/mol	Ortigueira et al. (2015)
Turkey	Effluent of sugar beet thick juice	Batch and continuous photobioreactors	Mutant strains of *Rhodobacter capsulatus*	34 % (0.49 mmol/(Lch)	Uyar et al. (2015)
France	NH_4^+ and glucose	Electrochemical and biological processes	Anaerobic bacteria (sludge from waste water treatment plant)	0.35 mol H_2 mol^{-1} glucose	Abdallah et al. (2015)
USA	Duckweed (fermentation feedstock)	Anaerobic fermentation	–	75.3 mL H_2 per g dry duckweed in 7 days	Xu and Deshusses (2015)
France	35 g/L NaCl	Microbial electrolysis cell (MEC)	*Proteobacteria*, *Bacteroidetes*, *Firmicutes* and *Spirochaetes* and *Actinobacteria*	201.1 ± 7.5 $L_{H2}/m^2_{Cathode}$·d	Carmona-Martinez et al. (2015)
Belgium	Fermentative media (LM-H, FHM, marine LB)	Hydrogen-driven remediation strategies	*Pseudoalteromonas* sp. BH11	2.01 ± 0.05 mmol L^{-1} day^{-1} g^{-1}	Hosseinkhani et al. (2014)

Turkey	Modified media	Photofermentation	*Rhodobacter capsulatus wild type* (DSM 1710)	0.326 mol H_2/mol substrate	Androga et al. (2014)
Belgium	Glucose	Anaerobic sequencing batch reactor	*C. butyricum* CWBI 1009	278 ml h^{-1}	Beckers et al. (2015)
Greece	Cottonseed cake	Dark fermentative hydrogen production	Thermophilic bacterium *Caldicellulosiruptor saccharolyticus*	84–112 %	Panagiotopoulos et al. (2013)
Sweden	Wheat straw	Anaerobic digestion	*Caldicellulosiruptor saccharolyticus*	5.2 L H_2/L/Day	Pwar et al. (2013)
Turkey	Thick juice dark fermentor effluent	Photofermentation (solar tubular photobioreactor)	*Rhodobacter capsulatus* (PNS)	0.4 mol H_2/mol acetic acid	Boran et al. (2012b)
Finland	Crude glycerol	Bioconversion	*Clostridium* species	1.1 ± 0.1 mol-H_2/mol-glycerol$_{consumed}$	Mangayil et al. (2012)
Italy	Crude glycerol	Dark fermentation	Microbial mixed culture	0.96 mol H_2/mol	Varrone et al. (2012)
Turkey	Thermophilic dark fermentor effluent of sugar beet thick juice	Photofermentation (solar fed-batch panel photobioreactor)	*Rhodobacter capsulatus* YO3 (hup^-)	1.12 mmol H_2/L_c./h	Ozkan et al. (2012)
UK	Propane	Sorption-enhanced steam reforming (SESR)	–	9.1 mol hydrogen per mole of propane	Wang et al. (2011)
Turkey	Molasses dark fermentation effluents	Sequential dark and photofermentation	*Rhodobacter capsulatus wild type* (DSM 1710) and *Rhodobacter capsulatus* hup^- (YO3)	0.50 mmol H_2/L_c.H	Avcioglu et al. (2011)
Denmark	Wheat straw (hemicellulose rich)	Upflow anaerobic sludge bed (UASB) reactor	–	212.0 ± 24.1 mL-H_2/g-sugars	Kongjan and Angelidaki (2010)

(continued)

Table 14.2 (continued)

Country	Substrate	Technology	Selected microbial utilisation	Hydrogen production	References
Turkey	Sugar beet molasses	Dark fermentation and sequential dark and photofermentation	Extreme thermophile (*Caldicellulosiruptor saccharolyticus*) and photosynthetic bacteria (*Rhodobacter capsulatus*, *Rhodobacter capsulatus* hup⁻ mutant and *Rhodopseudomonas palustris*)	4.2 mol H_2/mol sucrose and 13.7 mol H_2/mol sucrose	Ozgur et al. (2010)
Sweden	Potato steam peels	Dark fermentation	–	1.37–3.48 mmol H_2/mole glucose	Ljunggren and Zacchi (2010)
Turkey	Hydrolysed and untreated potato steam peels and glucose	Starch fermentation	Extreme thermophile (*Caldicellulosiruptor saccharolyticus* and *Thermotoga neapolitana*)	2.4–3.8 mol H_2/mol glucose	Mars et al. (2010)
Netherlands	Carrot pulp and (mixture of) glucose and d-fructose	Dark fermentation	(Thermophilic bacteria) *Caldicellulosiruptor saccharolyticus* and *Thermotoga neapolitana*	2.7–2.8 mol H_2 (mol hexose)$^{-1}$	De Vrije et al. (2010)
Turkey	Volatile fatty acids (VFAs) individually (malate, acetate, propionate, butyrate and lactate)	Photofermentation	*Rhodobacter sphaeroides* OU 001	$24ml_{hydrogen}/l_{reactor}h$	Uyar et al. (2009)
Hungary	Wheat straw, maize leaves, sweet sorghum, sugarcane, bagasse and silphium	Dark fermentation	*Caldicellulosiruptor saccharolyticus*	3.8 mol H_2/mol glucose	Ivanova et al. (2009)
Turkey	Olive mill waste water (OMW)	Column photobioreactor	*Rhodobacter sphaeroides* OU 001	13.9 l_{H2}/l_{OMW}	Eroglu et al. (2004)
Turkey	Pretreated sugar refinery waste water (SRWW)	Column photobioreactor	*Rhodobacter sphaeroides* OU 001	2.67 l_{H2}	Yetis et al. (2000)

from hydrogen. The power output using, for comparison, commercial Pd (0) powder and Bio-Pd made from *Desulfovibrio desulfuricans* was about 100 mW (Orozco et al. 2010). An approach to maximise the hydrogen yields from food waste in a three-step conversion scheme coupling dark fermentation, photofermentation and anaerobic digestion was investigated by researchers and reported appreciating results.

Effects of pre-culture conditions on hydrogen production and fermentation balance and modification of hydrogenase uptake genes were studied by Redwood (2008) and Redwood et al. (2009). Formate deletions of hydrogenase 3 abolish hydrogen production of aerobically pre-grown resting cell suspensions, whilst the uptake hydrogenase deletion improved the hydrogen production by 37% over the parent strain. Possibilities to use organic wastes, sunlight and CO_2 as substrates under moderate conditions provide additional preference to biological methods over chemical methods of hydrogen production (Singh et al. 2015).

Combining microorganisms of different capabilities may improve the strength and their weaknesses overcome. Chandrasekhar et al. (2015) described strategies for microbial integration and discussed the mechanisms of biohydrogen production after integration. Broadly, dual systems can be categorised into entirely light driven systems (with microalgae/cyanobacteria as the first stage) and partially light-driven systems (with a dark, fermentative initial reaction). Partially light-driven systems have advantage over fully light-driven systems as the latter required land area that is too large for both, centralised (macro) and decentralised (micro) system. Feasibility of industrial biohydrogen production by fully light-driven system was affected by its land requirement. The potential contribution to the hydrogen economy of partially light-driven dual systems is overviewed alongside that of other biofuels such as biomethane and bioethanol (Redwood 2008).

Physical characteristics of some industrial effluent barred the potentiality of fermentation regardless of its fermentation supporting chemical property. In such cases, combining two different effluents may resolve the problem. Photofermentation of palm oil mill effluent (POME) can be used to produce biohydrogen by bacterium *Rhodobacter sphaeroides* NCIMB8253. The dark colour of palm oil mill effluent hinders the light penetration which can cease the process of photofermentation. Mixing of pulp and paper mill effluent reduces the turbidity of palm oil mill effluent and improves light penetration. Budiman et al. (2015) reported biohydrogen yield of 4.670 mL H_2/mL medium using NS4 treatment containing 25% and 75% (v/v) of palm oil mill effluent and pulp and paper mill effluent, respectively. Greater than 25% of palm oil mill effluent concentration increases turbidity (>16 450 NTU) and reduces biohydrogen production.

Biohydrogen production was investigated by thermophilic enrichment culture from a Turkish hot spring using batch experiments. To select spore-forming bacterial enrichment, the culture was heat-treated at 100 °C for 10 min (Karadag et al. 2009; Nissila 2013). Masset et al. (2012) reported negatively affected hydrogen yield by accumulation of lactate and ethanol, whilst hydrogen production was associated by acetate–butyrate-type fermentation. Results of Karadag et al. (2009) showed highest production of hydrogen at temperature range from 49.6 to 54.8 °C with the optimum values for initial pH and 6.5, 40 mg/L and 4–13.5 g/L concentrations of iron, yeast extract and glucose, respectively.

Analysis of process kinetic for biohydrogen production enrichment cultures showed that substrate (glucose) concentration higher than 1 g/L is likely to inhibit hydrogen production. Biofilm reactors working on synthetic medium with glucose as the only carbon and energy source and operated at 70 °C were used to check the output of various start-up strategies for biohydrogen production (Zheng et al. 2008; Sung 2004). A biofilm reactor started up with carriers that were previously inoculated with the enrichment cultures required longer start-up time (about 1 month) but resulted in higher yield (2.21 mol H_2/mol glucose consumed), whilst a biofilm reactor directly inoculated with the enrichment cultures reached stable state much faster in about 8 days only but with very low hydrogen yield (0.69 mol

H_2/mol glucose consumed). On the basis of these results, Zheng et al. (2008) concluded that hydraulic pressure is necessary for successful immobilisation of bacteria on carriers, whilst there is the risk of washing out specific high-yielding bacteria.

14.5 USA

Despite its present nonenergy utilisation, hydrogen is considered as one of the most important future energy resources in the USA, although hydrogen is mainly produced by thermochemical and electrochemical processes using coal and natural gas as the most economical sources (EIA, US (energy information administration) 2008). Rachman and Prasetyo (2014) described biological hydrogen production processes as more environment friendly and less energy intensive as compared to thermochemical and electrochemical processes. Biological hydrogen is mostly controlled either by photosynthetic organism or by fermentative organisms. Two enzymes, nitrogenase and hydrogenase, play a very important role in biological hydrogen generation process (Das and Veziroglu 2001; Tamagnini et al. 2002; Zhang et al. 2014). In this study, Das and Veziroglu (2001) presented the microorganisms and biochemical pathways involved in biohydrogen generation processes and several developmental works. In a similar experiment, Das and Veziroglu (2001) explained that an immobilised system is suitable for the continuous hydrogen production, and fermentative hydrogen production processes have some edge over the other biological processes.

Davila-Vazquez et al. (2009) studied the fermentative biohydrogen production in a continuous stirred-tank reactor (CSTR) operated during 65.6 days with cheese whey as substrate. They concluded that continuous fermentative biohydrogen production from cheese whey can be significantly enhanced by an appropriate selection of parameters such as hydraulic retention time (HRT) and organic loading rate (OLR). After optimising HRT and OLR, cheese whey fermentation can be applied for sustainable and clean energy generation as it also showed significant enhancements in volumetric hydrogen production rate (VHPR) which is a critical parameter to determine the full-scale practical application of fermentation.

14.6 Canada

Rates and yields of hydrogen production, as for many other bioprocesses, are a function of several variables (Hallenbeck and Ghosh 2009). Modelling and optimisation have been carried out in attempts to improve biohydrogen production rate and yield. Bacterium *Enterobacter aerogenes* strain NRRL B-407 showed several advantages in biohydrogen production using crude glycerol as a substrate (Sarma et al. 2013). Therefore, different less expensive (or wastes) materials (e.g. slaughterhouse liquid waste, brewery waste biomass) have been evaluated as supplementary nutrient for H_2 production from bioconversion of crude glycerol. Sarma et al. (2013) succeeded in replacing the large amount (~5–6 g/L) of expensive nutrients/buffering agents by negligible amount (~10 mg/L) of different waste materials, without compromising the cumulative H_2 yield, and the strain used in the present study was found to grow at an acidic pH as low as 3.3, indicating its prospective application for dark fermentative H_2 production. In a study, Skonieczny and Yargeau (2009) reported an increasing hydrogen production rate directly proportionate with an increase in both substrate concentration and pH. In Canada, various techniques have been used for biohydrogen production (Table 14.3) using a different substrate as carbon source.

Baghchehsaraee et al. (2010) reported increased biohydrogen yield with increasing temperature. With activated sludge, hydrogen production yields were 0.56 and 1.32 mol H_2 mol^{-1} glucose at 37 ° C and 55 ° C, respectively, whilst with anaerobically digested sludge, hydrogen yields were 2.18 and 1.25 mol H_2 mol^{-1} glucose at 37 °C and 55 °C, respectively. Yield reduction using activated sludge was defended by the researchers as the instability of hydrogen

Table 14.3 Biohydrogen production trend in Canada

Country	Substrate	Technology	Selected microbial utilisation	Hydrogen production	References
Canada	Waste crude glycerol obtained from biodiesel production process	Anaerobic fermentation	*Clostridium butyricum* NRRL B-41122 and *Enterobacter aerogenes* NRRL B-407	0.95 mmol H_2/mol glycerol	Pachapur et al. (2016)
Canada	Apple pomace hydrolysate co-fermented with crude glycerol	Anaerobic fermentation	*Clostridium butyricum* NRRL B-41122 and *Enterobacter aerogenes* NRRL B-407	26.07 mmol H_2/L	Pachapur et al. (2015)
Canada	Sugar beet juice	Dark fermentation	–	6 mol H_2/mol hexose	Dhar et al. (2015)
Canada	NaCl and magnetite nanoparticles, alginic acid, chitosan	Dark fermentation in bioreactor	*Clostridium beijerinckii NCIMB8052*	2.1 mol H_2/mol glucose	Seelert et al. (2015)
Canada	Pharmaceutical waste water	Upflow anaerobic sludge blanket reactor	Pretreated mixed anaerobic sludge (for anaerobic bacterial culture)	2.95 mmol/day	Krishna et al. (2013)
Canada	Sugar industry waste (beet molasses and black strap)	Photofermentation	*Rhodobacter capsulatus* JP91 (purple non-sulphur photosynthetic bacterium)	10.5 mol H_2/mol sucrose	Keskin and Hallenbeck (2012)
Canada	RCV medium (glucose as a carbon source in different concentrations)	Photofermentation	*Rhodobacter capsulatus* JP91 (purple non-sulphur photosynthetic bacterium)	3 mol H_2/mol glucose	Abo Hashesh et al. (2011)

production with activated sludge during the repeated batches due to formation of lactic acid as the predominant metabolite in some batches.

14.7 Africa

Electricity Supply Commission (ESCOM), a parastatal power supplier of South Africa, has been unable to fulfil the expanding energy demand of the country, resulting into extensive and importunate power cuts throughout the country. The present situation of the country urges an urgent need for exploration and implementation of clean and sustainable energy. Feasibility of using biohydrogen as a potential and sustainable source of energy in South Africa was examined by Sekoai and Daramola (2015). The present scenario of biohydrogen production, however, is facing two major challenges, viz. high production cost and low yield. Recent researches have made some advances to overcome these challenges. Biohydrogen researches to improve the hydrogen yield through physiological manipulations, process modifications, use of metabolic and genetic engineering and use of nanomaterial are providing some promising results to overcome the conventional barriers.

Kumar et al. (2013) recently showed an increase in hydrogen production by cell immobilisation using nanoparticles in the bioreactor. Lignocellulosic biomass was considered as a promising bioprocess substrate available in sufficient quantity for the production of biofuels and biomaterials (Kang et al. 2014). Moodley and Gueguim Kana (2015) conducted an experiment with xylose and glucose of sugarcane leaves after pretreatment by HCl and moist heat and examined the dynamics of biohydrogen production from these substrates at a semi-pilot scale. They modelled and optimised the production. Response surface methodology was applied to optimise the hybrid pretreatment within the ranges of 4–10 % HCl, at 60–100 °C and 60–240 min of process time. The coefficients of determination (R^2) of 0.93 and 0.99 were obtained for glucose and xylose models, respectively. Obtained result indicated the suitability of the models to spin the optimisation space. Process optimisation predicted glucose and xylose yields of 1.68 g l^{-1} and 8.92 g l^{-1} on a dual pretreatment of 5.28 % HCl for 187 min at 94.94 °C. In a 13 L bioreactor, these optimised substrates showed a peak production of biohydrogen fraction by 26.73 % at the 30th hour and total yield of 248.05 ml H_2 g^{-1} of fermentable sugar. Findings of this research suggested that sugarcane leaves (which are not contributing in the economical yield and burnt out during harvesting) can be an excellent renewable source of fermentable sugars for the production of biofuels such as biohydrogen. The optimum inoculum pretreatment conditions predicted by Faloye et al. (2014) were pH 11 and 2 min microwave treatment at 860 W, and the validation experiments demonstrated 32.41 % increase on hydrogen yield. The optimally pretreated inoculum was processed in two semi-pilot-scale bioreactor (10 L) in batch mode with 7 L working volume. In the absence of pH control, 46 % of glucose was utilised corresponding to a molar hydrogen yield of 1.78 mol H_2/mol glucose and a maximum hydrogen fraction of 49.3 %, whereas, under a controlled pH environment, a twofold increase in glucose utilisation was obtained which corresponds to a molar hydrogen yield of 2.07 mol H_2/mol glucose and a maximum hydrogen concentration of 56.4 %. The controlled fermentation significantly improved biohydrogen production in the scaleup process, and methane production was completely suppressed due to effectiveness of the combined pretreatment to enrich hydrogen-producing bacteria (Mohan 2008; Faloye et al. 2014). Viable counts and microscopical analysis indicated the presence of hydrogen-producing endospore-forming presumptive *Clostridium* species (Faloye et al. 2014; Kumar et al. 2015).

Sekoai and Kana (2014) studied organic segment of municipal solid waste as a substrate for production of biohydrogen on a semi-pilot scale. Wen et al. (2010) and He et al. (2013) assessed the potential of process effluents for electricity generation in a two-chambered microbial fuel cell. Total biohydrogen yield of 246.93 ml H_2/g total volatile solids and a maximum hydrogen fraction of 46.7 % were obtained at optimum operational set points of 7.9 pH, 30.29 °C and 60 h hydraulic retention time (HRT). These experiments suggested that the energy generation from organic wastes could improve the process economics significantly by integrating a two-stage process: fermentative hydrogen production with electricity generation (Sekoai and Kana 2014). Two factors influence the hydrogen generation process goals: (1) to attain a net positive energy balance at thermophilic temperatures and high effluent recycle rates and (2) to attain volumetric hydrogen productivities that are sufficient to drive a 5 kW fuel cell when scaled up to 1 m^3. However, an increase in the hydrogen yield above the Thauer limit (75 % value or 3 mol H_2/mol glucose) was not attained (Obazu et al. 2015b). Pawar and van Niel (2013) and Obazu et al. (2015a) showed that undefined bacterial cultures from mesophilic sources rapidly generated thermophilic bacterial granules adapted to 70 °C which could achieve hydrogen production efficiencies.

Sweet sorghum (*Sorghum bicolor*, an annual tropical plant), adaptive in hot and dry season, could be a potential raw material for biohydrogen gas production (Rachman et al. 2015). Moreover, it has a high biomass production and can adapt to extreme and subtropical regions. Production of biohydrogen gas has been studied by researchers using sweet sorghum at packed-bed reactor by *Enterobacter aerogenes* ADH-43 and to get optimum dilution rate in order to increase gas H_2 production (Ntoampe 2012; Rachman et al. 2015). In the batch experiments, the fresh sorghum medium was fed into the reactor before 2 h of the stationary phase in order to achieve continuous culture. The steady-state condition showed that optimum dilution rate was 0.15 h^{-1} with biohydrogen production of 81.50 mmol L^{-1} h^{-1} and yields 0.87 mol H_2/mol total sugars (Rachman et al. 2015). Nasr et al. (2015) investigated the hydrogen production potential of waste water from starch industries in a sequential dark–photofermentation process using two anaerobic baffled reactors (ABRs) operated in parallel. OLR of the process was maintained at 8.11 ± 0.97 g COD L^{-1} day^{-1} with 15 h HRT. Pretreated sludge was used as inoculum for one reactor, whilst another reactor was inoculated with sludge immobilised on maghemite nanoparticles. Better hydrogen yield (104.75 ± 12.39 mL H_2/g-COD removed) was achieved in reactor inoculated with sludge immobilised on maghemite nanoparticles as compared to (66.22 ± 4.88 mL-H_2/g-COD removed) the reactor inoculated with pretreated sludge. The effluent of ABR inoculated with sludge immobilised on maghemite nanoparticles was used in the third ABR for further hydrogen production by photofermentation. At a total HRT of 30 h, hydrogen yield of 166.83 ± 27.79 mL-H_2/g-COD removed was achieved. Three hundred and forty photosynthetic bacteria utilise VFAs that enhance the pH to about 6.5 in the third ABR.

Theoretically, Thauer limit of 4.0 mol H_2/mol glucose could be achievable only under thermophilic temperatures, and bacterial granules from mesophilic inoculant were adapted to generate biohydrogen from sucrose under thermophilic temperature range (50, 55, 60, 65, 70 °C). However, present findings were not at all reached near to Thauer limit of biohydrogen generation. Obazu (2013) and Obazu et al. (2015b) explained this as the high H_2 yields may be obtained under thermophilic temperatures, but it was not a sufficient condition because possibly the yield lower than the theoretical value is by substrate utilisation for cell synthesis too. Some advanced research on hydrogen production of African countries is given in Table 14.4.

On the South African effort to develop reliable, clean and safe energy sources, the hydrogen economy is providing a promising option over and above fossil fuel. Hydrogen is considered as the most reliable, safe and clean energy carrier and can be used to store and distribute energy with fuel cell technologies to produce electricity (Raji and Kahn 2013; http://www.saasta.ac.za/). Platinum is the key catalytic material used in most fuel cells. The finding of platinum group metal (PGM) reserves in South Africa is another driving force behind the prevalence of biohydrogen technology in the country. More than 75 % of the world's known platinum reserves in South Africa provide a platform for socio-economic benefits and environmental clean energy (Sibanyoni 2012; Raji and Kahn 2013).

14.8 Australia

Per capita use of energy is highest in Australia. To fulfil its substantial energy requirements, the country is dependent on fossil fuels, which makes Australia stand in the row of the highest per capita polluters. Plentiful natural resources available in Australia make it fit for the utilisation of all the options available for a hydrogen economy (McLellan et al. 2005). At this stage, natural gas, biomass, water and coal are the most potential hydrogen sources, whilst due to current capability in utilising hydrogen-rich gases, molten carbonate and solid oxide fuel cells may hold the advantage for stationary power in Australia (McLellan et al. 2005).

In a batch process, Wang and Jin (2009) and Cai et al. (2013) optimise fermentation parameters for hydrogen production from molasses using newly isolated *Clostridium butyricum* W5. Researchers obtained 1.85 mol hydrogen/mol hexose with a total productivity of 17.38 mmol h^{-1} L^{-1} with initial cell concentration of 9×10^4 cell/ml under 39 °C temperature at pH 6.5 and 1.2 g/L NH_4NO_3, 100 g/L molasses. During the fermentation process, researchers detected propionic acid with very low level, whilst other solvents such as ethanol, butanol and acetone were nil detected. Analysis of data showed exponential increase of hydrogen yield with the increase in cell growth with no correlation between the

Table 14.4 Biohydrogen production trend in African countries

Country	Substrate	Technology	Selected microbial utilisation	Hydrogen production	References
Egypt	Pretreated sludge and sludge immobilised by nanoparticles of maghemite	Dark–photofermentation	*Clostridium* and *Rhodopseudomonas palustris*	166.83 ml H_2 /g COD removed	Nasr et al. (2015)
Egypt	Municipal solid waste (organic fraction)	Anaerobic digestion (upflow intermittently stirred-tank reactor)	Anaerobic consortia	2.05 ± 0.33 molH_2/mol carbohydrate	Elsamadony and Tawfik (2015a)
South Africa	Artificial neural networks with carbon source (glucose, acetate, glycerol, etc.)	Microbial electrolysis cell voltage values:	–	Mol H_2/mol acetate,	Sewsynker et al. (2015)
		1.2 V,		4.04 Mol	
		1.0 V		3.71 Mol	
		1.15 V		3.97 Mol	
Egypt	Paperboard mill waste water	Anaerobic baffled reactor	Anaerobic consortia	9.3 kg COD/Ld	Farghaly et al. (2015)
Egypt	Combined digestion of kitchen waste water and municipal food waste	Anaerobic baffled reactor	Anaerobic consortia	156–329 mL H_2/g VS removed/d	Tawfik et al. (2015)
Egypt	Sludge of paperboard industry and gelatin solid waste	Anaerobic fermentation	Thermophilic bacteria	4.5 ± 0.3 to 7.2 ± 0.6 lH_2/L substrate	Elsamadony and Tawfik (2015b)
Egypt	Petrochemical industrial effluents	Anaerobic baffled reactor	Sludge from aerobic treatment plant (for bacterial seed)	45.50–377.03 ml H_2/g COD removed	Elreedy and Tawfik (2015)
Egypt	Municipal solid waste (organic fraction) with Tween 80 and polyethylene glycol (PEG 6000)	Anaerobic digestion	*Enterobacter*, *Escherichia*, *Buttiauxella* and *Pantoea*	109.9 ± 7.1 and 113.8 ± 7.7 mlH_2/g Carb. initial	Elsamadony et al. (2015)
Nigeria	Palm oil mill effluents (POMEs)	Thermophilic fermentation	*Thermoanaerobacterium*, *Clostridium butyricum*, POME mixed culture	57 billion L	Ohimain and Izah (2015)
Egypt	Hydrolysed (alkali) rice straw	Anaerobic baffled reactor	*Paludibacter*, *Clostridium*, *Ensifer*, *Prevotella* and *Petrimonas* (from treated and untreated sludge)	0.97 and 1.19 mol H_2/mol (in two different reactors, respectively)	El-Bery et al. (2013)
South Africa	Modified Endo formulation media with C/N/P ratio of 334:42:1 in growth medium	Dark fermentation	*Citrobacter freundii* Cf1 and *E. cloacae* Ecl	2.33 mmol H_2 (mmol sucrose)$^{-1}$	Thompson et al. (2008)

hydrogen yield and ratio of acetic acid to butyric acid.

Rittmann and Herwig (2012) presented results of more than 2,000 conditions for quantitative normalised H_2 production and qualitative growth characteristics in a normalised and comparable format to the scientific community. Evidence based on statistical analysis showed that high substrate conversion efficiency was achieved by thermophilic strains, but high volumetric productivity was sustained by mesophilic strains. Furthermore, *Thermoanaerobacterales* is a group of microbes to be preferred to achieve high substrate conversion efficiency than the families *Clostridiaceae* and *Enterobacteriaceae* (Rittmann and Herwig 2012; Ciranna 2014). Available results suggested that the potential of bioprocessing application of dark fermentative biohydrogen production from fed-batch cultivations is underestimated. For efficient bioprocess development and optimisation of biohydrogen production, an experimental strategy should be designed primarily aiming at revealing cultivation conditions, improving medium and describing inhibitory effects. This strategy will facilitate the researchers to compare and optimise strains and processes independently from the initial conditions and scale (Rittmann and Herwig 2012). Researchers working on biohydrogen production concluded that biofilm reactors fed with glucose and filled with plastic carriers could efficiently produce biohydrogen at 70 °C (Qureshi et al. 2005; Zhang et al. 2008). Zheng et al. (2008) have conducted batch experiments in a biohydrogen CSTR with household solid waste to develop glucose-fed hyperthermophilic hydrogen-producing microorganisms. Kinetic analysis revealed that the substrate (glucose) is likely to inhibit the hydrogen production if its concentration was higher than 1 g/L in biohydrogen enrichment culture. To check the potential of biohydrogen production, biofilm reactors operated at 70 °C, provided with synthetic medium, and glucose as the sole carbon and energy source was tested at variable start-up strategies. A biofilm reactor, previously inoculated with the enrichment cultures, started up with plastic carriers and resulted in 2.21 mol H_2/mol glucose yield of hydrogen, although 1-month start-up time was required for this biohydrogen production approach. On the other hand, biofilm reactor directly inoculated with the enrichment cultures reached stable state in 8 days but came up with a very low hydrogen yield (0.78 mol H_2/mol glucose). Results of these experiments clearly indicated that, despite being the possibility of washing out specific high-yielding bacteria, hydraulic pressure is a specific requirement for successful immobilisation of bacteria (Zhang et al. 2007).

14.9 The Middle East

Air Liquide Arabia (a joint venture company) situated at the west coast of Saudi Arabia was incorporated with Saudi Arabia in 2008 and invested more than US$ 393 million to begin producing hydrogen from Yanbu Industrial City. Air Liquide Arabia made a long-term agreement with the new Yanbu Aramco Sinopec Refining (YASREF) to supply hydrogen. Air Liquide site has a total hydrogen capacity of 340,000 m^3 per hour provided by two global-scale hydrogen production units accompanied by one purification unit. The production can further be enhanced as per the requirement of YASREF (Oil Review Middle East 2015). Novel strains of microalgae *Chlorella vulgaris* (strain YSL01 and YSL16) upregulate the expression of the hydrogenase gene. It produces hydrogen under aerobic conditions during photosynthesis using CO_2 as the sole carbon source with continuous illumination (Hwang et al. 2014). Setting up an experimental condition corresponding to natural aquatic condition by employing dissolved oxygen in water clarifies that *C. vulgaris* YSL01 and YSL16 enzymatically produce hydrogen, even under atmospheric conditions by experimental expression of hydrogenase gene (Hwang et al. 2014).

Batch and continuous culture process of a column photobioreactor was studied by researchers of Middle East Technical University for hydrogen production potentiality of *Rhodobacter sphaeroides* OU 001. The study demonstrated that the cell concentration and the ratio of L-malic acid and sodium glutamate are significant aspects affecting hydrogen production rate. The study also suggested for the optimisation of amount of feed and time intervals of the dilutions in a continuous system for optimum hydrogen yield (Eroglu et al. 1998). Upon illumination, *Rhodobacter sphaeroides* OU 001 is also able to produce hydrogen anaerobically at the same time; the bacterium was also able to produce valuable by-products such as poly-β-hydroxybutyric acid (PHB) which can improve the feasibility of the process (Redwood and Macaskie 2006). Possibility of bacterial growth into waste water from a sugar refinery further enhances the feasibility of the system. Various researchers studied the growth of *Rhodobacter sphaeroides* OU 001 under aerobic conditions with medium containing 30% sugar refinery waste water and a standard growth media containing L-malic acid and sodium glutamate (Yigit et al. 1999; Schneider et al. 2012; Budiman et al. 2015). The PHB concentration was higher (0.3 g L^{-1}) in medium containing 30% waste water as compared to standard media (approximately 0.2 g L^{-1}). Under anaerobic conditions, the amount of hydrogen gas collected was 35 mL in 108 h, and the maximum concentration of PHB produced was around 0.5 g L^{-1} in the medium containing 30% waste water. The above-mentioned result suggested that the sugar refinery waste water could increase the yield of hydrogen, and PHB can be collected during hydrogen production (Yigit et al. 1999).

Kinetics study of hydrogen production by *Rhodobacter sphaeroides* OU 001 and the effects of various parameters on batch column photobioreactor were investigated by researchers (Koku et al. 2003; Waligórska et al. 2006). Their main focus was on the effect of the implementation of a light–dark cycle illumination scheme, inoculum age and replacement of vitamin mixture in the medium with yeast extract. High culture age of the initially inoculated bacteria decreases the hydrogen production. Furthermore, continuous illumination reduces the total yield of hydrogen compared to exposure of the bacterial culture to light–dark cycles. Replacing vitamins by yeast extract increased the growth and hydrogen production rates; however, total amount of gas produced and hydrogen fraction in the evolved gas are slightly reduced (Koku et al. 2003).

Studies were also carried out in a leaching-bed reactor to improve biohydrogen production by fermentation of food waste by heat-shocked anaerobic sludge and to investigate the effect of dilution rate (D) on the production rate of metabolites and hydrogen during fermentation (Han and Shin 2004; Nasr 2012). Adjustment of environmental conditions during the fermentation is a key limitation factor amongst various reaction constraints that can affect the fermentation of food waste. It is because of various components present in the food waste with variable characteristics and diverse degradation regime. This obstacle can be overcome by enhancing the degradation of slowly degradable complex material. Further, controlling the dilution rate could delay the shift of predominant metabolic flow from hydrogen and acid-forming pathway to solvent-forming pathway (Han and Shin 2004). Continuous pilot tubular photobioreactors with *R. capsulatus* hup^- mutant can achieve stable bacterial growth with hydrogen production (Boran et al. 2012a). Fed-batch operation for photofermentative hydrogen production by *Rhodobacter capsulatus* (hup^-) mutant in outdoor conditions by a solar tubular photobioreactor has been successfully executed on a pilot scale (Androga et al. 2012; Boran et al. 2012a). Increase in the wall thickness of the tubes increases reliability and the durability of the tubes which can provide a long-term operation, whilst decreased tube diameter can increase hydrogen productivity. It is suggested that the diameter of the tube and thickness of the tube wall should be optimised to provide a better light exposure to the cells (Boran et al. 2012a).

14.10 Future Perspectives

Globally, a momentum is developed towards hydrogen energy. Hydrogen production, distribution and use become an important aspect for research planning and policy making (McLellan et al. 2005; IPHE 2010). Primary energy sources can be used to produce hydrogen using various technologies (Balat and Kırtay 2010). High-throughput experimentations are pivotal in biohydrogen fermentation processes in order to obtain consistent data for scaleup studies. This will require novel reactor designs with high level of parallelisation combined with online computer systems to assess the critical process conditions during biohydrogen production. Application of mathematical and statistical tools in biohydrogen fermentation processes is also essential to facilitate investigation on the synergistic effects of various process parameters on the overall yield (Sekoai and Daramola 2015).

Production of hydrogen using biological tool is the predominant challenge for biotechnology concerning present and future environmental problems. Future of biological hydrogen production is not only determined by research advances (e.g. genetic engineering of microorganisms for efficiency improvement and designing complications of bioreactor) but also by fuel economic (cost of fuel), societal adaptation and the development of systems for hydrogen energy (FAO 1997). Thermocatalytic and gasification processes are contributing about half of current hydrogen production using naphtha, coal, natural gas and heavy oils. Most of the present research has been focused on environmentally sustainable energy from replacing conventional fuels to biomass. Hydrogen production from biomass is already economically competitive today along with major challenges, as completed technology demonstrations are not available (Balat and Kırtay 2010). Current strategies geared towards improving biohydrogen production include microbial culture immobilisation, bioreactor modifications, the optimisation of process conditions (temperature, pH, OLR and HRT), culture selection and enrichments, substrate choice and the metabolic engineering of biohydrogen specialists (Arimi et al. 2015; Soydemir et al. 2016).

Biohydrogen could be used for domestic, thermal, industrial and transport energy requirements. Development in fuel cell encourages hydrogen production for power generation. Potentiality of hydrogen gas for power generation with high efficiency without causing any pollution makes it an encouraging choice over other fuels. Hydrogen production through the biological route had gained importance as it could be produced through a variety of renewable resources (waste water/sewage/biomass) (Mallikarjun 2012). Although biological process has been recognised as a nascent approach for hydrogen production, production rate and cost are still hindering this technology on an industrial scale (Wang et al. 2008).

Energy systems based on hydrogen is emerging into an attractive proffer for providing a future replacement for the current fossil fuel-based energy systems. With current mixing ratio of about 510 ppb, hydrogen is an important trace component of the atmosphere. Hydrogen is an indirect greenhouse gas with a global warming potential of 5.8 over a 100-year time horizon due to of its reactivity to tropospheric hydroxyl radicals which can upset the distributions of important greenhouse gases (methane and ozone) (Derwent et al. 2006). According to Derwent et al. (2006), replacing the current fossil fuel-based energy to the hydrogen energy with a leakage rate of 1 % would produce a climate impact of 0.6 % of the current fossil fuel-based system. Therefore, careful attention should be provided to reduce the leakage of hydrogen from the synthesis, storage and usage in the future global hydrogen economy if the full climate benefits are to be realised.

14.11 Conclusion

Hydrogen has been perceived as a widespread, clean and eco-accommodating fuel which is relied upon to give energy for all our needs like transport, industrial and domestic energy requirements. It is considered as a standout amongst the most intriguing distinct options for non-renewable energy fuels. A lot of innovative work on production, storage and transportation and use of hydrogen is in advancement worldwide. A breakthrough can be achieved if the production of biohydrogen reaches the commercial scale by using waste (water, materials, garbage, etc.) as a raw material.

References

Abdallah R, Amrane A, Djelal H, Taha S, Fourcade F, Labasque T, Geneste F, Floner D (2015) Energetic valorization of ammonium resulting from nitrate electrochemical reduction-feasibility of biohydrogen production. Biochem Eng J 94:145–152

Abo-Hashesh M, Ghosh D, Tourigny A, Taous A, Hallenbeck PC (2011) Single stage photofermentative hydrogen production from glucose: an attractive alternative to two stage photofermentation or co-culture approaches. Int J Hydrog Energy 36(21):13889–13895

Akil K, Jayanthi S (2014) The biohydrogen potential of distillery wastewater by dark fermentation in an anaerobic sequencing batch reactor. Int J Green Energy 11(1):28–39

Akman MC, Erguder TH, Gündüz U, Eroğlu İ (2015) Investigation of the effects of initial substrate and biomass concentrations and light intensity on photofermentative hydrogen gas production by response surface methodology. Int J Hydrog Energy 40(15):5042–5049

Androga DD, Ozgür E, Eroglu I, Yücel M, Gündüz U (2012) Photofermentative hydrogen production in outdoor conditions. In: Minic D (ed) Hydrogen energy - challenges and perspectives. InTech.. doi:10.5772/50390, Available from: http://www.intechopen.com/books/hydrogen-energy-challenges-and-perspectives/photofermentative-hydrogen-production-in-outdoor-conditions

Androga DD, Sevinç P, Koku H, Yücel M, Gündüz U, Eroglu I (2014) Optimization of temperature and light intensity for improved photofermentative hydrogen production using Rhodobacter capsulatus DSM 1710. Int J Hydrog Energy 39(6):2472–2480

Arimi MM, Knodel J, Kiprop A, Namango SS, Zhang Y, Geißen SU (2015) Strategies for improvement of biohydrogen production from organic-rich wastewater: a review. Biomass Bioenergy 75:101–118

Armor JN (1999) The multiple roles for catalysis in the production of H_2. Appl Catal Gen 176(2):159–176

Atif AAY, Fakhru'l-Razi A, Ngan MA, Morimoto M, Iyuke SE, Veziroglu NT (2005) Fed batch production of hydrogen from palm oil mill effluent using anaerobic microflora. Int J Hydrog Energy 30(13):1393–1397

Avcioglu SG, Ozgur E, Eroglu I, Yucel M, Gunduz U (2011) Biohydrogen production in an outdoor panel photobioreactor on dark fermentation effluent of molasses. Int J Hydrog Energy 36(17):11360–11368

Babu ML, Sarma PN, Mohan SV (2013) Microbial electrolysis of synthetic acids for biohydrogen production: influence of biocatalyst pretreatment and pH with the function of applied potential. J Microbial Biochem Technol. S6: 003. doi:10.4172/1948-5948.S6-003

Baghchehsaraee B, Nakhla G, Karamanev D, Margaritis A (2010) Fermentative hydrogen production by diverse microflora. Int J Hydrog Energy 35(10):5021–5027

Balat H, Kırtay E (2010) Hydrogen from biomass–present scenario and future prospects. Int J Hydrog Energy 35(14):7416–7426

Beckers L, Masset J, Hamilton C, Delvigne F, Toye D, Crine M, Thonart P, Hiligsmann S (2015) Investigation of the links between mass transfer conditions, dissolved hydrogen concentration and biohydrogen production by the pure strain Clostridium butyricum CWBI1009. Biochem Eng J 98:18–28

Boran E, Ozgur E, Yucel M, Gunduz U, Eroglu I (2012a) Biohydrogen production by Rhodobacter capsulatus Hup– mutant in pilot solar tubular photobioreactor. Int J Hydrog Energy 37(21):16437–16445

Boran E, Özgür E, Yücel M, Gündüz U, Eroglu I (2012b) Biohydrogen production by Rhodobacter capsulatus in solar tubular photobioreactor on thick juice dark fermenter effluent. J Clean Prod 31:150–157

Brown MR, Chesbro W, Mule R, Peace C, Petersen G, Leary M (2008) Current anaerobic digestion technologies used for treatment of Municipal Organic Solid Waste. California Integrated Waste Management Board, Sacramento, CA

Budiman PM, Wu TY, Ramanan RN, Md. Jahim J (2015) Improvement of biohydrogen production through combined reuses of palm oil mill effluent together with pulp and paper mill effluent in photofermentation. Energy Fuels 29(9):5816–5824

Cai G, Jin B, Monis P, Saint C (2013) A genetic and metabolic approach to redirection of biochemical pathways of Clostridium butyricum for enhancing hydrogen production. Biotechnol Bioeng 110(1):338–342

Carmona-Martínez AA, Trably E, Milferstedt K, Lacroix R, Etcheverry L, Bernet N (2015) Long-term continuous production of H_2 in a microbial electrolysis cell (MEC) treating saline wastewater. Water Res 81:149–156

Chandrasekhar K, Lee YJ, Lee DW (2015) Biohydrogen production: strategies to improve process efficiency through microbial routes. Int J Mol Sci 16(4):8266–8293

Ciranna A (2014) Biohydrogen production in extreme conditions: a comprehensive study of the fermentative metabolism of a polyextremophilic bacterium. Tampere University of Technology. Publication:1228, Tampere 2014

Costa JAV, De Morais MG (2011) The role of biochemical engineering in the production of biofuels from microalgae. Bioresour Technol 102(1):2–9

Dahiya S, Sarkar O, Swamy YV, Mohan SV (2015) Acidogenic fermentation of food waste for volatile fatty acid production with co-generation of biohydrogen. Bioresour Technol 182:103–113

Das D, Veziroğlu TN (2001) Hydrogen production by biological processes: a survey of literature. Int J Hydrog Energy 26(1):13–28

Dasgupta CN, Suseela MR, Mandotra SK, Kumar P, Pandey MK, Toppo K, Lone JA (2015) Dual uses of microalgal biomass: an integrative approach for biohydrogen and biodiesel production. Appl Energy 146:202–208

Davila-Vazquez G, Cota-Navarro CB, Rosales-Colunga LM, de León-Rodríguez A, Razo-Flores E (2009) Continuous biohydrogen production using cheese whey: improving the hydrogen production rate. Int J Hydrog Energy 34(10):4296–4304

De Vrije T, Budde MA, Lips SJ, Bakker RR, Mars AE, Claassen PA (2010) Hydrogen production from carrot pulp by the extreme thermophiles Caldicellulosiruptor saccharolyticus and Thermotoga neapolitana. Int J Hydrog Energy 37:15616–15631

Derwent R, Simmonds P, O'Doherty S, Manning A, Collins W, Stevenson D (2006) Global environmental impacts of the hydrogen economy. Int J Nucl Hydrog Prod Appl 1(1):57–67

Dhar BR, Elbeshbishy E, Hafez H, Lee HS (2015) Hydrogen production from sugar beet juice using an integrated biohydrogen process of dark fermentation and microbial electrolysis cell. Bioresour Technol 198:223–230

Dwidar M, Park JY, Mitchell RJ, Sang BI (2012) The future of butyric acid in industry. Sci World J 2012:471417

EIA, US (energy information administration) (2008) The impact of increased use of hydrogen on petroleum consumption and carbon dioxide emissions report, SR-OIAF-CNEAF/2008-04

El-Bery H, Tawfik A, Kumari S, Bux F (2013) Effect of thermal pre-treatment on inoculum sludge to enhance bio-hydrogen production from alkali hydrolysed rice straw in a mesophilic anaerobic baffled reactor. Environ Technol 34(13,14):1965–1972

Elreedy A, Tawfik A (2015) Effect of hydraulic retention time on hydrogen production from the dark fermentation of petrochemical effluents contaminated with Ethylene Glycol. Energy Procedia 74:1071–1078

Elsamadony M, Tawfik A (2015a) Potential of biohydrogen production from organic fraction of municipal solid waste (OFMSW) using pilot-scale dry anaerobic reactor. Bioresour Technol 196:9–16

Elsamadony M, Tawfik A (2015b) Dry anaerobic co-digestion of organic fraction of municipal waste with paperboard mill sludge and gelatin solid waste for enhancement of hydrogen production. Bioresour Technol 191:157–165

Elsamadony M, Tawfik A, Suzuki M (2015) Surfactant-enhanced biohydrogen production from organic fraction of municipal solid waste (OFMSW) via dry anaerobic digestion. Appl Energy 149:272–282

Eroglu I, Aslan K, Gunduz U, Yucel M, Turker L (1998) Continuous hydrogen production by Rhodobacter sphaeroides O.U. 001. In: Zaborsky OR (ed) Biohydrogen. Plenum Press, New York, pp 143–149

Eroglu E, Gündüz U, Yücel M, Türker L, Eroğlu İ (2004) Photobiological hydrogen production by using olive mill wastewater as a sole substrate source. Int J Hydrog Energy 29(2):163–171

Faloye FD, Kana EG, Schmidt S (2014) Optimization of biohydrogen inoculum development via a hybrid pH and microwave treatment technique–Semi pilot scale production assessment. Int J Hydrog Energy 39(11):5607–5616

FAO (1997) Hydrogen production. Renewable biological systems for alternative sustainable energy production. Miyamoto, K (ed.) Food and Agriculture Organization of the United Nations. http://www.fao.org/docrep/w7241e/w7241e0g.htm

Farghaly A, Tawfik A, Eldin MG (2015) Continuous biological treatment of paperboard mill wastewater along with hydrogen production. Energy Procedia 74:926–932

Ghimire A, Valentino S, Frunzo L, Trably E, Escudié R, Pirozzi F, Lens PN, Esposito G (2015) Biohydrogen production from food waste by coupling semi-continuous dark-photofermentation and residue post-treatment to anaerobic digestion: a synergy for energy recovery. Int J Hydrog Energy 40(46):16045–16055

Hallenbeck PC, Ghosh D (2009) Advances in fermentative biohydrogen production: the way forward? Trends Biotechnol 27(5):287–297

Han SK, Shin HS (2004) Biohydrogen production by anaerobic fermentation of food waste. Int J Hydrog Energy 29(6):569–577

Hasyim R, Imai T, Sompong O, Sulistyowati L (2011) Biohydrogen production from sago starch in wastewater using an enriched thermophilic mixed culture from hot spring. Int J Hydrog Energy 36(21):14162–14171

Hawkes FR, Hussy I, Kyazze G, Dinsdale R, Hawkes DL (2007) Continuous dark fermentative hydrogen production by mesophilic microflora: principles and progress. Int J Hydrog Energy 32(2):172–184

He Z, Zhang F, Ge Z (2013) Using microbial fuel cells to treat raw sludge and primary effluent for bioelectricity generation: final report

Hosseinkhani B, Hennebel T, Boon N (2014) Potential of biogenic hydrogen production for hydrogen driven remediation strategies in marine environments. New Biotechnol 31(5):445–450

Hsiao CL, Chang JJ, Wu JH, Chin WC, Wen FS, Huang CC, Chen CC, Lin CY (2009) Clostridium strain co-cultures for biohydrogen production enhancement from condensed molasses fermentation solubles. Int J Hydrog Energy 34(17):7173–7181

http://www.marketsandmarkets.com/Market-Reports/hydrogen-generation-market-494.html. Accessed on 12 Jan 2016

Hwang JH, Kim HC, Choi JA, Abou-Shanab RAI, Dempsey BA, Regan JM, Kim JR, Song H, Nam IH, Kim SN, Lee W (2014) Photoautotrophic hydrogen production by eukaryotic microalgae under aerobic conditions. Nat Commun 5(3234):1–6

IPHE (2010) Hydrogen and fuel cell global commercialization & development update:1–10. http://www.iphe.net/docs/Resources/IPHE_FINAL_SON_press_quality.pdf

Ivanova G, Rákhely G, Kovács KL (2009) Thermophilic biohydrogen production from energy plants by Caldicellulosiruptor saccharolyticus and comparison with related studies. Int J Hydrog Energy 34(9):3659–3670

Kang Q, Appels L, Tan T, Dewil R (2014) Bioethanol from lignocellulosic biomass: current findings determine research priorities. Sci World J 2014. DOI: 10.1155/2014/298153

Karadag D, Mäkinen AE, Efimova E, Puhakka JA (2009) Thermophilic biohydrogen production by an anaerobic heat treated-hot spring culture. Bioresour Technol 100(23):5790–5795

Keskin T, Hallenbeck PC (2012) Hydrogen production from sugar industry wastes using single-stage photofermentation. Bioresour Technol 112:131–136

Koku H, Eroglu I, Gunduz U, Yucel M, Turker L (2003) Kinetics of biological hydrogen production by the photosynthetic bacterium Rhodobacter sphaeroides OU 001. Int J Hydrog Energy 28(4):381–388

Kongjan P, Angelidaki I (2010) Extreme thermophilic biohydrogen production from wheat straw hydrolysate using mixed culture fermentation: effect of reactor configuration. Bioresour Technol 101(20):7789–7796

Koutinas A, Kanellaki M, Bekatorou A, Kandylis P, Pissaridi K, Dima A, Boura K, Lappa K, Tsafrakidou P, Stergiou PY, Foukis A (2016) Economic evaluation of technology for a new generation biofuel production using wastes. Bioresour Technol 200:178–185

Krishna RH, Mohan SV, Swamy AVVS (2013) Bio hydrogen production from pharmaceutical waste water treatment by a suspended growth reactor using environmental anaerobic technology. Am Chem Sc J 3(2):80–97

Kumar A, Singh JS (2016) Microalgae and cyanobacteria biofuels: a sustainable alternate to crop-based fuels. In: Singh JS, Singh DP (eds) Microbes and environmental management. Studium Press, USA

Kumar GS, Kumari S, Reddy K, Bux F (2013) Trends in biohydrogen production: major challenges and state-of-the-art developments. Environ Technol 34(13–14):1653–1670

Kumar G, Bakonyi P, Sivagurunathan P, Nemestóthy N, Belafi-Bako K, Lin CY (2015) Improved microbial conversion of de-oiled Jatropha waste into biohydrogen via inoculum pretreatment: process optimization by experimental design approach. Biofuel Res J 2(1):209–2014

Lai Z, Zhu M, Yang X, Wang J, Li S (2014) Optimization of key factors affecting hydrogen production from sugarcane bagasse by a thermophilic anaerobic pure culture. Biotechnol Biofuels 7(1):1

Lay CH, Kuo SY, Sen B, Chen CC, Chang JS, Lin CY (2012) Fermentative biohydrogen production from starch-containing textile wastewater. Int J Hydrog Energy 37(2):2050–2057

Lee DH, Lee DJ (2008) Hydrogen economy in Taiwan and biohydrogen. Int J Hydrog Energy 33:1607–1618

Ljunggren M, Zacchi G (2010) Techno-economic analysis of a two-step biological process producing hydrogen and methane. Bioresour Technol 101(20):7780–7788

Mallikarjun Y (2012) Bio-hydrogen produced from wastewater, industrial effluents. The Hindu Hyderabad, 25 July 2012. http://www.thehindu.com/

Mangayil R, Karp M, Santala V (2012) Bioconversion of crude glycerol from biodiesel production to hydrogen. Int J Hydrog Energy 37(17):12198–12204

Marone A, Varrone C, Fiocchetti F, Giussani B, Izzo G, Mentuccia L, Rosa S, Signorini A (2015) Optimization of substrate composition for biohydrogen production from buffalo slurry co-fermented with cheese whey and crude glycerol, using microbial mixed culture. Int J Hydrog Energy 40(1):209–218

Mars AE, Veuskens T, Budde MA, Van Doeveren PF, Lips SJ, Bakker RR, De Vrije T, Claassen PA (2010) Biohydrogen production from untreated and hydrolyzed potato steam peels by the extreme thermophiles Caldicellulosiruptor saccharolyticus and Thermotoga neapolitana. Int J Hydrog Energy 35(15):7730–7737

Masset J, Calusinska M, Hamilton C, Hiligsmann S, Joris B, Wilmotte A, Thonart P (2012) Fermentative hydrogen production from glucose and starch using pure strains and artificial co-cultures of Clostridium spp. Biotechnol Biofuels 5(1):35

McLellan B, Shoko E, Dicks AL, Da Costa JD (2005) Hydrogen production and utilisation opportunities for Australia. Int J Hydrog Energy 30(6):669–679

Mohan VS (2008) Fermentative hydrogen production with simultaneous wastewater treatment: influence of pretreatment and system operating conditions. J Sci Ind Res 67:950–961

Mohan SR (2015) Structure and growth of research on bio hydrogen generation using wastewater. Int J Hydrog Energy 40:16056–16069

Moodley P, Kana EG (2015) Optimization of xylose and glucose production from sugarcane leaves (Saccharum officinarum) using hybrid pretreatment techniques and assessment for hydrogen generation at semi-pilot scale. Int J Hydrog Energy 40(10):3859–3867

Mullai P, Rene ER, Sridevi K (2013a) Biohydrogen production and kinetic modeling using sediment microorganisms of pichavaram mangroves. Ind Bio Med Res Int. doi:10.1155/2013/265618

Mullai P, Yogeswari MK, Sridevi K (2013b) Optimisation and enhancement of biohydrogen production using nickel nanoparticles–a novel approach. Bioresour Technol 141:212–219

Nasr NES (2012) Investigation of bio-hydrogen and biomethane production from thin stillage. electronic thesis and dissertation repository. Paper 491. http://ir.lib.uwo.ca/etd/491

Nasr M, Tawfik A, Ookawara S, Suzuki M, Kumari S, Bux F (2015) Continuous biohydrogen production from starch wastewater via sequential dark-photo fermentation with emphasize on maghemite nanoparticles. J Ind Eng Chem 21:500–506

Nath K, Das D (2004) Biohydrogen production as a potential energy resource-present state of art. J Sci Ind Res 63:729–738

National Hydrogen Energy Roadmap (2002) By United States Department of Energy. https://www.hydrogen.energy.gov/pdfs/national_h2_roadmap.pdf. Accessed 4 Mar 2016

Nissila M (2013) Biohydrogen, bioelectricity and bioalcohols from cellulosic materials. Tampere University of Technology. Publication 1103, Tempere. TUTCRIS Portal (http://www.tut.fi/tutcris)

Ntoampe M (2012) Microbial fuel and chemical production using sweet potatoes, PhD thesis submitted to the

Faculty of Engineering and the Built Envirnonment, University of the Witwatersrand Johannesnurg. Accessed through http://wiredspace.wits.ac.za/jspui/bitstream/10539/11742/2/THESIS%20final.pdf

Nyberg M, Heidorn T, Lindblad P (2015) Hydrogen production by the engineered cyanobacterial strain Nostoc PCC 7120 ΔhupW examined in a flat panel photobioreactor system. J Biotechnol 215:35–43

Obazu FO (2013) Development of bio-reactor for the production of hydrogen from plant biomass. Doctoral dissertation, Lund University

Obazu FO, Afolabi AS, Daramola MO, Gray VM (2015a) Biohydrogen production by bacterial granules adapted to grow at different thermophilic temperatures. In: Proceedings of the world congress on engineering, vol 2

Obazu FO, Afolabi A, Daramola MO, Gray VM (2015b) Sustenance of dark fermentative hydrogen production by an undefined bacterial culture. In: Proceedings of the world congress on engineering, vol 2

Ohimain EI, Izah SC (2015) Estimation of potential biohydrogen from palm oil mills' effluent in Nigeria using different microorganisms under light independent fermentation. J Environ Treat Technol 3(2):97–104

Oil Review Middle East (2015) Air liquide starts its largest hydrogen project in Saudi Arabia. Oil review Middle East 24 June 2015, http://www.oilreviewmiddleeast.com/petrochemicals/air-liquide-starts-its-largest-hydrogen-project-in-saudi-arabia. Accessed on 6 Feb 2016

Orozco RL, Redwood MD, Yong P, Caldelari I, Sargent F, Macaskie LE (2010) Towards an integrated system for bio-energy: hydrogen production by Escherichia coli and use of palladium-coated waste cells for electricity generation in a fuel cell. Biotechnol Lett 32(12):1837–1845

Ortigueira J, Pinto T, Gouveia L, Moura P (2015) Production and storage of biohydrogen during sequential batch fermentation of Spirogyra hydrolyzate by Clostridium butyricum. Energy 88:528–536

Özgür E, Mars AE, Peksel B, Louwerse A, Yücel M, Gündüz U, Claassen PA, Eroğlu İ (2010) Biohydrogen production from beet molasses by sequential dark and photofermentation. Int J Hydrog Energy 35(2):511–517

Ozkan E, Uyar B, Özgür E, Yücel M, Eroglu I, Gündüz U (2012) Photofermentative hydrogen production using dark fermentation effluent of sugar beet thick juice in outdoor conditions. Int J Hydrog Energy 37(2):2044–2049

Pachapur VL, Sarma SJ, Brar SK, Le Bihan Y, Buelna G, Verma M (2015) Biohydrogen production by co-fermentation of crude glycerol and apple pomace hydrolysate using co-culture of Enterobacter aerogenes and Clostridium butyricum. Bioresour Technol 193:297–306

Pachapur VL, Sarma SJ, Brar SK, Bihan YL, Buelna G, Verma M (2016) Hydrogen production from biodiesel industry waste by using a co-culture of Enterobacter aerogenes and Clostridium butyricum. Biofuels: 1–19

Panagiotopoulos IA, Pasias S, Bakker RR, de Vrije T, Papayannakos N, Classen PAM, Koukios EG (2013) Biodiesel and biohydrogen production from cotton seed cake in a biorefinery concept. Bioresour Technol 136:78–86

Parihar RK, Upadhyay K (2015) Production of biohydrogen gas from wastewater by anaerobic fermentation process: a review. Int J Chem Stu 3(3):7–14. SCIENCEDOMAIN. www.sciencedomain.org

Pattanamanee W, Chisti Y, Choorit W (2015) Photofermentive hydrogen production by Rhodobacter sphaeroides S10 using mixed organic carbon: effects of the mixture composition. Appl Energy 157:245–254

Patterson T, Esteves S, Dinsdale R, Guwy A, Maddy J (2015) Life cycle assessment of biohydrogen and biomethane production and utilisation as a vehicle fuel. Bioresour Technol 131:235–245

Pawar SS, van Niel EW (2013) Thermophilic biohydrogen production: how far are we? Appl Microbiol Biotechnol 97(18):7999–8009

Pawar SS, Nkemka VN, Zeidan AA, Murto M, van Niel EW (2013) Biohydrogen production from wheat straw hydrolysate using Caldicellulosiruptor saccharolyticus followed by biogas production in a two-step uncoupled process. Int J Hydrog Energy 38(22):9121–9130

Penfold DW, Forster CF, Macaskie LE (2003) Increased hydrogen production by Escherichia coli strain HD701 in comparison with the wild-type parent strain MC4100. Enzym Microb Technol 33(2):185–189

Qureshi N, Annous BA, Ezeji TC, Karcher P, Maddox IS (2005) Biofilm reactors for industrial bioconversion processes: employing potential of enhanced reaction rates. Microb Cell Fact 4(1):24

Rachman MA, Prasetyo J (2014) Electricity from biowaste-based gas hydrogen as a source of renewable energy biomass. In: Informatics, Electronics & Vision (ICIEV), International conference on IEEE 2014:1–5

Rachman MA, Eniya LD, Widyastuti ET (2015) Biohydrogen production from fermentation of sweet Sorghum (Sorghum Biocular L) by interobacter aerogenes Adh-43 in the packed-bed reactor. J Technol 75(6): 67–72

Raji A, Kahn MTE (2013) Efficient and effective domestic energy generation. Energize 19–21. http://www.ee.co.za/wp-content/uploads/2014/06/energize-oct-2013-pgs19-21.pdf

Redwood MD (2008) Bio-hydrogen and biomass-supported palladium catalyst for energy production and waste-minimisation. Doctoral dissertation, University of Birmingham

Redwood MD, Macaskie LE (2006) A two-stage, two-organism process for biohydrogen from glucose. Int J Hydrog Energy 31(11):1514–1521

Redwood MD, Paterson-Beedle M, Macaskie LE (2009) Integrating dark and light bio-hydrogen production strategies: towards the hydrogen economy. Rev Environ Sci Biotechnol 8(2):149–185

Reungsang A, Sangyoka S, Imai T, Chaiprasert P (2004) Biohydrogen production from cassava starch manufacturing wastewater. In: The joint international conference on "Sustainable Energy and Environment (SEE):1–3

Rittmann S, Herwig C (2012) A comprehensive and quantitative review of dark fermentative biohydrogen production. Microbial Cell Fact 11(1):1

Salleh SF, Kamaruddin A, Uzir MH, Karim KA, Mohamed AR (2015) Investigation of the links between heterocyst and biohydrogen production by diazotrophic cyanobacterium A. variabilis ATCC 29413. Arch Microbiol:101–113. doi: 10.1007/s00203-015-1164-6

Sarma SJ, Brar SK, Le Bihan Y, Buelna G, Rabeb L, Soccol CR, Naceur M, Rachid B (2013) Evaluation of different supplementary nutrients for enhanced biohydrogen production by Enterobacter aerogenes NRRL B 407 using waste derived crude glycerol. Int J Hydrog Energy 38(5):2191–2198

Schneider T, Graeff-Hönninger S, French WT, Hernandez R, Claupein W, Holmes WE, Merkt N (2012) Screening of industrial wastewaters as feedstock for the microbial production of oils for biodiesel production and high-quality pigments. J Comb. http://dx.doi.org/10.1155/2012/153410

Seelert T, Ghosh D, Yargeau V (2015) Improving biohydrogen production using Clostridium beijerinckii immobilized with magnetite nanoparticles. Appl Microbiol Biotechnol 99(9):4107–4116

Seengenyoung J, Sompong O, Thong TI, Prasertsan P (2011) Effect of inoculum size for biohydrogen production from palm oil mill effluent. TIChE international conference, 2011

Sekoai PT, Daramola MO (2015) Biohydrogen production as a potential energy fuel in South Africa. Biofuel Res J 2(2):223–226

Sekoai PT, Kana EBG (2014) Semi-pilot scale production of hydrogen from Organic Fraction of Solid Municipal Waste and electricity generation from process effluents. Biomass Bioenergy 60:156–163

Sen B, Chou YP, Wu SY, Liu CM (2016) Pretreatment conditions of rice straw for simultaneous hydrogen and ethanol fermentation by mixed culture. Int J Hydrog Energy 41(7):4421–4428

Sewsynker Y, Kana EBG, Lateef A (2015) Modelling of biohydrogen generation in microbial electrolysis cells (MECs) using a committee of artificial neural networks (ANNs). Biotechnol Biotechnol Equip 29(6):1208–1215

Show KY, Lee DJ, Tay JH (2012) Biohydrogen production: current perspectives and the way forward. Int J Hydrog Energy 34(17):7349–7357

Sibanyoni JM (2012) Nanostructured light weight hydrogen storage materials, PhD thesis

Sinbuathong N, Kanchanakhan B, Leungprasert S (2015) Biohydrogen production from normal starch wastewater with heat-treated mixed microorganisms from a starch factory. Int J Glob Warming 7(3):293–306

Singh A, Sevda S, Abu Reesh IM, Vanbroekhoven K, Rathore D, Pant D (2015) Biohydrogen production from lignocellulosic biomass: technology and sustainability. Energies 8(11):13062–13080

Skonieczny MT, Yargeau V (2009) Biohydrogen production from wastewater by Clostridium beijerinckii: effect of pH and substrate concentration. Int J Hydrog Energy 34(8):3288–3294

Sompong O, Hniman A, Prasertsan P, Imai T (2011) Biohydrogen production from cassava starch processing wastewater by thermophilic mixed cultures. Int J Hydrog Energy 36(5):3409–3416

Soydemir G, Keris-Sen UD, Sen U, Gurol MD (2016) Biodiesel production potential of mixed microalgal culture grown in domestic wastewater. Bioproc Biosyst Eng 39(1):45–51

Sung S (2004) Biohydrogen production from renewable organic wastes, DOE Award Number: DE-FC36-00G010530, Iowa State University 2207 Pearson Hall, Room 15 Ames, IA 50011-2207 PhD thesis

Tamagnini P, Axelsson R, Lindberg P, Oxelfelt F, Wünschiers R, Lindblad P (2002) Hydrogenases and hydrogen metabolism of cyanobacteria. Microbiol Mol Biol Rev 66(1):1–20

Tang GL, Huang J, Sun ZJ, Tang QQ, Yan CH, Liu GQ (2008) Biohydrogen production from cattle wastewater by enriched anaerobic mixed consortia: influence of fermentation temperature and pH. J Biosci Bioeng 106(1):80–87

Tawfik A, El-Qelish M, Salem A (2015) Efficient anaerobic co-digestion of municipal food waste and kitchen wastewater for bio-hydrogen production. Int J Green Energy 12(12):1301–1308

Thompson LJ, Gray VM, Kalala B, Lindsay D, Reynolds K, von Holy A (2008) Biohydrogen production by Enterobacter cloacae and Citrobacter freundii in carrier induced granules. Biotechnol Lett 30(2):271–274

Uyar B, Eroglu I, Yücel M, Gündüz U (2009) Photofermentative hydrogen production from volatile fatty acids present in dark fermentation effluents. Int J Hydrog Energy 34(10):4517–4523

Uyar B, Gürgan M, Özgür E, Gündüz U, Yücel M, Eroglu I (2015) Hydrogen production by hup– mutant and wild-type strains of Rhodobacter capsulatus from dark fermentation effluent of sugar beet thick juice in batch and continuous photobioreactors. Bioproc Biosyst Eng 38(10):1935–1942

Varrone C, Giussani B, Izzo G, Massini G, Marone A, Signorini A, Wang A (2012) Statistical optimization of biohydrogen and ethanol production from crude glycerol by microbial mixed culture. Int J Hydrog Energy 37(21):16479–16488

Vatsala TM, Raj SM, Manimaran A (2008) A pilot-scale study of biohydrogen production from distillery effluent using defined bacterial co-culture. Int J Hydrog Energy 33(20):5404–5415

Waligórska M, Seifert K, Szymańska K, Łaniecki M (2006) Optimization of activation conditions of Rhodobacter sphaeroides in hydrogen generation process. J Appl Microbiol 101(4):775–784

Wang X, Jin B (2009) Process optimization of biological hydrogen production from molasses by a newly isolated Clostridium butyricum W5. J Biosci Bioeng 107(2):138–144

Wang X, Jin B, Mulcahy D (2008) Impact of carbon and nitrogen sources on hydrogen production by a newly isolated Clostridium butyricum W5. Int J Hydrog Energy 33(19):4998–5005

Wang X, Wang N, Wang L (2011) Hydrogen production by sorption enhanced steam reforming of propane: a thermodynamic investigation. Int J Hydrog Energy 36(1):466–472

Wei J, Liu ZT, Zhang X (2010) Biohydrogen production from starch wastewater and application in fuel cell. Int J Hydrog Energy 35(7):2949–2952

Wen Q, Wu Y, Zhao LX, Sun Q, Kong FY (2010) Electricity generation and brewery wastewater treatment from sequential anode-cathode microbial fuel cell. J Zhejiang Univ Sci B 11(2):87–93

Winter CJ (2005) Into the hydrogen energy economy—milestones. Int J Hydrog Energy 30(7):681–685

Won SG, Lau AK (2015) Effects of manipulating cyclic duration and pH on fermentative hydrogen production in an anaerobic sequencing batch reactor. J Environ Sci Health (A) 50(7):750–756

Wu TY, Hay JXW, Kong LB, Juan JC, Jahim JM (2012) Recent advances in reuse of waste material as substrate to produce biohydrogen by purple non-sulfur (PNS) bacteria. Renew Sustain Energy Rev 16(5):3117–3122

Xia A, Jacob A, Tabassum MR, Herrmann C, Murphy JD (2016) Production of hydrogen, ethanol and volatile fatty acids through co-fermentation of macro-and micro-algae. Bioresour Technol 205:118–125

Xu J, Deshusses MA (2015) Fermentation of swine wastewater-derived duckweed for biohydrogen production. Int J Hydrog Energy 40(22):7028–7036

Yetis M, Gündüz U, Eroglu I, Yücel M, Türker L (2000) Photoproduction of hydrogen from sugar refinery wastewater by Rhodobacter sphaeroides OU 001. Int J Hydrog Energy 25(11):1035–1041

Yigit DO, Gunduz U, Turker L, Yucel M, Eroglu I (1999) Identification of by-products in hydrogen producing bacteria; Rhodobacter sphaeroides OU 001 grown in the waste water of a sugar refinery. J Biotechnol 70(1):125–131

Zhang ML, Fan YT, Xing Y, Pan CM, Zhang GS, Lay JJ (2007) Enhanced biohydrogen production from cornstalk wastes with acidification pretreatment by mixed anaerobic cultures. Biomass Bioenergy 31(4):250–254

Zhang X, Sherman DM, Sherman LA (2014) The uptake hydrogenase in the unicellular diazotrophic cyanobacterium Cyanothece sp. strain PCC 7822 protects nitrogenase from oxygen toxicity. J Bacteriol 196(4):840–849

Zheng H, Zeng RJ, Angelidaki I (2008) Biohydrogen production from glucose in upflow biofilm reactors with plastic carriers under extreme thermophilic conditions (70 C). Biotechnol Bioeng 100(5):1034–1038 http://www.saasta.ac.za/images/stories/HydrogenF.pdf

Index

A. Singh, D. Rathore (eds.), *Biohydrogen Production: Sustainability of Current Technology and Future Perspective*, DOI 10.1007/978-81-322-3577-4